SUSTAINABLE DESIGN AND BUILD

SUSTAINABLE DESIGN AND BUILD

Building, Energy, Roads, Bridges, Water and Sewer Systems

Md. FARUQUE HOSSAIN

Adjunct Professor, Department of Civil and Urban Engineering, New York University, Brooklyn, New York;
CEO, Green Globe Technology, Inc., Flushing, New York

ELSEVIER

Butterworth-Heinemann
An imprint of Elsevier

Butterworth-Heinemann is an imprint of Elsevier
The Boulevard, Langford Lane, Kidlington, Oxford OX5 1GB, United Kingdom
50 Hampshire Street, 5th Floor, Cambridge, MA 02139, United States

Library of Congress Cataloging-in-Publication Data
A catalog record for this book is available from the Library of Congress

British Library Cataloguing in Publication Data
A catalogue record for this book is available from the British Library

ISBN: 978-0-12-816722-9

For information on all Butterworth-Heinemann publications visit our website at https://www.elsevier.com//books-and-journals

Publisher: Matthew Deans
Acquisition Editor: Ken Mc Combs
Editorial Project Manager: Gabriela D. Capille
Production Project Manager: R.Vijay Bharath
Designer: Matthew Limbert

Typeset by TNQ Technologies

To

Nowshin, Shafin, and Faria

I love them more than their imagination.

I hope one of them will fulfill my dream of changing the world by building a sustainable earth.

CONTENTS

ABOUT THE AUTHOR

Dr. Md. Faruque Hossain has more than 20 years of industry experience in the field of sustainable research, design and build for large-scale buildings and civil, energy, environmental, and critical Infrastructure projects. He worked and consulted in diverse small companies to fortune-listed companies and managed as less as million-dollar to over billion-dollar projects. Faruque also worked for the New York City as the member of senior management team and interacted with the highest level government officials of local, state, federal, and international agencies for collaborating to build a sustainable Earth. During his tenure in New York City agency, Faruque managed a world-class team of scientists, architects, engineers, contractors, and consultants from top tier engineering firms and maintained highest level professional relationship to conduct sustainability practices on assigned projects. Hossain received his PhD from Hokkaido University, did postgraduate research in engineering at the University of Sydney, and executive education in architecture at Harvard University. He is a LEED-certified professional and editor of several international building and sustainable engineering related journals. Dr. Hossain, of Bangladesh origin, is renowned as the industry leader and notable scientist in the field of sustainable research, development, and project management for building a better planet. He has dozens of world class publications in very high-impact journals, and he wrote two books (Elsevier), and four book chapters (Francis and Taylor, and Springer) in the field of sustainable design and build. Currently, he is working at the Department of Civil and Urban Engineering at New York University as an adjunct professor and simultaneously running his own company "Green Globe Technology" with the motto to practice sustainability for Building a Cleaner and Greener World.

PREFACE

What world are we living in? Indeed, a terrible one with the crisis of environment, energy, shelter, infrastructure, transportation, and water. The fact that anthropogenic conventional practice to consume the natural resources is putting mother Earth in a dangerous level results in natural resource getting finite level and causing a deadly unequilibrium environment to survive any living being on it near future. Simply, the total system on Earth is getting malfunctioned silently, which needs to be fixed by giving immediate attention to utilize the wise consumption of natural resource and the implementation of sustainable technology to secure a balanced Earth. The notion of sustainability might be wide, but if we think for the practical solution to the survival of the Earth, it might have many innovative ideas to secure the Earth in a balanced environment. Therefore, sustainability research, development, and application must be practices by best scholastic approach of science and technology to build a clearer and greener world for our survival.

This book necessarily focuses on describing the holistic approaches to the technical aspects of sustainability for commercial applicability in society and the environment to achieve a best-balanced Earth. Therefore, this book describes the root cause of global sustainability problems and sets out a series of solutions by presenting innovative ideas through research and application. Simply, the goal of this book is to define the context of instigation to think through the scientific theories and practical applications with the holistic approach of both problems and solutions of sustainable mechanism by the presentation of the main seven themes (1. Introduction, 2. Environment, 3. Energy, 4. Advanced Building Design Technology, 5. Infrastructure and Transportation, 6. Water, and 7. Best Management Practices), which are very much interconnected to the global environment and our daily lives. Consequently, the importance of sustainability is also discussed considering the wise application of science and technology by trickling down the advancement thoughts, research, and application to achieve a broader goal of sustainability for building a better planet for the future generation.

Md. Faruque Hossain
Department of Civil and Urban Engineering
New York University

CHAPTER ONE

Introduction

1.1 ENVIRONMENT

Sustainability, within the environmental sector, means that the biological systems must remain productive and diverse for an indefinite period of time. One example of biological systems that are considered sustainable is long-lived and healthy ecosystem. Generally, sustainability can be defined as the durability of processes and systems including the interrelated domains of culture and politics, economics, and ecology to acquire healthy environments that will support human survival and that of other creatures of planet Earth [1,2]. Consequently, in preserving the natural resources, sustainability encounters social challenge that involves ethical consumerism, individual and local lifestyle, urban transportation and planning, as well as national and international laws. While sustainable development must be adopted as the holistic method to acquiring a greener and cleaner planet Earth, it is essential to view sustainability as the target goal of humanity to confirm an equilibrium ecosystem [3,4]. As a result, it is essential for sustainability to be majorly concerned with the commitment of policymakers, investors, engineers, architects, and scientists to administer and promote necessary planetary environmental resources essential in securing resilience and attain sustainability of these important resources for benefits of forthcoming generations [5,6].

Environmental resiliency, and thus environmental sustainability, is usually measured by occurrences or junctures where the mixing of naturally befalling regenerative forces, such as biomass, vegetation, atmosphere, soil, water, and solar energy, intermingle with their underlying forces into the environment. Human activities are the major drivers to destructing the systems of the Earth as well as its biophysical mechanism [7–9]. Therefore, the impact of a community on the environment is instigated by a single person or the available population, which in turn relies on complex ways on exactly what natural resources are being utilized, whether those natural resources are renewable, as well as the human activity scale in comparison to the ecosystems' carrying capacity. Accordingly, the resource consumption pattern by a population of individuals within all sectors is generating adversative effect on

Sustainable Design and Build
ISBN: 978-0-12-816722-9
https://doi.org/10.1016/B978-0-12-816722-9.00001-X

biodiversity, conservation biology, environmental science, and Earth science. Progressively, the biodiversity loss within the environment mainly from the habitat fragmentation and loss generated by human land appropriation for agriculture, forestry, and development as natural capital is rapidly changing all over the globe [10,11]. As a result, this change in land use plays a major part to the operations of the changes in the biosphere in relative magnitudes of land devoted to grassland, woodland, forest, agriculture, and urbanization. All of which have significant impact on global nitrogen, carbon, and water biogeochemical cycles. Basically, product consumption at all scales via the consumption chain, beginning with the economic sectors' impact via national economies to the international economies and with the impacts of personal spending patterns and lifestyle choices via demands of resources of particular services and goods, is seriously impacting on the environment [12,13]. To maintain the resource consumption, resource productivity, as well as resource intensity, it is necessary to investigate the pattern of consumption that is associated with resources to the economic, social, and environmental effects at the context or scale. The initial world scientific evaluation on the effects of production and consumption was published in 2010 by the international Resource Panel of United Nations Environment Program (UNEP), which recognized the priority actions for both developing and developed countries [14—16]. The findings of the study indicated that consumption by household associated with energy-using, food, shelter, and mobility products is the major cause of life cycle effects of consumption and generating limited level of the existing natural resources. As a result, to safeguard these natural resources through the implementation of basic principles of complex ecological issues and formulation of operational solutions, it is necessary to undertake advanced research to encounter challenges generated internationally through increasing urbanization, ecological degradation, and population growth.

1.2 ENERGY

The global atmosphere is now getting seriously dangerous as increment of all aspects of the carbon cycle into the atmosphere has become the major crisis throughout the world due to the usage of conventional energy [17,18]. Air toxicants such as volatile organic compounds, sulfur oxide, nitrogen oxide as well as airborne pollutant substances that generate acid rain and photochemical smog, air pollution and the deadly chlorofluorocarbon are causing severe effects on the Earth's atmosphere and environment

[19,20]. Therefore, sustainable energy is an urgent demand to serve the needs of the present without compromising the ability of future generations to meet their needs. Whereas renewable energy refers to the energy that is naturally refilled on a human being's timescale, sustainable energy is the energy whose usage will not jeopardize the system within which it is implemented to an extent of being unfit to offer the needs for the coming days. The principle of sustainability encompasses having the four interrelated domains: (1) culture, (2) politics, (3) economics, and (4) ecology. Technologies promote sustainable energy including renewable energy sources, such as hydroelectricity, solar energy, wind energy, wave power, geothermal energy, bioenergy, tidal power, and technologies designed to improve energy efficiency.

Significant advancement is being carried out in transition of energy from fossil fuels to environmentally sustainable systems, and finally to the point where 100% renewable energy is being supported by several studies. As a result, changes that need to be made on the present-day conventional energy consumption will not only be on how energy is supplied but also on how it is used, and it is important to reduce the volume of energy needed to deliver different goods and/or services. Stabilizing and decreasing the emissions of CO_2 simply requires energy efficiency and renewable energy to remain as the "twin pillars" of environmental sustainability. Based on the current historical examination, the growth rate in demand of energy has generally overtaken the rate of enhancements in energy efficiency [21—23]. This is because of the ongoing population and economic growth. Consequently, aggregate use of energy as well as correlated emissions of carbon have constantly increased, which ultimately causes deadly climate changes. In consequence, supplies of renewable and sustainable (clean) energy are an exigent demand to alleviate global energy demand and mitigate climate change crisis. Therefore, clean and renewable energy (and energy efficiency) are no longer niche sectors that are promoted only by governments and environmentalists but also by private sector by increasing the levels of investment for confirming a clean and green Earth.

So, to finally achieve a clean world, it is essential to look after the sustainable and renewable energy sector through application technology, carrying out advanced research, as well as via commercial application. Essentially, much focus must be directed toward renewable power system planning, design and building, and sustainable application of energy within all sectors of infrastructure and building to approve sustainable energy system construction and design, and control.

1.3 BUILDING

All over the world, buildings have been identified to consume very large portions of natural resources including water and energy. In the present day, buildings are responsible for 40% of the worldwide CO_2 emissions, which is equal to 9 billion tons of carbon dioxide yearly, and by 2050, these emissions are likely to double [24–26]. It is essential for one to think through the clean building to attain ecologically friendly building and energy efficiency, which eventually will combine a massive collection of skills, approaches, and practices to cut and finally eliminate the adverse environment impacts. Sustainable building refers to both the structure and the application of processes that are environmentally responsible and resource efficient throughout a building's life cycle from planning to design, construction, operation, maintenance, renovation, and demolition. This requires close cooperation of the contractor, the architects, the engineers, and the client for the entire project life cycle to achieve the benefit economy, durability, and comfort. So, the sustainable building development has several drives including social, economic, and environmental benefits. As such, the initiatives of sustainability require a synergistic and an incorporated design in both the existing building retrofitting and new construction to support environment and energy. Considering energy efficiency, toxic and waste reduction, maintenance and operations optimization, interior environmental quality improvement, material efficiency, design efficiency, and water efficiency, the technologies or practices applied in sustainable building essentially require to be focused, so as to generate a larger amassed impact.

Making the most of the renewable resources is frequently stressed in sustainable building. Examples may include utilizing sunlight via photovoltaic equipment; active and passive solar; and trees and planets via rainwater runoff reduction, rain gardens, and green roofs [27,28]. In addition, utilizing low-effect constructing materials and using permeable concrete or packed gravel rather than asphalt or conventional concrete to improve groundwater replenishment are other approaches that are being used [29,30]. Moreover, having appropriate synergistic design in place may enable individual green building technologies to join forces to generate increasing impact. Clean design or green architecture on the artistic side is the philosophy of planning a construction that is in line with the resources nearby the site and natural feature. Designing clean building involves many key steps: identifying "green" materials of building from indigenous sources, reducing loads, optimizing systems, and finally producing on-site renewable energy.

Consequently, it is necessary for the building sector to confirm that a dynamic clean research, management, and development are all in stages of the expansion of any amenities within the rural and urban regions. Essentially, emphasis should be directed to the assessments of project life cycle, solicitation, preconstruction, design development, project planning, project ecology, site selection, methods and application of construction materials to ultimately confirm a sustainable building for the civil and rural community.

1.4 INFRASTRUCTURE AND TRANSPORTATION

Traditional infrastructure is not only causing trillions of dollars every year but also playing vital role to losses of lands and creating adverse environmental and climate perplexity [31,32]. Sustainable infrastructure is a network providing the "ingredients" for solving urban and climatic challenges by building with nature, thus it would be a best option to secure a resilient community. The main components of this approach are roads, highways, bridges, tunnel management to achieve climate adaptation, less heat stress, more biodiversity, food production, better air quality, sustainable energy production, clean water, and healthy soils, as well as the more anthropocentric functions such as increased quality of life through recreation and provision of shade and shelter in and around towns and cities [33—35]. Subsequently, sustainable infrastructure serves to provide an ecological framework for social, economic, and environmental health of the surroundings. Thus, sustainable infrastructure is considered as the best engineering practice that would achieve the construction and management of more holistic roads, highways, bridges, tunnels to smoothly run daily life.

On the other hand, traditional transport systems have significant impacts on the environment, accounting for nearly 28% of world energy conventional consumption, and it is causing proportional climate change and adverse environmental impact [36—38]. The sustainable transportation refers to the broad subject of transport that should be environmentally benign in the sense of social, environmental, and climate impacts and the ability to mitigate the environmental pollution indefinitely. Components for evaluating sustainable transport include the advanced vehicle technology to be used for road, water, or air transport by using renewable and clean energy, where the infrastructure should be able to accommodate the clean fuel—operated transport for roads, railways, airways, waterways, canals, and terminal pathways to mitigate energy and traffic jam crisis. Simply sustainable transport systems will make a positive contribution to the environmental, social, and economic sustainability of the communities by binding social and economic connections where people can be quickly benefitted by this sustainable mobility such as zero-emission vehicle and flying transportation technology. Necessarily, the promotion of incremental improvement in zero-emission fuel vehicle and clean and renewable energy—operated flying transportation vehicle technology migrating from fossil-based transportation system would be the best option to measure the sustainability measurement and optimization.

Therefore, the sustainable infrastructure system and advanced transportation vehicle are needed urgently to have the better, safer, and faster mobility and less environmental impact comparing traditional infrastructure and conventional vehicles. Therefore, the main research and development must be focused to practice sustainable infrastructure development and advanced zero emission and advanced technology for flying vehicles for building cleaner and greener world.

1.5 WATER

Global environment has been impacted by the global water cycle by the acceleration of evaporation which creates a cooling effect on Earth and climate change [14,39]. Water covers 75% of the Earth's surface, and out of this, 97% is the salty water of the oceans and only 3% freshwater, most of which is locked up in the Antarctic ice sheet [8,40]. Most of it is in icecaps, glaciers (69%), and groundwater (30%), whereas all lakes, rivers, and swamps combined only account for a small fraction (0.3%) of the Earth's total freshwater reserves [1,41]. The main source of our daily usage of water is either a freshwater or groundwater reserve. Although, the supply of freshwater is constantly replenished through precipitation because of the water cycle, within the last several decades, groundwater strata have been getting lower nearly 10 m, scaring the groundwater finite level in the near future [31,42]. In contrast, rising urbanization contaminates freshwater supplies, thereby triggering adverse environmental impact, which eventually alarms the survival of all living beings on Earth due to the shortage of water near future. Because water has distinctive features that are important for life proliferation to respond in ways that eventually permit replication, it is very important to all living organisms for their own existence. Simply water is essential to all living beings for their survival because it has many distinct properties that are critical for the proliferation of life to react in ways that ultimately allow replication. It is, therefore, vital both as a solvent in which many of the body's solutes dissolve and as an essential part of many metabolic (catabolism and anabolism) processes within the living body [39,43]. In catabolism, water is utilized to break the bonds within large molecules to create smaller molecules, and in anabolism, water is detached from molecules to create larger molecules. Both these processes of anabolism and catabolism cannot exist without water [8,18]. In the plant kingdom, water is the fundamental element for photosynthesis and respiration where photosynthetic cells use the sun's energy to split off hydrogen from oxygen in water. Afterward, hydrogen is mixed with carbon dioxide (absorbed from air or water) to form glucose that is utilized as their food and release oxygen to balance the ecosystem.

Within the world's economy, water naturally plays a significant role, with almost 70% of the freshwater being utilized by humans going to agricultural sector, which has a larger contribution to global economy [21,40]. For several parts of the globe, fishing in freshwater and saltwater bodies is a key food source and an important part of the globe's economy referred to as Blue and Brown Economy. In homes and industries, huge amounts of steam, ice, and water are utilized for heating and cooling. For an extensive chemical substances variety, water is a tremendous solvent, and as a result, it is extensively utilized for washing and cooking, and in industrial processes.

Unfortunately, this natural resource is becoming scarcer globally where certain places are in a vulnerable condition. In the developing world, 90% of all wastewater still goes untreated into local rivers and streams, causing dangerous water environment in the water world. In developed countries, usage of conventional treatment processes causes severe environmental pollution [30,31]. Some 50 countries, with roughly a third of the world's population, also suffer from medium- or high-water stress, and 17 of these extract more water annually than that is recharged through their natural water cycles [8,14]. The strain not only affects surface freshwater bodies such as rivers and lakes but also degrades groundwater resources. Currently, about a billion people around the world routinely drink unhealthy water, resulting in some 5 million deaths each year caused by polluted drinking water.

Thus, advance research and development of water is needed to sustain this natural resource where primary focus needs to be given by conducting emerging distributed systems for water supply and wastewater treatment to confirm a sustainable Earth. Consequently, much more scientifically and technologically advanced research and development must be applied, considering environmentally friendly (1) physical and chemical treatment processes for water and wastewater treatment process, (2) environmental biotechnology for use in water resource management and bioremediation, and utilize wastewater into useful product, (3) watershed and wetland management to reduce the water loss, (4) advanced environmental engineering design to mitigate the groundwater, (5) sustainable water resource development as a new source of water supply.

1.6 CONCLUSION

The UNEP estimates that each year 2.4 million people die from air pollution and nearly 5 million people die because of water pollution and nearly 10 million people die due to other environmental problems caused by humans. The most dangerous, particularly hazardous to health, are emissions of black carbon, a component of particulate matter, which is a known cause of respiratory and carcinogenic diseases and the main contributor to global climate change. Currently the atmospheric CO_2 is 400 ppm, and it is increasing at the rate of 2.11% yearly, which is heading to reach the toxic level of CO_2 concentration into the air of 60,000 ppm where all living beings will die in 30 s. If the current level of CO_2 emission is not stopped, all human race on Earth will be extinct $\left[\int_{400^{(2.11\%)}}^{60,000}(2017)\right]$ in 121,017,712 years, and thus, it would be the end of the story of human civilization on Earth. Simply, sustainable science and technology implementation is an urgent need in all sectors of environment, energy, building, infrastructure, transportation, water for all living beings and the human race on Earth.

ACKNOWLEDGMENTS

This research was supported by Green Globe Technology under the grant RD-02018-03. Any findings, conclusions, and recommendations expressed in this paper are solely those of the author and do not necessarily reflect those of Green Globe Technology.

REFERENCES

[1] Garcia-Robledo E, Padilla CC, Aldunate M, Stewart FJ, Ulloa O, Paulmier A, Gregori G, Revsbech NP. Cryptic oxygen cycling in anoxic marine zones. Proc Natl Acad Sci USA 2017;114(31):8319–24.

[2] Hossain MF. Invisible transportation infrastructure technology to mitigate energy and environment. Energy Sustain Soc 2017;7:27 [Springer].

[3] Arnell N, et al. Climate change 1995. In: Watson RT, Zinyowera MC, Moss RH, editors. Impacts, adaptations, and mitigation of climate change. Cambridge University Press; 1996. p. 325–63.

[4] Davis SJ, Caldeira K. Consumption-based accounting of CO2 emissions. Proc Natl Acad Sci USA 2010;107:5687–92.
[5] Li W, et al. Reducing uncertainties in decadal variability of the global carbon budget with multiple datasets. Proc Natl Acad Sci USA 2016;113:13104–8.
[6] Pickup G, Bastin GN, Chewings VH. Identifying trends in land degradation in non-equilibrium rangelands. J Appl Ecol 1998;35:365–77.
[7] Ballantyne AP, Alden CB, Miller JB, Tans PP, White JWC. Increase in observed net carbon dioxide uptake by land and oceans during the past 50 years. Nature 2012;488: 70–2.
[8] Denman KL, et al. Couplings between changes in the climate system and biogeochemistry. Cambridge University Press; 2007.
[9] Duce RA, et al. Impacts of atmospheric anthropogenic nitrogen on the open ocean. Science 2008;320:893–7.
[10] Bauer JE, et al. The changing carbon cycle of the coastal ocean. Nature 2013;504: 61–70.
[11] Erb K-H, et al. Bias in the attribution of forest carbon sinks. Nat Clim Change 2013;3: 854–6.
[12] Helbling DE, Hollender J, Kohler H-PE, Singer H, Fenner K. High-throughput identification of microbial transformation products of organic micropollutants. Environ Sci Technol 2010;44:6621–7.
[13] Hossain MF. Photonic thermal energy control to naturally cool and heat the building. Adv Thermal Eng 2018;131:576–86 [Elsevier].
[14] Betts RA, Jones CD, Knight JR, Keeling RF, Kennedy JJ. El Nino and a record CO_2 rise. Nat Clim Change 2016;6:806–10.
[15] Hossain MF. Transforming dark photon into sustainable energy. Int J Energy Environ Eng 2018. https://doi.org/10.1007/s40095-017-0257-1 [Springer].
[16] Hossain MF. Solar energy integration into advanced building design for meeting energy demand. Int J Energy Res 2016;40:1293–300 [Wiley].
[17] Canadell JG, et al. Contributions to accelerating atmospheric CO_2 growth from economic activity, carbon intensity, and efficiency of natural sinks. Proc Natl Acad Sci USA 2007;104:18866–70.
[18] Ciais P, Sabine C. Chapter 6: carbon and other biogeochemical cycles in climate change 2013. In: Stocker T, Qin D, Platner GK, editors. The physical science basis. Cambridge University Press; 2013.
[19] Postel SL, Daily GC, Ehrlich PR. Human appropriation of renewable fresh water. Science 1996;271:785.
[20] van der Werf GR, et al. Climate regulation of fire emissions and deforestation in equatorial Asia. Proc Natl Acad Sci USA 2008;105:20350–5.
[21] Fairhead J, Leach M. Misreading the African landscape: society and ecology in a forest-Savanna mosaic. Cambridge: Cambridge University Press; 1996.
[22] Hossain MF. Green science: independent building technology to mitigate energy, environment, and climate change. Renew Sustain Energy Rev 2017;73:695–705 [Elsevier].
[23] Hossain MF. Design and construction of ultra-relativistic collision PV Panel and its application into building sector to mitigate total energy demand. J Build Eng 2017; 9:147–54 [Elsevier].
[24] Bao H, Niggemann J, Luo L, Dittmar T, Kao S-J. Aerosols as a source of dissolved black carbon to the ocean. Nat Commun 2017.
[25] Hossain MF. Green science: advanced building design technology to mitigate energy and environment. Renew Sustain Energy Rev 2018;81(2):3051–60 [Elsevier].
[26] Hossain MF, Fara N. Integration of wind into running vehicles to meet its total energy demand. Energy Ecol Environ 2017;2(1):35–48 [Springer].

[27] Hossain MF. Green science: decoding dark photon structure to produce clean energy. Energy Rep 2018. https://doi.org/10.1016/j.egyr.2018.01.001 [Elsevier].
[28] Liverman D, Moran EF, Rindfuss RR, Stern PC, editors. People and pixels: linking remote sensing and social science. Washington, D.C.: National Academy Press; 1998.
[29] Prietzel J, Zimmermann L, Schubert A, Christophel D. Organic matter losses in German Alps forest soils since the 1970s most likely caused by warming. Nat Geosci 2016;9:543–8.
[30] Scherr SJ. Meeting household needs: farmer tree-growing strategies in Western Kenya. In: Arnold JEM, Dewees PA, editors. Tree management in farmer strategies. Oxford, U.K.: Oxford University Press; 1995.
[31] Hossain MF. Application of advanced technology to build a vibrant environment on planet Mars. Int J Environ Sci Technol 2017;14(12):2709–20 [Springer].
[32] Hossain MF, Mukai H. Importance of nutrients (N, P, and NO_3+NO_2) in growth of the surfgrass, *Phyllospadix iwatensis* Makino. 2000.
[33] Hossain MF. Production of clean energy from cyanobacterial biochemical products. Strat Plann Energy Environ 2016;3:6–23 [Taylor and Francis].
[34] Hossain MF. Theoretical modeling for hybrid renewable energy: an initiative to meet the global power. J Sustain Energy Eng 2016;4:5–36 [Wiley].
[35] Stephens BB, et al. Weak northern and strong tropical land carbon uptake from vertical profiles of atmospheric CO_2 science. Science 2007;316:1732.
[36] Hossain MF. In situ geothermal energy technology: an approach for building cleaner and greener environment. J Ecol Eng 2016;77:49–55 [Polish Society of Ecological Engineering].
[37] Hossain MF. Theory of global cooling. Energy Sustain Soc 2016;6:24 [Springer].
[38] Schwietzke S, et al. Upward revision of global fossil fuel methane emissions based on isotope database. Nature 2016;538:88–91.
[39] Liu Z, et al. Reduced carbon emission estimates from fossil fuel combustion and cement production in China. Nature 2015;524:335–8.
[40] Mason Earles J, Yeh S, Skog KE. Timing of carbon emissions from global forest clearance. Nat Clim Change 2012;2:682–5.
[41] McAvoy DC, Schatowitz B, Jacob M, Hauk A, Eckhoff WS. Measurement of triclosan in wastewater treatment systems. Environ Toxicol Chem 2002;21:1323–9.
[42] Parazoo, et al. Contrasting carbon cycle responses of the tropical continents to the 2015–2016 El Niño. Science 2017.
[43] Borrelli P, Robinson DA, Fleischer LR, Lugato E, et al. An assessment of the global impact of 21st century land use change on soil erosion. Nat Commun 2017.

CHAPTER TWO

Environment

2.1 AIR

At present, global carbon emission and sequestration rates are not in equilibrium level, therefore, the global environment is currently under a severe vulnerable condition, which threatens the survival of all living beings on Earth. The annual global carbon emission considering all forms of CO_2 emissions, including the burning of fossil fuel and terrestrial land-use change, along with other related factors and the global carbon sequestration by all sources, such as the ocean sink, the earth sink, absorption by terrestrial vegetation, and other associated factors, has been estimated in this report to determine the acceleration of atmospheric CO_2 concentration. The estimate shows that the atmospheric CO_2 concentration has been increasing at a rate of 2.11% ppm per year for the past several years, and it continues to rapidly increase. It is well established that the toxic level of atmospheric CO_2 concentration is 60,000 ppm at which all living beings will die within 30 min [1,2]. If the current level of CO_2 emission is not stopped, all human race on Earth will be extinct $\left[\int_{400(2.11\%)}^{60,000}(2017)\right]$ in 121,017,712 years because the toxic level of CO_2 will reach 60,000 ppm within this time, and thus, it would be the end of the story of human civilization on Earth.

2.1.1 Background

The annual average atmospheric CO_2 concentration is currently 400 ppm [1,3]. Because of the industrial revolution of the early 1990s, fossil fuels, deforestation, land-use change, and external forces have become the key sources of anthropogenic CO_2 that have caused the concentration of CO_2 in the atmosphere to rapidly increase and change the global climate (Fig. 2.1).

In this paper, I have calculated the "global carbon budget," the "global CO_2 concentration," and, by determining the amount of CO_2 being input to the atmosphere relative to the global CO_2 balance, the rate of

Sustainable Design and Build
ISBN: 978-0-12-816722-9
https://doi.org/10.1016/B978-0-12-816722-9.00002-1

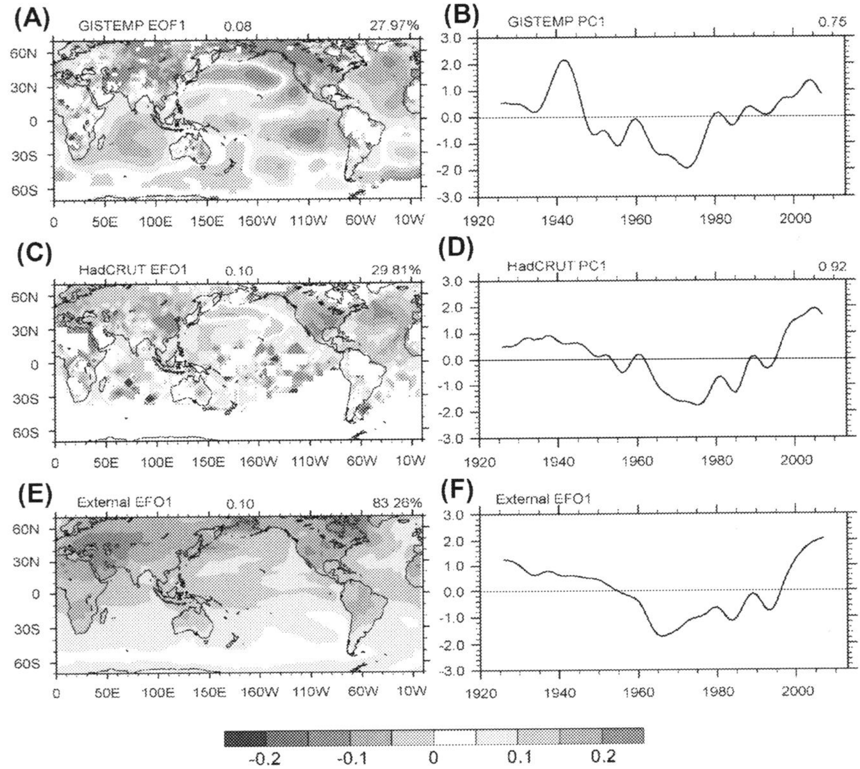

Figure 2.1 The leading EOF pattern (in °C) and associated standardized principal component (PC) of global climate change based on 8-year low-pass filtered annual anomalies after the long-term trend for 1920–2015 is removed from (*A*) and (*B*) GISTEMP; (*C*) and (*D*) HadCRUT; and (*E*) and (*F*) external forcing based on 18 CMIP5 models. Values shown at the top-right of the left panels represent the percentage variance explained by each mode, and those at the right panels represent the correlation coefficient with PC1 of external forcing. Values shown on in the middle of the left panels.

acceleration of atmospheric CO_2 emissions annually. Furthermore, I have prepared detailed data sets using MATLAB software and have calculated the annual total global carbon emissions from the preindustrial period (1750) to the modern period (2015). All CO_2 emissions from the burning of fossil fuel and land-use change from 1750 to the industrial revolution and from 1870 to the modern era have been analyzed in detail by incorporating data from the Fifth Assessment Report of the Intergovernmental

Panel on Climate Change (IPCC AR5) into MATLAB software to ensure the most accurate estimates of CO_2 emissions into the atmosphere because this gas is the major driver of climate change. The estimated CO_2 emissions from burning fossil fuel are referred to as E_{FF}; $Gt(yr^{-1})$, and the emissions resulting from land-use change are E_{LUC}; $Gt(yr^{-1})$. The sequestration and the absorption of CO_2 by the oceans, which is referred to as S_{OCEAN}; $Gt(yr^{-1})$, and that by the terrestrial vegetation, which is referred to as S_{LAND}; $Gt(yr^{-1})$, have also been estimated to determine the growth rate of the atmospheric CO_2 concentration (G_{ATM}; $Gt(yr^{-1})$) by using the following equation describing the CO_2 balance among the atmosphere, ocean, and land:

$$E_{FF} + E_{LUC} = G_{ATM} + S_{OCEAN} + S_{LAND} \tag{2.1}$$

$$G_{ATM} = E_{FF} + E_{LUC} - S_{OCEAN} - S_{LAND} \tag{2.2}$$

The atmospheric G_{ATM} is calculated in parts per million per year (ppm yr^{-1}), which can be converted into the total mass of carbon per year (GtC yr^{-1}). Assessments of environmental vulnerability have been conducted globally for the past several decades, and numerous studies have been performed [2–4]. To date, little research has been done on the adverse impacts of toxic levels of CO_2 in the atmosphere on life on Earth. Thus, the purpose of this report is to identify the yearly growth rate of the CO_2 concentration in the atmosphere and its impact on the future state of the global environment to determine the survival period of humanity and other living organisms.

2.1.2 Methods and Approach

Global CO_2 emissions, absorption, and sequestration were analyzed by interpreting reports from several organizations (CDIAC, IEA, UNEP, USDoE, ECE, EIA, PBL, NEAA, NEDO, NOAA, and NASA), and the data were incorporated into MATLAB software to develop the data set. To accurately calculate annual global carbon estimates, I considered all data up to the year 2015 and the projected fossil energy emissions for 2016, and from the projected total carbon estimates for 2016, the annual growth rate of the atmospheric concentration of CO_2 was determined [5,6].

2.1.2.1 CO_2 Emissions From Fossil Fuel

The yearly growth rate in CO_2 emissions was estimated from the difference between two consecutive years, which was divided by the first-year emissions per the following equation:

$$\left[\frac{E_{FF(t_0+1)} - E_{FF(t_0)}}{E_{FF(t_0)}}\right] * 100\%\ \text{yr}^{-1} \tag{2.3}$$

In general, a simple calculation can characterize the yearly CO_2 emission growth rate. However, to accurately determine the growth rate over multiple decades, I applied a leap year factor to confirm the net annual growth rate of carbon (E_{FF}) by using its logarithm equivalent in the following equation:

$$\frac{1}{E_{FF}} \frac{d\,(\ln E_{FF})}{dt} \tag{2.4}$$

Here, I calculated the pertinent CO_2 emission growth rates considering multidecadal periods by implementing a nonlinear drift into $\ln(E_{FF})$ in Eq. (2.4) and by calculating the yearly growth percentage. Thus, I fitted the logarithm of E_{FF} into the equation rather than directly using E_{FF} to ensure an accurate growth rate estimate and satisfy Eq. (2.3) [1,7,8].

2.1.2.2 CO_2 Emissions From the Land-Use Change (E_{LUC})

The emissions reported here (E_{LUC}) include CO_2 fluxes from deforestation, forest degradation, and the abandonment of agricultural land associated with modern civilization and were calculated by implementing dynamic global vegetation modeling (DGVM) bookkeeping simulations in MATLAB [9—11]. The simulations were prepared with DGVMs, in which I initially clarified the historical changes in land use followed by the atmospheric CO_2 concentrations [12—14]. Therefore, I implemented a time series of the distribution of preindustrial land cover by allocating the estimated variance into the first simulation and the dynamic evolution of biomass soil carbon to the prescribed land-cover change [15—17]. All the DGVMs here represent complete vegetation growth and decay processes as well as the decomposition of dead organic matter to determine the response to increasing atmospheric CO_2 levels [18—20].

2.1.2.3 Ocean CO_2 Sink

The CO_2 sequestered by the ocean from 1959 to 2015 was calculated by combining seven global oceans biogeochemical cycle models, and this approach can be used to comprehensively analyze the physical, chemical, and biological processes that are directly impacted by the concentration of CO_2 at the ocean surface as well as the air–sea CO_2 fluxes [21,22]. Thus, the ocean CO_2 sequestration is normalized by accurate observational values by dividing the individual yearly values by the modeled average for 1990–99 and then multiplying the result by an observation-based calculation of 2.2 GtC yr^{-1} [16,23,24]. Therefore, the oceanic CO_2 sequestration per year (t) in GtC yr^{-1} is calculated as follows:

$$S_{OCEAN}(t) = \frac{1}{n}\sum_{m=1}^{m=n}\frac{S^{m}_{OCEAN}(t)}{S^{m}_{OCEAN(1990-1999)}} * 2.2 \tag{2.5}$$

where n represents the number of variables; m represents the factors; and t represents the period. The normalization is considered when the ratio or the sequestration of CO_2 in the 1990s are assumed to be underestimated and is initially triggered by diffusion, which relies on the CO_2 gradient. Therefore, the ratio is considered a naturally appropriate approach that accounts for the time dependence of the CO_2 gradient in the oceans [25–27].

2.1.2.4 CO_2 Absorption by Terrestrial Vegetation and the Earth

The variations in the CO_2 emissions from fossil fuels (E_{FF}) and land-use change (E_{LUC}) as well as the growth rate of the atmospheric CO_2 concentration (G_{ATM}) and ocean CO_2 sequestration (S_{OCEAN}) can be accounted for to determine the net sequestration of CO_2 by the terrestrial vegetation (S_{LAND}) by considering Eq (2.1). Therefore, this type of sequestration can be computed as the CO_2 remaining from the mass balanced budget, which is expressed as follows:

$$S_{LAND} = E_{FF} + E_{LUC} - (G_{ATM} + S_{OCEAN}) \tag{2.6}$$

Here, S_{LAND} is computed from the remainder of the estimates and includes all perturbed carbon from fossil fuels, land-use change, and the CO_2 atmospheric growth rate. The computation of S_{LAND} in Eq. (2.6) with the budget from the DGVMs can be used to calculate E_{LUC} by subtracting the impact of land-use changes, which will provide an independent calculation of a consistent S_{LAND}. Thus, it can represent an appropriate understanding of the role of the terrestrial vegetation in determining the response to CO_2 and climate variability (Fig. 2.2).

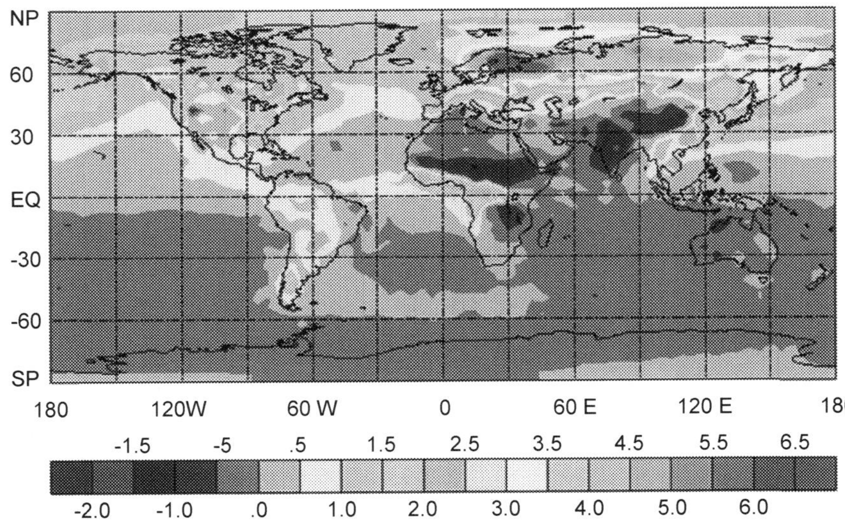

Figure 2.2 Annual mean mixing ratio of CO_2 (ppmv), according to the PBL, as simulated in MATLAB using terrestrial biosphere surface fluxes. The concentration was initialized to be globally uniform, and the model was run on a coarse 7.2 x 9 grid with 9 levels for 10 years. The end of this "*spin up: run" was then used as the initial condition for a further 4-year integration on a 4 x 5 grid with 17 levels. All results presented here are for the final 3 years of the 4 x 5 run.

2.1.2.5 Calculation of the Growth Rate of the Atmospheric CO_2 Concentration (G_{ATM})

The total growth rate of the concentration of atmospheric CO_2 was calculated for 1956–2016, that is, the annual increase in the concentration of atmospheric CO_2 was calculated by comparing records of CO_2 emissions and sequestration from multiple sources (DEP, NASA, NOAA/ESRL, 2015; Scripps, NEDO, and UNDP), considering that the ocean boundary layers and the air are very well mixed and very much in a functional band with time period and latitude [28–30]. The growth rate unit, ppm yr^{-1}, is then converted from GtC yr^{-1} to be consistent with the other components.

2.1.3 Results and Discussion

The average global carbon emissions from 1875 to 2016 have been estimated by numerous scientists from NASA, NOAA, DEP, NASA, NOAA/ESRL, Scripps, NEDO, UNDP and several other agencies (Fig. 2.2). Over this timescale, 91% of the total emissions ($E_{FF} + E_{LUC}$) was from the burning of fossil fuels and other industrial activities, and 9% resulted from land-use change [31,32]. These emissions were captured by sinks within the atmosphere (44%), the oceans (26%), and the land (30%), which accounts for the annual growth of the CO_2 concentration in the atmosphere [18,28]. Here, except for the land-use component, carbon emissions have grown since 1959 with important interannual differences in the growth rate of atmospheric CO_2 concentration and CO_2 sequestration by the terrestrial vegetation (Fig. 2.4) and considering decadal differences in all terms (Table 2.1).

2.1.3.1 CO_2 Emissions From Fossil Fuels, Land-Use Change, and Other Factors

Global CO_2 emissions from burning fossil fuels and industrial activities increased at an average of 3.1 ± 0.2 GtC yr^{-1} per decade in the 1960s and by an average of 9.3 ± 0.5 GtC yr^{-1} from 2006 to 2015 (Table 2.1 and Fig. 2.3). The emissions growth rate decreased in the 1960s and 1990s being 4.5% per year in the 1960s (1960–69), 2.8% per year in the 1970s (1970–79), 1.9% per year in 1980s (1980–89), and 1.1% per year in the 1990s (1990–99). Growth started to increase again in the 2000s at an average rate of 3.5% per year, but it has decreased to 1.8% per year in recent decades (2006–2015). However, the CO_2 emissions from land-use change have remained nearly constant at approximately 1.3 ± 0.5 GtC yr^{-1} during 1960–2015. The reduction in emissions from land conversion suggests that significant emissions occurred in the 1990s and 2000s as calculated by bookkeeping and DGVM methods in MATLAB software (Table 2.1). A change in tropical deforestation does not account for this reduction, but the flux in the 1990s was like that in the 2000s, suggesting a decrease in E_{LUC} from 2011 to 15 relative to that from 2001 to 2010.

2.1.3.2 Ocean and Terrestrial Vegetation CO_2 Sinks

The global CO_2 sequestration by the ocean was estimated based on the mean CO_2 sequestration value for the 1990s combined with the differences in ocean CO_2 sequestration for 1959–2015 including all seven global ocean biogeochemical cycles. For the 1990s, the mean ocean CO_2 sequestration was estimated to be approximately 2.2 ± 0.4 GtC yr^{-1} based on indirect observational data from the IPCC and considering the timescale for CFC penetration [33–35]. This CO_2 sequestration value is comparable with that of 2.0 ± 0.5 GtC yr^{-1} calculated for the 1990s and 1.9 to 2.5 GtC yr^{-1} for 1990–2009[9,13] with a ± 0.3 to ± 0.7 GtC yr^{-1} range of uncertainty. Therefore, this method can accurately calculate observation-based ocean

Table 2.1 The results implemented from using bookkeeping technology and budgeted residuals from DGVMs implemented in MATLAB software and its inverse calculations for 1960–69, 1970–79, 1980–89, 1990–99, and 2000–09 shown in $GtCyr^{-1}$

Mean ($GtC\ yr^{-1}$)	1960–69	1970–79	1980–89	1990–99	2000–09	2006–2015	2016
Carbon Emissions from Land-Use Change (E_{LUC})							
Bookkeeping	1.5 ± 0.5	1.3 ± 0.5	1.4 ± 0.5	1.6 ± 0.5	1.0 ± 0.5	1.0 ± 0.5	1.3 ± 0.5
DGVMs[2]	1.2 ± 0.3	1.2 ± 0.3	1.2 ± 0.2	1.2 ± 0.2	1.1 ± 0.2	1.3 ± 0.3	1.2 ± 0.4
Residual Terrestrial Sink (S_{LAND})							
Budget Residual	1.7 ± 0.7	1.7 ± 0.8	1.6 ± 0.8	2.6 ± 0.8	2.6 ± 0.8	3.1 ± 0.9	1.9 ± 0.9
DGVMs[a]	1.2 ± 0.5	2.2 ± 0.5	1.7 ± 0.6	2.3 ± 0.5	2.8 ± 0.6	2.8 ± 0.7	1.0 ± 1.4
Total Land Fluxes ($S_{LAND}-E_{LUC}$)							
Budget	0.2 ± 0.5	0.4 ± 0.6	0.1 ± 0.6	1.0 ± 0.6	1.6 ± 0.6	2.2 ± 0.7	0.6 ± 0.7
($E_{EF} - G_{ATM} - S_{ocean}$) DGVMs[a]	−0.2 ± 0.7	1.1 ± 0.5	0.4 ± 0.5	1.1 ± 0.3	1.8 ± 0.4	1.7 ± 0.5	−0.1 ± 1.4
Inversions (CTE2016- FT/Jena CarboScope/(CAM)[b]	–/–/-	–/–/-	-/0.2[b]/0.9[b]	-/1.0[b]/1.9[b]	1.5[b]/1.6[b]/2.5[b]	2.2[b]/2.3[b]/3.4[b]	1.9[b]/2.6[b]/2.6[b]

DVGMs, dynamic global vegetation modeling.

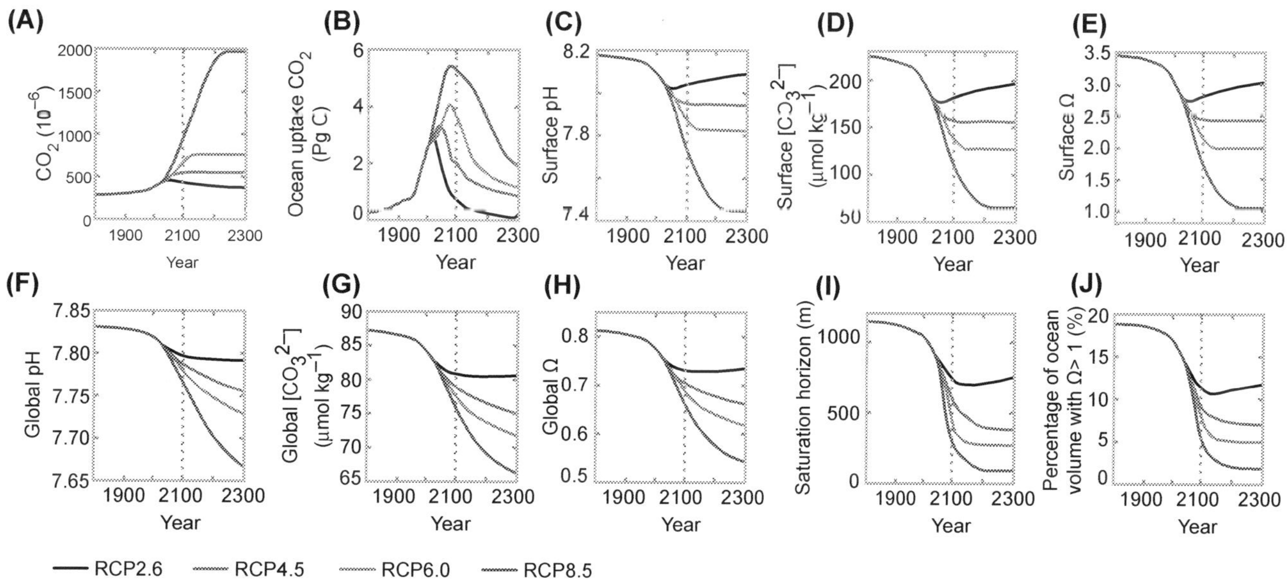

Figure 2.3 The impact of global scale carbon emission (A) on earth, (B) ocean, (C—E) earth surface in various parameters, (F—J) ocean pH, and various parameters of RCP2.6, RCP4.5, RCP6.0, and RCP8.5.

sequestration by implementing the sea—air pCO_2, (sea—air pCO_2 variable) ⋆ (gas transfer coefficient), and the results are completely consistent with the observed results [36,37].

Ultimately, the calculation shows that residual terrestrial CO_2 sequestration increased from 1.7 ± 0.7 GtC yr^{-1} in the 1960s to 3.1 ± 0.9 GtC yr^{-1} from 2006 to 2015 with an interannual variability of up to 2 GtC yr^{-1}. It also indicates reduced land sequestration during El Niño and overaccumulation by ocean sequestration, which accounts for the increasing growth rate of atmospheric CO_2 (Fig. 2.4). The high CO_2 sequestration by the land from 2006 to 2015 is comparable with that of the 1960s as determined by DGVMs by incorporating the estimated budget residual in response to the increase in cumulative atmospheric CO_2 (Fig. 2.4). This DGVM mean of 2.8 ± 0.7 GtC yr^{-1} is technically the observational average for 2006—15, which is estimated from the budget residual (Fig. 2.5).

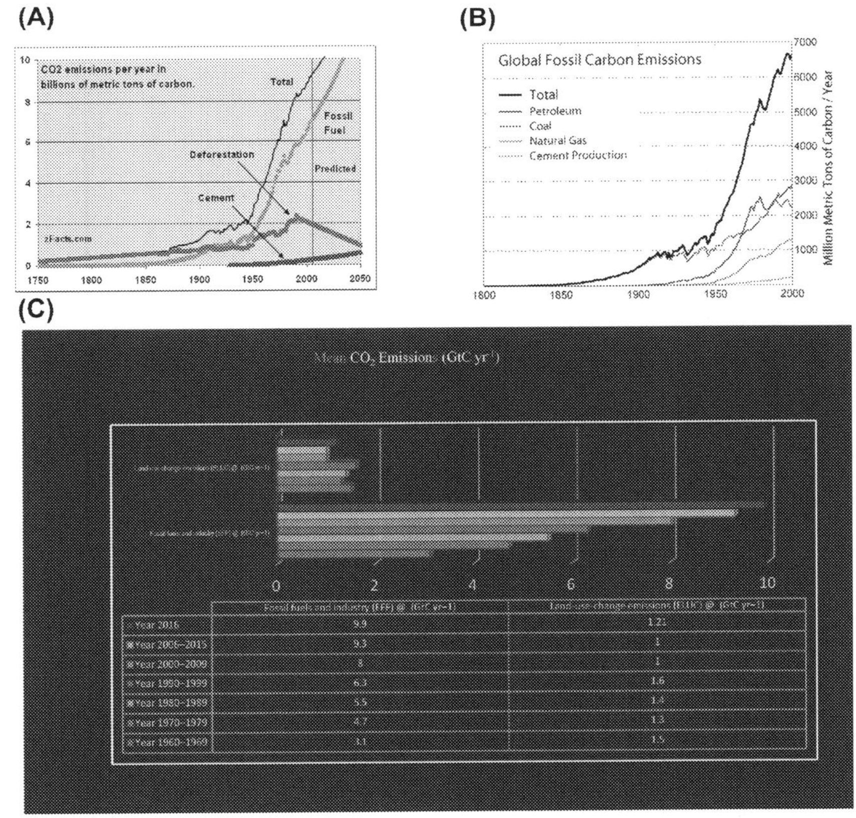

Figure 2.4 (A) Total CO_2 emission of earth, (B) CO_2 budget from burning fossil fuels, e.g., gas, oil, and coal. (C) Total estimated CO_2 emissions from both fossil fuels and land-use change from 1960—69 through 2016.

(A)

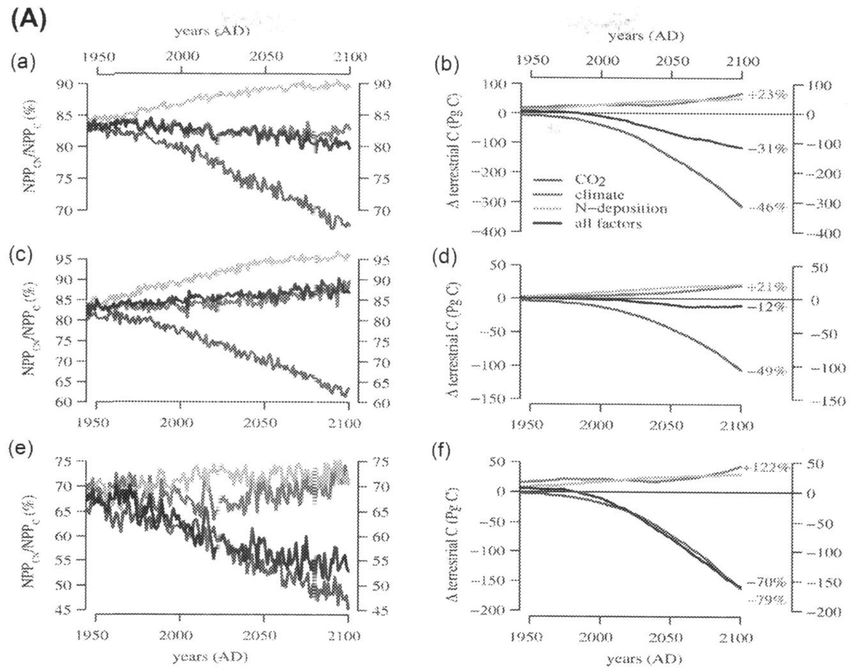

(B)

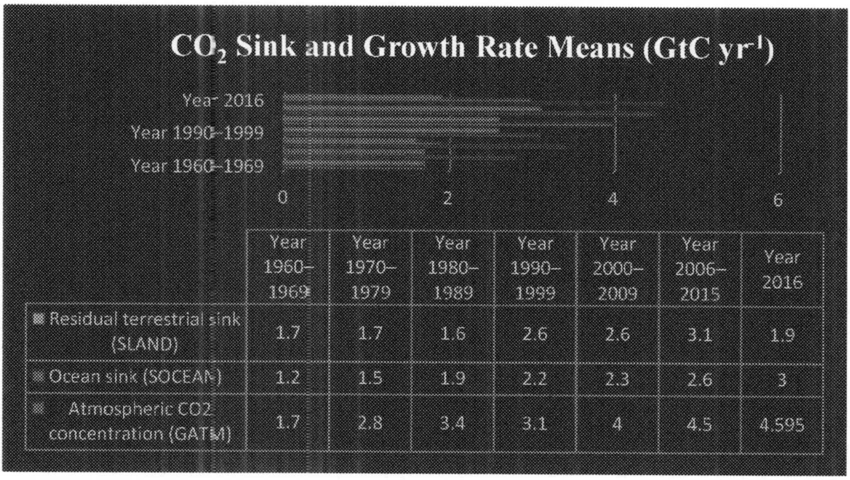

	Year 1960–1969	Year 1970–1979	Year 1980–1989	Year 1990–1999	Year 2000–2009	Year 2006–2015	Year 2016
Residual terrestrial sink (SLAND)	1.7	1.7	1.6	2.6	2.6	3.1	1.9
Ocean sink (SOCEAN)	1.2	1.5	1.9	2.2	2.3	2.6	3
Atmospheric CO2 concentration (GATM)	1.7	2.8	3.4	3.1	4	4.5	4.595

Figure 2.5 (A) CO_2 sink, N-depletion, climate dynamics, and all other factors in terrestrial environment, (B) CO_2 sinks and annual growth rate of CO_2 emissions in $GtCyr^{-1}$ from 1960–69 through 2015. The calculation $\left[\left(\frac{4.595-4.500}{4.500}\right)\times 100\% = 2.11\%\right]$ revealed that atmospheric CO_2 concentration increasing at a rate of 2.11% annually.

Subsequently, the entire CO_2 land flux (S_{LAND} - E_{LUC}), which is extended by atmospheric inversion, is greatly beneficial to the calculation of the global budget, G_{ATM}, along with the assumption of relative uncertainty in S_{OCEAN} and E_{FF} due to the inversions. The total land flux has the same magnitude considering the decadal mean, which for 2006–15, accounted for three inversions of 2.2, 2.3, and 3.4 GtC yr^{-1} relative to the total flux calculation of 2.1 ± 0.7 GtC yr^{-1} using Eq. (2.1). The total land sequestrations by these inversions are 1.8, 1.8, and 3.0 GtC yr^{-1} including the average of all reverse flux variables, 0.45 GtC yr^{-1}. Interestingly, the interannual variation within the inversions matched the residual-based S_{LAND} very closely (Fig. 2.5). Therefore, the total land fluxes determined by DGVM confirm that the calculation of the carbon budget and its inversion in the atmosphere has a decadal mean of 1.7 ± 0.5 GtC yr^{-1} (Table 2.1 and Fig. 2.5).

2.1.3.3 Cumulative CO_2 Emissions and Atmospheric Impact

In this study, the total emission from fossil fuel burning and land-use change from 1870 to 2015 was calculated as 555 ± 55 GtC. Within the atmosphere, the emissions are partitioned among the atmosphere (235 ± 5 GtC per the atmospheric CO_2 concentrations of 288 ppm in 1870 and 399.1 ppm in 2015), the ocean (160 ± 20 GtC), and the land (160 ± 60 GtC). The total emissions during the preindustrial development period of 1750–1869 were 3 GtC for E_{FF} and approximately 45 GtC for E_{LUC} (rounded to the nearest 5), of which 10 GtC were emitted in the 1850–70 period and 30 GtC in the 1750–1850 period for an average of 25 GtC, which revealed that the atmospheric CO_2 concentration growth rate during this period was 25 GtC and that of the ocean sequestration was 20 GtC and terrestrial land sequestration was 5 GtC [38–40].

Based on the calculations in this study, the growth of the atmospheric CO_2 concentration (G_{ATM}) is projected to increase in 2016 to 6.7 ± 1.1 GtC (3.15 ± 0.53 ppm). Combining the projected E_{FF} and G_{ATM} reveals that summing the sequestration by the land and ocean and subtracting the emissions from land-use change ($S_{LAND} + S_{OCEAN}$ - E_{LUC}) only yields approximately 3 GtC. S_{OCEAN} accounted for 3.0 GtC in 2015 and is expected to increase in 2016 [41–43], and E_{LUC} accounted for 1.3 GtC in 2015 with a decadal average of 1.0 GtC yr^{-1}. Here, the remaining terrestrial sequestration, S_{LAND}, for 2016 is, as expected, far below the mean balance of E_{LUC} for 2006–15 [44,45]. Therefore, to determine the total carbon accumulated and its rate of increase in the atmosphere since

1750 or 2016, I have used the atmospheric CO_2 concentrations of 288 ± 3 or 399 ± 3 ppm, respectively. The variation of ± 3 ppm (converted to $\pm 1\sigma$) was obtained from IPCC data, and the rate of growth of the atmospheric CO_2 concentration was calculated using MATLAB software by comparing the decadal and individual annual values for the past 10 years. The concentration of CO_2 in the atmosphere was shown to be increasing at a rate of 2.11% ppm annually. Because the atmospheric concentration of CO_2 was 399.4 ± 0.1 ppm in 2016 and it has been growing at a rate of 2.11% per year, a toxic CO_2 concentration of 60,000 ppm will result in 121,017,712 years.

2.1.4 Conclusion

Global CO_2 emissions and their redistribution within the atmosphere, ocean, and terrestrial biosphere through absorption are calculated for the past several decades to determine the future atmospheric CO_2 concentration. The annual growth rate of the global atmospheric CO_2 concentration (G_{ATM}) (ppm) is computed considering the global carbon budget of the atmosphere, the ocean, and the land over the last several years and was shown to currently be 2.11% annually. If we cannot reduce this current level (400 ppm) of annual CO_2 growth in a timely manner, the atmospheric CO_2 concentration will eventually reach a toxic level (60,000 ppm), at which time the entire human race and all other living beings on the planet will perish in 30 min in 121,017,712 years.

2.2 WATER

Water is the vital element in our daily life which needs to be conserved and should be free from contaminants to keep the global environment and living beings safe. This report thus, identifies the emerging contaminants (ECs) in water environment and recommends the direction for future monitoring for contaminant-free water for our daily needs. These ECs are ubiquitous in the aquatic environment, mainly derived from the discharge of municipal wastewater effluents. Their presence is of concern due to the possible ecological impact (e.g., endocrine disruption) to biota within the environment. To better understand their fate in water and in the environment, a standardized approach is needed to ensure the representative data are attained and facilitates a better understanding of spatial and temporal trends of ECs occurrence. This report thus, describes the impact of traditional wastewater treatment into the environment. Because conventional wastewater treatment has the lack of suspended particulate matter analysis due to further preparation requirements and a lack of good analytical approaches which results in aquatic environment got pollutant. In the traditional technology, most of the treated sludge is applied directly to agricultural land without analysis for ECs which results in soil contamination and consequently, enforcing to get toxicity into the exposed terrestrial organisms. Therefore, this report suggested abandoning the traditional wastewater treatment mechanism and finding better technology to obtain a world of pollution-free water world to secure a better environment Earth.

2.2.1 Introduction

In surface waters, a broad range of inorganic and organic contaminants primarily have traditionally been originated from industrial or agricultural chemicals wastes that cause environmental pollution. Besides, a greater number of municipal wastewater treatment chemicals and pharmaceutical chemicals described as ECs such as 17β-estradiol (E2), 17α-ethinylestradiol (EE2), and diclofenac are also part of causing water contamination. To date, millions of different industries alone have been responsible for water contamination and toxicity globally, with concentrations up to a maximum of 6.5 $mg\ L^{-1}$ for the antibiotic ciprofloxacin [46–48]. In this report, single compound acute toxicity testings (including crustaceans, algae, and bacteria) conducted under controlled laboratory conditions have found median effective concentrations (EC_{50}'s—concentration at which the toxicological response to an organism is halfway between a normal and maximum response for a preset time period) for a number of these ECs to be $< 1\ mg\ L^{-1}$ [43,49,50]. Such effect concentrations classify the chemical as potentially very toxic to aquatic organisms [39,41,51]. The presence of these chemicals in the environment is more concerning considering that they do not appear individually, but as a complex mixture, which could lead to unwanted synergistic effects. The ubiquity of a high number of potentially toxic ECs in the environment underpins the need to better understand their occurrence, fate, and ecological impact. This report describes current knowledge on the occurrence of ECs in wastewaters and surface waters using the standard data set as a model to represent global water contamination. From the data set and wider literature, areas of concern considered to be understudied are discussed, considering spatial and temporal variability of ECs in wastewater and river water, partitioning of ECs to solid matter during wastewater treatment, fate of ECs in environmental waters and toxicological impact of ECs within the environment. Finally, recommendation for future water environmental monitoring approaches is proposed for contaminant and toxic free water world.

2.2.2 Current Knowledge of Emerging Contaminant Occurrence in Wastewaters and Surface Waters

The presence of ECs in the water environment is mainly attributed to the discharge of treated wastewater from wastewater treatment plants (WwTPs), which are contaminated by chemicals, where the aquatic environment gets polluted and that has an impact on dissolved oxygen and chlorophyll production of aquatic plants in the body of water and causes rapid growth of deadly microorganism (Fig. 2.6). Conventional secondary processes (activated sludge and trickling filters) represent the most extensively used and studied processes and are therefore focused on in this report. However, this report is not designed to remove ECs resulting in their discharge to receiving surface waters including rivers, lakes, and coastal waters but describes the impact of chemicals such as chlorine, NSAIDs, β-blockers, antidepressants, and the antiepileptic carbamazepine into influent wastewaters [29,39,52]. Reports suggested that removal varies broadly from low (<50%) to high (>80%) due to their different physicochemical properties such as chlorophyll and susceptibility to microbiological attack into the water environment by receiving surface waters in the ng to mg l^{-1} range which leads surface water getting contaminated [51–53].

2.2.3 Factors Which Influence Receiving Wastewater Concentration

A large variation in influent wastewater concentrations (equivalent to more than an order of magnitude) has been observed for some ECs where chlorine and acetaminophen were found within influent wastewater at mean concentrations ranging from 6924 to 492,340 ng L^{-1} [54–57]. This indicates spatial and/or temporal variations in their usage. However, dilution from industrial inputs, degradation within the upstream sewer, rainfall, and sampling mode will all contribute to this variability. The use of inappropriate sampling strategies is considered the greatest weakness of reported occurrence data. Ort et al. gave an excellent account of uncertainties associated with different sampling modes [58,59]. Current approaches tend to use sampling of low interday frequency and often no intraday repetition. This approach has

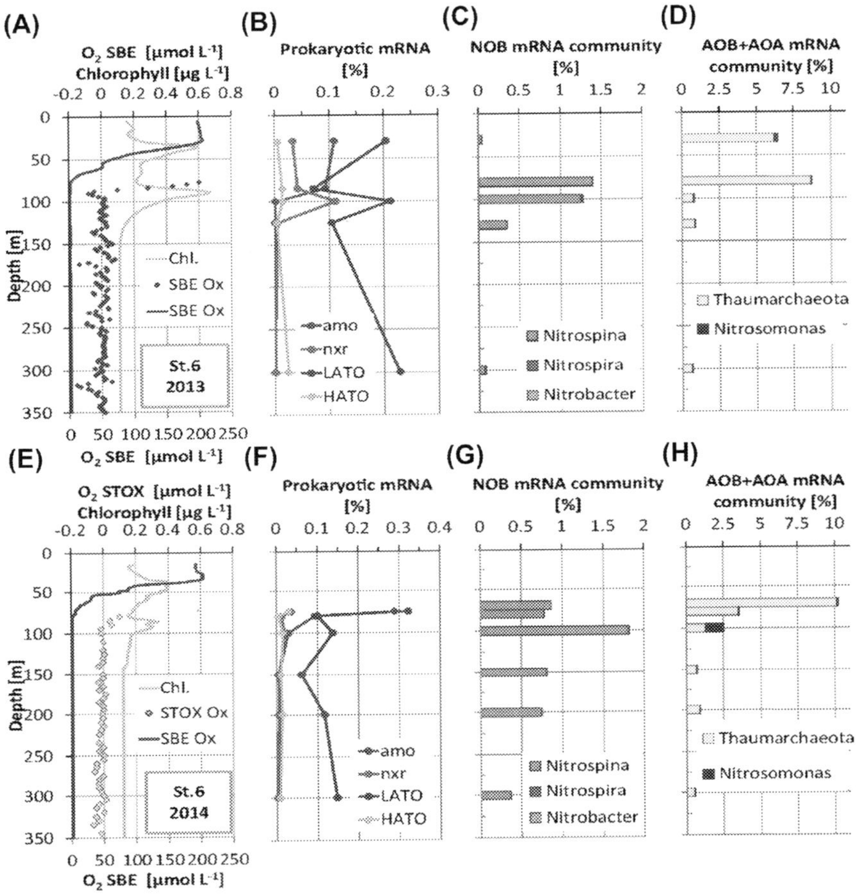

Figure 2.6 Water column dissolved oxygen (O_2), chlorophyll concentrations (Chl.), and microbial transcript gets abundances at station T6 in the ETNP in (*A–D*) 2013 and (*E–H*) 2014. (*A* and *E*) O_2 based on SBE and STOX sensor measurement and chlorophyll inferred from in vivo fluorescence. (*B* and *F*) Abundances of transcripts encoding LATO and HATO, *amo*, and *nxr*, as a percentage of total prokaryotic mRNA. (*C* and *G*) Taxonomic classification of total mRNA affiliated with NOB and (*D* and *H*) AOB (Nitrosomonas) and AOA (Thaumarchaeota), as a percentage of total prokaryotic RNA. (Source: Emilio Garcia-Robledo et al. PNAS (2017); 114 [31] 8319–8324).

limitations as it only yields a snap shot of EC concentration for a specific point in time. Time and volume/flow proportional composite samplers are also reported, but to a lesser extent [60–62]. The latter being preferred as it accounts for fluctuations in flow. The relatively high cost and logistical issues associated with this sampling mode have resulted in very few studies applying flow proportional sampling report. Also, a major uncertainty of

24-h composite sampling is chemical stability. This is not often investigated but is known to be significant for some chemicals [63,64]. Due to uncertainties of existing sampling methods, there is lack of understanding on spatial and temporal variations in EC concentrations.

2.2.4 Spatial Distribution

Assessing spatial distribution of EC contamination is notoriously difficult. The collation and interpretation of literature data from a variety of sources has obvious limitations. On the other hand, studies have been able to tentatively assess spatial distribution within a single catchment [65–67]. Such studies must rely heavily on sampling due to the large number of sites which are monitored at approximately the same time. Despite the uncertainties associated with sampling report, information attained from such studies is extremely valuable. It was used as a primary indicator of sites within the catchment that requires more detailed study. These can then be subject to more robust sampling protocols to facilitate greater understanding of EC fate and occurrence during wastewater treatment and in the environment at these predetermined sites of interest.

2.2.5 Intraday Variation

It is anticipated that concentrations of ECs in receiving wastewater vary throughout the day. Coutu et al. collected hourly composite samples (i.e., one sample every 15 min to create an hourly composite) for influent wastewater over 24-h sampling periods [68,69]. This was used to investigate within day variability of antibiotic concentrations. Samples were chilled upon collection but their stability at 4°C was not investigated or referenced to. However, significant degradation (>15%) has been observed for some ECs stored at 4°C for only 12 h (Baker and Kasprzyk-Hordern, 2013). Despite this uncertainty, the six antibiotics studied (trimethoprim, norfloxacin, ciprofloxacin, ofloxacin, clindamycin, metronidazole) showed variation in concentration throughout the day [70–72]. A peak in receiving load was observed between 07:00 and 9:00 h (Fig. 2.7). This is explainable by posology and the accumulation of administered drugs within urine during sleep [73–75]. Therefore, following the first toilet flush of the day an increase in load of antibiotics is observed. Other urine-derived ECs are expected to behave in the same way. Yet it is unknown how the performance of WwTPs responds to this daily spike in receiving concentration and volumetric load. Collection of corresponding final effluent samples would help address this.

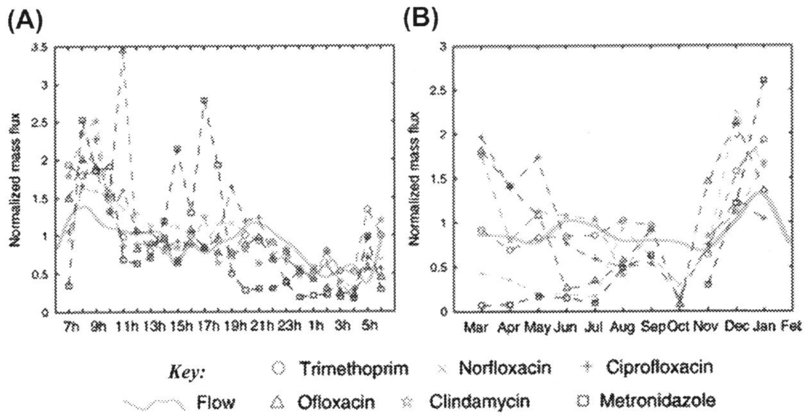

Figure 2.7 Fluctuations of mass flux for selected antibiotics in wastewater throughout a 1-day period (A) and during a year (B). Adapted from Coutu et al. (2013).

2.2.6 Interday Variability

Wide reports of 19 countries using a variety of time, flow, and volume proportional sampling attempted to investigate interday variability (i.e., over a 1-week period) of chemicals in influent wastewater [76–78]. Collated reports revealed a trend of recreational use for some compounds [79,80]. Both organic and inorganic chemicals showed elevated levels of water contaminants. In contrast, it is anticipated that also hospital-dispensed chemicals such as X-ray contrast media and anticancer chemicals are observed at higher concentrations into the wastewater body. Interday variability of EC concentration is difficult to fully appreciate without conducting flow proportional sampling. This is most significant when comparing weekday and weekend periods where wastewater flows can vary notably. Differences in weekday/weekend flow will be specific for the catchment in question. For example, a residential (commuter) or tourist area will see increased flow during weekends. On the other hand, a catchment containing industrial works may observe a reduction in flow and dilution during weekends. Rainfall will also result in changes to wastewater flow between days. This illustrates that reporting load (mass per day) over the traditional approach of concentration (mass per liter) to overcome temporal variations in flow is more suited to describe findings.

2.2.7 Partitioning of Emerging Contaminants to Solid Matter during Wastewater Treatment

EC analysis in wastewaters tends to be undertaken on the aqueous phase only (i.e., a prefiltered sample). Particulate phase analysis is not routinely undertaken because up to 1 g of dry solids is often required for each analysis and an additional sample extraction is needed [76,81,82]. However, here it is essential as some chemicals have a high affinity to particulate matter. Numerous chemicals including amitriptyline, EMDP, dosulepin, fluoxetine, norfluoxetine, triclosan, ofloxacin, and ciprofloxacin have been found to be within the particulate phase of influent wastewater at significant concentrations (>20% of the total concentration). Consequently, particulate phase determination is necessary to correctly report influent concentration for some chemicals. This is also essential for back calculations of population usage by wastewater-based epidemiology [83,84]. Partitioning of ECs to suspended matter in final effluents is even less studied due to the very low solid concentrations encountered (typically 10–30 mg L^{-1} for secondary processes). Nevertheless, some researchers found that final effluent of secondary processes had >20% of the total triclosan, ofloxacin, and ciprofloxacin concentration to be within the particulate phase, despite very low suspended solids concentrations [85–87]. Particulate phase concentrations were equivalent to and in the range of 26–296 ng L^{-1}. This provides a route for their release into the environment which goes unmonitored, and the environmental fate of these particulate bound chemicals is unknown.

2.2.8 River Sediments and Amended Soils

ECs have also been reported in riverine sediments due to the rapid growth of bacterial contamination (Fig. 2.8). Determining toxicological impact is essential as sediments can act as a sink for their accumulation. However, this is notoriously difficult to ascertain due to the complexity of the system. Benthic organisms can be exposed to ECs within the sediment itself, in interstitial water, and in overlying water [88–90]. This makes experimental design and set up critical for reproducing representative conditions. A study investigating toxicity of diclofenac to the crustacean *Hyalella azteca* prepared synthetic sediment at a 1:3 ratio with water [72,91]. The authors failed to report or investigate aqueous/liquid partitioning of diclofenac, whether equilibrium conditions were present or if desorption occurred throughout the test. Consequently, reporting toxicological effect concentrations for sediments can have considerable uncertainties. An improved study of biological response information with chemical analysis for both aqueous and particulate phases has been performed [57,92,93]. Their findings suggested that sorption to sediments resulted in a reduction of bioavailability and toxicity. On the other hand, the accumulation (and increased concentration) of readily sorbed compounds in sediments within

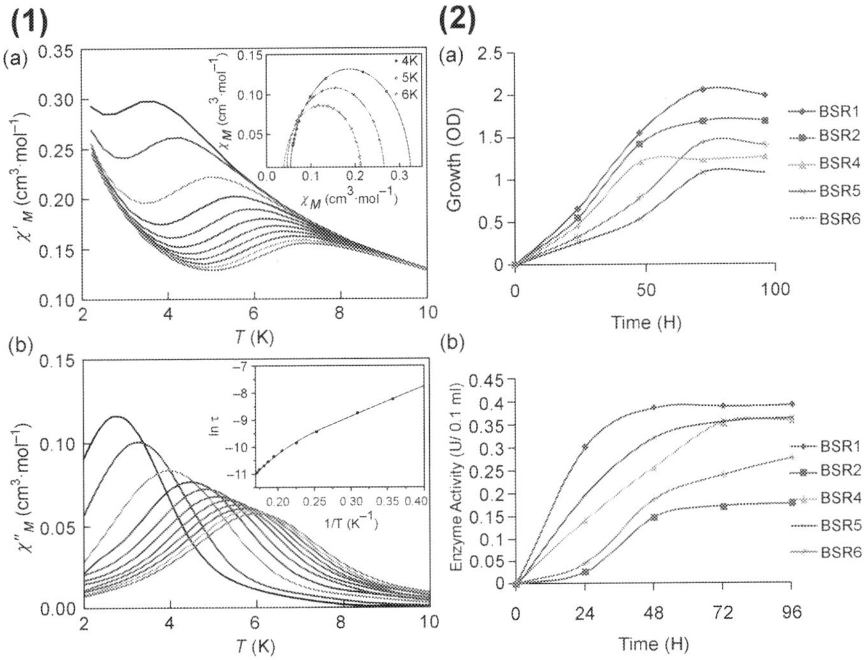

Figure 2.8 (1) The contaminant bacterial ignition in response to contaminant water temperature per unit areas; (2) the contaminant bacterial growth during the certain period of time where enzyme activities were initiated by the water contamination.

the environment could compensate for this reduction in toxicity. Interestingly, the activity of benthic invertebrate also resulted in increased desorption, leading to improved bioavailability [75,86]. Numerous mechanisms take place, illustrating the complexity of ascertaining sediment toxicity. Studies over long-time periods which simulate representative steady-state riverine sediment conditions are required for a range of indicator species. These firstly need to be supported with chemical analysis to assess the behavior of the EC(s) in question. Such studies would help assess chronic and multigeneration impact of sediment contamination to benthic organisms which is ultimately impacting global ecology and finally affecting the human being.

2.2.9 Diagnosis of Emerging Contaminant Removal During Wastewater Treatment

To better understand their pathways of removal during wastewater treatment, particulate phase analysis is needed as well as analysis of the biomass (either suspended or attached) of the process [94–96]. Ideally, corresponding aqueous and particulate determinations should be undertaken for each sampling point such that a complete process mass balance is attained. This can give valuable information on the dominant mechanisms which govern removal [97–99]. Removal can vary greatly between ECs from a physically driven process (adsorption) to biologically mediated enzymatic reactions (biodegradation) [99–101]. Their identification also needs to be supported with information of process conditions and operation, nutrient removal, and complimentary analysis of the physical/biological characterization of biomass [102,103]. With this knowledge, the operation of the process could be adjusted to favor their removal [59,79]. Such information can also be used to identify where further research efforts may be needed. For example, those chemicals removed by biodegradation suggest further investigation of possible biotransformation products in final effluents is needed, and those removed by adsorption require further understanding of their fate during and following sludge treatment where ECs undergo microbially mediated reactions during secondary wastewater treatment [104,105] and in the environment can be effective technology. Biodegradation is often referred to as the dominant fate pathway for the removal of some ECs from the aqueous phase of wastewaters and surface waters [61,64]. However, this can result in the formation of numerous degradation or transformation products. Therefore, the ECs in wastewater, it is expected that a great number of transformation products (of unknown toxicity and persistence) exist in final effluent and receiving surface waters need to be removed to avoid chronic impact to the water environment and living being on Earth [87,105].

2.2.10 Conclusions

Global water is getting contaminated due to the lack of knowledge for water conversation and the application of conventional wastewater treatment process, which needs to be resolved urgently to protect the environment and living being. Currently, the application of conventional technology for the treatment of wastewater and water management is backdated, causing severe environmental pollution and loses of water severely. Therefore, removal performance of different WwTPs process types at various operational conditions needs to be reevaluated to conserve and treat the water to mitigate the water perplexity throughout the world. Necessarily, further research and sustainable technology are needed to determine fate and removal of ECs during treatment sophisticatedly, considering its impact across the complete ecosystem and living being on Earth.

2.3 LAND

A vital component in comprehending the human activities and interactions with the atmosphere is land use and land cover (LULC). Therefore, having the capacity to stimulate changes is essential. Because the land surface heats up and cools down faster compared with water or air, the planet land interacts with and affects climate heavily [39,62]. Land use, reflectivity, topography, elevation, and latitude all have variable impacts. The land latitude is capable of effecting, however, considerable radiations that reach the surface. The planetary albedo reflexivity as well as the sort of land cover that obtains energy from the Sun influences the number of energy that is mirrored to planet Earth. As a result, land management and land use practices possess a substantial influence on natural resources comprising animals, plants, nutrients, soil, and water.

2.3.1 Background and Introduction

In international environment change, LULC change is a key issue. During the 1972 Stockholm Conferences on the Human Environment and the 1992 United Nations Conferences on Environment and Development (UNCED) the community of scientific research called for practical study of land-use changes [106,107]. Simultaneously, the International Geosphere and Biosphere Program (IGBP) as well as the International Human Dimension Program (IHDP) coarranged a working group promoted research activity to set up research agenda for LULC changes. Where other factors are incorporated on the necessity basis to drive different developmental index for water and land resources, land cover/land use mapping is a vital element. Naturally, based on NRSA (1989), land cover refers to artificial covers, rock/soil, water bodies, natural vegetation, and others found on the land. On the other hand, land use refers to the activities of human beings and various uses which are conducted over land. One of the most essential earth system properties is land cover which is defined as biotic and abiotic component assemblage on the surface of the earth. While land cover refers to that which covers the earth surface, land use defines the way in which land cover is transformed. Land cover is comprised of bare soil, forest, grassland, snow, and water. Conversely, land use includes wildlife management areas, recreation area, built-up land, and agricultural land. Land use describes the activities of man on land, which are directly associated to the land [39,48,51]. Land cover mirrors the biophysical condition of the surface of the earth and the immediate subsurface, hence embracing water, vegetation, and the soil material. Land cover and land use are self-motivated, and transformations may include the intensity of nature of change, although it may as well involve spatial and aspects of time. In addition, land cover and land use involve the alteration either indirect or direct of natural habitations and theirs influence on the area ecology. Moreover, population pressure which results in intense use of land minus proper management practices is mainly the cause of land degradations.

Because of overpopulation, people tend to shift toward sensitive areas on earth, use land without taking into consideration the slope as well as the erodibility, which eventually results to severe soil erosion. Road construction effect as well other similar landscape disturbances on landslides and erosion, and other mass movements on earth are well recognized [51,74]. Evidently, the extent of the land underneath forest is getting reduced due to human activities. Similarly, the land being utilized for cultivation is decreasing as well. On the other hand, instantaneously, land underneath built-up region is constantly increasing. The function of the property promoters and the real estate individuals has in the recent time been bringing a severe tragedy to agricultural land and forest area. Clearly, this is a harmful land management condition. Therefore, understanding the existing state is essential, and this kind of study on LULC change detection, to make future (Fig. 2.9).

Use of land cover or land-use change investigation, as well as human activities assessments and biophysical capacities would greatly increase the method to comprehending the key causes and extent of land degradation as well as loss of variety [3,4,28]. Land use, however, is dynamic and essential data for planning of operative land measurement program, thus requiring more practical to deal with degradation's root cause. This report therefore offers an overall framework and guide for this analysis. The report will discuss the exact information that can be acquired from land use investigation and provide a theoretical framework for recognizing the key causes of land-use change. Thus, it will offer a set of common problems as well as research questions and finally will discuss the analytical framework implications for the choice of approaches and results interpretations. This is based on intuitions as well as capabilities acquired from enduring research and development activities carried out by team members of LUCID project crossways

(1)

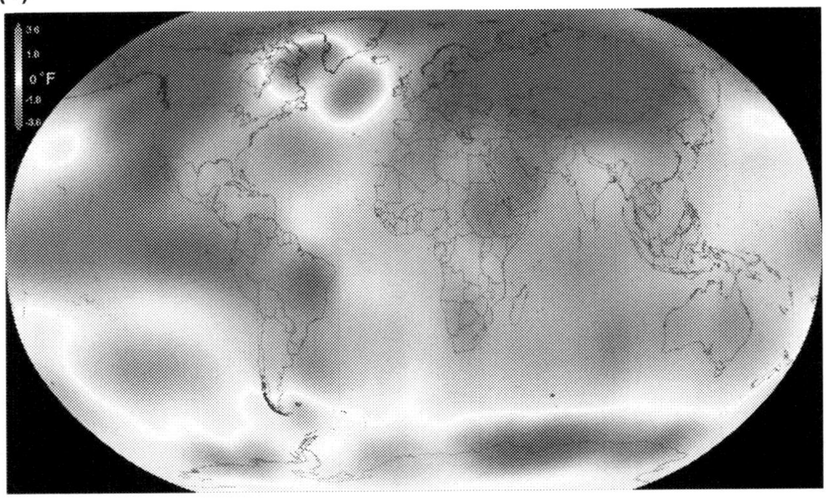

(2)

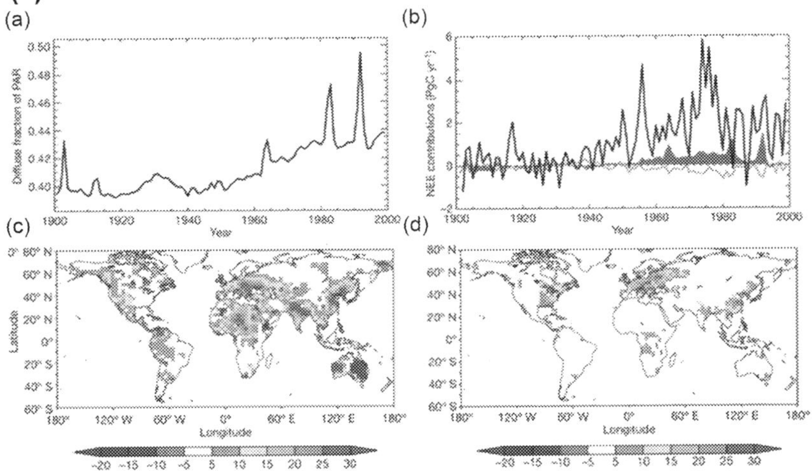

Figure 2.9 (1) Global temperature anomalies for 2015 compared with the 1951–80 baseline. 2015 was the *warmest year* in the NASA/NOAA temperature record, which starts in 1880. It has since been superseded by 2016 (NASA/NOAA; 20 January 2016). (2) Simulated global mean annual diffuse fraction of the PAR, based on aerosol optical depth from volcanic and anthropogenic sources, as simulated by.

numerous sites within east Africa, and from other land use transformations and key causes research [77,78,101]. Through offering an overall analytical framework for research design of land use and the gathering and analysis of data, this report matches other LUCID project procedural papers.

2.3.2 Root Cause Analysis for Land-Use Change

While related root cause investigation illuminates the way in which trends in land management are affecting land biodiversity and degradation, land use spatial analysis makes available rich societal and environmental information. The following are the examples of data that root cause and land use analyses can make available: leverage points within the system to enhance land management; identification of serious conditions, groups, and locations; scenarios of the influence of likely program or policies; determination of the paste program and policies effect on land management as well as environment so as to acquire lessons learned from cautionary and positive experiences; and recognition of the elements of landscape that are changing, key reasons for those particular changes; and their influence on land biodiversity and degradation. To completely understand trends as well as their causes, a multiscale method is essential because environmental and societal processes work at varied scales, and different information types exist at different scales. For instance, depending on the analysis scale, the part of land that is being utilized particularly for cultivation in a semiarid area is likely to appear different. This has been demonstrated by several researchers while plot level analysis discloses that farmers have in fact left several of their fields uncultivated because of low productivity resulting to those fields reverting to bush [39,51]. Usually, at the plot and household level, fieldwork is carried out to acquire particular information. Often, samples are utilized as representation of wider areas or small area is deeply inspected [51,77]. Land management practices such as gender of head, ethnicity, labor availability, income activities, household assets, indicator plants, yield, fuelwood collection or fertilizer application, and soil conservation; household and individual information such as wealth, ethnic group, or farm size; animal species counts; water quantity and quality; erosion estimates; soil properties; and plant species composition are the examples of the information that may be hence acquired and mapped.

2.3.3 Spatial Processes and Spatial Pattern Identification

Spatial processes and spatial patterns identification, like the ones mentioned above, can make available answers to "where" kind of questions and can contribute to the comprehending of "why that is happening there." Nevertheless, in case-specific study, site is not placed within its comprehensive context, and in case the higher-level forces influence at international, regional, and national levels disturbing the local land use as well as management factors is left out, these answers are often incomplete. Different kinds of information regarding land biodiversity and degradation that are of global interest are provided by higher level analyses. Root cause and land use analyses at both the global and regional levels may make available information on, for instance, (1) status of wildlife migration corridors that pass through national borders, (2) land use/land cover change impact on the local to international climate. Conservation of land use such as afforestation or deforestation affects the quantity of atmospheric dust within the atmosphere, as well as the carbon amount stocked in vegetation within the soil. In addition, the parameters of land surface such as surface roughness, soil moisture, leaf area index, and albedo are extremely sensitive to the conservations of land cover and affect the regional and local climate directly. (3) Root cause and land use comparison amid sites make available new information on the root causes of land use modification, the recognition of common driving forces, and on common patterns and land-use change process. These kinds of analyses allow the development of upcoming land-use change scenarios as well as its effect on the environment. (4) Private and public investments, governance, markets and prices, agricultural or land policies, (5) trends in internationally important biodiversity, including the changing of habitats, specific ecosystems, or flora, (6) the vegetation removal off-site impact. For instance, watershed deforestation may cause surface water siltation downstream and erosion (Fig. 2.10).

Although the framework proposes a linear, simplex causative chain, it does not embrace the connections amid variables within boxes. Furthermore, it does not expressly embrace temporal trends or spatial scale. There

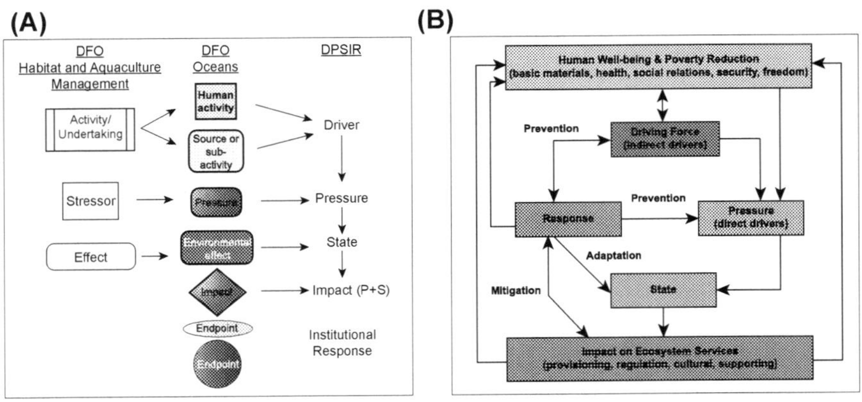

Figure 2.10 (A) Terminology and labels used in conceptual modeling of DFO to get results of DPSIR, (B) the DPSIR Framework.

is an additional assumption that indicators same as a political action as a response indicator, trail from the past indicator even though not different elements live. In summary, it makes available a logical framework for demonstrating presumed factors, then offer very little assistance in the initial recognition of the significant variables, their relation, or whichever way they connect to the issue as well as its achievable resolution.

2.3.4 Intensification

Intensification theory refers to subevolution in the use of land and agricultural techniques adoption in the direction of higher application of capital and labor to the land in response to increasing demand for agricultural products and population densities. Within the continent, several farming systems are intimated with rising population densities as well as modifications in their system of farming, as indicated by intensification, same as decreasing usage of fallow, as well as embracing upper yielding of stringent crops [39,61]. Nevertheless, intensification studies, do not usually adopt factors same as those of land distribution amidst households or the wider socioeconomic framework in which agricultural change occurs. The analysis of these human measurements, such as differentiation amid households and financial achievement diversification ways, disclose the forces behind dynamic management of soil and land (Fig. 2.11).

As a result, the motivating force of land use modification is many-faceted. As time goes by, these motivating forces may change in relative impact, which will differ change in the context. The land use modification examination demands theoretical contexts as well as analytical approaches that are both broad enough to accommodate socioenvironment interaction dynamics at different scales and also be flexible enough to capture the temporal underlying forces of these procedures [39,41,54].

2.3.5 Critical Processes Affecting Land Use and Land Management

Therefore, it is essential for the causes of land management and land-use changes to be analyzed in the course of a multiscale, multidisciplinary framework applying a style of nonspatial and spatial methods. Similar to social differentiation, as well as the role of power and politics, the analytical context put more emphasis on bound factors that need to be expressly thought of. Moreover, the past examination into the root causes as well as environmental effect of land management and land-use changes and that of the LUCID project has revealed significant procedures that play an

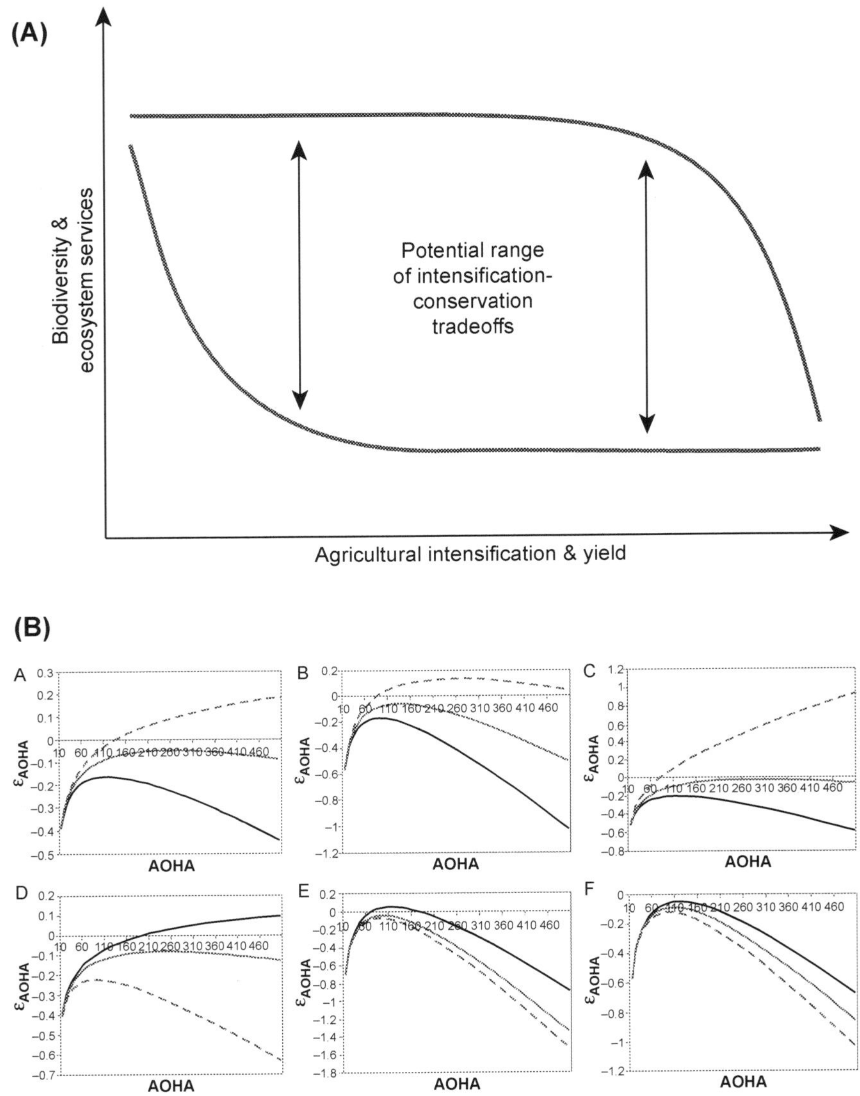

Figure 2.11 (A) The elasticity of agricultural land with respect to agricultural intensification εAOHA as estimated with model 1 (GOV = CORC in A), model 2 (GOV = ROL in B), model 3 (GOV = ACC in C), model 4 (GOV = PA in D), model 5 (GOV = ESI in E), and model 6 (GOV = EPI in F). (B) In computing the elasticity, the governance scores are set at high (sample maximum, red dashed line), moderate (sample mean, blue solid line), and low (sample minimum, black solid line) values. When conventional governance scores are high, intensification leads to agricultural expansion (thus signaling a Jevons paradox), whereas when conventional governance scores are low or moderate, intensification leads to agricultural contraction (thus signaling land sparing). Conversely, when the environmental governance indicator takes high or moderate values, land sparing occurs. For low values of the environmental governance indicator, either a Jevons paradox occurs (D and E) or the intensity of the land-sparing effect is reduced (F).

oversized role of shifting land connections to land degradation and variety [7,29]. Although, these procedures play very essential part in managing selections of shifting land management and land use, in broad land-use changes analyses, this is typically unnoticed. Diversification and urbanization economic process, financial gain, civil strife and insecurity, and national policies are the four essential processes/issues known by this analysis.

2.3.6 National Policies

How national governments have greatly influenced land use by basically changing who can access what land, is one of the striking finding from the LUCID root cause examinations. As the national governments have centralized and decentralized the management of protected areas and communal land; inspired or discouraged migration via settlement schemes developments or through permitting individuals from other parts to access land or possess land use rights; altered the regulations of land tenure, such as privatizing past public land, changing traditional land tenure systems that lead to changed rights over land or group ranches delimitation; and gazetted or degazetted as secured areas, changing who gets access to what land has been the case elsewhere.

2.3.7 Civil Strife and Insecurity

Often, civil strife, war, and overall insecurity are regarded temporary phenomena in root causes and land use analyses, despite the fact that they may have a key effect on land use and land management. This unfortunately, is necessarily not the case, and long-term and regional analyses must often consider their impacts. The effect on land use as well as management in the LUCID research relied on the length and severity of the unrest, but it included the reduced investment of household in developing land as well as delay in government investment in the construction of roads and other infrastructure causing slower economic growth and land use modification than expected; farmers out-migration to capital city and local urban areas; and uncertainty of trade in agricultural products as well as other commodities.

2.3.8 Income Diversification and Urbanization

Within rural areas, household often get involved in earning income through nonagricultural sources. Usually, this involves the adults, especially men leaving their farms to find work in other places such as large farms or urban centers or family members engaging in local off-farm work such as in service, commerce, or small-scale manufacturing. This can have a great impact on the local land use given that labor is pulled out from farms, and as a result, cultivation is less intense or extensive, as limited labor is invested in the farm as well as in the management of soil. For poor households with very small farms or the ones in marginalized environments, it is specifically a critical approach to continue farming where it otherwise might be trivially productive to sustain a family [49,102]. Out-migration of men can result in altered gender responsibilities and roles. On the contrary, richer households with additional income from nonagricultural income-earning activities and can manage their farms with a large amount of capital inputs including hiring labor (Olson et al., 2004). Given that people respond to outside and local opportunities and constraints differently, the extent at which they get involved in nonfarm income varies across time and crossways regions.

2.3.9 Environmental Change Impact

Debatably, the most ubiquitous socioeconomic driving force of ecosystem degradation and changes is land-use change. The Earth's landscape has considerably been changed by a number of human activities including agriculture, urban development, and deforestation [18,32,62]. This kind of land disturbance affects essential ecosystem services and processes, which eventually might impact severely on environment causing calamities such as draught.

Engaging in exhaustive agriculture can possibly cause severe environmental consequences. For instance, for a long time, agricultural practices and land use have been recognized to be a likely cause of water population, and the government policies influences this effect. Both in coastal and inland waters, a leading water pollution source is run off from agricultural lands. Because of wetlands conversions to irrigation water diversions and crop production, a number of wildlife species have been brought to the edge of extermination (Fig. 2.12).

Forests play a major role within ecosystem balance [51,74,101]. They reduce flooding and soil erosion, slow down surface runoff, intercepts precipitation, removes CO_2 from the atmosphere, provide critical habitat for wildlife, and support biodiversity. If forests are transformed to urban development of agriculture, the abovementioned essential ecosystem services will be destroyed or reduced. For instance, deforestation together with agriculture, urban sprawl as well as other human activities has considerably changed and broken the vegetative cover of the Earth. This

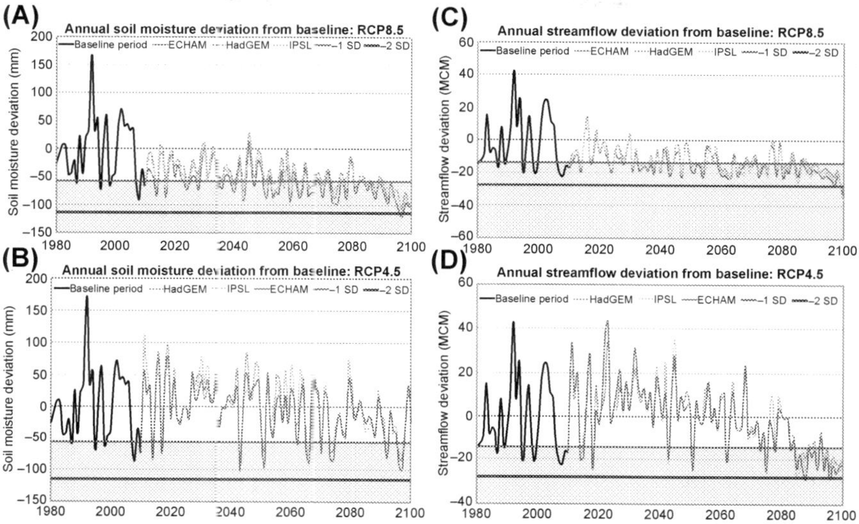

Figure 2.12 Global draught analysis and project from 1980 to 2100. (A) Represents deviation of soil moisture versus annual deviation of soil moisture from baseline Representative Concentration Pathway of RCP8.5; (B) Show deviation of soil moisture versus annual deviation of soil moisture from baseline Representative Concentration Pathway of RCP4.5; (C) Streamflow deviation versus annual deviation of soil moisture from baseline Representative Concentration Pathway of RCP8.5; (D), Streamflow deviation versus annual deviation of soil moisture from baseline Representative Concentration Pathway of RCP4.5.

kind of disturbance is likely to alter the worldwide atmospheric concentration of the principal heat-trapping gas; CO_2; and affect the international, regional, and local climate through altering the energy equilibrium on the surface of the Earth.

Metropolitan development has been connected to several environmental issues. These issues include wildlife habitat loss, pollution of water, and pollution of air. In addition to nutrients sediments, urban runoff often has toxic contaminants which cannot only pollute water but also cause variation in temperatures and streamflow. Several research studies indicated that fragmentation, habitat destruction as well as alteration related with urban development have been recognized as the primary cause of species extinction and biodiversity decline [7,65,77]. All over the world, exhaustive agriculture and urban development in coastal areas and further inland are a key threat to marine environment biodiversity, productivity, and health.

2.4 DISCUSSION

Based on the above research, it is evident that this report commended the wide selection choice of approaches accessible to carry out root cause and land use analyses regarding dynamic land variety and degradation. As a result, the analysis findings offered a detailed data and sources of data through the examination of the complicated systems same as atmosphere/society association causing land use and management change. Table 2.1 shows the primary data variability as well as information that revealed the concerns in dynamic land degradation, variety, and land use. In the examination, secondary information and information from, for instance, literature review, different government statistics, and population censuses, could complement the above. In addition, the information is slightly different from every other source. Several spatial sampling methods were urged by the spatial distribution data to be useful, particularly, in representing the ecological variability that differs crossways house more probably than the socioeconomic characteristics [7,41]. Line transect that must cut crossways into different applicable variable or several ecosystems is one method that is usually used in geographical and ecological field analysis. To acquire information on the phenomena distribution, the systematic or random purpose samples will afterward be analyzed on the transect line. Conversely, a random points distribution crossway a neighborhood offers a less prejudiced demonstration of the real, nonetheless, transports to every other point is tough as well if protection is disturbed (Fig. 2.13).

A difference in this is likely to be clustered sampling, in which the points to look are easier to attain because they are clustered, although, the clusters are chosen haphazardly. An associate method to cluster sampling, stratified sampling, captures each socioeconomic or ecological variability purpose location, same as farms or villages, and the sample are haphazardly selected within a spatial institution of interest. Same as in agroecosystem, sample represents those within associate system, and in a systematic way, valid similarities and dissimilarities might be generated amid ecosystems [18,51,67]. Therefore, each contextualize level data within the regional, nation, and global frameworks for spatial distribution is analyzed by micro-level if data on higher level root causes of land use change is shifting land managers to the native level. By their nature, land use analysis and root cause analysis are multifaceted and complicated reasons, and as such, the taken analyzed and conferred data may essentially remove this complexness to determine the real root causes of land-use change.

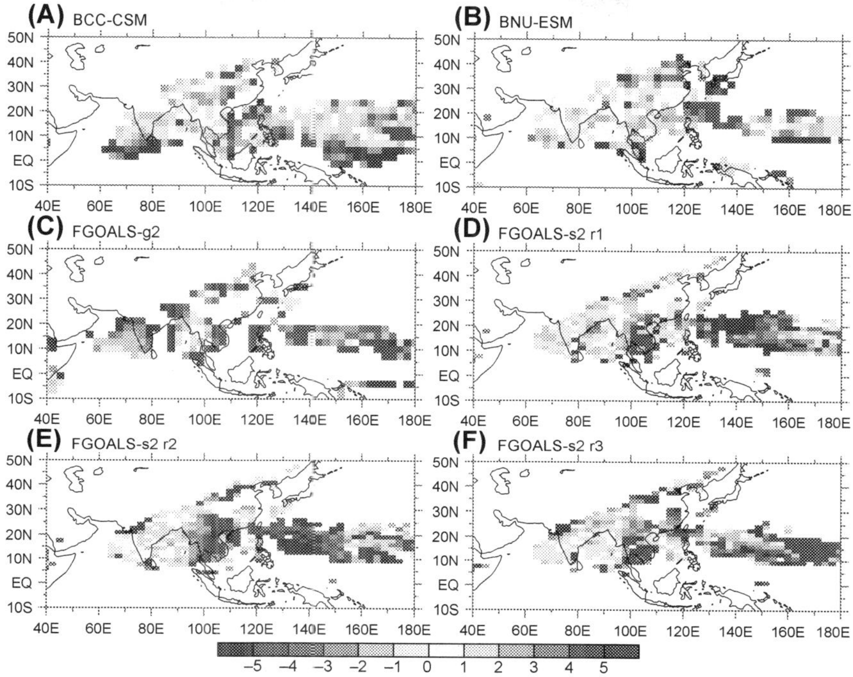

Figure 2.13 Spatial distributions of the global land use derived from (A) BCC-CSM, (B) BNU-ESM, (C) FGOALS-g2, (D) FGOALS-s2 r1, (E) FGOALS-s2 r2, and (F) FGOALS-s2 r3. "r1," "r2," and "r3" denote the three realizations of FGOALS-s2, respectively. The data of coupled models are derived from the 20th century historical climate simulations during 1981−2015. The results are presented with the grid of individual model/observation. Note that pentad 30 is the end of May. The numbers shown in the upper-right corner of each plot are the spatial pattern correlation coefficients with GPCP land-use change data and with CMAP data.

These opposing information convenience types at different scale shall therefore represent the detailed interpretations and processes of land-use change all over the world. Lastly, soil erosion impact represented an important force driving the LULC modification, where social action, power, social scale, and above all, human activities are liable for worldwide land use alteration. As a result, the motivating forces to alter soil erosion in numerous scales were identified to a chief objective of land use management. These findings disclosed a significant empirical goal of providing corrected interpretations as well new information of soil erosion which eventually is part of land use modification (Fig. 2.14).

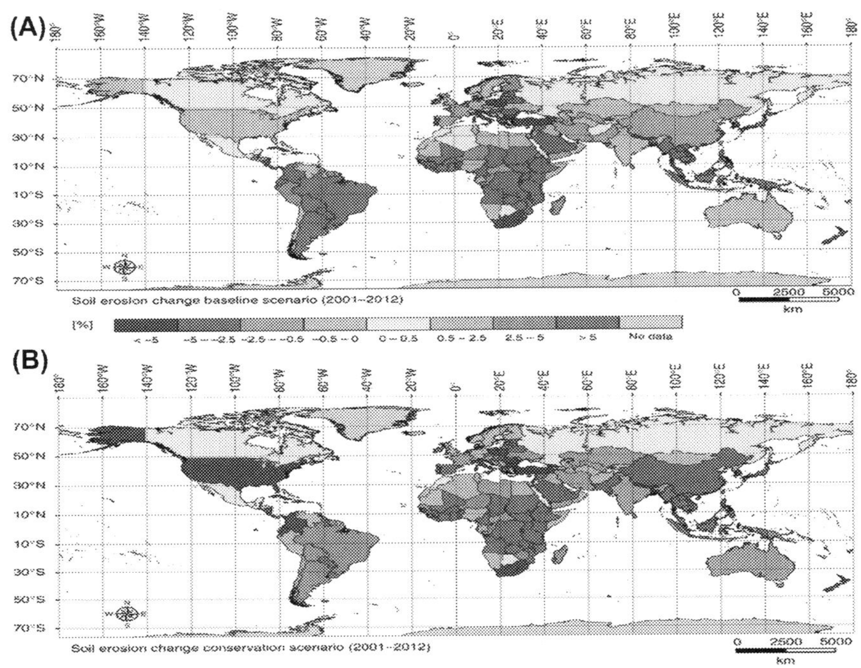

Figure 2.14 Country-specific changes of the annual average soil erosion. Panels (A) and (B) have a common legend, with the chromatic scale representing the decrease or increase of the yearly average rates of soil erosion in percentage. The rates are acquired through comparing the pixel-based values in each of the 22 countries that are under study. Panel (A) shows soil erosion change amid 2001 and 2012 based on the baseline scenario. The existing data amid the two periods being observed, relies only on LULC change outline joining the agricultural inventory information with the satellite-derived LULC data. Panel (B) displays soil erosion alteration amid 2001 and 2012 based on the conservation scenario. The percentage change of soil erosion, in this instance, is caused by the extenuating impact of soil preservation practices as well as the joined impact of LULC modification. LULC, land use and land cover.

2.5 CONCLUSION

In conclusion, in several developing countries, modification of land use trends is very rapid. The course of modification and rates are fluctuating, and before corrective program and policies can be effective in the long-term crossways wide areas, it is a must for the root cause pattern of these trends to be identified, and their effect on land biodiversity and degradation understood. Research of this study is being based on one or the other aspects of social or ecological system to make analysis of the ecological or social trends crossways space to allow for root causes, processes, and patterns generalization, and also permit the learning of general lessons. Considering the basic modification processes as well as the effect of their being effectual into the environment, much attention has indeed been directed toward corrective approaches to address land-use change concerns such as land degradation. As a result, this specific objective is significantly improved through the adoption of a mutual analytical framework to carry out such cross-site comparisons and analysis. This report has therefore, sketched one such framework, which requires the adoption of a system method to understand the connection of environmental and social processes over space and time, and eventually cause climate changes.

2.6 CLIMATE CHANGE

Science has made enormous inroads in understanding climate change and its causes and is beginning to help develop a strong understanding of current and potential impacts that will affect people today and in coming decades. Given that this understanding enables decision makers to place climate change in the framework of additional encounters facing nations and the world as whole, it (understanding) is crucial. In understanding a difficult system such as the Earth's climate, there are, and there always will be some uncertainties. However, based on several research lines, there exists a strong and credible body of proof, indicating that climate is changing and human activities are key causes of these changes.

2.6.1 Introduction

The climate change mechanism refers to a change in weather pattern's statistical distribution when the changes last for a long period of time. Therefore, climate change can be defined as a change in average weather conditions, or a change in the time weather variation in the context of long-lasting average conditions. In simple words, climate change results from factors such as volcanic eruptions where human activities have been recognized as the key cause of current change of climate, plate tectonics, variations in solar radiation acquired by the Earth, and biotic processes. Based on scientific facts, climate is changing and increase in the levels of carbon dioxide (CO_2) is a primary cause of this changes [1,2]. This is particularly because of the releases from fossil fuel burning, after that aerosol, and then finally the CO_2 released by manufacture of cement. Additional factors including deforestation, animal farming, ozone depletion, and land use, are also of great concern in the part they play both distinctly and when combined with other elements in affecting climate variable measures, micro-climate, and climate itself. Through the use of the theoretical models, and observations, scientist aggressively work to comprehend past and forthcoming climate. Based on geological proof from past sea level records, stable-isotope and other analyses of sediment layers, glacial and periglacial processes, faunal and floral records, cores removed from deep accumulations, and profiles of boreholes temperature acclimate record prolonging deep into past of the Earth has been collected and keeps on being built up. The instrumental record provides the most recent information. Based on the physical science, overall circulation models are frequent in theoretical methods to connect effects and causes in climate change, make forthcoming projections, and match older climate information [18,32]. A modification in the climate system arithmetical properties when considering a prolonged period, mainly its spread and mean, is however, the most common climate change terminology and is mostly caused by human activities. In addition, while climatic forces have a considerable effect to change the climate, many other natural processes cause climate modification.

2.6.2 Climatic Forces

Based on the broadest scales, the rates at which the Sun radiates energy and loses it to space determine the equilibrium climate and temperature of Earth. By ocean currents, winds, and other mechanisms, this energy is circulated all over the world to affect the climates of different areas. Forcing mechanisms or climate forcing refer to factors that are likely to shape the climate. Processes such as changes in greenhouse concentrations;, continental drift and mountain building; variations in the albedo or reflectivity of the oceans, atmosphere, and continents; variation in orbit of the Earth; and variations in solar radiations are examples of forcing mechanisms. There exist a range of climate change responses that are likely to either diminish or amplify the original forcing. Just like ice caps and the oceans, some climate system parts react more gradually in response to forcing mechanism, while others' response is quicker. Besides, there exist major threshold elements which are likely to produce fast change when surpassed. There are either external or internal forcing mechanisms. External climate forcing can either be natural, for example, volcano eruption, the Earth's orbit or changes in solar output, anthropogenic, for instance, greenhouse dust and gases increased emissions [41,48]. Internal climate forcing are natural procedures, such as, thermohaline circulation, that exist within the climate system itself. Whether it is an external or internal forcing mechanism, the climate system reaction can be fast, slow, or a combination. As a result, the reaction of climate system can be abrupt, although the complete reaction to forcing both the internal and external forcing mechanisms may not be completely developed for ages or even extended periods.

2.6.3 Internal Forcing Mechanisms

In general, according to scientists, biosphere, lithosphere, cryosphere, hydrosphere, and atmosphere are the five elements of the climate system of the Earth. Internal forcing mechanisms, which involve natural modifications in the climate system, cause internal "climate changeability" changes; circulations of ocean atmosphere, and species type and distribution are examples of internal climate variations (Fig. 2.15).

It is likely for the atmosphere and ocean to work in partnership to impulsively create internal climate changeability that may stay for a very long period, years even to decades. The Atlantic Multidecadal Oscillation, the Pacific Decadal Oscillation, and the El-Nino-Southern Oscillation are examples of this kind of changeability [77,101,102]. It is likely for these variabilities to affect the worldwide average surface temperature through the redistribution of heat amid the atmosphere and the deep ocean and/or via changing the water/cloud sea/vapor distribution of ice, which is likely to

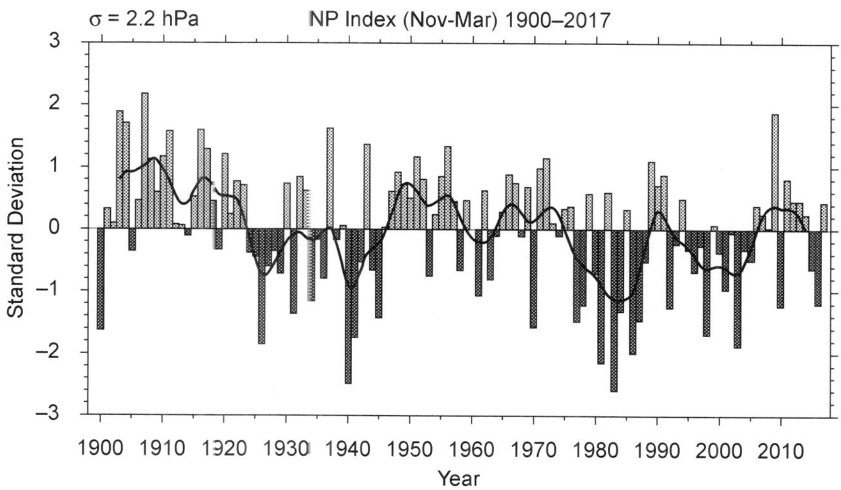

Figure 2.15 Internal force from the period of 1900–2100 that causes and predicted climate variations globally.

impact on the Earth's total energy budget. Because the ocean has hundreds of times more mass than the atmosphere, hence, extremely high thermal inertia, it is possible for the oceanic facets of these particular circulations to create changeability on centennial timescales. For instance, modification to ocean processes such as circulation of thermohaline plays a major part in the redistribution of heat within the world's oceans, which as well takes a significant role in climate alteration.

2.6.4 Thermohaline Circulation

The term thermohaline is derived from the word "thermo," which refers to temperature, and "haline," which refers to the content of salt, the components which together determine the seawater density. Thermohaline circulation is usually a part of the extensive ocean circulation that is driven by international density gradients generated by freshwater and surface heat fluxes. Surface currents such as the Gulf Stream, usually wind-driven, move poleward from the equatorial Atlantic Ocean, then the cooling end direction, and ultimately sinking at very high latitudes, thereby creating *The North Atlantic Deep Water* which then flows into the basins of the ocean [4,32]. As a result, widespread mixing occurs amidst the ocean basins, causing the reduction in the differences that exist amidst them and eventual makes a global system for the Earth's oceans. During the journey, the masses of water transport both matter, as solids, gases, and dissolved matters, and energy in form of the heat all over the world. Accordingly, the circulation condition has a huge effect in the Earth's climate (Fig. 2.16).

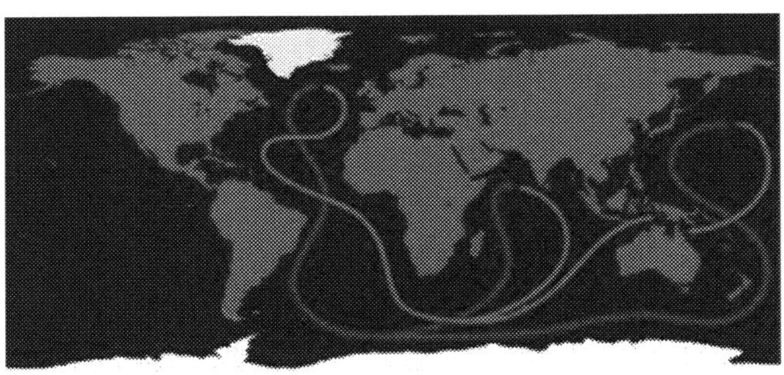

Figure 2.16 A summary of the path of the thermohaline circulation. Blue paths represent deepwater cold currents, while red paths represent surface warm currents.

While the red paths indicate surface warm currents, the blue paths indicate deepwater cold currents. Thermohaline circulation can in details be described that, the surface currents movement being pushed by the wind is impartially intuitive, as a result, the deep ocean wind devoid was presumed to be impeccably static. In a nutshell, the velocities of ocean water range from fractions of cms (centimeters)/sec to occasionally more than 1 m/s in the surface currents such as the Kuroshio and Gulf Stream. Differences in density, resulting from temperature and salinity variations (decreasing the temperature and increasing salinity of water together rises its density) is the major motivating force in the ocean. Moreover, tidal currents can be important in the deep ocean because ocean current because of tides are also important in several areas, most noticeable in moderately shallow coastal places. It is believed that tidal current in the deep ocean enable mixing processes, particularly, diapycnal mixing. Although ocean water density is not internationally similar, it varies discretely and significantly. There exist sharply demarcated boundaries amidst the masses of water which are formed at the surface and consequently preserve their own self inside the ocean. These sharp borders, however, are to be imagined in $T-S$ diagram, where the masses of water are distinguished, and not spatially. The boundaries usually place themselves below or above each other based on their own density, which relies on both salinity and temperature.

Because the warm seawater inflates, it is less dense compared to cooler seawater. Also, as the liquefied salts fill the intersect amid the molecules of water, causing increased mass per unit volume, as compared with fresher water, saltier water is denser [10,37]. Just like a piece of ice of wood floats on water lighter masses of water float over denser masses of water. This is referred to as "stable stratification" versus unstable stratification where more dense water masses are positioned over lighter masses of water. Dense masses of water are unstably stratified when they are first created, therefore, they seek to position themselves within the correct vertical position based on their own density. This kind of motion is referred to as convection, which through gravitation, orders the stratification. Motivated by density inclines, this forms the major motivating forces behind deep ocean currents such as the DWBC (Deep Western Boundary Current)

Equivalent water quantities rising somewhere else must offset the great dense water quantities sinking at very high latitudes. It is essential to be aware that, while warm water in tropical and temperate zones rise more slowly crossways a much bigger area, the cold water in the polar zones sink comparatively speedily over a very small area. Afterward, it returns gradually poleward nearby the surface to repeat the same cycle. The repeated deepwater diffuse upwelling sustains the permanent thermocline presence found in all places at a small and midlatitudes. This model which is called the Stommel–Arons box model for the MOC was explained in 1960 by Arnold B. Arons and Henry Stommel. Over most the ocean, this slow upward shifting is estimated to be almost 1 cm/day (0.5 inch/day). In case this rise stops, downward heat movement would trigger the descending of the thermocline decreasing its steepness. The formation of deep masses of water inside the North Atlantic and the Southern Ocean, brought about by variations in salinity and temperature in water, is what mainly drives this thermohaline circulation.

2.6.5 Effects on Global Climate

Even if poleward heat transport outdoor the tropics is significantly larger in the atmosphere than in the ocean, the thermohaline circulation plays a vital role of delivering heat to the polar areas, and hence in controlling the sea ice amount in these areas. It is believed that modifications in the thermohaline circulation significantly effects on the radiation budget of the Earth [11,39,42]. Thermohaline circulation may as well play a significant role in CO_2 circulation in the atmosphere because it controls the rate of exposure of deepwaters to the surface. Although it is frequently indicated that the thermohaline circulation is primarily the reason as to why Western Europe is so warm, suggestion have been made indicating that this is highly incorrect, as Europe is being found to be generally temperate because it is located downwind of an ocean basin. Another reason is the atmospheric wave effect that brings warm air from the subtropics to the north. Nevertheless, critics have also been directed toward the underlying presumptions of this specific analysis. It is assumed that the deglaciation in North America as well as the huge influxes of low-density meltwater from Lake Agassiz resulted in deepwater formation shifting and sinking in the extreme North Atlantic, which eventually caused Younger Dryas (climate time in Europe).

2.6.6 External Climate Forcing

Modifications in the periodic supply of sunlight reaching the surface of the Earth and the way it is supplied across the world are caused by orbital differences in the motion of the Earth. Although there exists very slight change to the area-averaged yearly averaged sunshine, strong alterations within seasonal and geographical distribution can exist. Earth's axis procession, modifications in the Earth's axis tilt angle of rotation, and differences in Earth's eccentricity are the three kinds of kinematic change [2,41]. If these three kinds of kinematic change are joined together, they will produce Milankovitch cycles, which have great effect on the Earth's climate and are visible for their appearance in the stratigraphic record, their connection with the Sahara retreat and advance, and for their connection to interglacial and glacial periods. According to the IPCC notes, Milankovitch cycles drove the ice age cycle, carbon dioxide came after change in temperature, "with some hundreds of years lag," and a response that amplified the change in temperature. The ocean depths a gap period in changing temperature, carbon dioxide solubility, and other factors influencing air—sea carbon dioxide exchange changed upon the change of the seawater temperature (Fig. 2.17).

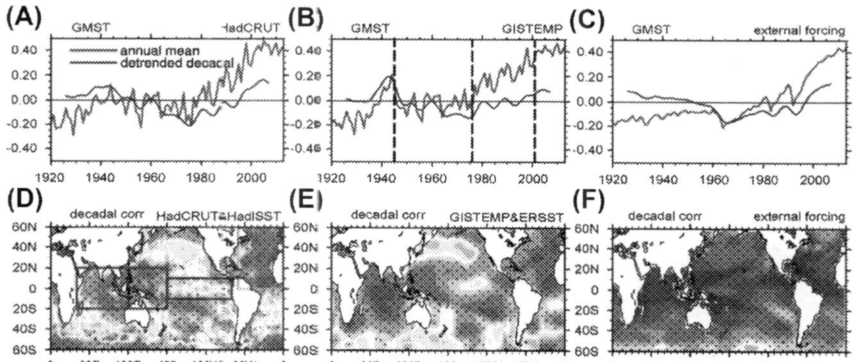

Figure 2.17 Time series (in °C) for annual GMST anomalies (red lines) and detrended 8-year low-pass filtered GMST (blue lines) from (*A*) HadCRUT, (*B*) GISTEMP, and (*C*) external forcing from 18 CMIP5 models. Correlation coefficients between SST and GMST for detrended 8-year low-pass anomalies during 1920—2013 from (*D*) HadISST and HadCRUT, (*E*) ERSST and GISTEMP, and (*F*) external forcing. Stippling indicates regions exceeding the 95% statistical significance. Standard deviation of detrended 8-year low-pass filtered SST from (*G*) HadISST, (*H*) ERSST, and (*I*) external forcing.

2.6.7 Solar Output

Heat from radioactive compound decay, tidal energy from the moon, and geothermal energy from the Earth's core are other sources of energy input to the Earth besides the Sun, which is the leading source. Together, the short-term and the long-term differences in solar intensity are recognized to influence the international climate. The Sun emitted just 75% as much as it emits in the present days, 3 to 4 billion years ago [39,64,77]. There should be no existence of liquid water on Earth in case the atmospheric composition had been similar as in the current days. A massively different atmosphere, having a greater greenhouse gases (GHGs) concentration than presently exist, is the assumed answer to this paradox. Beyond, the subsequent nearly 4 billion years, the atmospheric composition transformed as the Sun energy output increased.

2.6.8 The Great Oxygenation Event

The most prominent change about 2.4 billion years ago was the atmosphere oxygenation. From the present day, over the next 5 billion years, the final death of the Sun as it continues to become a red giant, followed by a white dwarf, will have huge effects on the Earth's climate, with the red giant stage probably finishing any life on Earth that lives till that estimated time (Fig. 2.18).

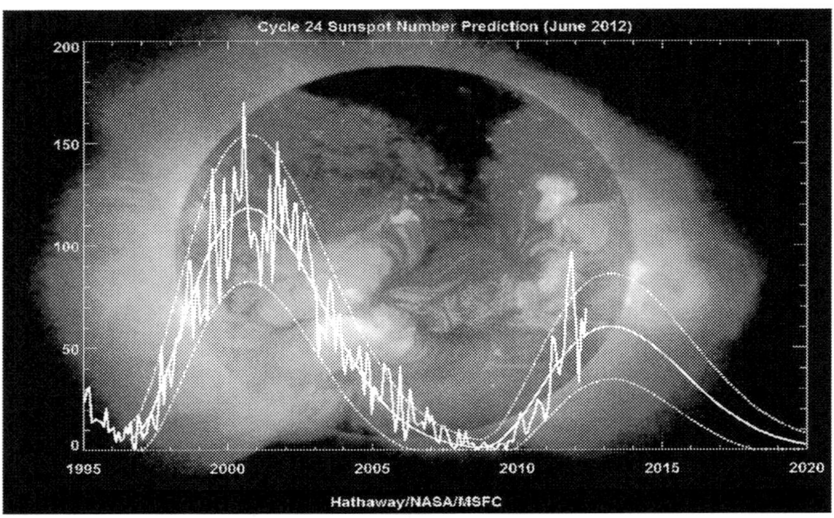

Figure 2.18 The above figure shows the differences in solar activities during the past many eras based on the beryllium isotopes and sunspots observations. The time of strangely few sunspots during the 17th century was the Maunder Minima.

Solar output differs on a temporary timescale as well as the long-term modulations and the 11-year solar cycle. Differences in solar intensity, probably caused by Maunder Minima, Sporer, and Wolf, are regarded to have been very powerful initiating the Little Ice Age. This occurrence stretched from 1550 A.D to 1850 A.D and was identified by relative cooling as well as larger glacier degree than earlier of subsequent centuries [51,77]. Moreover, solar variations may have influenced some of the heating witnessed between 1900 and 1950. The energy out of the Sun cyclical nature is not up till now completely understood; it varies from the extremely slow alteration that is occurring in the Sun as it gets old and evolves. Some research studies indicate that solar radiation rises from the activity of cyclical sunspot disturbing the global warming. The total effects, which includes the anthropogenic radiative forcing and solar variations, finally impacts on the Earth's climate. According to the 2010 study, the solar variability effect on the temperature all over the atmosphere might be in contradiction of the present expectations. An announcement on the publication in the *nature* paper in the preliminary findings from the CLOUD experiment was made by CERN, in a press release of August 2011 [31,32]. The findings show that in presence of water and sulfuric acid ionization from cosmic rays, the aerosol formation is considerably improved. Although, within the lower atmosphere where it is also necessary for ammonia to be present, this is inadequate to count, as aerosol creation as well as more trace vapors have been involved. Determining whether trace vapors are of human or natural origin is the subsequent step of finding more information of trace vapors (Fig. 2.19).

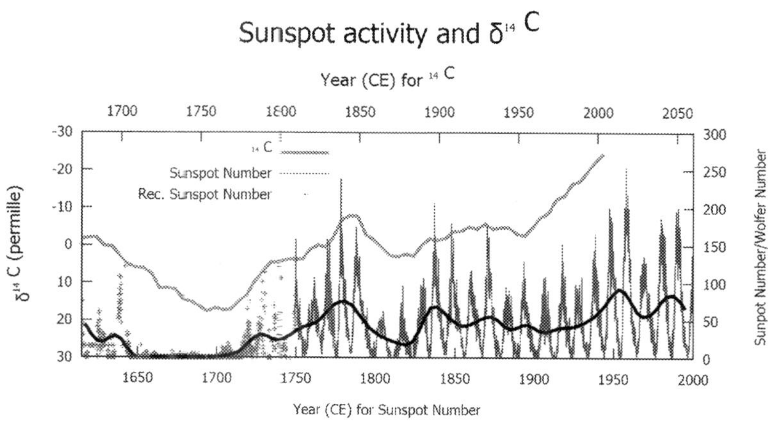

Figure 2.19 Solar activity events recorded in radiocarbon. Values since 1950 are not shown.

2.6.9 Volcanism

The 1991 volcano eruption affected the Earth's climate well, subsequently globe temperature attenuate via concerning zero 5 °C for up to 3 years. As a result, in 1991–93, the cooling over enormous planet components reduced the surface temperatures to appreciate a discount in Internet radiation of 4 W/square measures (Fig. 2.20).

The eruptions that are capable of injecting more than 100,000 tons of SO_2 into the stratosphere are well-thought-out to be huge enough to disturb the climate of the Earth on a scale of over a year. This is because of the optical features of sulfate aerosol and SO_2, which powerfully scatter or absorb the solar radiation, thereby forming a worldwide sulfuric acid mist layer [39,51]. Averagely, these kinds of eruptions take place many times per century at a period of at least a few years.

Mount Pinatubo eruption in 1991, recorded as the second biggest terrestrial eruption of the twentieth century, significantly affected the Earth's climate and consequently caused the worldwide temperatures to decrease by almost 0.5 °C for a period of up to 3 years. As a result, beginning 1991 to 1993, the cooling over of big sections of the earth decreased the overall surface temperature, which was equal to a decrease in net radiation of 4 W/m^2. In 1815, the eruption of mount Tambora resulted in a year without a summer season. Considerably larger eruptions referred to as igneous provinces take place just a few times in every 5100 hundred million via flood basalt, as well as caused by the earlier mass extinctions and global warming of the Earth. Given that changes in temperature can be compared with natural changeability, small eruption, having for instance, less than

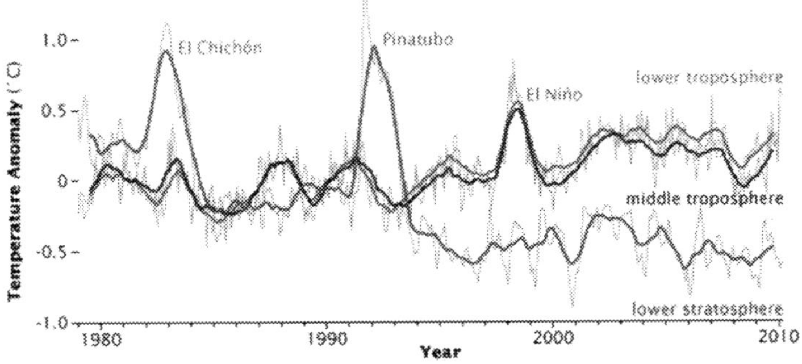

Figure 2.20 In atmospheric temperature from 1979 to 2010, determined by MSU NASA satellites, effects appear from aerosols released by major volcanic eruptions (El Chichón and Pinatubo). El Niño is a separate event, from ocean variability.

0.1 Mt of SO_2 into the stratosphere, impacts only slightly on the atmosphere [31,39,77]. They as well have considerable effect on the atmosphere of the earth because smaller eruptions take place at a considerably higher frequency. Besides mapping the present and forthcoming trends in volcanic activities, seismic monitoring also attempts to create early warning systems. Studying the volcanic forcing response and physical mechanisms is the key aim in climate modeling. Likewise, volcanoes are part of the prolonged carbon cycle. For extremely long period of times, volcanoes release CO_2 from the mantle and crust of the Earth, offsetting the uptake through sedimentary rocks as well as geological CO_2 sinks. Based on the estimates of the United States Geological Survey, volcanic emissions are at a considerable lower level than the present human activities effects, which usually creates between 100 and 300 times the CO_2 amount released by volcanoes. According to an analysis of published research studies, the yearly volcanic CO_2 releases, which include amounts emitted from hot spot volcanoes, volcanic arcs, and midocean ridges, are just equal to 3–5 days of output caused by human activities.

2.6.10 Climate Change Evidence

There is proof that the impact of Chicxulub some 66 million years ago harshly affected the climate of the Earth. Worldwide temperatures were decreased by up to 26°C by the huge amounts of aerosol and GHGs spread into the atmosphere [3,9,31]. This also caused subfreezing of temperature for a period of 3 to 16 years. This involved both the internal and external driving forces, and hence time sequence for yearly means of climate dynamics running in extremely considerable flexural changeability (Fig, 2.21).

2.6.11 Temperature Measurements and Proxies

Widespread atmospheric monitoring by the mid-20th century, radiosonde balloons, and the global satellite data from the 1970s onwards supplemented the instrumental temperature records from the surface stations. While the 17th and 19th centuries were cool, a larger part of the 20th century was observed to be unprecedentedly warm. The $^{18}O/^{16}O$ ratio in the core samples of calcite and ice utilized to infer the temperatures of the ocean in the distant past, represents an example of a temperature proxy approach, just like other climate metrics recognized in consequent classes. Corresponding modifications in agricultural and settlement patterns can be utilized to detect

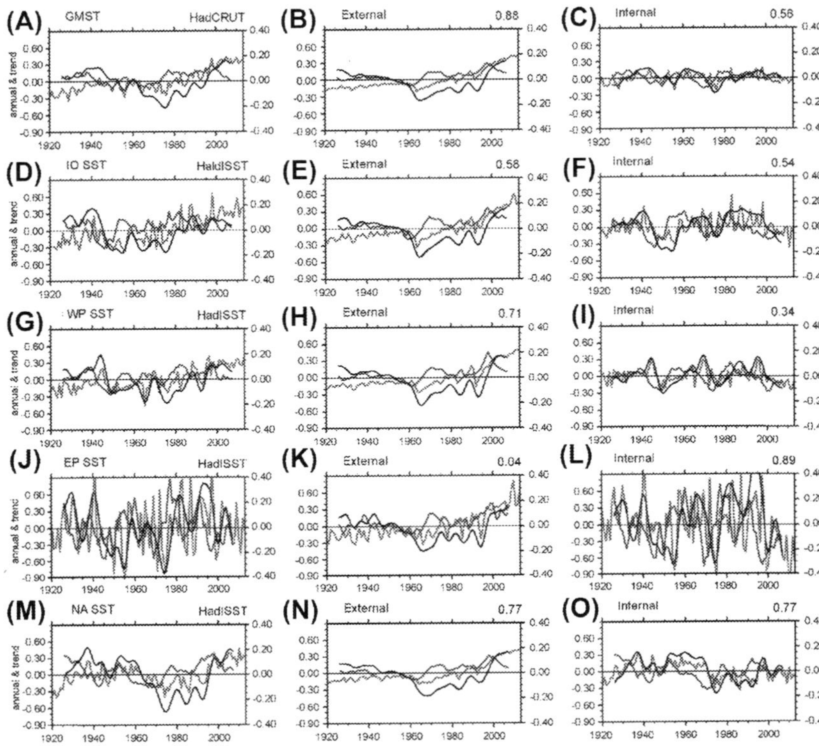

Figure 2.21 Time series for annual means (red; in °C), time sequence for yearly means (in red), detrended 8-year low-pass filtrated yearly means (in black) versions and the centered 13-year running yearly means trends (in blue) for *(A)*–*(C)* GMST, *(D)*–*(F)* Indian Ocean SST (20°S–20°N, 40–120°E), *(G)*–*(I)* tropical western Pacific SST (20°S–20°N, 120–180°E), *(J)*–*(L)* eastern equatorial Pacific SST (10°S–10°N, 180°E–80°W), *(M)*–*(O)* North Atlantic SST (0–60°N, 70–180°W) indicated as boxes in Fig. 1(*d*), from the observations HadISST and HadCRUT, in the left column, internal changeability accessed from the variations amid externally forced variation and observation (right column), and external forcing based on 18 CMIP5 model which is in the middle column [48,51]. The numbers which are at the top right signify the correlation coefficient amidst the internally created or externally forced differences on decadal timescale, and the observation. Sensibly, it is from the beginning of the midlate 19th century that full international records of surface temperature exist. For periods earlier than the midlate 19th century, a great deal of the proof is indirect climatic modifications inferred from the alterations in representations; indictors that represent the Earth's climate include glacial geology, change of sea level, dendrochronology, ice cores, and vegetation.

climate change in current past. Historical document, oral history, and archaeological proof may provide insight into the earlier climate changes. The rise and collapse of different civilizations have related to the effects of climate change (Fig. 2.22).

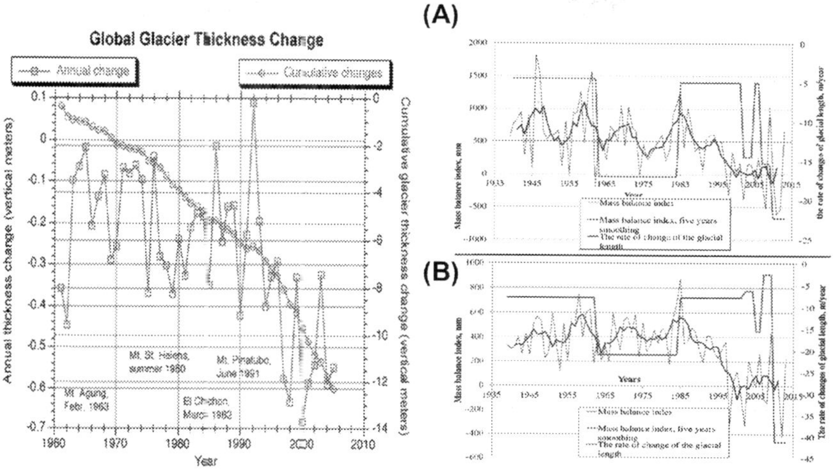

Figure 2.22 Illustration of decrease in glacier thickness over the last half-century.

2.6.12 Glaciers

Among the most sensitive climate change indicators are glaciers, whose size is calculated by a mass balance amid snow input and the melt output. Glaciers always retreat as temperatures become warm except in the case where the participation of snow rises to make up for added melt; the opposite is as well true. Due to both external forcing and natural changeability, glaciers develop and also shrink. Glacier evolution within a specific season can strongly be determined by subglacial and glacial hydrology, precipitation, and changeability in temperature. As a result, to smooth out the indigenous temporary changeability and acquire a history of a glacier that is associated to the climate, it is essential for one to find the average over a decade or a lasting timescale and/or over several individual glaciers. Since 1970s, a globe glacier stock has been accumulated, at first based majorly on maps and aerial photographs, but currently based more on satellite.

The compiling trails over 100 thousand glaciers covering an overall area of nearly 240,000 km^2, and initial estimates show that the surplus ice cover is almost 445,000 km^2. Information on glacier mass balance and glacier retreat is collected yearly by the World Glacier Monitoring Services (78.88.99). Based on these collected data, glaciers globally have been recognized to be shrinking considerably. Strong glacier retreats occurred in the 1940s, with the stable or developing state during the 1920s and 1970s. Retreat again occurred from the middle of 1980s to the current day. Ever since the mid- to late-Pliocene, the glacial as well as the interglacial cycles have been the most important climate processes, with the current interglacial period lasting for almost 11,700 years. Feedback such as fall and rise of mainland ice sheets, as well the substantial sea level modifications, shaped by variations in the Earth's orbit, assisted in forming the climate. However, additional modifications comprising Younger Dryas, Dansgaard—Oeschger occurrences, and Heinrich occurrences demonstrate the way in which variations in glaciers might as well impact on the climate minus the orbital forcing. Moraines that has a fortune of material documenting the periods within which a glacier improved and retreated is left behind by the glaciers. In the same way, the absence of glacial cover can possibly be determined by the volcanic tephra horizon or soil presence, through tephrochronological methods. Since 2002, mass has continually been lost by the land ice sheets in both Greenland and Antarctica, and since 2009, both sheets of ice have experienced ice mass loss acceleration.

2.6.13 Arctic Sea Ice Loss

Over the last many years, the decrease in Arctic sea ice, both in thickness and extent, is more proof for rapid change in the Earth's climate. Sea ice covers millions of km^2 in the polar areas, changing with seasons. While nearly all sea ice within the Antarctic sea or Southern Ocean melt down and reforms yearly, some sea ice inside the Arctic remains unchanged year after year [2,48,77]. Based on the satellite observation, presently sea ice is decreasing at a rate of 13.2% for each decade. The retreat of the Arctic summer sea ice in 2007 was exceptional. Arctic sea ice has been placed in a risky position by the years of thinning and shrinking in a temperate climate, presently it is susceptible to atmospheric modifications. Both the volume and the extent anomaly vary slightly from the first month of the year to the seventh month and subsequently decline steeply between the eighth month of the year and the ninth month. Lessened production of ice caused by strangely high SAT is reason for this decrease.

2.6.14 Vegetation

Given the change in climate, there is also a likely occurrence in the change in vegetation type, coverage, and distribution. Increased warmth and precipitation may be caused by some alterations in climate, which eventually lead to enhanced growth in plants and subsequent airborne CO_2 sequestration

[31,44]. Earlier flowering and finally fruiting times is caused by steady increase in warmth of an area, which motivates an alteration in timing-dependent organisms' life cycle. On the contrary, cold climate causes plant biocycle to lag.

2.6.15 Forest Genetic Resources

Over the coming 50 years, climate change is expected to impact on the forest genetic resources diversity and thus on the forest composition and forest tree species distribution, despite the fact that this field is of several uncertainties. Forest genetic resource diversity allows the prospective for a species to adapt to changes in climate as well as associated forthcoming challenges, such as forest fire, diseases, pests, and temperature changes [54,74]. Species, however, are naturally incapable of adapting in the pace at which the climate is changing, and increase in temperature is most likely to facilitate spreading of diseases and pests, generating more threat to trees in the forest as well as their populations. Human interventions, such as forest reproductive material transfer, may be required to prevent these issues.

2.6.16 Cloud Cover and Precipitation

The global network of precipitation can be utilized to estimate past precipitation in the contemporary era. Although surface coverage over remote regions and oceans is comparatively sparse, since 1970s, decreasing dependence on satellite cloud, interpolation, and precipitation data have existed. Although precipitation climatological variation measurement in earlier eras and epoch is incomplete, it can be estimated using proxies such as tree rings, cave stalagmites, ice cores, and marine sediments. Based on the scientists, precipitation and cloud cover substantially get affected by climatological temperatures [21,32].

2.6.17 Sea Level Change

For much of the past century, tide gauge measurements organized over a prolonged period to a certain long-term average have been utilized to generally estimate the global sea level change. In the most recent days, altimeter measurements have offered an enhanced measurement of international seal level change. Scientists have currently dated reefs that are growing nearby, for example, coastal sediments, oceans, and marine terraces surfaces, to measure sea level before instrumental measurements

[52,78]. Radiocarbon and uranium series are the major dating approaches utilized. Although in the early Pliocene sea level was 15–25 higher than that in the present, the global temperatures were identified to be warmer than the current temperature.

2.7 CONCLUSION

Despite some impact by natural process, the evidence stated that climate change is mainly caused by humane activities, primarily burning fossil fuel, land-use changes, and misuse of natural resources. During the past several decades, there has been serious concern over the dangers associated with the accumulation of GHGs in the atmosphere. Hazardous air pollutants are usually emitting from fossil fuel and gas extraction and processing industries; hydrocarbon and dioxin emissions are caused by the use and combustion of petrol and diesel oil for transport; and radionuclides are released during the burning/combustion of gas, petroleum, coal, and heavy fuel oil. Excessive concentrations of these pollutants have demonstrated health, welfare, and ecological effects felt globally in addition to climate change. The changing climate is thus impacting the global environment severely. In a broad way such as causing flood or draught, the influence of agricultural crop yields, affect human health, and destroy ecosystems are causing due to this climate change. So, what can we do because we are the first generation with the tools is to see how global environmental system is running toward danger due to the conventional system of natural resource consumption, at the same time, we are the last generation with the tools to prevent these dangers. Indeed, we can individually do something great by applying sustainable technologies in every sector and collectively helping to create global view for the development of the sustainable technology for confirming a cleaner and greener earth.

ACKNOWLEDGMENTS

This topic is written by Bruce Petrie,[a] Ruth Barden,[b] Barbara Kasprzyk-Hordern,[a] at [a]Department of Chemistry, University of Bath, Bath BA2 7AY, UK; [b]Wessex Water, Bath BA2 7WW, UK. This paper was originally published as "A review on emerging contaminants in wastewaters and the environment: Current knowledge, understudied areas and recommendations for future monitoring" in journal "Water Research" 72 (2015); 3–27. The report has been modified from its original research to fit the publication in this book chapter under the topic "Water."

REFERENCES

[1] Andres RJ, Boden TA, Higdon D. A new evaluation of the uncertainty associated with CDIAC estimates of fossil fuel carbon dioxide emission. Tellus B Chem Phys Meteorol 2014;66:23616.

[2] Arnell N, et al. Climate change 1995. In: Watson RT, Zinyowera MC, Moss RH, editors. Impacts, adaptations, and mitigation of climate change. Cambridge University Press; 1996. p. 325−63.

[3] Andreozzi R, Caprio V, Ciniglia C, De Champdore M, Lo Giudice R, Marotta R, Zuccato E. Antibiotics in the environment: occurrence in Italian STPs, fate, and preliminary assessment on algal toxicity of amoxicillin. Environ Sci Technol 2004;38: 6832−8.

[4] Bagnall JP, Malia L, Lubben AT, Kasprzyk-Hordern B. Stereoselective biodegradation of amphetamine and methamphetamine in river microcosms. Water Res 2012b;47:5708−18.

[5] Achard F, et al. Determination of tropical deforestation rates and related carbon losses from 1990 to 2010. Glob Change Biol 2014;20:2540−54.

[6] Ciais P, Sabine C. In: Stocker T, Qin D, Platner GK, editors. Chapter 6: carbon and other biogeochemical cycles in climate change 2013 Phys Sci Basis. Cambridge University Press; 2013.

[7] Chevallier F. On the statistical optimality of CO_2 atmospheric inversions assimilating CO_2 column retrievals. Atmos Chem Phys 2015;15:11133−45.

[8] Mills GA, Gravell A, Vrana B, Harman C, Budzinski H, Mazella N, Ocelka T. Measurement of environmental pollutants using passive sampling devices e an updated commentary on the current state of the art. Env Sci Process Impact 2014; 16:369−73.

[9] Campbell DJ, Lusch DP, Smucker TA, Wangui EE. Root causes of land use change in the Loitokitok area, Kajiado District, Kenya. Nairobi: International Livestock Research Institute; 2004.

[10] Cleaver K, Schreiber G. Reversing the spiral: the population, agriculture and environment Nexus in sub-Saharan Africa. Washington, D.C.: World Bank; 1994.

[11] Conroy MJ. Mapping of species richness for conservation of biological diversity: conceputal and methodological issues. Ecol Appl 1996;6(3):763−73.

[12] Le Quéré C, Andrew RM, Friedlingstein P, Sitch S, et al. Global carbon budget 2017. Earth Syst Sci Data Discuss 2017.

[13] Davis SJ, Caldeira K. Consumption-based accounting of CO_2 emissions. Proc Natl Acad Sci USA 2010;107:5687−92.

[14] Rapport DJ, Singh A. Framework for the state of environment reporting [Paper read at UNEP- SCOPE workshop on "Making science more policy relevant", at Prague]. 2002.

[15] Betts RA, Jones CD, Knight JR, Keeling RF, Kennedy JJ. El Nino and a record CO_2 rise. Nat Clim Change 2016;6:806−10.

[16] Campbell DJ. Towards an analytical framework for land use change. In: Bergstrom, Kirschmann, editors. Carbon and nutrient dynamics in tropical agro-ecosystems. Wallingford, UK: CAB International; 1998.

[17] Prietzel J, Zimmermann L, Schubert A, Christophel D. Organic matter losses in German Alps forest soils since the 1970s most likely caused by warming. Nat Geosci 2016;9:543−8.

[18] Bebbington AJ, Batterbury SPJ. Transnational livelihoods and landscapes: political ecologies of globalization. Ecumene 2001;8(4):369−80.

[19] Berardi G. Commentary on the challenge to change: participatory research and professional realities. Soc Nat Resour 2002;15:847−52.

[20] Dietzenbacher E, Pei J, Yang C. Trade, production fragmentation, and China's carbon dioxide emissions. J Environ Econ Manage 2012;64:88–101.
[21] Bridges EM, Oldeman LR. Global assessment of human-induced soil degradation. Arid Soil Res Rehabil 1999;13(4):319–25.
[22] Campbell BM, Sayer JA. Integrated natural resources management: linking productivity, the environment and development. Wallingford, UK: CABI Publishing; 2003.
[23] Denman KL, et al. Couplings between changes in the climate system and biogeochemistry. Cambridge University Press; 2007.
[24] Dlugokencky E, Tans P. Trends in atmospheric carbon dioxide. National Oceanic & Atmospheric Administration, Earth System Research Laboratory (NOAA/ESRL). http://www.esrl.noaa.gov/gmd/ccgg/trends/global.
[25] Duce RA, et al. Impacts of atmospheric anthropogenic nitrogen on the open ocean. Science 2008;320:893–7.
[26] de Sherbinin A. Land-use and land-cover change. In: A CIESIN thematic guide. Palisades, NY: Center for International Earth Science Information Network of Columbia University; 2002.
[27] Marti E, Variatza E, Balcazar JL. The role of aquatic ecosystems as reservoirs of antibiotic resistance. Trends Microbiol 2014;22:36–41.
[28] Barbier EB. The economic determinants of land degradation in developing countries. Phil Trans R Soc London 1997;352:891–9.
[29] Diouf A, Lambin EF. Monitoring land-cover changes in semi-arid regions: remote sensing data and field observations in the Ferlo, Senegal. J Arid Environ 2001;48:129–48.
[30] Lambin EF, Rounsevell MDA, Geist HJ. Are agricultural land-use models able to predict changes in land use intensity? Agric Ecosyst Environ 2000;82:321–31.
[31] Ballantyne AP, Alden CB, Miller JB, Tans PP, White JWC. Increase in observed net carbon dioxide uptake by land and oceans during the past 50 years. Nature 2012;488:70–2.
[32] Ballantyne AP, et al. Audit of the global carbon budget: estimate errors and their impact on uptake uncertainty. Biogeosciences 2015;12:2565–84.
[33] Biot Y, Blaikie PM, Jackson C, Palmer-Jones R. Rethinking research on land degradation in developing countries. Washington, D.C.: World Bank; 1995.
[34] Evans TP, Moran EF. Spatial integration of social and biophysical factors related to landcover change. Popul Dev Rev 2002;28(Supplement).
[35] Erb K-H, et al. Bias in the attribution of forest carbon sinks. Nat Clim Change 2013;3:854–6.
[36] Box P. Spatial units as agents: making the landscape an equal player in agent-based simulations. In: Gimblett RH, editor. Integrating geographic information systems and agent-based modeling techniques for simulating social and ecological processes. New York: Methuen; 2002.
[37] Bauer JE, et al. The changing carbon cycle of the coastal ocean. Nature 2013;504:61–70.
[38] Freire P. Pedagogy of the oppressed. New York: Herder and Herder; 1970.
[39] Gonzalez-Gaya B, et al. High atmosphere-ocean exchange of semivolatile aromatic hydrocarbons. Nat Geosci 2016;9:438–42.
[40] Bao H, Niggemann J, Luo L, Dittmar T, Kao S-J. Aerosols as a source of dissolved black carbon to the ocean. Nat Commun 2017.
[41] Donald CE, Elie MR, Smith BW, Hoffman PD, Anderson KA. Transport stability of pesticides and PAHs sequestered in polyethylene passive sampling devices. Environ Sci Pollut Control Ser 2016.

[42] Garcia-Robledo E, Padilla CC, Aldunate M, Stewart FJ, Ulloa O, Paulmier A, Gregori G, Revsbech NP. Cryptic oxygen cycling in anoxic marine zones. Proc Natl Acad Sci USA 2017;114(31):8319–24.
[43] Hyland KC, Dickenson ERV, Drewes JE, Higgins CP. Sorption of ionized and neutral emerging trace organic compounds onto activated sludge from different wastewater treatment configurations. Water Res 2012;46:1958–68.
[44] Campbell DJ, Olson JM. Environment and development in Kenya: flying the kite in Kajiado District. Centen Rev 1991;2:295–314.
[45] Canadell JG, et al. Contributions to accelerating atmospheric CO_2 growth from economic activity, carbon intensity, and efficiency of natural sinks. Proc Natl Acad Sci USA 2007;104:18866–70.
[46] Ewel KC. Natural resource management: the need for interdisciplinary collaboration. Ecosystems 2001;4:716–22.
[47] Fairhead J, Leach M. Misreading the African landscape: society and ecology in a forest-Savanna Mosaic. Cambridge: Cambridge University Press; 1996.
[48] Hossain MF, Nowshin F. Integration of wind into running vehicles to meet its total energy demand. Energy Ecol Environ 2017;2(1):35–48 [Springer].
[49] Irwin EG, Geoghegan J. Theory, data, methods: developing spatially explicit economic models of land use change. Agric Ecosyst Environ 2001;85:7–23.
[50] McConnell WJ, Moran EF. Meeting in the Middle: the challenge of meso-level integration. In: An international workshop, October 17–20, 2000, Ispra, Italy. LUCC report series. Bloomington, IN: LUCC Focus 1 Office; 2000.
[51] Jeon J, Kurth D, Hollender J. Biotransformation pathways of biocides and pharmaceuticals in freshwater Crustaceans based on structure elucidation of metabolites using high resolution mass spectrometry. Chem Res Toxicol 2013.
[52] Karl DM, Church MJ. Microbial oceanography and the Hawaii ocean time-series programme. Nat Rev Microbiol 2014;12:699–713.
[53] Maitima JM, Olson JM. Guide to field methods for comparative site analysis for the land use change, impacts and dynamics project. Nairobi: International Livestock Research Institute; 2001.
[54] Kinzig AP. Bridging disciplinary divides to address environmental and intellectual challenges. Ecosystems 2001;4:709–15.
[55] Parazoo, et al. Contrasting carbon cycle responses of the tropical continents to the 2015–2016 El Niño. Science 2017.
[56] Petrie B, Barden R, Kasprzyk-Hordern B. A review on emerging contaminants in wastewaters and the environment: current knowledge, understudied areas and recommendations for future monitoring. Water Res 2015.
[57] Postel SL, Daily GC, Ehrlich PR. Human appropriation of renewable fresh water. Science 1996;271:785.
[58] Runnstrom MC. Rangeland development of the Mu us sandy land in semiarid China: an analysis using landsat and NOAA remote sensing data. Land Degrad Dev 2003;14:189–202.
[59] Vörösmarty C, Fekete B, Meybeck M, Lammers R. Global system of rivers: its role in organizing continental land mass and defining land to ocean linkages. Global Biogeochem Cycles 2000;14:599–621.
[60] Kok K, Farrow A, Veldkamp A, Verburg PH. A method and application of multi-scale validation in spatial land use models. Agric Ecosyst Environ 2001;85:223–38.
[61] Liverman D, Moran EF, Rindfuss RR, Stern PC, editors. People and pixels: linking remote sensing and social science. Washington, D.C.: National Academy Press; 1998.
[62] Li W, et al. Reducing uncertainties in decadal variability of the global carbon budget with multiple datasets. Proc Natl Acad Sci USA 2016;113:13104–8.

[63] Pickup G, Bastin GN, Chewings VH. Identifying trends in land degradation in non-equilibrium rangelands. J Appl Ecol 1998;35:365–77.
[64] Qi J, Weltz M, Huete AR, Sorooshian S, Bryant R, Kerr YH, Moran MS. Leaf area index estimates using remotely sensed data and BRDF models in a semiarid region. Remote Sens Environ 2000;73(1):18–30.
[65] Hossain MF. Green science: advanced building design technology to mitigate energy and environment. Renew Sustain Energy Rev 2018;81(2):3051–60 [Elsevier].
[66] Hossain MF. Application of advanced technology to Build a vibrant environment on planet Mars. Int J Environ Sci Technol 2017;14(12):2709–20 [Springer].
[67] Hossain MF. Invisible transportation infrastructure technology to mitigate energy and environment. Energy Sustain Soc 2017;7:27 [Springer].
[68] Lambin EF, Turner II BL, Geist HJ, Agbola SB, Angelsen A, Bruce JW, Coomes O, Dirzo R, Fischer G, Folke C, George PS, Homewood K, Imbernon J, Leemans R, Li X, Moran EF, Mortimore M, Ramakrishnan PS, Richards JF, Skånes H, Steffen W, Stone G, Svedin U, Veldkamp TA, Vogel C, Xu J. The causes of land-use and land-cover change: moving beyond the myths. Global Environ Change 2001;11(4):5–13.
[69] Lambin EF, Geist HJ, Lepers E. Dynamics of land-use and land-cover change in tropical regions. Ann Rev Environ Resources 2003;28:205–41.
[70] Hossain MF. Photonic thermal energy control to naturally cool and heat the building. Adv Therm Eng 2018;131:576–86 [Elsevier].
[71] Hossain MF. Solar energy integration into advanced building design for meeting energy demand. Int J Energy Res 2016;40:1293–300 [Wiley].
[72] Hossain MF. In situ geothermal energy technology: an approach for building cleaner and greener environment. J Ecol Eng 2016;77:49–55 [Polish Society of Ecological Engineering].
[73] Geist HJ, Lambin EF. What drives tropical Deforestation? A meta-analysis of proximate and underlying causes of deforestation based on subnational case study evidence. Louvain-la-Neuve, Belgium: LUCC International Project Office, University of Louvain; 2001.
[74] Gibson CC, Ostrom E, Ahn TK. The concept of scale and the human dimensions of global change: a survey. Ecol Econ 2000;32:217–39.
[75] Ogunlaja OO, Parker WJ. Impact of activated sludge process configuration on removal of micropollutants and estrogenicity. Water Sci Technol 2015.
[76] Hossain MF. Transforming dark photon into sustainable energy. Int J Energy Environ Eng 2018. https://doi.org/10.1007/s40095-017-0257-1 [Springer].
[77] Liu Z, et al. Reduced carbon emission estimates from fossil fuel combustion and cement production in China. Nature 2015;524:335–8.
[78] Mason Earles J, Yeh S, Skog KE. Timing of carbon emissions from global forest clearance. Nat Clim Change 2012;2:682–5.
[79] Pound B, Snapp S, McDougall C, Braun A. Managing natural resources for sustainable livelihoods. Sterling, VA: Earthscan Publications Ltd.; 2003.
[80] Scherr SJ. Meeting household needs: farmer tree-growing strategies in Western Kenya. In: Arnold JEM, Dewees PA, editors. Tree management in farmer strategies. Oxford UK: Oxford University Press; 1995.
[81] Le Quéré C, et al. Global carbon budget 2016. Earth Syst Sci Data 2016;8:605–49.
[82] Le Quéré C, et al. Global carbon budget 2015. Earth Syst Sci Data 2015.
[83] McAvoy DC, Schatowitz B, Jacob M, Hauk A, Eckhoff WS. Measurement of triclosan in wastewater treatment systems. Environ Toxicol Chem 2002;21:1323–9.
[84] Metzger KL, Boone RB, Coughenour MB, Reich RM. Effects of seasonal grazing on plant species diversity and vegetation structure in a semi-arid ecosystem. J Arid Environ 2005;61(1):147–60.

[85] Maitima JM, Reid R, Gachimbi LN, Majule A, Lyaruu H, Pomery D, Mugatha S, Mathai S, Mugisha S. A methodological guide on how to identify trends and linkages between changes in land use, biodiversity and land degradation. Nairobi: International Livestock Research Institute; 2004.
[86] Milliman J, Mei-e R. In: Eisma D, editor. Climate change: impact on coastal habitation. CRC Press; 1995. p. 57–83.
[87] Schwietzke S, et al. Upward revision of global fossil fuel methane emissions based on isotope database. Nature 2016;538:88–91.
[88] Hossain MF. Production of clean energy from cyanobacterial biochemical products. Strat Plann Energy Environ 2016;3:6–23 [Taylor and Francis].
[89] Hossain MF. Theoretical modeling for hybrid renewable energy: an initiative to meet the global power. J Sustain Energy Eng 2016;4:5–36 [Wiley].
[90] Zou L, Zhou T. Asian summer monsoon onset in simulations and CMIP5 projections using four Chinese climate models. Adv Atmos Sci 2015;7:924–34.
[91] Houghton RA. Balancing the global carbon budget. Annu Rev Earth Planet Sci 2007.
[92] Hossain MF. Theory of global cooling. Energy Sustain Soc 2016;6:24 [Springer].
[93] Pijanowski BC, Brown DG, Manik G, Shellito B. Using neural nets and GIS to forecast land use changes: a land transformation model. In: Computers, environment and urban systems; 2002.
[94] FAO. Data sets, indicators and methods to assess land degradation in drylands: report of the LADA e-mail Conference 9 October–4 November 2002. In: World soil resources reports. Rome: Food and Agriculture Organisation; 2003.
[95] Fox J, Rindfuss RR, Walsh SJ, Mishra V. People and the environment: approaches for linking household and community surveys to remote sensing and GIS. Boston, MA: Kluwer Academic Publishers; 2003.
[96] Helbling DE, Hollender J, Kohler H-PE, Singer H, Fenner K. High-throughput identification of microbial transformation products of organic micropollutants. Environ Sci Technol 2010;44:5621–7.
[97] Hossain MF. Green science: independent building technology to mitigate energy, environment, and climate change. Renew Sustain Energy Rev 2017;73:695–705 [Elsevier].
[98] Hossain MF. Design and construction of ultra-relativistic collision PV panel and its application into building sector to mitigate total energy demand. J Build Eng 2017; 9:147–54 [Elsevier].
[99] Hossain MF, Mukai H. Importance of nutrients (N, P, and NO_3+NO_2) in growth of the surfgrass, *Phyllospadix iwatensis* Makino. 2000.
[100] Houghton R. Balancing the global carbon budget. Annu Rev Earth Planet Sci 2007; 35:313–47.
[101] Harini N, Munroe DK, Southworth J. From pattern to process: landscape fragmentation and the analysis of land use/land cover change. Agric Ecosyst Environ 2004; 101:111–5.
[102] Moran MS, Kerr Y, Hymer DC, Qi J. Comparison of ERS-2 SAR and Landsat TM imagery for monitoring agricultural crop and soil conditions. Remote Sens Environ 2002;79(2–3):243–52.
[103] Borrelli P, Robinson DA, Fleischer LR, Lugato E, et al. An assessment of the global impact of 21st century land use change on soil erosion. Nat Commun 2017.
[104] Stephens BB, et al. Weak northern and strong tropical land carbon uptake from vertical profiles of atmospheric CO_2 science. Science 2007;316:1732.
[105] Yin Y, et al. Variability of fire carbon emissions in equatorial Asia and its nonlinear sensitivity to El Niño. Geophys Res Lett 2016;43. 10,472–410,479.

[106] Tengberg A, Ellis-Jones J, Kiome R, Stocking M. Applying the concept of agrodiversity to indigenous soil and water conservation practices in eastern Kenya. Agric Ecosyst Environ 1998;70:259–72.
[107] van der Werf GR, et al. Climate regulation of fire emissions and deforestation in equatorial Asia. Proc Natl Acad Sci USA 2008;105:20350–5.

CHAPTER THREE

Energy

3.1 CONVENTIONAL ENERGY

Wood, the most traditional conventional energy source since ancient time, is being widely used which indeed involves a complete destruction mechanism that eliminates the natural resources and disturbs environmental balance. Moreover other conventional energy sources such as coal, petroleum, and natural gases are fossil fuels, the primary sources of energy used in our daily life, causing a deadly climate change.

3.1.1 Wood

Wood, as an energy source, in the form of firewood, charcoal, chip, sheet, and pellet, is being used significantly in both developing and developed nations. Particularly, wood is being used hugely as the essential wellspring of energy for the rural people in developing countries principally to cook either in a mud boiler or an open-let stove. In developed nations, wood is also being used for warming and sporadically used in a heater, stove, or chimney or outside in a heater, pit fire, or blaze as a mold. Besides, it is also being used as the primary energy for the industrial process such as smoking meat and producing maple syrup (Fig. 3.1).

Figure 3.1 Stapled firewood is being used for fireplace, which is normally cut from forest and being sold by stere (1 $m^3 = 0.276$ cords, relating to a woodpile as 8′ wide × 4′ high log) as a leading traditional heating supply for houses.

Sustainable Design and Build
ISBN: 978-0-12-816722-9
https://doi.org/10.1016/B978-0-12-816722-9.00003-3

3.1.1.1 Energy Science

Burning wood is a simple combustion science. The energy limit of wood is nearly equivalent to the moist content of the species and/or the varieties' biomass configuration. However, a standard hardwood such as red oak contains energy limit (thermal amount) of 14.9 MJ/Kg (6388 BTU/LB), of which 10.4 MJ is recombustible once burned at 70% efficiency, which is equivalent to nearly the energy of 4.5 kWh/kg [7,8].

3.1.1.2 Environmental Impact

To make fire by consuming wood leads to colossal side effects that are terribly undesirable and irritating, which emits H_2O vapor, CO_2, and CO aerosol particulates, alongside caustic alkali fly ash that are indeed dangerous and account for causing deadly human health and climate change [9,10].

3.1.1.3 Human Health

Burning wood substantially contribute to air pollution, particularly particulates. These particulate air pollutants do cause human health problems such as asthma and heart diseases. Besides, wood burning by-products contain toxic and carcinogenic substances such as formaldehyde, polycyclic aromatic hydrocarbons (PAHs), dioxins, and furans, which can cause birth defect [11,12].

3.1.1.4 Climate Change

Wood burning also creates CO and CO_2 by combustion process, and this process results in elevated levels of greenhouse gases in the atmosphere [13,14]. Numerous research has been performed and revealed that particles formed by combustion of woods and other biomass create "black carbon," which is the primary player for climate change. The International Panel on Climate Change (IPCC) has, therefore, ruled out a group of projections to mitigate the climate change by eliminating the use of wood and other biomass burning and to curb the destruction of natural resources. Which means, eliminating our reliance on wood for heating and cooking by breaking our conventional habit to burn wood, thereby making a significant contribution to reduce Green House Gases (GHGs), black carbon, and then finally climate change.

3.1.2 Coal

Coal originated from remains of plants and ferns that grew in swamp area millions of years ago and is considered as the fossil fuel. Microorganisms and natural chemical compounds played a vital role to decompose those tree debris buried beneath water or mud and formed peat, which mainly consists of cellulose $(C_6H_{10}O_5)_n$, and finally peat turned into coal. The top 10 [3] coal producers of 2012 in the world were (in million metric tons = mmt): (1) China, 3,621 mmt; (2) the United States of America, 922 mmt; (3) India, 629 mmt; (4) Australia, 432 mmt; (5) Indonesia, 410 mmt; (6) Russia, 351 mmt; (7) South Africa, 261 mmt; (8) Germany, 196 mmt; (9) Poland, 144 mmt; and (10) Kazakhstan, 122 mmt [15,16].

3.1.2.1 Formation

The formation of coal occurred at about 500 million of years ago when the earth was concealed mostly by huge swampy forests consisting of dense jungle, large ferns, reeds, and giant mosses. During their life cycle process, vegetations grew, died, and fell into the swamp waters and again new vegetation grew on it, died, fell into it, and the process has been going on since. Naturally a deep layer of dead vegetation had been rotten in the swamp. In a period of millions of years, layers got much thicker and placed one on top of others. The weight of the top layers and the water and dirt created a force to pack down the lower layers of the vegetations debris. Then the bacteria and chemical components started to decompose the layer. Furthermore, the internal and external heat of the debris and the weight of debris forced the O_2 out and left rich C deposits to conduct a major physical change in those vegetations which leads to form coal in a sequence as follows: (1) peat, (2) lignite, (3) bituminous coal, and (4) anthracite coal. By the reformative disintegration and thermal and weight pressure, the cellulose lost moisture, H_2, and O_2 and got converted into coal as per the sequences shown in Fig. 3.2.

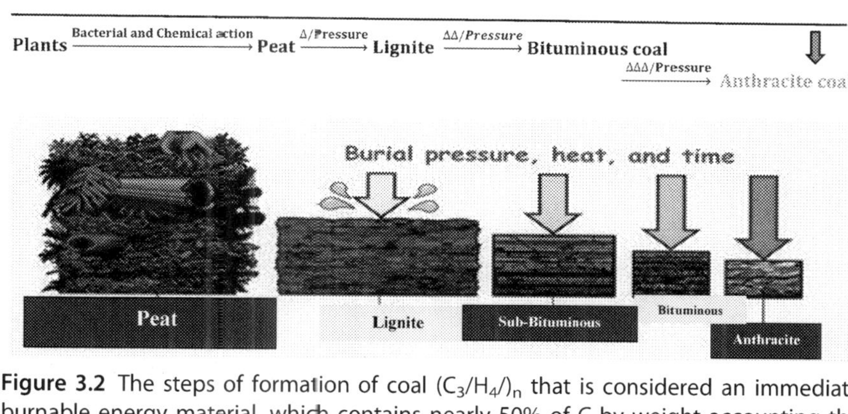

Figure 3.2 The steps of formation of coal $(C_3/H_4/)_n$ that is considered an immediate burnable energy material, which contains nearly 50% of C by weight accounting the amount of carbon and hydrogen present.

3.1.2.2 Energy Science

Though there are various kinds of coal, the main components of coal consist primarily of carbon, water, hydrogen, and some oxygen. Besides, a significant amount of N_2, S, and some other minerals are also present in coal. Nearly all C in coal is bonded, with only one C−C bond present at every C atom. Therefore the formation of energy calculation $C + O_2 \rightarrow CO_2$ can be assumed to be from the breakdown of C−C bond as follows.

C (1 mol)	+	O_2 (1 mol)	→	CO_2 (1 mol)
C−C	+	O = O	→	O = C = O
Energy requirement for breaking the bond:				
83 kcal/mol	+	119 kcal/mol =		202 kcal/mol
Energy formation:				

2×177 kcal/mol = 354 kcal/mol

Net release = 152 kcal/mol of C

Therefore 12 g of C yields 152 kcal of energy, provided sufficient O_2 is available to complete combustion.

1 kg of C thus gives 11,000 kcal of energy

$\frac{152 \text{ kcal}}{12 \text{ g}} \times 1000 \text{ g} = 12{,}666 \text{ kcal}$

Since coal contains other ingredients, the net energy calculation yield of 1 kg of coal is determined as 700 kcal.

The aforementioned equation shows that CO_2 is the primary by-product of coal combustion (44g of CO_2) and forms for every 12 g of carbon burnt. Inadequate O_2 supply during the process of combustion, for example, a room without sufficient ventilation or a fireplace without a sophisticated chimney, forms CO by depleting O_2 ($C + CO_2 = 2CO$), which has no color and odor and is indeed a very dangerous gas [17,18].

3.1.2.3 Production Engineering

The most efficient technique for coal extraction relies on the profundity and nature of the creases and the geological and environment factor components. Most of the coal obtained from surface and subterranean mines is transferred to plants for washing and stockpiling. In more detail it can be stated that it is the way toward extricating coal. Starting from the earliest stage, it has had numerous advancements over the years, that is, from the beginning of men burrowing and physically removing the coal on trucks by conducting substantial open-cut excavation and creating long divider mines. Necessarily coal mining requires appropriate assessment of locations and utilization of depth, dragging, trucking, transportation, and pressure-driven jacks and shearers. Coal mining can be constructed either on surface or underground or through other appropriate processes, and these are classified into categories discussed in the following section.

3.1.2.4 Surface Mining

At the point when coal creases are close to the exterior, it might be efficient to obtain the coal by using open-pit mining. Open-pit coal mining recuperates a more significant coal store than subsurface techniques because a higher amount of the coal crease in the layers might be used. Such technology might be incorporated by an accompanying Dragline that is operated by expelling the overburden. Control scoops in vast trucks transport overburden and coal, and then basin wheel excavators extract the mineral. In this mining technology, use of explosives is the first step to remove the of the mining zone. Then, the overlaying soil and rock is removed by heavy excavators. When the coal crease is uncovered, it is bored and then cracked completely and mined into strips. The coal is then stacked on vast conveyances for transporting it to the coal arrangement plant or directly to the customer.

3.1.2.5 Strip Mining

Strip mining uncovered coal by excavating dirt or bedrock over coal crease. This excavation work is used to remove unwanted material, and thus it is evacuated as much to expose the coal seams. This unwanted material from the excavation is stored near the mining zone temporarily, and eventually it is dumped in an appropriate landfill from a notable distance of the mining zone. Once all the unwanted excavated material is dumped, coal seams will be visible as the blocks of coals and layers. Providing sophisticated support of excavation, these block and layers of coal are to be drilled and broken into pieces by using large earth-moving front-loader excavator for dragging, scooping, and trucking. The coal shall be then stacked on site for in-site processing or can be transported to the coal planning plant for commercial use.

3.1.2.6 Contour Mining

This technique involves excavating unwanted materials from the coal crease in a sequence and dumping the unwanted material along an edge or around the slope of mine. The exposed coals are drilled and cut inside the mining zone by using heavy earth-moving equipment and stored inside the mining zone. In this technology, a path is being created to move trucks from surface to inside the mining zone back and forth by creating appropriate slope support. Once coals are stored in large amounts inside the mine, the truck gets inside the mining zone, gets loaded by coal, and transports the coal to the coal planning plant.

3.1.2.7 Underground Mining

Maximum coal creases are so deep underneath the earth that open-pit mining becomes difficult and requires underground mining, a technique used currently for around two-third of world coal mining. In profound mining, the room and pillar strategy advances along the crease, whereas columns and timber are left in place to support the rooftop. When room and column mines are restricted at a certain point (restricted by topography, ventilation, or financial matters), an accompanying rendition of room and column extraction, named second mining or withdraw digging, is normally begun. Mineworkers expel the coal in the columns, accordingly recouping, however, much coal from the coal crease as could reasonably be expected. A work territory engaged with column extraction is known as a column area. The column areas use remote-controlled gear, including expansive pressure-driven versatile rooftop underpins, which can avert collapses until the point that the diggers and their hardware have left a work region. The portable rooftop underpins are like an extensive lounge area table, however, with water-powered jack support. Once the substantial mainstays of coal are excavated, the portable rooftop bolster's props abbreviate and are dragged back to the protected zone. The mine rooftop regularly falls after the portable rooftop underpins leave a region where primarily the foremost strategies for underground mining are used by Longwall digging. Longwall digging represents around half of the underground generation, where the longwall power loader has a face of 1000 ft (300 m) or more. It is an advanced machine with tractive motive units to move over a coal crease. The extricated coal falls on a protected chain transport or container line that transports the coal to the transmit line for removal from the coal face. Longwall structures have self-propelled water-driven rooftop bolsters that progress with the machine as mining progresses. As the longwall mining gear pushes ahead, overlying rock that is never again bolstered by coal is allowed to fall behind the power loader in a controlled way. Using automated systems with sensors improves productivity along with safety, as coal crease has substances such as methane. Longwall frameworks have coal recuperation rate efficiency between 60% and 80% while encompassing geography permits their utilization [19,20]. Once the coal is evacuated, three-fourth of the segment by rule, the rooftop is permitted to crumple in a controlled way and then is stacked on conveyances for carrying to the coal planning unit (Fig. 3.3).

3.1.2.8 Environmental Impact

The coal industry has some negative impacts on the environment, including air and water contamination caused during coal extraction, processing, and consumption. Apart from air contamination, consumption of coal product produces myriad amounts of hazardous waste products every year, such as fly ash, hot debris, and flue gas that contain radioactive material and other

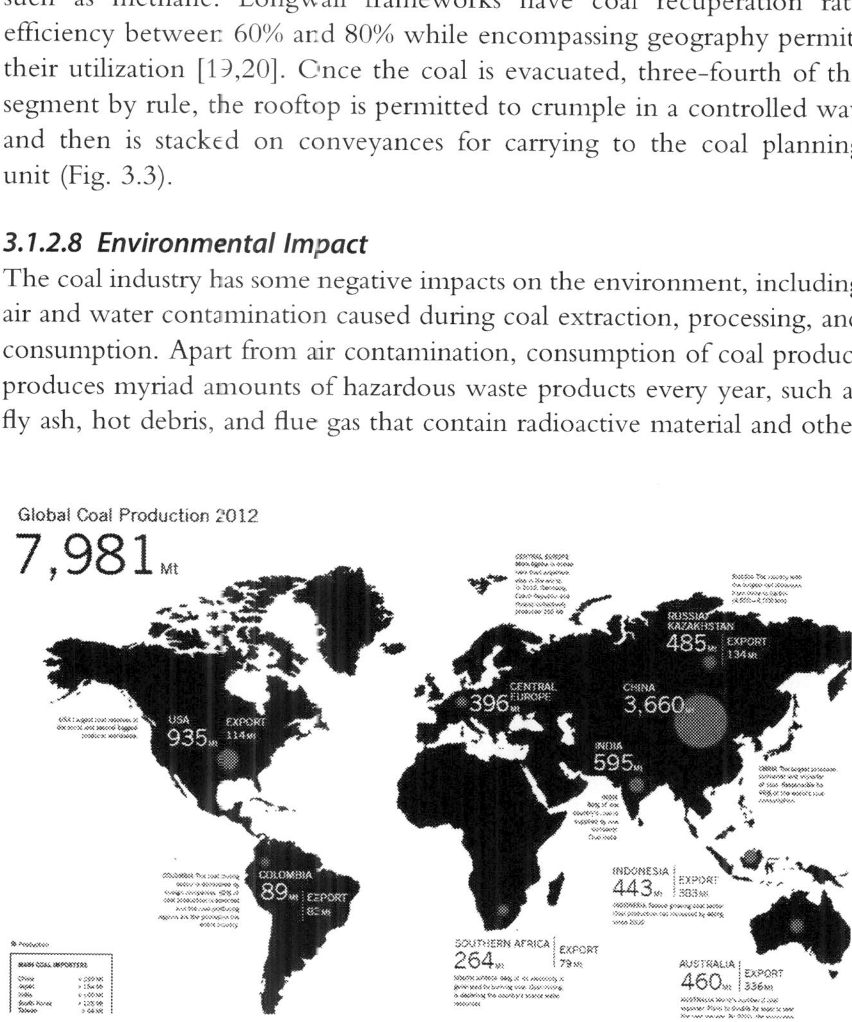

Figure 3.3 Coal generation patterns (1980–2012) in the main five coal-creating nations (US Energy Information Administration [EIA]) such as China, India, the United States of America, Australia, and Russia. Coal is mined for financial purposes in more than 50 nations, and more than 7981 Mt/yr of coal was mined in 2012, which is a considerable increment over the past 25 years. The accompanying figures demonstrate the gauge for the best coal-delivering nations in the planet where China is the biggest producer followed by India and the United States.

substantial metals. Surface mining of coal totally wipes out existing plants, wrecks the traditional soil contour, disturbs natural life and living space, corrupts air quality, changes land utilizes, and, to some extent, changes whole geology of the region.

3.1.2.9 Land and Surroundings

Strip mining is extremely dangerous because it decreases the natural habitat in the surrounding land as it takes out existing vegetation, annihilates the genetic soil profile, dislodges or destroys wildlife and territory, and adjusts current land utilizes. Adverse effects on topographical highlights of human intrigue may happen in a coal strip mine. Aimless mining might result in geomorphic and geophysical stresses and relinquish amazing picturesque assets. Study of prehistoric life forms and social and other memorable esteem might be endangered due to the troublesome actions involved in unearthing coal. Archeological and notable highlights are wrecked by the process of undressing soil and rocks, unless they are shifted to some other place before mining.

Surface mining upsets all tasteful components of the scene fundamentally. Alteration of landforms frequents new and intermittent setups. Further direct instances show up as the material is extricated and squander stacks are created. Assorted hues and surfaces are exposed as the vegetative cover is expelled, and overburden is dumped to the site. Tidy, quivering, and diesel fume scents are made (affecting vision, hearing, and smell). Nearby occupants often find such effects frustrating or obnoxious. During heavy excavation, pinnacles of mountains or slopes are damaged to explore thick coal face lying underneath. The dust and shake evacuated is stored in close by valleys, hollows, and discouragements, clogging and polluting waterways.

Expulsion of soil and rock covering the coal face may cause detention and loss of topsoil and uncovered parent material, making land unfertile. Soil unsettling is influenced, and the related compaction results in conditions helpful for disintegration. Earth expulsion from the surface mining region harms many natural soil properties and diminishes its biodiversity and profitability for agriculture. Soil structure might be exasperated by pounding or total breakdown.

3.1.2.10 GroundWater

Surface mining may weaken groundwater from various perspectives: (1) by wasting usable water from shallow aquifers; (2) bringing down water levels in neighboring territories and altering the stream course inside aquifers; (3) pollution of usable aquifers underneath mining operations because of invasion (permeation) of low-quality mine water; and (4) expanded penetration of precipitation on ruin heaps. Where coal or carbonaceous shale is available, expanded invasion takes place and it may bring about the following: (1) expanded overflow of low-quality water and disintegration from ruin heaps, (2) reviving low-quality water to shallow groundwater aquifers and low-quality water stream to close by streams. The defilement of both groundwater and adjacent streams might happen for drawn out stretches of time. Disintegration of stream quality is an outcome from corrosive mine seepage, lethal follow components, high amount of broken solids in mine wastewater, and expanded silt loads released to streams. At the point when coal surfaces are uncovered, pyrite meets water and air and structures sulfuric corrosive. As water channels from the mine, the corrosive material moves into the conduits; insofar, as rain falls on the mine tailings, the sulfuric corrosive generation proceeds, regardless of whether the mine is yet working or not. Also, squander heaps and coal stockpiling heaps can yield silt to streams. Surface waters might be rendered unfit for agribusiness, human utilization, washing, or other family uses.

3.1.2.11 Surface Water

Coal-fired boilers influence plants to release lignite rich in limestone that contains quicklime (CaO). It rapidly breaks down in water to form calcium hydroxide ($Ca(OH)_2$) and is carried by rainwater to the water bodies from the slag dump zones. Lime-softening procedure accelerates Ca and Mg particles, which gives temporary hardness to the water and then forms $NaHCO_3$ and white soda. Sodium carbonate reacts with the remaining calcium and magnesium ions in the water to provide permanent hardness. Moreover, water-soluble sodium salt in the fiery debris increases the sodium content in water. Therefore, water is changed into the subtle water by substituting Ca and Mg particles with Na particles released by coal-terminated boilers. Application of sodium-influenced water into the soil makes land infertile. Stream water alkalinity as a result of collecting salts in the challenge of meeting different transpiration and vanishing misfortunes ends up plainly severe when many coal-terminated containers and power stations are introduced in a waterbody.

3.1.2.12 Wildlife

Surface mining of coal brings immediate and aberrant harm to untamed life. The effect on natural life stems principally from aggravating, evacuating, and redistributing the land surface. The most direct impact on natural life is demolition or relocation of species in the ranges of uncovering and ruin heaping. Pit and ruin ranges are not fit for giving sustenance and cover to most types of untamed life. Versatile untamed life species such as diversion creatures, fowls, and predators leave these territories. More stationary creatures such as spineless creatures, reptiles, tunneling rodents, and little warm-blooded animals might be crushed. The people group of microorganisms and supplement cycling forms are furious about development, stockpiling, and redistribution of soil. Debasement of amphibian territories is a noteworthy effect by surface mining and might be clear numerous miles from a mining site. Silt defilement of surface water is regular with surface mining. Silt yields may expand a 1000 times of their previous level because of strip mining. The impacts of silt on sea-going untamed life change with the species and the measure of defilement. High residue levels can kill angle straightforwardly, cover upcoming beds, lessen light transmission, change temperature slopes, fill-in pools, spread streamflow over more extensive shallower territories, and decrease generation of amphibian creatures used as nourishment by different species. These progressions crush the living space of esteemed species and may upgrade natural surroundings for less-alluring species.

3.1.2.13 Food Chain

Coal power plants are in charge of half of the mercury discharges all through the world. Mercury is part of the natural way of life, as it is changed over into methylmercury, as a lethal compound which hurts both untamed life and individuals who expend freshwater angle. The mercury is devoured by worms, which are eaten by angle. Individuals come across methylmercury by eating tainted fish that are at the highest point of sea-going natural ways of life where sea angle represents the vast majority source of methylmercury introduced into humans, which can cause cancer.

3.1.2.14 Radiation Introduction

Coal additionally contains low levels of uranium, thorium, and other normally happening radioactive isotopes, which, if discharged into the earth, may prompt radioactive sullying. Coal plants discharge radiation as radioactive fly fiery remains, which is inhaled in and ingested by neighbors and fuses into crops. Barrings containing waste and accidental discharges from coal plants convey more radioactive squanders into the earth than atomic plants per unit of created vitality. Plant-discharged radiation discharged by coal-inferred fly cinder conveys 100 times more radiation to the encompassing condition than does the typical operation of a comparably gainful atomic plant.

3.1.3 Petroleum

The utilization of petroleum was first invented in ancient China over 2000 years ago. In I Ching, one of the most punctual Chinese compositions refers to the use of oil in its crude form without refining, which was first found, removed, and used in parts of China in the first century BCE. Pechelbronn (Pitch wellspring) is known to be the primary European location where oil was investigated and used since 1498. The interest in oil as a fuel for illumination in North America and all over the globe rapidly sprouted since the eighteenth century. Today, around 90% of fuel demands are fulfilled by petroleum. The three leading oil-producing nations are Russia, Saudi Arabia, and the United States.

3.1.3.1 Formation

Petroleum is a derivative formed from old fossilized natural materials, for example, zooplankton and algae. It is a unique blend of hydrocarbons, for the most part, alkanes and cycloalkanes, originated from the vast amounts of these remaining parts settled to ocean or base of lake, blending through dregs in the absence of oxygen. As more layers sank into the ocean or seafloor, with severe heat and pressure, organic material began to convert into a waxy substance known as kerogen, which is found in different oil reserves worldwide, and subsequently more heat, time, and pressure are given into the fluid and vaporous hydrocarbons through a process called catagenesis. The arrangement of oil happens from hydrocarbon decomposition due to variety of primarily endothermic reactions at high temperature as well as pressure. There were certain supplement-rich locations where a lot of natural material settling to the sea bottom surpassed the rate of its decomposition. This brought about huge loads of natural material being covered beneath the consequent stocks, for example, shale shaped from mud. This vast natural stock later ended up plainly warmed and changed underweight into petroleum (Fig. 3.4).

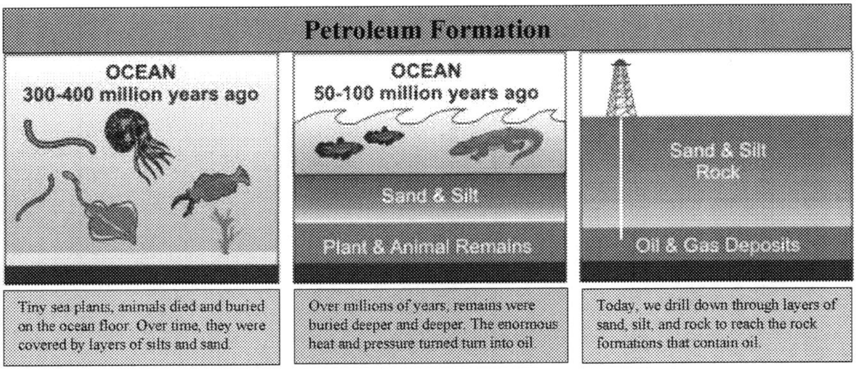

Figure 3.4 The formation of petroleum by the process that started from 300 to 400 million years ago when sea plants and animals died and buried, and then 50–100 million years later, with enormous pressure and heat, they deepened down and turned into oil. Today, we drill down through the sand, silt, and rock to pump up the raw petroleum.

3.1.3.2 Energy Science

Petroleum blend consists mainly of hydrocarbons such as alkanes, cycloalkanes, and different fragrant hydrocarbons, whereas the other natural mixes contain gases such as nitrogen, oxygen, and sulfur, and some amount of metals such as nickel, copper, and vanadium are also present. The correct subatomic arrangement of raw petroleum fluctuates approximately from development to development, yet the number of compound components differs in range of 83%–85% for carbon, 10%–14% for hydrogen, 0.1%–2% for nitrogen, 0.05%–1.5% for oxygen, 0.05%–6.0% for sulfur, and for metals <0.1%. Four unique sorts of hydrocarbon atoms show up in raw petroleum. The comparative level of every shift from oil to oil decides the properties of each type of oil. There are alkanes for around 15%–60%, naphthenes at 30%–60%, aromatics at 3%–30%, and asphaltics at 6% or for leftover portion. Raw petroleum differs enormously in appearance relying on its creation [21,22].

Because oil is a blend of a substantial number of various hydrocarbons, the most ordinarily discovered atoms are alkanes (paraffin), cycloalkanes (naphthenes), aromatic hydrocarbons, or more complex chemicals such as asphaltenes. Every fuel mixture has a one of a kind blend of particles, which characterizes its physical and synthetic properties, similar to shading and consistency. The alkanes, otherwise called paraffin, are immersed hydrocarbons with linear or expanded series, which contain just carbon and hydrogen and have the general formula C_nH_{2n+2}. They have carbon atoms in the range of 5−40 for each hydrocarbon, in spite of the fact that following measures of smaller or larger particles might be available in the blend [23,24].

The alkanes consisting five to eight carbon atoms, i.e., pentane (C_5H_{12}) to octane (C_8H_{18}), are refined into gas, the ones having nine to 16 carbon atoms, i.e., from nonane (C_9H_{20}) to hexadecane ($C_{16}H_{34}$), and later into diesel fuel, lamp oil, and fly fuel. Higher order alkanes which have more than 16 carbon atoms can be refined into fuel oil and greasing up oil [25,26]. At the larger end of the range, paraffin wax is an alkane with around 25 carbon atoms, whereas black-top has 35 and up, in spite of the fact that present-day refineries break them into more profitable items. The shortest particles consisting four or fewer carbon atoms are in a vaporous state at room temperature. They are called oil gasses. Contingent upon request and the expenditure of recuperation, these gasses are either burnt off and sold as condensed oil gas under pressure or used in controlling the refinery's particular burners. During the winter, butane (C_4H_{10}) is mixed into the fuel pool at high frequencies because its high vapor pressure helps when winter begins [27]. When Butane is liquefied under pressure just above barometrical pressure, it is used for controlling cigarette lighters, yet it is likewise a principle gas hotspot for some generating nations. Propane can be liquefied under low pressure and is expended for almost every function depending on oil for vitality, from cooking to heating or even as a fuel for transportation.

The naphthenes are soaked hydrocarbons which consists of only hydrogen and carbon atoms arranged in single ring by the formula C_nH_{2n}. Cycloalkanes have similar properties to alkanes but have higher breaking points [4].

The aromatic hydrocarbons are unstable hydrocarbons which contain one or more benzene rings, to which hydrogen molecules are attached with the formula C_nH_{2n-6} [28,29]. They tend to consume with a dingy fire, and many have a sweet fragrance. Several aromatic hydrocarbons cause cancer.

Aromatic hydrocarbons are isolated by fragmentary refining at an oil refinery to create gas, stream fuel, lamp oil, and different hydrocarbons [22,30]. For example, isooctane, broadly used as a part of fuel, has a molecular equation of C_8H_{18}, and it reacts with oxygen.

$$2C_8H_{18(l)} + 25O_{2(g)} \rightarrow 16CO_{2(g)} + 18H_2O_{(g)}$$
$$(\Delta H = -5.51\ \text{MJ/mol. of octane}) \quad (3.1)$$

The quantity of different atoms in an oil test can be detected by research center examination. The atoms ordinarily extricated in a dissolvable form, at that point isolated in a gas chromatograph, lastly decided with an appropriate indicator, for example, a fire ionization locator or a mass spectrometer. As oil consists of large number of coeluted hydrocarbons, they cannot be settled by conventional gas chromatography and thus regularly show up as a mound in the chromatogram. This uncertain complex blend (UCM) of hydrocarbons is especially evident while examining weather-beaten oils and concentrates from tissues of organisms present in oil. Inadequate ignition of oil or gas brings about dangerous results. When amount of oxygen is very less in ignition, it results in emission of carbon monoxide. Owing to high temperature and pressure included, fumes and gasses from fuel burning in automotors as a rule incorporate nitrogen oxides which are in charge of formation of photochemical exhaust cloud. At a consistent volume, the warmth required for burning an oil-based good can be estimated as:

$$Q_v = 12,400 - 2100\, d^2 \tag{3.2}$$

where the unit of Q_v is calories per gram and d is the specific gravity at 60°F (16°C). The thermal conductivity of petroleum-based liquids can be stated using the following formula:

$$K = \frac{1.62}{\text{API}}\left[1 - 0.0003(t - 32)\right] \tag{3.3}$$

where K is measured in BTU. $\text{F}^{-1}\text{hr}^{-1}\text{ft}^{-1}$, t is measured in °F, and API in degrees American Petroleum Institute (API) gravity. The specific heat of petroleum oils can be stated using the following formula:

$$c = \frac{1}{d}\left[0.388 + 0.00046t\right] \tag{3.4}$$

where c is measured in BTU/lb°F, t is measured in °F, and d is the specific gravity at 60°F (16°C). If c is measured in kcal/(kg °C), the formula can be given as:

$$c = \frac{1}{d}\left[0.4024 + 0.00081t\right] \tag{3.5}$$

where t is in degree Celsius and d is the specific gravity at 15°C. The latent heat of vaporization can be stated using the following formula:

$$L = \frac{1}{d}\,[110.9 - 0.09t] \tag{3.6}$$

where L is measured in BTU/lb, t is in °C, and d is the specific gravity at 60°F (16°C).

If L is in kcal/kg, the formula can be as follows:

$$L = \frac{1}{d}\,[194.4 - 0.162t] \tag{3.7}$$

where t is in degree Celsius and d is the specific gravity at 15°C.

3.1.3.3 Production Engineering

Latest technologies such as seismic mapping and imaging innovation have helped researchers for signaling out gas and petroleum fields. Seismologists can expose supply of hydrocarbons by using a supersensitive sound-discharging device such as geophones, which gives bobbing movement of shake development beneath the ground. By recording the echoes and then analyzing 3D maps of them through supercomputers, it helps in finding the location of petroleum and gas fields in less time and also avoids the cost of standard exploration. Out of all the technologies available to extract energy out of the petroleum mining, drilling is the most popular technology in practice. It is done after the study of supplementary geography, sedimentary bowl examination, and supply portrayal. After this, makers can reach to oil and gas stocks with the help of powered cracking and level penetrating, which was almost impossible beforehand with immense cost.

3.1.3.4 Horizontal Drilling

Horizontal drilling and penetrating help oil and gas producers to limit the adverse effects on surface and also help to bore various wells from a single cushion. Horizontal drilling may be a debatable issue for mineral rights, as drillers can drill into neighboring mineral deposits, yet it is one of the most renowned technologies in the petroleum industry. It came into existence in 1930, and collaborating with water-driven breaking, it has brought a revolution in worldwide energy production. Because of this combination, oil production has increased significantly, and this is termed as "Shale Revolution."

3.1.3.5 Hydraulic Fracturing

Oil and gas producers are taking benefit of hydraulic fracturing due to the water-driven cracking system. This has been in practice since 1947, and it has brought shale revolution all over the world by using a combined method of hydraulic fracturing and horizontal drilling. According to the national petroleum council, this innovation will take over 70% of petroleum gas production in the United States. For breaking the well efficiently, makers introduce pressurized fluid-containing water, chemicals, and proppant into well, which splits stone opening and thus makes it easy for petroleum gas to stream openly. Pressure-driven cracking has some adverse ecological effects such as surface contamination and excessive water consumption and violation, the potential for initiated quakes, and so use of it has always been arguable. These potential natural dangers could delay the spreading of water-powered cracking to different nations, although all these difficulties are inclined toward state and neighborhood level in the United States.

3.1.3.6 Deepwater Drilling

The Deepwater Horizon is a semisubmersible ultra—deepwater-penetrating unit which is designed for working in hard surface conditions and great depth up to 10,000 feet with a group of 135 faculties. The vessel does not fix its position to any particular point but instead keeps on varying the position according to the well, using thrusters and propellers. Structure of semisubmersible platforms is designed in such a way using ballasted boats that they can easily resist waves and can remain more steady than ordinary boats. Also, they do not have substantial deck territories with control and operations space, helipads, and stacking docks.

3.1.3.7 Tar Sands

Petroleum can be generated from tar sands, but it has wide ramifications for worldwide petroleum advertise. Tar sands are blend of mud, sand, water, and bitumen (heavy type of petroleum). They are often referred as petroleum sand. First, tar sands are mined and bitumen is extracted from it, which is later on refined into oil. This procedure is more tedious and expensive than traditional oil extraction process as only one barrel of oil can be extracted from two tons of tar sands. There are two famous extraction techniques for tar sands, such as mining and in situ. Regarding mining, open-pit mining is the most famous type of extraction technique in practice, in which extensive pressure is required to remove the tar sand and load them onto trucks carrying around 320 tons in each load. In the end, oil is separated from bitumen using blend of warmth, water, chemicals, and steady development. Regarding in situ, this technique is used when bitumen stocks are too deep to extract ordinarily. This technique relies on introducing steam to covered tar sands and extracting bitumen through wells. The advancement of tar sands far and wide may lead to ecological issues.

3.1.3.8 Utilization

Petroleum is very crucial to several projects, and for industrialized human development, it is of vital importance. It is also a primary concern for many countries that out of all the energy sources, oil represents a massive part in worldwide energy consumption, ranging 32% for Europe and Asia, 53% for the Middle East, 44% for South and Central America, 41% for Africa, and 40% for North America. Every year, around 30 billion barrels of oil are consumed across the globe, and most of the oil purchasers are the countries who extract the oil. In 2004, United States alone consumed 24% of worldwide consumption. However, it dropped to 21% in 2007. According to the report of US EIA, 2011, the world consumes 87.421 million barrels of oil every day. This table shows that oil was consumed in 1000 barrels per day (1000 bbl/day) and 1000 cubic meters per day (1000 m^3/day) in the year 2011 [31,32] (Tables 3.1 and 3.2).

3.1.3.9 Environmental Impact

Humans use petroleum for fueling transportation, for keeping houses warm, and also in plastic and drug industry. Although petroleum-based products make human life much comfortable, the whole process of extracting to delivering them can cause the air and water contamination to the earth. Harmful substances are radiated from burning the petroleum. Gases such as carbon dioxide (CO_2), carbon monoxide (CO), sulfur dioxide (SO_2), nitrogen oxides (NO_x), and volatile organic compounds (VOCs), particulate matter (PM), lead, and other air toxins, for example, benzene, acetaldehyde, and 1,3-butadiene are emitted, which have deadly effects on earth and human welfare.

3.1.3.10 Harmfulness

A thin shiny layer is formed on the surface of water due to the presence of oil distillates called interphase. Raw petroleum is an amalgam of many natural mixes, most of which are hazardous and cancer-causing. Even it can cause problem to child birth. Benzene present in petroleum can cause leukemia

Table 3.1 Top petroleum-consuming countries (million barrels per day) among which the United States of America, China, Japan, India, Russia, and Saudi Arabia are in the top of the list (in 2016)

Consuming nation 2011	1000 bbl/day	1000 m^3/day	Population in millions	bbl/year, per capita	m^3/year, per capita	National production/ consumption
United States	18,835.5	2994.6	314	21.8	3.47	0.51
China	9,790.0	1556.5	1345	2.7	0.43	0.41
Japan	4,464.1	709.7	127	12.8	2.04	0.03
India	3,292.2	523.4	1198	1	0.16	0.26
Russia	3,145.1	500.0	140	8.1	1.29	3.35
Saudi Arabia (OPEC)	2,817.5	447.9	27	40	6.4	3.64
Brazil	2,594.2	412.4	193	4.9	0.78	0.99
Germany	2,400.1	381.6	82	10.7	1.70	0.06
Canada	2,259.1	359.2	33	24.6	3.91	1.54
South Korea	2,230.2	354.6	48	16.8	2.67	0.02
Mexico	2,132.7	339.1	109	7.1	1.13	1.39
France	1,791.5	284.8	62	10.5	1.67	0.03
Iran (OPEC)	1,694.4	269.4	74	8.3	1.32	2.54
United Kingdom	1,607.9	255.6	61	9.5	1.51	0.93
Italy	1,453.6	231.1	60	8.9	1.41	0.10

OPEC, American Petroleum Institute.

Table 3.2 Top petroleum-producing countries (million barrels per day) in the world (2016)

Rank	Country	Oil production (bbl/day, 2016)
1	Russia	10,551,497
2	Saudi Arabia (OPEC)	10,460,710
3	United States	8,875,817
4	Iraq (OPEC)	4,451,516
5	Iran (OPEC)	3,990,956
6	China	3,980,650
7	Canada	3,662,694
8	United Arab Emirates (OPEC)	3,106,077
9	Kuwait (OPEC)	2,923,825
10	Brazil	2,515,459

in people. It can also reduce the tally of white platelet in the human body and thus decrease human's immunity against diseases. Studies have found out that even presence of small amount of benzene in range of parts per billion can cause deadly diseases such as leukemia, Hodgkin's lymphoma, and other blood and invulnerable framework within 5–15 years.

3.1.3.11 Air Contamination

Whenever petroleum distillates are burned and burning has not finished, alternate mixes are also formed instead of water and carbon dioxide. These mixes such as carbon monoxide and methanol are very dangerous to life. Fine particulates of soot can blacken human and animal lungs and can even cause heart diseases. Residue from the burning of petroleum causes cancer.

3.1.3.12 Corrosive Rain

Nitrogen gas in the atmosphere gets oxidized, making nitrous oxides due to a high temperature by burning of petroleum. Nitrous oxides along with sulfur dioxide from sulfur present in oil mix with water in the atmosphere and cause acid rain. Acid rains can cause problems such as dead trees and fermented lakes with dead fish. Acid rain slaughters coral reefs.

3.1.3.13 Petroleum Slicks

Petroleum or oil slick is a contact of oil with nature, especially with marine, due to human activity and is a type of contamination. The term is generally connected to marine oil slicks, where oil is discharged into sea or river and sometimes ashore due to spills. These slicks can occur because of leaking of petroleum from tankers, pipelines, railcars, seaward stages, boring apparatus, and wells or due to spills from refined oil-based products. It can also be due to bulkier fills used by ships or spill of any waste oil. Split oil infiltrates into feathers of bird or fur of animals, reducing its capacity for protection against extreme temperature and making them more cumbersome in water. Oil slicks offshore are more harmful than ashore because it can spread over a greater distance, even a small oil spill, which can easily cover even a shoreline with a thin layer of oil. This layer of oil can affect ocean flying creatures, warm-blooded animals, shellfish, and other living beings.

3.1.3.14 Unstable Natural Mixes

Unstable natural mixes (VOCs) are gasses or vapors produced by various solids and liquids and most of which have harmful effects on human society and earth. VOCs from oil are toxic and pollute the air, and some such as benzene are deadly to living beings. Aliphatic unstable vapors are more critical from spills of diesel and raw petroleum. "Less poisonous" particles than benzene, for example, albeit in abundance, can cause well-being concerns although when benzene levels are moderately low. Inhaling petroleum hydrocarbons, for example, gas, diesel, and fly fuels from underground storage tanks can affect human health adversely.

3.1.3.15 Environmental Change

Consumption of oil emits carbon dioxide, an ozone-depleting substance. Along with coal consumption, oil burning might be the biggest factor contributing to climatic CO_2. The content of CO_2 has risen to 390 ppmv in recent course of 150 years, which was about 180–300 ppmv in earlier 800–1000 years. This temperature variance may have reduced the Arctic ice top to 1,1,00,000 miles^2 (2,800,000 km^2). According to the report of International Energy Agency, around 13% of world's oil resides in the Arctic.

3.1.4 Natural Gas

Natural gas is a similar product used for generation of energy for warming, cooking, and power purposes. It is also used as a fuel and in plastic and chemical industry. Natural gas was first found in old China when they were boring for salt water. Natural gas was used in the production of salt. It was first used by Chinese in around 500 BCE. They found out a method to transfer gas leaking from the ground in bamboo pipes to Ziliujing District of Sichuan, where it was used in the withdrawal of salt by boiling salt water. The extraction of natural gas first began in the year 1825 at Fredonia, New York, the United States. By the year 2009, approximately 66 000 km^3 had been consumed out of the total recoverable reserve of 850 000 km^3 of natural gas. In the 19th century, natural gas was acquired as a by-product of delivering oil. The volume of rare gas is assessed to be 660,000 km^3, such as, shale gas, out of the total volume of 180,000 km^3 that might be extractable for use. Researchers from Massachusetts Institute of Technology (MIT), Black and Veatch, and the Department of Energy (DOE) have predicted that the natural gas will represent a great share of power and heat source in coming time. Russia has the greatest demonstrated reserve (476,000 km^3), whereas Iran, Qatar, Turkmenistan, and the United States has 33,100, 25,100, 17,500, and 8500 km^3, respectively [33–35] (Fig. 3.5).

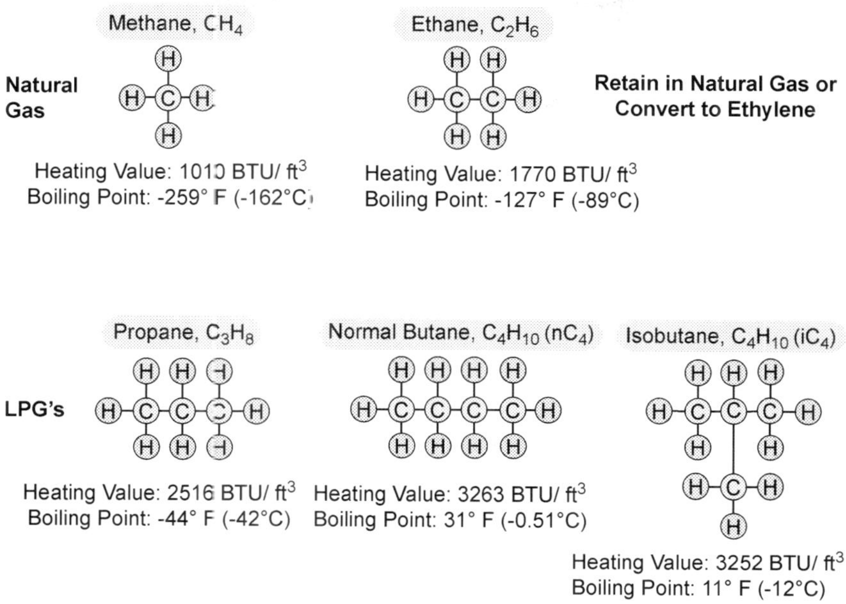

Figure 3.5 Methane can be considered as the direct result of gaseous petrol, whereas propane, normal butane, and isobutene are results of common gasses.

According to the estimated 2015 world's annual consumption rate of 3400 km^3 of the natural gas, the aggregate estimated left stocks of natural gas which are economically recoverable would last for 250 years at the current consumption rate. An yearly increment of 2%–3% in the consumption rate could possibly lower down the estimated reserves by 80–100 years [36,37].

3.1.4.1 Formation

Natural gas is a typical blend of hydrocarbon gas comprising methane, yet sometimes it has some content of other alkanes and little level of carbon dioxide, nitrogen, hydrogen sulfide, or helium. When layers of decaying plant and animal matter remain under the surface of the earth with high pressure and heat for a long time, natural gas is formed. Plants gain energy from the sun, which is put away as a synthetic bond in the gas. Most of the natural gas is developed by two systems, namely biogenic and thermogenic. Methanogenic living organisms in swamps, landfills, and shallow residue make biogenic gas, whereas thermogenic gas is produced from covered natural material at greater depth with prominent heat and pressure (Fig. 3.6).

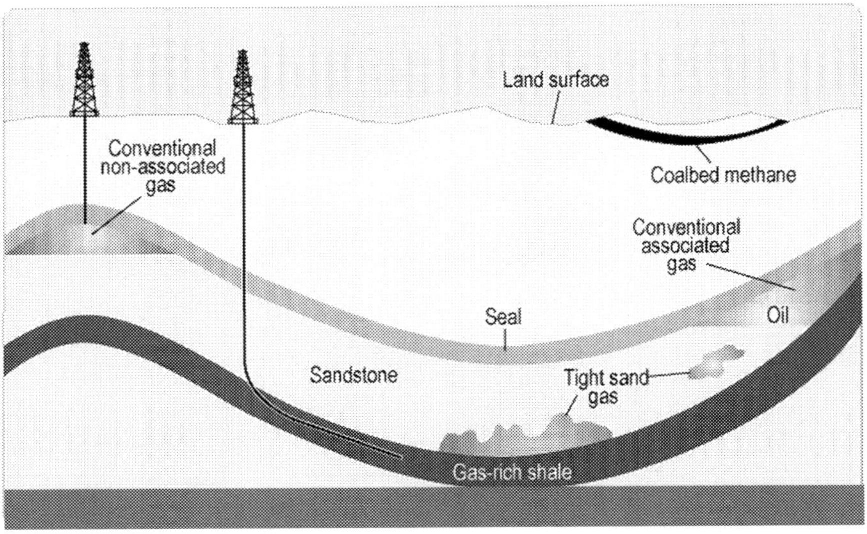

Figure 3.6 Schematic geography of gaseous petrol deposition in the layers of ocean floors which is formed in long period of geological time scale. *U.S. Energy Information Administration and U.S. Geological Survey.*

Natural gas is referred just as a "gas," particularly when contrasted with other energy sources of gas, such as shale gas and town gas.

3.1.4.2 Shale Gas

As the name suggests, shale gas is natural gas supplied from shale. Because shale has a low surface absorptivity, shale gas wells rely on cracks to stream the gas. Early shell gas wells relied on natural crack for streaming gas, but nowadays shale gas wells require water-driven cracking to stream the gas. Since the year 2000, shale gas became a unique resource of flammable gas in the United States and Canada. Owing to the efficiency of shell gas in the United States, its investigation has started in countries such as Poland, China, and South Africa. The United States is the leader in shale gas production in the planet.

3.1.4.3 Town Gas

Town gas is a combustible vaporous fuel made my destructive refining of coal. It contains a mixture of calorific gases such as hydrogen, carbon monoxide, methane, and other hydrocarbons with few noncalorific gases such as carbon dioxide and nitrogen and is used as a substitute for flammable gas. The gas liberated from coal is conveyed to living rooms and different units through the system of channels. (gas warming came into use in the latter half of 20th century.)

3.1.4.4 Energy Science

Natural gas primarily consists of methane. The gross heating value of the natural gas is the total heat acquired by complete combustion at a unit weight of unit volume of a gas in air and heat liberated by gathering water vapor in the combustion products (gas, air, and ignition items taken at standard weight and temperature), which can be calculated by following equation.

$$\text{Energy absorbed by water (J)} = \text{mass of the water (g)} \times \Delta T\ (^{\circ}\text{C}) \times \left(\frac{4.18\ \text{J}}{\text{g}\cdot{}^{\circ}\text{C}}\right)$$

$$\text{Energy released by burning natural gas (J)} = \text{time (s)} \times \left(\frac{\text{vol gas (L)}}{\text{time (s)}}\right) \times \left(\frac{1\ \text{mol gas}}{22.4\ \text{L}}\right) \times \left(\frac{\Delta H^{\circ}_{\text{rxn}}\ (\text{J})}{1\ \text{mol gas}}\right) \quad (3.8)$$

$$\%\ \text{Efficency} = \left(\frac{\text{Energy absorbed by water (J)}}{\text{Energy released by burning natural gas (J)}}\right) 100\%$$

The gross heat of combusting 1 m^3 of business-quality gaseous petrol is around 39 MJ ($\approx$10.8 kWh), although this value can differ by a few percent. This comes to around 49 MJ ($\approx$13.5 kWh) for 1 kg of gaseous petrol (expecting a thickness of 0.8 kg/m^3, a rough estimate).

The gross heating value of the natural gas is the total heat acquired by complete combustion at a unit weight of unit volume of a gas in air and heat liberated by gathering water vapor in the combustion products (gas, air, and ignition items taken at standard weight and temperature). The average heating value of the natural gas is given in Table 3.3.

3.1.4.5 Production Engineering

Natural gas production streaming is done by transmission pipe drilling, and lines controlled by motors and wellhead are regularly monitored to control gas production. There are compressors which regulate pressure of gaseous petrol in a conveyance line when gas transmits. Gaseous petrol conveyance channels reach out to the natural gas—handling unit, and from there the higher atomic weight flammable gas hydrocarbons are discharged to deliver a BTU estimate in the range of 950—1050 BTUs. Natural gas separated from oil wells is called a casing head gas or a related gas. The natural gas industry is liberating a rising amount

Table 3.3 Average heating value of the natural gases

Component	Typical analysis (mole %)	Range (mole %)
Methane	93.9	87.0–97.0
Ethane	4.2	1.5–9.0
Propane	0.3	0.1–1.5
Isobutane	0.03	0.01–0.3
Normal butane	0.03	0.01–0.3
Isopentane	0.01	Trace: 0.04
Normal pentane	0.01	Trace: 0.04
Hexanes plus	0.01	Trace: 0.06
Nitrogen	1.0	0.2–5.5
Carbon dioxide	0.5	0.05–1.0
Oxygen	0.01	Trace: 0.1
Hydrogen	Trace	Trace: 0.02
Specific gravity	0.59	0.57–0.62
Gross heating value (MJ/m^3), dry basis	38.7	36.0–40.2

The properties shown in the table are a total average when following conditions are given: ignition point, 593°C; flammability limits, 4%–16% (volume % in air); theoretical flame temperature, 1960°C (3562°F); and maximum flame velocity, 0.3 m/s. The typical sulfar content is 5.5 mg/m^3 inclusive of 4.9 mg/m^3 of sulfar added for safety reasons. The water vapor content of natural gas is typically 16–32 mg/m^3 [38–40].

of gas from testing asset sorts such as sharp gas, tight gas, coalbed methane, and shale gas (Fig. 3.7).

Because natural gas is an adulterated item when nonrelated gas is liberated from a field under supercritical conditions, repository mass drops down, and the higher subatomic weight portions may mix upon an

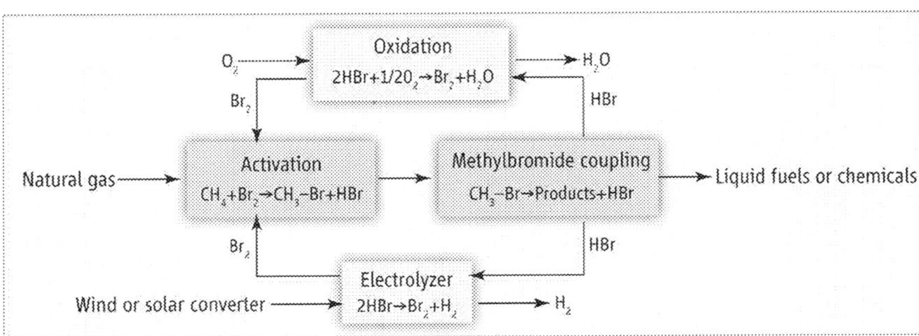

Figure 3.7 Schematic stream outline of an ordinary gaseous petrol preparing plant to refined nature for the use as energy source.

isothermic depressurizing as an impact called retrograde build-up. Thus the fluid along these lines is responsible for managing this issue by reinfusing dry gas to maintain the underground weight and to permit removal of condensates. The fluid concentrates at the surface and gas plant collect this condensate in standard cubic meters (at 0°C and 101.325 kPa) or in standard cubic feet (at 60°F [16°C] and 14.73 psia).

3.1.4.6 Environmental Impacts

Because natural gas primarily consists of methane, after discharging it to the atmosphere, it is removed by steady oxidation to carbon dioxide and water by hydroxyl radicals (•OH) formed in the atmosphere, $CH_4 + 2O_2 \rightarrow CO_2 + 2H_2O$. Compared to carbon dioxide, methane has a very short lifespan of 7 years, but as it is more capable of absorbing warmth and given the amount of methane, it has 84 times the impact on temperature as of carbon dioxide in a 20-year span and 28 times in a 100-year span. Therefore natural gas is much more intense at depleting ozone layer than carbon dioxide because of the high content of methane emission.

3.1.4.7 CO_2 Discharges

Natural gas is often pictured as purest among all the petroleum products. It generates around 29% less CO_2 and 44% less CO_2 for each joule consumed than oil and coal, respectively, and is less toxic compared with any other hydrocarbon fuel. It includes a moderate amount of human carbon emissions, which is anticipated to develop. As per the IPCC Fourth Assessment Report, 2004, gaseous petrol emitted around 5.3 billion tons CO_2 per year, on the other hand, coal and oil contributed 10.6 billion tons and 10.2 billion tons per year to CO_2 emissions, respectively. Petroleum gas would add to 11 billion tons of CO_2 annually as per Special Report on Emissions Scenarios by 2030, and coal and oil would contribute 8.4 and 17.2 billion tons individually with an expansion rate of 1.9% a year [41,42].

3.1.4.8 Air Contamination

The combustion of natural gas produces a negligible amount of sulfur, mercury, and particulates. Combusting natural gas produces nitrogen oxides (NOx), which are responsible for brown haze, although at much lower level than fuel and diesel used for running engines. DOE assessment states that if each of 10,000 homes of United States is using natural gas instead of coal, then the yearly discharge of 1900 tons of NOx, 3900 tons of SO_2, and 5200 tons of particulates can be avoided [43]. Reduction in these emissions can help in improving human health as these poisons are connected with diseases such as asthma, bronchitis, and lung cancer.

3.1.5 Hydropower

Hydropower is the most reliable, developed, and practical energy source innovation available to meet the daily energy demand [44,45]. Hydropower schemes have critical flexibility in their plan and are intended to meet urgent stack requests with moderately high limit factors or have higher introduced limits with lower limit factors but meet the more significant offer of a top request. However, hydropower plans can have essential visual and spatial impression. One of the main obstacles in developing hydropower is assuring that development and plan of undertaking this project are sensible. In spite of having financial approval, if there is any adverse effect of the project on the environment, living beings, and biodiversity, then it must be eliminated before undertaking the design.

3.1.5.1 Energy Science

The key elements of hydropower are rate of flow of liquid and water head. The static head refers change in the height of the gradient through which the water falls. Dynamic head acknowledges with speed of moving water. To maximize the output of energy generation, maximum head and flow should be there. Every unit of water can do a work equivalent to product of weight and head. The power generated from falling water can be computed using the following formula:

$$P = \eta \rho Q g h \tag{3.9}$$

where P is control in watts, η is the dimensionless productivity of the turbine, ρ is the thickness of water in kilograms per cubic meter, Q is the stream in cubic meter every second, g is the speeding up because of gravity, and h is the height difference among bay and outlet in meter.

To define, control is ascertained for a turbine that is 80% productive, with water at 1000 kg/cubic meter (62.5 pounds/ft^3) and a stream rate of 80 cm/s (2800 ft^3/s), gravity of 9.81 m for each second squared, and a net head of 145 m (480 ft). In SI units, the vitality can be figured as:

$$\text{Power (W)} = 0.80 \times 1000 \times 80 \times 9.81 \times 145 = 91 \text{ MW} \tag{3.10}$$

For computing productivity, administrators can differentiate the actual electrical energy that is generated and hypothetical potential energy of the water flowing through the turbine. The richness of the hydropower turbine represents the head lost because of stream grinding, the rise in water level, the area of the station, impact of changing gravity, the temperature, and the barometric weight of the air, the thickness of water, and heights above forebay and tail bay. Hydropower can be computed for few GW for large-scale ventures. Smaller projects may depend on the only single turbine, and it can be calculated by the following formula of levelized cost of energy generation:

$$\mathrm{LCOE} = \frac{\sum_{t=1}^{n} \frac{I_t + M_t + F_t}{(1+r)^t}}{\sum_{t=1}^{n} \frac{E_t}{(1+r)^t}} \tag{3.11}$$

where LCOE = the average lifetime levelized cost of electricity generation; I_t = investment cost in the year t; M_t = operations and maintenance cost in the year t; F_t = fuel cost in the year t; E_t = electricity production in the year t; r = discount rate; and n = economic life of the system.

Large hydropower frameworks tend to associate with a unified structure, although assuring that in the end, demand will meet their age limit. Hydropower plants can be used in scaled down networks or as a separated network framework. If huge storage is impractical, then hydropower plants may require joining with other age sources to assure constant supply in dry periods. The two biggest hydropower ventures on the earth are 14 GW Itaipu in Brazil and Three Gorges in China with 22.4 GW. These two units alone deliver 80−100 TWh/year according to the IPCC, 2011 [46,47].

3.1.5.2 Production Engineering

There are various possible factors to be considered when building up an LCOE approach for hydropower. The model displayed here is very simplified, given that it should be connected with a wide variety of innovations in various nations. Similarly a detailed study of LCOE brings about a substantially higher overhead as far as the granularity of presumptions is required. When it is not possible to run the model with much hypotheses and suspiciousness cannot be separated, given genuineness, "exactness" of the approach can delude. Hydropower plants can be built in various sizes with different traits, and instead of the importance of head and flow rate, it can be put into categories:

1. Run-of-stream hydropower: These plants can have negligible or no storage capacity, and their age are determined from planning and size of flowing streams.
2. Reservoir hydropower: These plants are having storage capacity behind the dam with the specific goal to separate age from water inflow. Storage limit is subjected to the site and financial aspect of the project.
3. Pumped storage hydropower: These plants pump water off top from the storage situated at a lower elevation, and during the time of high electricity demand, energy can be generated by releasing water through turbines.

These are the three most famous hydropower plants in practice and can produce a wide range of size and capacity depending on geography and hydrology of the waterbody. The Hydropower plants consist of following fundamental elements:

Dam: A dam can store water, which can be supplied when needed. Also, there may be a desilter plant to treat the residue developed behind the barrier.

Intake, penstock, and surge chamber: When gates of dam open, water gravitates through the penstock to turbines. There is a head race before reaching to the penstock. A surge chamber controls the flow of water, as excessive rush can harm the turbines.

Turbine: The water strikes the turbine and rotates the blades of a turbine. There is an arrangement of the attached generator with the pole above or beside the turbine. Francis turbine is the most common type of turbine in practice, which presents a side-by-side design of generator.

Generators: When blades of turbine rotate, similarly, the rotor inside the generator rotates to generate electric current (AC) through magnets pivoted inside the generator.

Transformer: The transformers inside the plant take the AC voltage and transform it into high-voltage current for more efficient transmission of current.

Transmission lines: These are used for transmitting the power directly to the purchaser, where the high-voltage current is transferred back to low-voltage current and nourished into appropriate arrangement. In remote areas, setting up new transmission lines can be costly and complex job.

Outflow: In the end, the used water is facilitated through channels, called tailraces, and reenters the downstream. There may be an arrangement of spillways in framework, which controls the excess water and works as a passage for it to flow.

The water used to run the turbines is not wasted but it comes back to the waterway framework. It may be some distance away or further downstream and without any significant effect on water framework. Hydropower framework can also be used for other purposes such as for surge control, irrigation system, water system, and drinking supplies. Along with this, it can also improve the conditions for route, tourism, and leisure parks.

The dam, water entrance, head race, surge chamber, penstock, tailrace, and the powerhouse are the major time-consuming segments of hydropower project. The penstock can be made of steel, plastics, cement, or wood. They may be lined or unlined depending on the shake.

The powerhouse also contains mechanical and electrical gear and is made out of the ordinary building materials. The most important mechanical and electrical parts of a small hydropower plant are the turbines and generators. Turbines are devices that generate the energy by falling water into pivoting shaft control. There are two types of turbines, namely "reaction" and "impulse." Impulse turbine uses force of streaming water to generate electricity, instead of weight of water. Reaction turbines generate energy from weight of water head. The flow rate of water needs to be computed, which is possible if measure of precipitation and catchment range of a waterway is known. Catchment regions are regions from which water streams into waterway. The aforementioned strategy is for run-of-stream type of hydro framework and can be calculated by the formula:

$$Q_{site} = k\left[\frac{A_{site}}{A_{guage}}\right]Q_{gauge} \tag{3.12}$$

where A_{site} is the catchment range of the power plant in (m^2), A_{guage} is the catchment territory of the gauge in (m^2), Q_{site} is the release at the site in (m^3/s), Q_{gauge} is the release at the gauge in (m^3/s), and k is a scaling steady or capacity. At that point, the power produced by the turbine can be computed as per the given formula:

$$p = \eta_{total}\rho g\, QH \tag{3.13}$$

where p is the mechanical power yield delivered at the turbine, η total is the pressure-driven effectiveness of the turbine, ρ is the thickness of water

(1000 kg/m^3), g is the speeding up caused by gravity (9.81 m/s^2), and H is the compelling weight head (m) (Fig. 3.8).

Here hydropower can deliver energy considering turbine limit, i.e., $P_{wg}(t) \geq$ Pload (t), and can be used to store power. The capacity limit at any given time (t) is given by the formula:

$$C_{bat}(t) = C_{bat}(t-1) + (P_{pv}(t) + P_{wg}(t) - P_{load}(t)\eta_{cad})\Delta t \eta_{cha} \quad (3.14)$$

where $C_{bat}(t)$ and $C_{bat}(t-1)$ are the accessible energy limits at time (t) and $(t-1)$, respectively. P_{pv} is the power produced by the PV, P_{wg} is the power produced by the turbine generator, $P_{load}(t)$ is the power expended at stack t, t is the reenactment time step ($\Delta t = 1$ h), η_{cad} is the proficiency of the AC/DC converter, and η_{cha} is the battery charging effectiveness, which relies upon the charging current and may range from 0.65 to 0.85 (Fig. 3.9).

In the end, this hydroelectricity can be evaluated by its accessible power. Power is an element of water-powered head and rate of flow of stream. The head is the energy per unit weight (or unit mass) of water. To outline, control can be ascertained for a turbine that is 80% productive, with water at

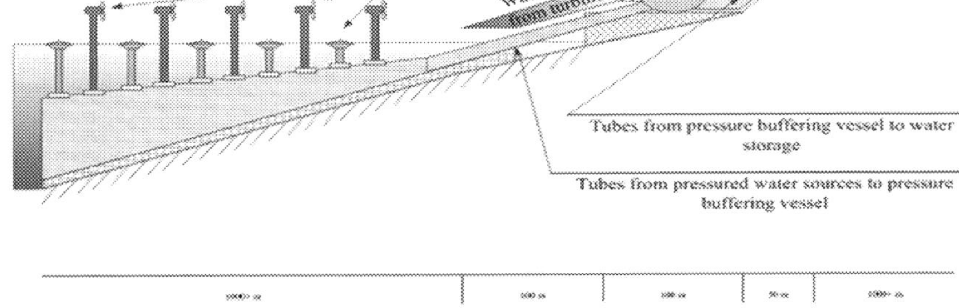

Figure 3.8 An image of a typical hydropower plant that uses weight buffering and assemblage of runoff waterway and pump for streaming water to rotate turbine and generate mechanical energy for driving a generator and deliver power.

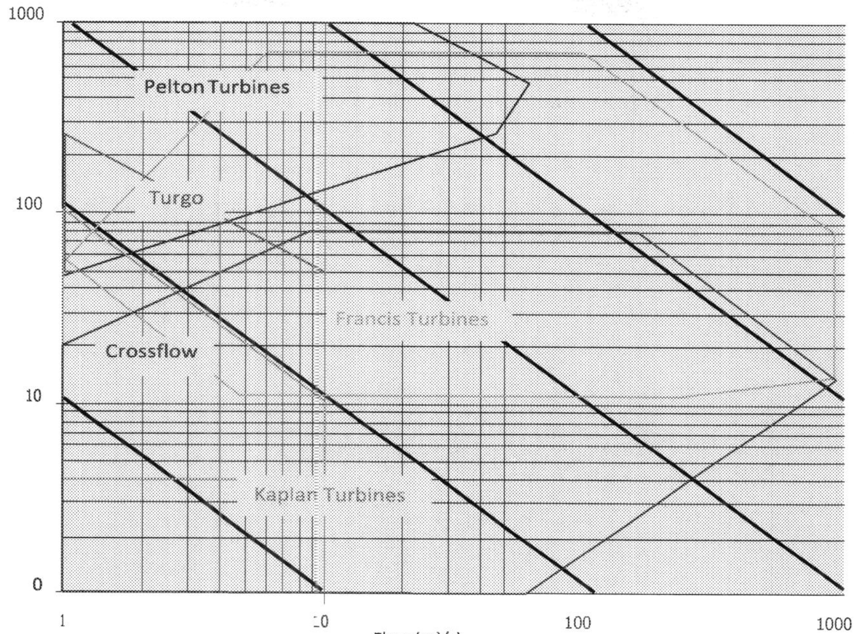

Figure 3.9 Working areas of various turbine types. There are two types of generators which can be used as a part of small hydropower plants: offbeat enlistment and synchronous machines (NHA and HRF, 2010). Offbeat generators are mostly used for the small-scale hydroventures.

1000 kg/m^3 (62.5 lb/ft^3), a stream rate of 80 m^3/s (2800 ft^3/s), gravity of 9.81 m for each second squared, and a net head of 145 m (480 ft) in SI units which gives 91 MW [48–50].

3.1.5.3 Environmental Impact

Hydropower is one of the clean energy sources, yet there are many negative impacts of hydropower on environment. Development of dam and power plant imposes many biological/physical changes on earth. Oceanic and earthbound biological system nourishment web gets destroyed by operating hydropower plant. Subsequently, potential effects must be analyzed, and tools such as Hydropower Sustainability Assessment Protocol and Environmental and Social Impact Assessment must be used to assess the impacts of the project.

3.1.6 Nuclear Energy

Radioactive element like U^{235} can go through process of nuclear fission and generate abundant amount of nuclear energy. For instance, energy liberated from a kilogram of uranium-235 (U-235) undergoing nuclear process would be approximately three million times more than that released by burning a kilogram of coal traditionally (7.2×10^{13} J/kg of uranium-235 compared to 2.4×10^{7} J/kg of coal) or similar to 12 million barrels of oil [51—53]. Nuclear power is an efficient way of generating electricity in nuclear power plant by rotating turbines through steam induced by nuclear energy during nuclear reactions taking place in the reactor. The term incorporates nuclear decay, nuclear fission, and nuclear fusion.

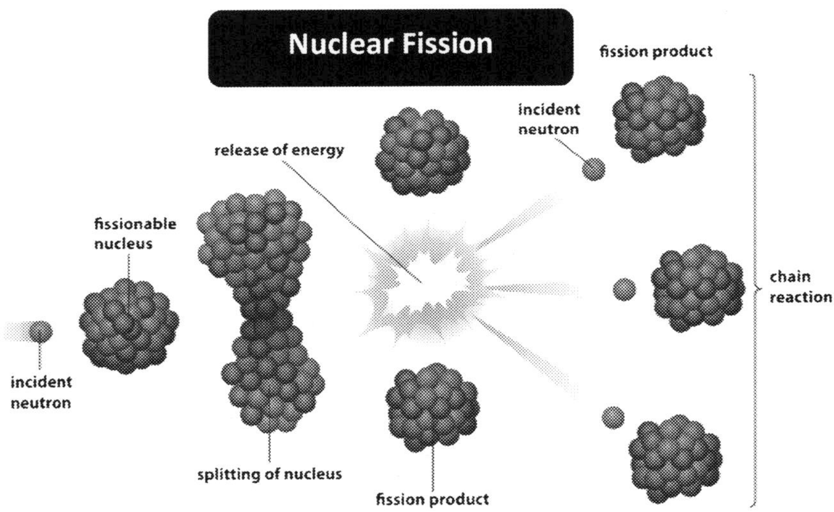

In 1932, physicist Ernest Rutherford found that enormous amount of energy was released from lithium atoms in accordance with the principle of mass-energy equivalence, when they were "split" into alpha particles by bombarding protons from a proton accelerator. Though Rutherford and his fellow nuclear physicists Niels Bohr and Albert Einstein thought that the idea of using atomic power for feasible purposes in near time was absurd, Rutherford even labeled such expectations "foolish."

To further examine Fermi's claims, in 1938, German chemists Otto Hahn and Fritz Strassman, with Austrian physicist Lise Meitner and

Meitner's nephew Otto Robert Frisch carried out several experiments by bombarding uranium with neutron atoms. They discovered that contradicting to Fermi's claims, nucleus of heavy uranium was split into two equal parts by a comparably tiny neutron. Before this discovery, all nuclear decays indulged only limited variation to the proton and neutron number of the nucleus, but this operation connected "nuclear fission" with the biology. Many scientists including Leo Szilard affirmed that independent nuclear chain reaction could occur if additional neutrons were to be released by a fission reaction. When Frederic Joliot-Curie declared this in 1939, after a functional validation, scientists across the globe appealed their government for cooperation in the in-depth analysis of the nuclear fission, on the edge of World War II, for the evolution of a nuclear weapon.

Emigration of Fermi and Szilard to the United States became reason for the invention of the first artificial reactor, known as Chicago Pile-1, which then gained importance in late 1942 and subsequently became a part of the Manhattan project, which produced enhanced uranium and constructed considerable reactors to cultivate plutonium for using it for defense purposes as a weapon, which were used to bomb Cities of Hiroshima and Nagasaki.

Within one decade from 1960 to late 1970s, installed nuclear capacity intensified relatively sharply, growing from 1 to 100 GW and as much as 300 GW in late 1980s. But then all over the world, there was significant fall in growth rate of nuclear capacity, reaching 366 GW in 2005. More than 50 GW of capacity was getting assembled between around 1970 and 1990 (assembling over 150 GW of capacity between late 1970s and early 1980s); again in 2005, another 25 GW of capacity was intended. Later on, 66% of nuclear plants ordered after 1970 got canceled. Between 1975 and 1980, 63 nuclear units got withdrawn in the United States of America [54–56].

3.1.6.1 Energy Science

Activities linked with generation of electricity from nuclear reactions are termed together as nuclear fuel cycle. Uranium is to be treated before it can be used as a fuel for nuclear reactor. Before using uranium in nuclear reactor, it undergoes steps of mining, enrichment, and fuel fabrication and then nuclear cycle begins. After using uranium in reactor, the used fuel is sent for reprocessing or recycling before final geological deposition. Of the used fuel, 95% can be recycled in reprocessing stage to use it again in reactor. A nuclear Plant is just one phase of nuclear power life cycle. First, Uranium is extracted from mines, which are underground, open pit, or in situ leach, and converted into an oxide of uranium such as yellowcake. Later on, it is transported to treating facility for converting it into uranium hexafluoride and enriching by different methods. Now enriched uranium, which contains more 0.7% U-235 than the natural, is used for making fuel rods for the particular reactor. The fuel rod will typically spend around 6 years (3 operational cycles) inside reactor until concentration of fissile U-235 increases to 3% and expressed as the following equation:

$$\boldsymbol{u}\cdot\boldsymbol{v} = u_i v_j \delta_{ij}$$

The following shows equivalence:

$$\begin{aligned}
&\boldsymbol{u} = u_x\boldsymbol{i} + u_y\boldsymbol{j} + u_z\boldsymbol{k}\\
&u_1\boldsymbol{e}_1 + u_2\boldsymbol{e}_2 + u_3\boldsymbol{e}_3 = \sum_{i=1}^{3} u_i\boldsymbol{e}_i\\
&\boldsymbol{v} = \sum_{j=1}^{3} v_j\boldsymbol{e}_j\\
\boldsymbol{u}\cdot\boldsymbol{v} &= \sum_{i=1}^{3} u_i\boldsymbol{e}_i\cdot\sum_{j=1}^{3} v_j\boldsymbol{e}_j\\
&= u_i\boldsymbol{e}_i\cdot v_j\boldsymbol{e}_j\\
&= u_i v_j(\boldsymbol{e}_j\cdot\boldsymbol{e}_i)\\
&= u_i v_j \delta_{ij}
\end{aligned} \tag{3.15}$$

where $\delta_{ij} = 1$ if $i = j$ and 0 otherwise.

Subsequently, the released energy moves to used fuel pool, and these unstable isotopes produced by fission reaction can perish away. The spent fuel is cooled off thermally and become radioactively stable enough to handle, after spending around 5 years in spent-fuel pool.

3.1.6.2 Production Engineering

A nuclear chamber is a device used for initiating and controlling constant nuclear chain reaction. Nuclear reactors are used for production of energy and in forward thrust of boats. The heat generated by nuclear reaction are used for running steam turbines, which either rotate shafts of ship or electrical generators. The steam generated can also be used for industrial heating. Some of the reactors are also used for producing different isotopes for industrial or medical purposes, and few are also used for production of weapons-grade plutonium. Across the globe, there are total 435 nuclear reactors in 31 countries, as per the report of International Atomic Energy Association (IAEA), April 2014. By absorbing neutrons, nucleus of uranium-235 splits into lighter elements with higher velocity and lighter neutrons. This event is called induced nuclear fission.

3.1.6.3 Fission

When a heavy atomic nucleus such as uranium-235 or plutonium-239, capable of undergoing fission reaction, absorbs a neutron, nuclear fission reaction starts. The nucleus splits into two or more lighter and faster nuclei, releasing kinetic energy and gamma radiations along with some free neutrons. Out of these free neutrons, some are again absorbed by other fissile atomic nucleus, starting more fission reactions, which frees more neutrons, and the process continues. This is called nuclear chain reaction.

$$ {}^{235}_{92}U + {}^{1}_{0}n \longrightarrow {}^{236}_{92}U \rightarrow \begin{cases} {}^{140}_{56}Ba + {}^{93}_{36}Kr + 3\,{}^{1}n0 \\ {}^{144}_{54}Xe + {}^{90}_{38}Sr + 2\,{}^{1}n0 \\ {}^{144}_{55}Cs + {}^{90}_{37}Rb + 2\,{}^{1}n0 \end{cases} \quad (3.16) $$

Neutron absorber or neutron moderators are used for controlling nuclear chain reaction by varying the number of neutrons that will strike further fissile atoms. Nuclear reactors, either manually or automatically, shut down the fission reaction by using moderators such as light water (74.85% of world's reactors), solid graphite (20% of reactors), and heavy water (5% of reactors).

3.1.6.4 Heat Energy Generation

The core of reactor generates heat mainly by (1) converting the kinetic energy of fission products to thermal energy when these nuclei collide (H_1) with nearby atoms and (2) the reactor absorbs some of the gamma rays produced during fission and converts their energy into heat (H_2), which expressed by the following equation:

$$H_1 = \pi^2 n_1^2\left(1 - \frac{4n_1}{9|c|} + \frac{8n_2}{|c|} + \frac{4n_3}{|c|} + \frac{12}{c^2}(2n_2 + n_3)^2 - \frac{8n_1}{3c_2}(2n_2 + n_3)\right)$$

$$-\frac{\pi^2 n_3^2}{9}\left(1 + \frac{4n_1}{3|c|} + \frac{32n_2}{9|c|} + \frac{4n_3}{3|c|} + \frac{(6n_1 + 16n_2 + 9n_3)^2}{27c^2} - \frac{2n_3(6n_1 + 16n_2 + 9n_3)}{9c^2}\right)$$

$$+\frac{10\pi^2 n_2^3}{27|c|}\left(1 + \frac{(6n_1 + 3n_2 + 8n_3)}{|c|}\right) + \frac{2c^2}{3},$$

$$H_2 = \frac{\pi^2 n_2^2}{2}\left(1 + \frac{4n_1}{|c|} + \frac{40n_2}{27|c|} + \frac{16n_3}{3|c|} + \frac{(6n_1 + 3n_2 + 8n_3)^2}{3c^2} - \frac{14n_2(6n_1 + 3n_2 + 8n_3)}{27c^2}\right)$$

$$-\frac{2\pi^2 n_3^2}{9}\left(1 + \frac{4n_1}{3|c|} + \frac{32n_2}{9|c|} + \frac{8n_3}{9|c|} + \frac{(6n_1 + 16n_2 + 9n_3)^2}{27c^2} - \frac{10n_3(6n_1 + 16n_2 + 9n_3)}{27c^2}\right)$$

$$+\frac{16\pi^2 n_1^3}{9|c|}\left(1 + \frac{6(2n_2 + n_3)}{|c|}\right) + \frac{5c^2}{6} \tag{3.17}$$

The heat generation mechanism provides higher order corrections to those derived from dressed energy equations. We also need energy-field transfer relation between the paired and unpaired phases $(H_1 - H_2)/2$ for determining the full-phase boundaries. These equations define the complete phase diagram and the critical fields activated by the fields H_1 and H_2. The decay heat source will remain activated for some time even after the reactor is shut down, as the heat is produced by the radioactive decay of the fission products and materials activated by absorbing a neutron.

3.1.6.5 Cooling

Often water is used as a coolant in a nuclear reactor, although sometimes a gas, liquid metal, or molten salt is used for heat-absorbing purpose. Later on, this heat is used to generate steam. In most of the reactor system, cooling units are different than the one used for boiling water, for producing pressurized steam for turbines. However, sometimes water is directly boiled by reactor core for steam generation, for example, the boiling water rector.

3.1.6.6 Reactivity Control

A fission reaction rate is controlled by regulating the number of neutrons that will stimulate additional fission reactions. Nuclear reactors use various approaches to manage neutron and reactor's productivity. Some arise naturally from radioactive decay and thus considered as a part of the operation, whereas some are mechanically introduced in the reactor plan for a specific objective. The best technique for neutron control is using control rods. They are made of neutron-absorbing materials and tend to attract fission-inducing neutrons. When a control rod is deeply placed in the reactor, it consumes more neutrons than the volume of moderator it replaces, which results in a reduction in further neutron fission reactions as well as the reactor's power output. By removing the control rod, again the number of neutrons and power output can be increased.

The radioactive decay also disturbs neutron population in a reactor. Delayed neutron emission by some neutron-rich fission isotopes is a technique using the concept of radioactive decay. About 0.65% of total emitted neutrons are delayed neutrons, and the fission products which produce this are called delayed neutron precursors. The delayed neutron is emitted anytime from some milliseconds to some minutes, and so, for determining critical time of the reactor, significant time is required. Critical mass of the delayed neutrons is required to keep the reactor in zone of chain reactivity, and enough time would be available for a mechanical device or human operator to control a chain reaction. The presence of delayed neutrons is not required to preserve criticality in the ultimate phase known as the prompt critical point. In reactors, sometimes even coolant acts as a moderator. A moderator is a substance which increases power of the reactor by converting fast neutrons emitted from fission to thermal neutrons. Thermal neutrons have more potential to cause fission reactions. If coolants are used as a moderator, moderator density and thus power output can be changed by temperature variance.

Most of the reactors are subtle to xenon poisoning or the iodine pit. Xenon-135 is a common fission product which acts as a neutron absorber and tends to shut down the reactor. Xenon-135 accumulation can be controlled by increasing the power of the reactor. The iodine pit is used to temporarily shut down a reactor due to accumulation of neutron-absorbing element such as xenon-135, which are mainly produced by beta decay (with a half-life of 6.57 h) of iodine-135. Xenon-135 decays to cesium-135 (with a half-life of 9.2 h). By absorbing xenon-135 and burning it to xenon-136, which is more stable and absorbs fewer neutrons, rector can be restarted. Control rods must be reinserted for replacing neutron absorption of lost xenon-135.

Reactors used in nuclear marine propulsion for generating heat cannot be run at same power and in the same manner all the time as land-based power reactors run, and also, they must have prolonged reactor life without the need for refueling. Therefore many reactors use highly enriched uranium fuel with flammable neutron absorber in the fuel rods and can have an excess of fissionable material. So, when the fuel becomes less reactive, this burnable neutron poison starts draining and thus compensates for the fuel and increase the operating life of the reactor.

3.1.6.7 Environmental Impact

The radioactive waste stream from nuclear power plant consists of mainly unconverted uranium, other transuranic actinides (mostly plutonium and curium), and fission products. It accounts for 3% of volume, which is deadly. The actinides and fission products both are responsible for radioactivity causing severe earth and human health issues.

3.1.6.8 High-Level Radioactive Waste

High-level radioactive waste is produced during production of nuclear power. It consists of both short-lived and long-lived materials. The technical issues in dealing with long-lived radioactive wastes are challenging, and they remain deadly to living organisms. Out of all the fission products, technetium-99 (with a half-life of 220,000 years) and iodine-129 (half-life of 15.7 million years) having long half-life are of a significant concern, as they remain radioactive for few thousand years, which is dangerous for surrounding environment and living beings.

3.1.6.9 Low-Level Radioactive Waste

The atomic power plant also generates a bulk quantity of low-level radioactive waste that comprises substances such as cloths, hand tools, and filters. In the United States, the Nuclear Regulatory Commission has often treated this type of radioactive waste as a regular waste, which is hazardous to the environment.

3.1.6.10 Deadly Accidents

The most harmful effect of nuclear energy is a fatal accident that deteriorates the environment and causes loss of lives. From 1952 to 2016, 99 accidents have been reported worldwide at the nuclear-powered plant, which has caused property damage of around 20.5 billion USD along with taking hundreds of lives and affected thousands of people with diseases such as cancer. Fifty-seven accidents have been reported since Chernobyl disaster, and almost two-third of all atomic accidents happened in the United States alone.

3.2 SUSTAINABLE ENERGY

Sustainable energy is the type of energy that meets the current demand for energy with manageable environmental impacts. In other words, it is the energy framework that satisfies present requirement without compensating needs of future generation. The primary principle of sustainability is the possible improvement, which incorporates the interconnected fields of ecology and culture. Sustainable science is the exploration of economic advancement and ecology. Thus the sustainable energy is, in fact, the energy that can be generated and consumed without significant harm to the earth. Maintainable power is a component of the sustainable power source and is one of those sustainable technologies that give fantastic natural benefit. The sustainable energy sources are the pillars of sustainable energy, and use of them decreases the reliance on fossil fuels and gives a chance to alleviate greenhouse gases.

3.2.1 Solar

Solar-based energy is using technologies to harness sun's light and energy to provide heat, electricity, and cooling. The normal power density of sun-based radiation outside the atmosphere of earth is 1366 W/m^2, which is known as solar constant (Fig. 3.10).

The Earth gets 174 petawatts (PW) of incoming sun radiation at the upper atmosphere. Out of which 30% radiations are reflected back to space and the rest are absorbed by mists, seas, and land masses. The solar light spectrum is mostly spread over close to the infrared region with a small part in the ultraviolet region. Every day, most of the world's population gets 150–300 W/m^2 or 3.5–7.0 kWh/m^2 exposure to the sun's rays.

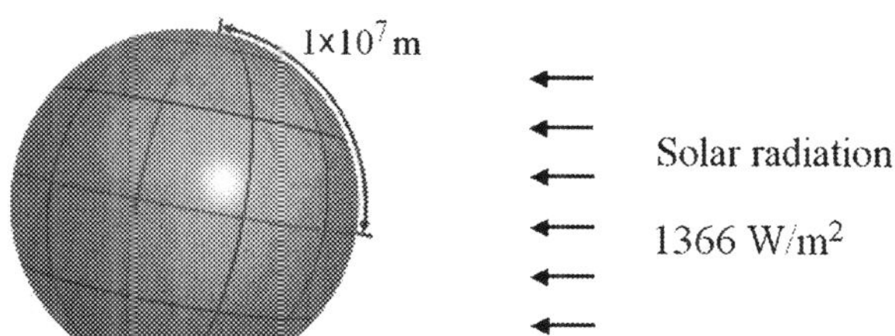

Figure 3.10 Annual solar energy reaching the surface of Earth. The average solar power on Earth is 1366 W/m^2 The length of the meridian of the earth is 10,000,000 m. The total solar energy that strikes at the surface of Earth per year is 5,460,000 EJ [57,58].

3.2.1.1 Energy Science and Technology

In this section, advance research on potential of solar energy has been described to garner solar energy more efficiently. The detail of the model is reflected in its correct prediction of performance. The performance of the individual components is modeled using either deterministic or probabilistic approaches, and thus the general approach for modeling the energy system is described in the following sections.

3.2.1.2 Photovoltaic Assessment

The total power yield of a Photovoltaic (PV) cluster depends on solar power received per unit area from the sun and surrounding temperature, and thus the model can be calculated as

$$P_{\mathrm{pv}} = \eta_{\mathrm{pvg}} A_{\mathrm{pvg}} G_t \tag{3.18}$$

where η_{pvg} is the PV-age effectiveness, A_{pvg} is the PV generator range (m^2), and G_t is the sun-oriented radiation in a tilted module plane ($\mathrm{W/m}^2$). The PV-age effectiveness (η_{pvg}) can be additionally characterized as

$$\eta_{\mathrm{pvg}} = \eta_r \eta_{pc}[1 - \beta(T_c - T_{\mathrm{cref}})] \tag{3.19}$$

where η_{pc} is the power-molding proficiency, which is equivalent to 1 when maximum power point tracking (MPPT) can be used; β is a temperature coefficient (0.004–0.006 for each °C); η_r is the reference module effectiveness; and T_{cref} is the reference cell temperature in °C. The reference cell temperature (T_{cref}) can be taken from the connection

$$T_c = T_a + \left(\frac{\mathrm{NOCT} - 20}{800}\right) G_t \tag{3.20}$$

where T_a is the encompassing temperature in °C, Nominal Operating Cell Temperature (NOCT) is the ostensible working cell temperature in °C, and G_t is the sun-powered radiation in a tilted module plane ($\mathrm{W/m}^2$). The aggregate radiation in the sun-oriented cell, considering both typical and diffuse sun-oriented radiation, can be evaluated as

$$I_t = I_b R_b + I_d R_d + (I_b + I_d) R_r \tag{3.21}$$

3.2.1.3 System Modeling

The sunlight-based cell, which is one of the components of sun-oriented cluster, is basically a P–N intersection semiconductor fit for generating power by means of the PV impact. PV cells are connected in parallel to frame a PV exhibit, and graphene is coordinated in the PV module to enhance the productivity of the subsequent PV (Fig. 3.11).

Using an ideal single diode, as shown in Fig. 3.2, for an array with N_s series-connected cells and N_p parallel-connected cells, the array current can be related to the array voltage as

$$I = N_p\left[I_{ph} - I_{rs}\left[\exp\left(\frac{q(V + IR_s)}{AKTN_s} - 1\right)\right]\right] \tag{3.22}$$

where

$$I_{rs} = I_{rr}\exp\left(\frac{T}{T_r}\right)^3\left[\frac{E_G}{AK}\left(\frac{1}{T_r} - \frac{1}{T}\right)\right] \tag{3.23}$$

Furthermore, q is the electron charge (1.6×10^{-9} C), K is Boltzmann's consistent, A is the diode ideality factor, and T is the cell temperature (K). I_{rs} is the cell turn-around immersion current at T, T_r is the cell-alluded temperature, I_{rr} is the invert immersion current at T_r, and E_G is the band hole vitality of the semiconductor used as a part of the cell. The photograph current I_{ph} differs with the cell's temperature and radiation as takes after

$$I_{ph} = \left[I_{SCR} + k_i(T - T_r)\frac{S}{100}\right] \tag{3.24}$$

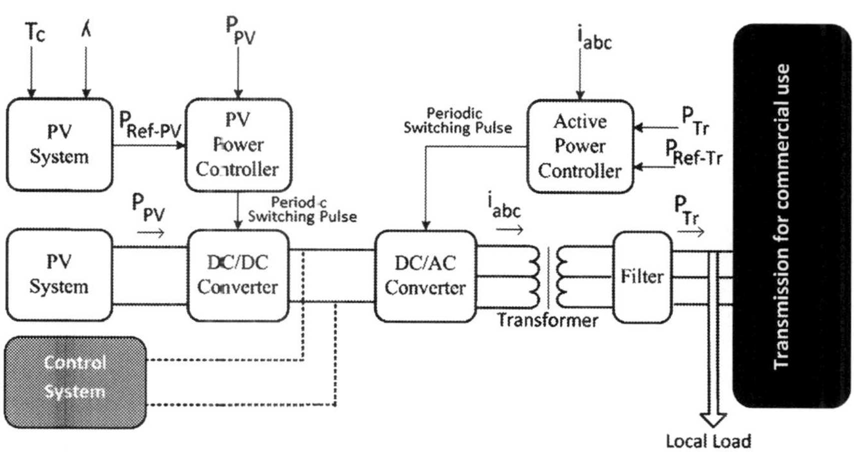

Figure 3.11 Modeling and simulation of a PV device and system, showing the multijunction PV system with a different efficiency perspective.

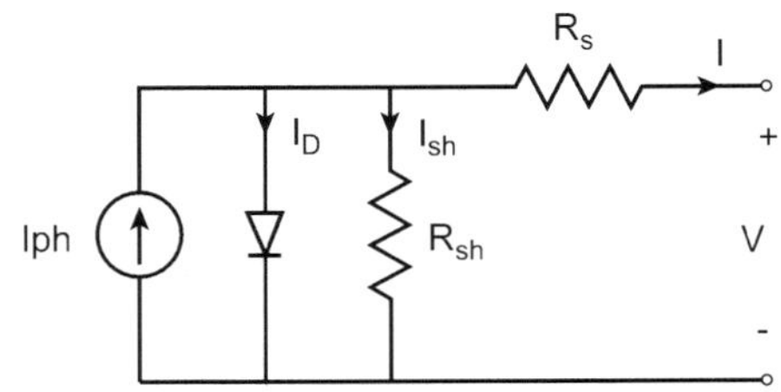

Figure 3.12 Single-diode equivalent circuit of a PV cell model module or array as a continuous function for a given set of operating conditions.

where I_{SCR} is the cell short-circuit current at the reference temperature and radiation, k_i is the short-circuit current temperature coefficient, and S is the solar radiation (mW/cm^2).

Solar cells can be generally modeled using single-diode (Fig. 3.2) or double-diode circuit models (Fig. 3.3) (Figs. 3.12 and 3.13).

The single-diode model includes an additional shunt resistance in parallel to the ideal shunt diode model. The $I-V$ characteristics of the PV cell can be derived using a single-diode model, as follows:

$$I = I_{ph} - I_D \tag{3.25}$$

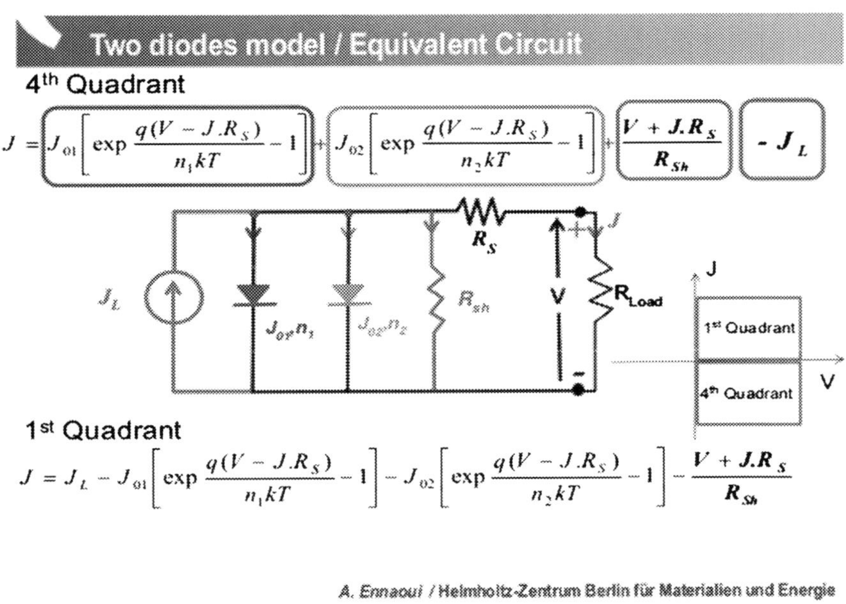

Figure 3.13 Equivalent circuit of the two-diode model with a parasitic series of R_s and shunt R_{sh} resistances.

$$I = I_{ph} - I_0 \left[\exp\left(\frac{q(V + R_s I)}{AKT} - 1 \right) - \frac{V + R_s I}{R_{sh}} \right] \tag{3.26}$$

where I_{ph} is the photograph current (A_n), I_D is the diode current (A_n), I_0 is the backward immersion current (A_n), A_n is the diode steady, q is the charge of the electron (1.6×10^{-9} C), K is Boltzmann's consistent, T is the cell temperature (°C), R_s is the arrangement protection (ohm), R_{sh} is the shunt protection (Ohm), I is the cell current (A_n), and V is the cell voltage (V).

The yield current of the PV cell using the two-diode model can be portrayed as,

$$I = I_{PV} - I_{D1} - I_{D2} - \left(\frac{V + IR_S}{R_{SH}} \right), \tag{3.27}$$

where

$$I_{D1} = I_{01} \left[\exp\left(\frac{V + IR_s}{a_1 V_{T1}} \right) - 1 \right] \tag{3.28}$$

$$I_{D2} = I_{02} \left[\exp\left(\frac{V + IR_s}{a_2 V_{T2}} \right) - 1 \right] \tag{3.29}$$

I_{01} and I_{02} are the reverse saturation currents of diode 1 and diode 2, respectively, and V_{T1} and V_{T2} are the thermal voltages of the respective diodes. The diode ideality constants can be represented by a_1 and a_2.

The simplified model of the PV system model proposed by Zhou et al. can be presented as follows:

$$v_{oc} = \frac{V_{oc}}{cKT/q} \tag{3.30}$$

$$P_{max} = \frac{\dfrac{V_{oc}}{cKT/q} - \ln\left(\dfrac{V_{oc}}{cKT/q} + 0.72 \right)}{\left(1 + \dfrac{V_{oc}}{nKT/q} \right)} \left(1 - \frac{V_{oc}}{\dfrac{V_{oc}}{I_{SC}}} \right) \times \left(\frac{V_{oc0}}{1 + \beta \ln \dfrac{G_0}{G}} \right) \left(\frac{T_o}{T} \right)^{\gamma} I_{sc0} \left(\frac{G}{G_o} \right)^{a} \tag{3.31}$$

where V_{oc} is the standardized estimation of the open-circuit voltage. V_{oc} concerning the warm voltage $V_t = nkT/q$, n is the ideality factor ($1 < n < 2$), K is Boltzmann's steady, T is the PV module temperature in kelvin, q is the electron charge, α is the factor in charge of all the nondirect impacts on which the photocurrent depends, β is a PV module innovation particular dimensionless coefficient, and γ is the factor speaking to all the nonstraight temperature–voltage impacts. Condition [59] gives the most extreme power yield of a solitary PV module. A genuine framework incorporates various PV modules associated in arranged and parallel connection. The total power yield for an arrangement with N_s cells connected in arrangement and N_p cells connected in parallel with control PM for every module would be

$$P_{\text{array}} = N_s N_p P_M \tag{3.32}$$

The power produced by solar device depends on exposure to the sun's rays and temperature. The voltage–current relationship and the voltage–control relationship are nondirect, as appeared in Fig. 3.6. The most extreme power point (MPP) needs to be followed for the proficient extraction of sun-based energy in a PV framework among the varying exposure condition. Maximum power can be extracted from a PV framework when MPPT is led for every asset. Singular following framework ought to be actualized for each as the variable used for MPPT is diverse for wind and PV frameworks. MPPT enhances the framework's productivity while limiting the arrival time of the establishment cost.

3.2.1.4 MPPT in a PV System

A few other new calculations are proposed in the writing. A few researchers have used fluffy rationale control, and a counterfeit neural system has been picturized for MPP following wind energy framework. Solar clusters rely on insolation, temperature, and voltage exhibit. Accordingly, MPPT must be actualized to keep the working voltage near the MPP under varying climatic conditions. MPPT is crucial in solar-based devices as it reduces the exhibit cost and gives maximum yield (Fig. 3.14).

The V–I characteristics of the solar array, neglecting the internal shunt resistance, are given by

$$I_0 = I_g - I_{\text{sat}}\left\{\exp\left[\frac{q(V_0 + I_o R_s)}{AKT}\right] - 1\right\} \tag{3.33}$$

where A is a dimensionless factor. The output characteristics of the PV array are described in Fig. 3.14, which shows MPP tracking under varying weather conditions.

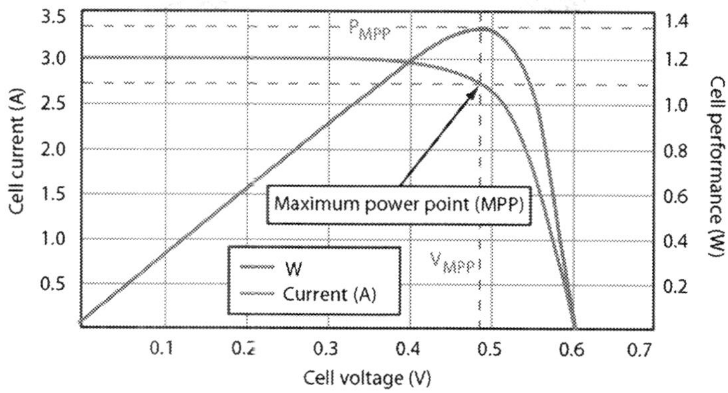

Figure 3.14 MPPT using control rationale, in which a PV cell's exponential connection among current and voltage amplifies.

In any case, integrated circuit (IC) strategy is more confusing than the P&O technique. Certainly, it requires four sensors to do calculations, it is costly, and it requires more time to calculate. In spite of the fact that IC technique is quicker than P&O, it is not much efficient when solar devices insolation changes quickly.

3.2.1.5 Query Table and Curve-Fitting Method

In this strategy, the PV-produced voltage and current are related, and they both are put away in a controlled framework under a given arrangement of climatic condition at MPP. It uses calculation similar to PSF examined under MPPT for the wind framework. In a bend fitting technique, a numerical approach is used instead of a query table. The use of a fitted condition to characterize the power—voltage (P—V) qualities could be examined in this work:

$$P_{\text{pv}} = aV_{\text{PV}}^3 + bV_{\text{PV}}^2 + cV_{\text{PV}} + d \tag{3.34}$$

here, a, b, c, and d are coefficients determined by the sampling values of the PV voltage. The voltage at maximum power is obtained using the formula

$$V_{\text{MPP}} = \frac{-b \pm \sqrt{b^2 - 3ac}}{3a} \tag{3.35}$$

This procedure can be repeated every millisecond for locating precise MPP. The detriment of this method is that it requires learning of physical parameters identifying with the cell material for every single climatic condition. Accordingly, a large memory capacity may be required for the estimations.

3.2.1.5.1 Open-Circuit Voltage-Based MPPT Technique

The power at the MPP can be characterized as

$$P_{\mathrm{MPP}} = I_{\mathrm{MPP}} V_{\mathrm{MPP}} \tag{3.36}$$

Because a single point exists at which the maximum power is obtained, if the estimation of either Current Maximum Power Point (IMPP) or Voltage at Maximum Power Point (VMPP) is known, the Power at Maximum Power (PMPP) can be followed. This system accepts that the proportion of the MPP voltage and the open-circuit voltage can be weakly influenced by insolation and temperature as follows:

$$V_{\mathrm{MPP}} \approx K_{oc} V_{oc} \tag{3.37}$$

where K_{oc} is the voltage factor, which can be calculated by examining the PV framework over an extensive variety of insolation and temperatures. Mainly it fluctuates between 0.71 and 0.92. The main issue with this technique is estimation of V_{oc}, which often requires detachment of the PV module. To measure V_{oc}, the PV framework circuit is opened momentarily. VMPP is then calculated using condition. This whole procedure is repeated at regular interval to refresh the estimation of VMPP. Another approach is to acquire the estimation with a different PV cell, called a pilot cell, which can state the attributes of the PV cluster.

3.2.1.5.2 Short-Out Current-Based MPPT Technique

This method takes an approach similar to the open-circuit voltage procedure accepting a straight support between the MPP current and the short-out current:

$$I_{\mathrm{MPP}} \approx K_{sc} I_{sc} \tag{3.38}$$

where K_{sc} is the present factor, which can be computed by examining the PV framework over an extensive variety of insolation and temperatures, and most of the part fluctuates in the range of 0.64 and 0.85 (Fig. 3.15).

To establish a relation between the number of light quanta at steady state and the intensity of radiation, a finite volume enclosure containing radiation is considered. The number of stationary states of light quanta of a certain type of polarization whose frequency is in the range of ν_r to $\nu_r + d_{\nu r}$ and maximum solar radiation can be achieved at 1.4 eV with an energy value of 27.77 MW/m^2 eV [60–62].

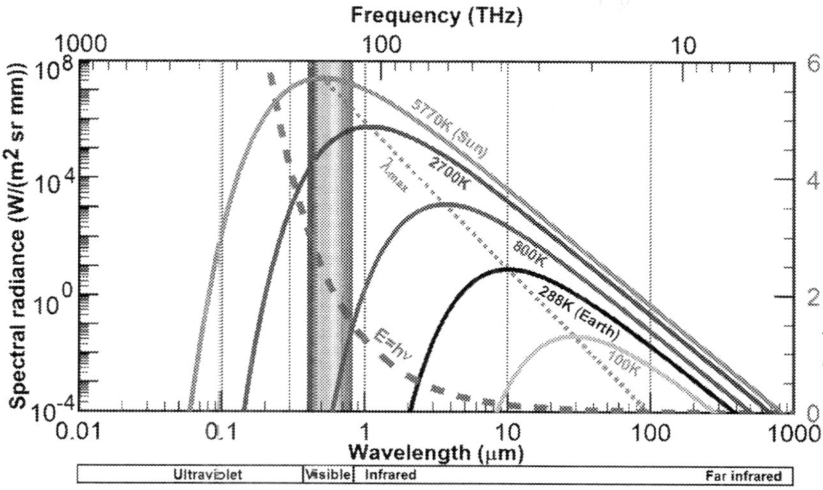

Figure 3.15 The figure depicts the blackbody radiation in various temperatures. At temperature 5800K, power is 6.31×10^7 (W/m^2), peak E is 1.410 (eV), peak λ is 0.88 (μm), peak u is 2.81×10^7 (W/m^2 eV).

3.2.1.6 Sustainability

Solar energy is pollution-free, renewable source of energy. It does not cause any harm to the environment and is a sustainable power source. Unlike fossil fuels, it does not emit any pollutant such as carbon dioxide, nitrogen oxide, or sulfur into the atmosphere. In this way, solar energy does not cause any global warming, corrosive rain, or brown haze.

3.2.2 Wind Energy

Wind energy is utilization of wind to rotate the blades of turbine which is connected to generator to produce electricity. It is an alternative renewable source of energy to fossil fuels. Wind energy is a clean, cheap, and free of greenhouse gases. The impacts on the environment are negligible compared to the nonrenewable sources of energy. Wind power gives variable energy which is steady most of the times, although sometimes there are significant variations for shorter period of time. Hence it is used along with other energy sources to give reliable supply. When there is a high wind power, on-grid system needs to be upgraded, and conventional energy use needs to be lowered. Power management techniques such as distributing power to neighboring areas, if having enough capacity, can overcome these problems.

3.2.2.1 Energy Research and Technology

The flowing modeling has been proposed by assessment of wind potential to capture wind power more efficiently. The critical condition representing the mechanical energy of wind turbine is given by

$$P_{\mathrm{MPP}} = \frac{1}{2} C_P(\lambda, \beta) \rho A V^3 \tag{3.39}$$

where ρ represents air density (kg/m^3), C_p represents power coefficient, A_n is the catching range of the rotor sharp edges (m^2), V is the average wind speed (m/s), and λ is the tip speed ratio. The hypothetical highest estimation of the power coefficient C_p is 0.593, and it is otherwise called Betz's coefficient. The tip speed ratio for a wind turbine can be characterized as the ratio of the rotational speed of the tip of a cutting edge to the wind speed. Scientifically,

$$\lambda = \frac{R\omega}{V} \tag{3.40}$$

where R is the range of the turbine (m), ω is the precise speed (rad/s), and V is the normal wind speed (m/s). The power generated by the wind can be calculated by the condition

$$Q_w = P \times (\mathrm{Time})(\mathrm{kWh}) \tag{3.41}$$

The wind speed at certain height cannot be measured, so different wind speed is taken as a reference and extrapolated to get wind speed at a certain height. However, care should be taken because wind speed at lower heights can have errors due to vegetation, shading, and restrictions. The wind speed can be figured out as

$$v(z)\ln\left(\frac{Z_r}{Z_0}\right) = v(z_r)\ln\left(\frac{z}{z_0}\right) \tag{3.42}$$

where Z_r is the reference tallness (m), Z is the height at which the wind speed is to be calculated, Z_0 is an estimation of the surface harshness, $v(z)$ is the wind speed at stature z (m/s), and $v(z_r)$ is the wind speed at the reference height Z (m/s). The cut-in speed is a low wind speed at which the turbine initially starts to pivot and produce power. The cut-out speed is a high wind speed at which the powers on the turbine structure are high, and accordingly, there is a danger of harm to the rotor. At this speed, a slowing mechanism can be used to stop the rotor and prevent any damage to the farm.

$$P_w(v)\begin{cases} \dfrac{v^k - v_C^k}{v_R^k - v_C^k}\cdot P_R & v_C \le v \le v_R \\ P_R & v_R \le v \le v_F \\ 0 & v \le v_C \text{ and } v \ge v_F \end{cases} \tag{3.43}$$

where P_R is the evaluated control, v_C is the cut-in wind speed, v_R is the appraised wind speed, v_F is the appraised removed speed, and k is the Weibull shape factor [5,63,64].

Power generation depends on wind speed, and thus for a fixed size of a rotor blade, energy output varies with the wind speed. Thus as the wind speed varies, ideal generator speed should be determined for maximum power. The angular velocity of the generator must be varied for maximum power output, and this process is known as MPPT. When the cutting-edge pitch angle is zero, the power coefficient is maximized for an ideal Total Solar Resource (TSR).

3.2.2.2 Drivetrain Modeling

The drivetrain transmits high streamlined turning from the propeller to the low-speed shaft of the generator using the gears. Sometimes, generators are directly attached with a propeller to avoid the complication in the designing. There are two ways to display drivetrain, namely one-mass model or a two-mass model. Built a scientific model on the basis of torsional multibody dynamic model.

3.2.2.3 One-Mass Model

If a perfectly rigid low-speed shaft is assumed, a single-mass model for the turbine can be represented by

$$J_t\dot{\omega}_t = T_a - K_t\omega_t - T_g \tag{3.44}$$

and

$$\begin{aligned} J_t &= J_r + n_g^2 J_g \\ K_t &= K_r + n_g^2 K_g \\ T_g &= n_g T_{em} \end{aligned} \tag{3.45}$$

where J_t is the turbine rotor moment of inertia in kg m^2; ω_t is the low-shaft precise speed in rad/s^2; K_t is the turbine damping coefficient in Nm/rad s, denoting the streamlined protection; and Kg is the generator damping coefficient in Nm/rad s, denoting the mechanical contact and windage [65].

3.2.2.4 Two-Mass Model

Bati and Brennam presented a two-mass model of a wind turbine in which the rotor side inertia J_r can be given by

$$J_t \frac{d\omega_t}{dt} = T_m - T_{ls} - K_t\omega_t, \tag{3.46}$$

the low speed shaft torque can be given by

$$T = B_{ls}(\theta_t - \theta_{ls}) + K_{ls}(\omega_t - \omega_{ls}), \tag{3.47}$$

and the generator inertia J_g is driven by the high-speed shaft, which can be braked by the electromagnetic torque T_g of the generator which is given as

$$J_g \frac{d\omega_g}{dt} = T_{hs} - K_g\omega_g - T_g \tag{3.48}$$

If we assume an ideal gearbox with ratio n, then

$$n = \frac{T_{ls}}{T_{hs}} = \frac{\omega_g}{\omega_t} = \frac{\theta_g}{\theta_{ls}} \tag{3.49}$$

where all the symbols are same as those of one-mass model. K_{ls} is the low-speed shaft damping coefficient in (Nm/rad s), ω_g is the fast shaft angular speed in (rad/s^2), T_m is the turbine torque in (Nm), T_{ls} is the low-speed shaft torque in (Nm), J_g is the generator rotor moment of inertia in (kg m^2), and T_{hs} is the rapid shaft torque in (Nm). By eliminating T_{ls} time subordinate from (3.47) and using (3.46) and (3.48), following dynamic framework is obtained

$$\frac{dT_{ls}}{dt} = \left(B_{ls} - \frac{K_{ls}K_t}{J_t}\right)\omega_t + \frac{1}{n}\left(\frac{K_{ls}K_t}{J_g} - B_{ls}\right)\omega_g - K_{ls}\left(\frac{J_t + n^2 J_g}{n^2 J_t J_g}\right)T_{ls} + \frac{K_{ls}}{J_t}T_\alpha + \frac{K_{ls}}{nJ_g}T_g, \tag{3.50}$$

where

$$K_{ls} = IG/L_{ls},$$

$$D_{ls} = \xi D_s,$$

$$\xi = \sqrt{1 - \left(\frac{\omega}{\omega_n}\right)^2}, \tag{3.51}$$

$$D_s = \sqrt[2]{K_{ls} m} \tag{3.52}$$

in which ω/ω_n is the ratio of the shaft frequency of oscillation to the undamped natural frequency of the shaft, m is the mass of the shaft, I is the second momentum of the area about the axis of rotation, L_{ls} is the shaft length, G is the modulus of rigidity, D_s is the critical damping of shaft, and ξ is the damping ratic of the shaft (Fig. 3.16).

3.2.2.5 MPPT in Wind

A wind turbine can work in following modes: (1) settled speed and (2) variable speed. Settled speed turbine generators can be controlled with a load, in contrast to which, an electric device is needed to control the fluctuating frequency and energy in variable-speed turbine. The utilization of a variable-speed wind turbine permits persistently adjusting the revolving velocity of the wind turbine to the wind speed with the goal to achieve more productivity and reduced power variations. Compared to fixed turbine, annual power production rises by 5%–10% in variable-speed turbine [66,67]. The main target of MPPT is to find extreme C_p value under changing wind speeds. The MPP could be found under differing wind speed by maintaining the total speed fraction at its ideal value in variable-speed turbine system. It is difficult to maintain maximum power yield at all wind speeds due to the nonlinear attributes of wind turbines. Thus several control techniques have been adapted to track the maximum power of wind turbine framework. Some of the strategies observe maximum power by keeping record of changes in generated power, whereas some use mechanical sensors to calculate generator speed that compels it to work in MPP by measuring wind speed.

3.2.2.6 Wind Power Generation

Either an acceptance or synchronous generator can be used for wind turbine system. Perpetual magnet synchronous generators are used as a part of wind turbine framework because they are more efficient, simple to control, and light weight, and they do not require receptive and charging current. Gearbox in wind turbine increases cost and weight in structure.

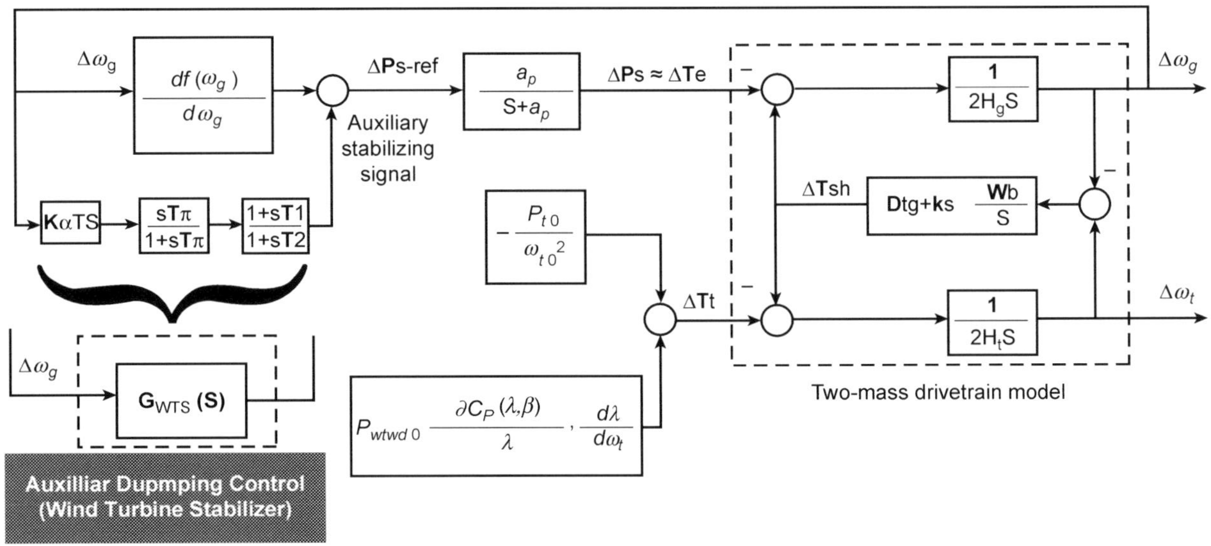

Figure 3.16 Two-mass model of a wind turbine framework. Schematic graph of a full-scale converter wind turbine with a synchronous machine.

The model for a permanent magnet synchronous generator (PMSG) depends on d-q synchronous reference outline. The PMSG voltage condition can be given by

$$V_d = -R_s i_d - L_d \frac{di_g}{dt} + \omega L_q i_q \tag{3.53}$$

$$V_q = -R_s i_q - L_q \frac{di_q}{dt} - \omega L_d i_d + \omega \lambda_m \tag{3.54}$$

The electronic torque can be given by

$$T_e = 1.5p[\lambda i_q + (L_d - L_q) i_d i_q] \tag{3.55}$$

where L_q is the q-axis inductance, L_d is the d-axis inductance, i_q is the q-axis current, i_d is the d-axis current, V_q is the q-axis voltage, V_d is the d-axis voltage, ω_r is the angular velocity of the rotor, λ is the amplitude of flux induced, and p is the number of pairs of poles. In the case of a squirrel cage induction generator (SCIG), the following equation can be used with a stationary $d-q$ frame of reference for dynamic modeling:

$$\begin{bmatrix} V_{qs} \\ V_{ds} \\ V_{qr} \\ V_{dr} \end{bmatrix} = \begin{bmatrix} R_s + pL_s & 0 & pL_m & 0 \\ 0 & R_s + pL_s & 0 & pL_m \\ pL_m & -\omega_r L_m & R_r + pL_r & -\omega_r L_r \\ \omega_r L_m & pL_m & \omega_r L_r & R_r + pL_r \end{bmatrix} \begin{bmatrix} i_{qs} \\ i_{ds} \\ i_{qr} \\ i_{dr} \end{bmatrix} \tag{3.56}$$

From the stator side, the equations are

$$\begin{aligned} \lambda_{ds} &= L_s i_{ds} + L_m i_{dr} \\ \lambda_{qs} &= L_s i_{qs} + L_m i_{dr} \\ L_s &= L_{is} + L_m \\ L_r &= L_{lr} + L_m \\ V_{ds} &= R_s i_{ds} + \frac{d}{dt}\lambda_{ds} \\ V_{qs} &= R_s i_{qs} + \frac{d}{dt}\lambda_{qs} \end{aligned} \tag{3.57}$$

and from the rotor side, the equations are

$$\begin{aligned}
\lambda_{dr} &= L_r i_{dr} + L_m i_{ds} \\
\lambda_{qr} &= L_r i_{qr} + L_m i_{qs} \\
V_{dr} &= R_r i_{dr} + \frac{d}{dt}\lambda_{dr} + \omega_r \lambda_{qr} \\
V_{qr} &= R_r i_{qr} + \frac{d}{dt}\lambda_{qr} - \omega_r \lambda_{dr}
\end{aligned} \tag{3.58}$$

For the air gap flux linkage, the equations are

$$\begin{aligned}
\lambda_{dm} &= L_m(i_{ds} + i_{dr}) \\
\lambda_{qr} &= L_m(i_{qr} + i_{qs})
\end{aligned} \tag{3.59}$$

where R_s is stator-winding resistance; R_r is motor-winding resistance; L_m is the magnetizing inductance; L_{ls} is stator leakage inductance; L_{lr} *is* rotor leakage inductance; ω_r is electrical rotor angular speed; i_d and i_q are current; V_d and V_q are voltage; λ_d and λ_q are the fluxes of the $d-q$ model. The output power and torque of the turbine (T_t) in terms of the rotational speed can be obtained by substituting (3.40) in (3.39):

$$P_w = \frac{1}{2}\rho A C_p(\lambda, B)\left(\frac{R\omega_{\text{opt}}}{\lambda_{\text{opt}}}\right)^3 \tag{3.60}$$

$$T_t = \frac{1}{2}\rho A C_p(\lambda, B)\left(\frac{R}{\lambda_{\text{opt}}}\right)^3 \omega_{\text{opt}} \tag{3.61}$$

The power coefficient (C_p) is a nonlinear function, which can be expressed by a fitting equation in the form

$$C_p(\lambda,\ \beta) = c_1\left(c_2\frac{1}{\lambda_i} - c_3\beta - c_4\right)e^{-c_5\frac{1}{\lambda_i}} + c_6\lambda \tag{3.62}$$

with

$$\frac{1}{\lambda_i} = \frac{1}{\lambda + 0.08\beta} - \frac{0.035}{\beta^3 + 1} \tag{3.63}$$

Here the value of constants $c_1 - c_6$ is considered induction generators, which can be used as a requirement of reactive power for excitation so that substations used in wind power generation system includes substantial capacitor banks for power factor correction (Fig. 3.17).

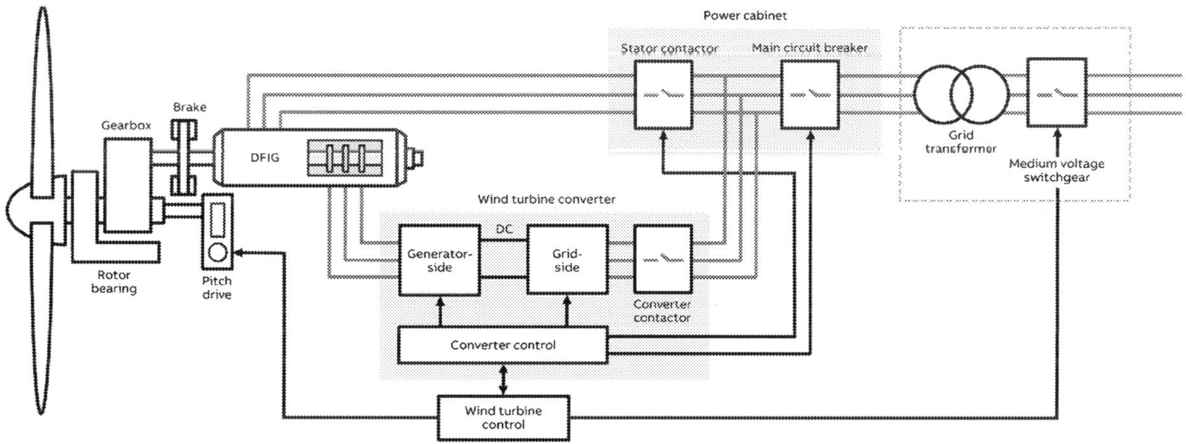

Figure 3.17 Wind energy generation using a standard high-speed drivetrain to confirm high-efficiency energy production. *DFIG*, doubly fed induction generator.

3.2.2.7 MPPT

Solar-based energy and wind energy are very different from other energy sources when we consider power control. Unlike others, it cannot be operated by controlling the fuel inflow rate or measuring the power connected to the propeller. Hence solar-based energy and wind energy requires proper control framework. The total extractable energy from wind relies on wind speed as well as on TSR. For obtaining the highest amount of energy from varying angular velocity, the rotational speed must be fluctuated to maintain the ideal estimation of TSR consistently. The voltage—current relationship and the voltage—control relationship are nonlinear. The MPP must be followed to extract maximum sun power in a PV framework among changing insolation conditions. The maximum power from a wind PV cross-breed framework can be obtained by directing MPPT to every asset. MPPT amplifies the framework's effectiveness as well as limits establishment cost (Fig. 3.18).

3.2.2.8 Optimum Torque Control

As in ideal TSR control, in which the ideal TSR (λopt) assures the maximum misuse of the accessible wind energy, this technique alters the PMSG torque with respect to the highest reference rotational force of the wind turbine at a given wind speed.

Eq. (3.40) can be recomposed as

$$V = \frac{\omega R}{\lambda} \tag{3.64}$$

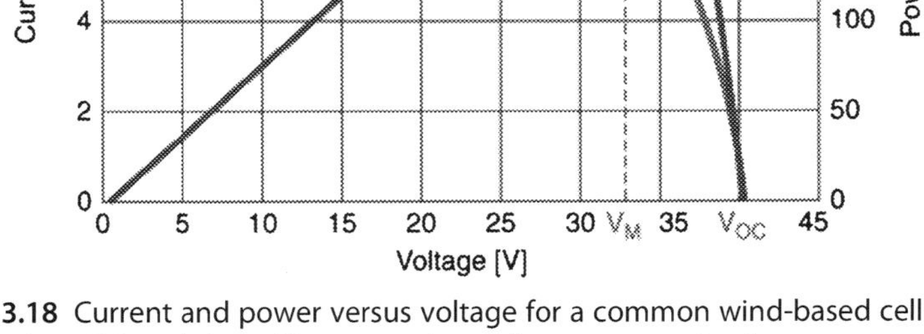

Figure 3.18 Current and power versus voltage for a common wind-based cell demonstrating the MPP and the cell's present bend for power production.

Using Eq. (3.64) in Eq. (3.39) gives

$$P = \frac{1}{2}\rho\pi R^5 \frac{\omega^3}{\lambda^3} C_P \tag{3.65}$$

when the rotor is rotating at λ_{opt}, $C_p = C_{pmax}$. Therefore, Eq. (3.65) becomes

$$P_{opt} = \frac{1}{2}\rho\pi R^5 \frac{C_{pmax}}{\lambda_{opt}^3}\omega^3 = K_{opt}\omega^3 \tag{3.66}$$

Power can also be defined as

$$P = \omega T \tag{3.67}$$

Rearranging Eq. (3.67) gives

$$T = \frac{P}{\omega} \tag{3.68}$$

and using Eqs. (3.68) and (3.66), we can obtain

$$T_{opt} = \frac{1}{2}\rho\pi R^5 \frac{C_{pmax}}{\lambda_{opt}^3}\omega^2 = K_{opt}\omega^2 \tag{3.69}$$

The value derived from the aforementioned equation can be given as the reference torque to the wind turbine. This approach is rapid and easy, although it has lower productivity than TSR control because it does not record the wind speed specifically. Thus instant reflection of varying wind speed in the reference flag is not possible. Maximum power following in this technique is performed by present power yield and control system to generate optimum power from a wind turbine. The energy coefficients (c_1−c_6) and parameters (λ_i) used in the studies by Banos, Imam et al., and Kalantar and Mausavi [14,47,65] are different, and the energy can be computed as

$$E = \frac{1}{\lambda_i} = \frac{1}{\lambda - 0.02\beta} - \frac{0.03}{1 + \beta^3} \tag{3.70}$$

3.2.2.9 Sustainability

Wind energy is a sustainable, pollution-free energy source. It will reduce our dependence on the fossil fuel for the power because it does not emit any pollutants. Out of all the renewable energy sources available, wind energy would be the most efficient and most used source of energy across the globe after the solar energy. In contrast to the solar power, one of the most significant advantage in wind energy is that the technology can be developed such that small wind turbine would give higher efficiency. The world is constantly going for advancement of an efficient technique to extract wind energy, and it will lead to more sustainability.

3.2.3 Geothermal Energy

The geothermal energy is energy stored inside the earth, originated from the planet and the radioactive waste. The geothermal gradient called magma is the difference in temperature between the earth's surface and center of the earth, and it drives a continues conduction of heat from the center of the planet to its cover. The temperature at the core-mantle may reach as high as 4000°C (7200°F). Owing to such high heat and pressure in earth's core, rock inside the core melts, and solid mantle acts like plastic, hence mantle moves slightly upward being lighter than the surrounding rock. The Earth's geothermal energy is hypothetically 10 times more than the energy needs of humankind, though only a small percentage may be available for use. Because the demand for electricity is rapidly increasing in the world, utilization of the geothermal energy would indeed be an alternative source of energy. Thus in this unit, much advanced technology has been proposed to extract geothermal energy as a source of the renewable energy.

3.2.3.1 Energy Science and Technology

The most economical in situ method to extract around 800 kpa of heat energy from geothermal energy is to explore naturally occurring "hydrothermal convection." In this system, cold water seeps into the hot earth's crust, heats up, and rises to the earth's surface as steam. Steam is used to run the generators to generate the energy. Geothermal power plant drills their holes into the rocks to capture the steam more efficiently up to 300 m. The technology is to extract geothermal energy by using heat pumps for serving the purpose of heating and cooling throughout the year. Air or antifreeze liquid is recirculated into the building by pumping through the pipes buried underneath. In the summer, it transfers heat from the building to the ground. In contrast to that, it does opposite in winter by providing prewarmed air and water to the heating system of the building. The following figure shows the direct and practical use of in situ ground energy source for heating and cooling. A tube runs from outside, under the ground, and into building's ventilation system to maximize the heat transfer.

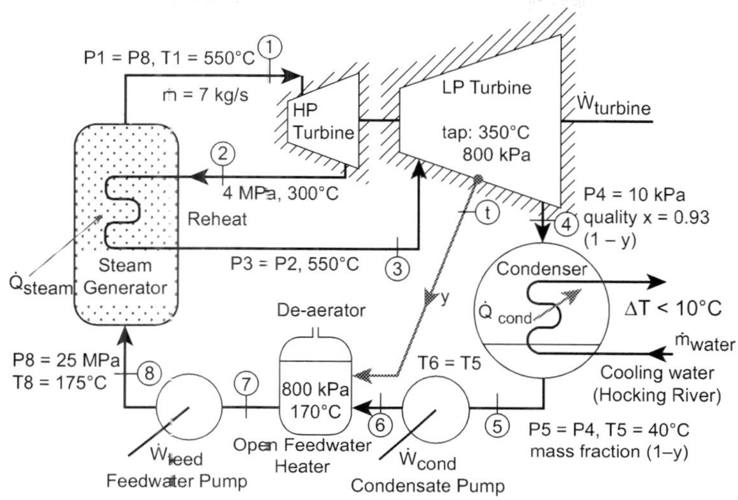

This system can be referred as a regenerative heat cycle, and the system will bring best thermal energy out of the geothermal plant. Before doing any experiment, it is necessary that entire cycle is plotted on P-h graph in coordination with weight, temperature, and the quality information. The following chart depicts it.

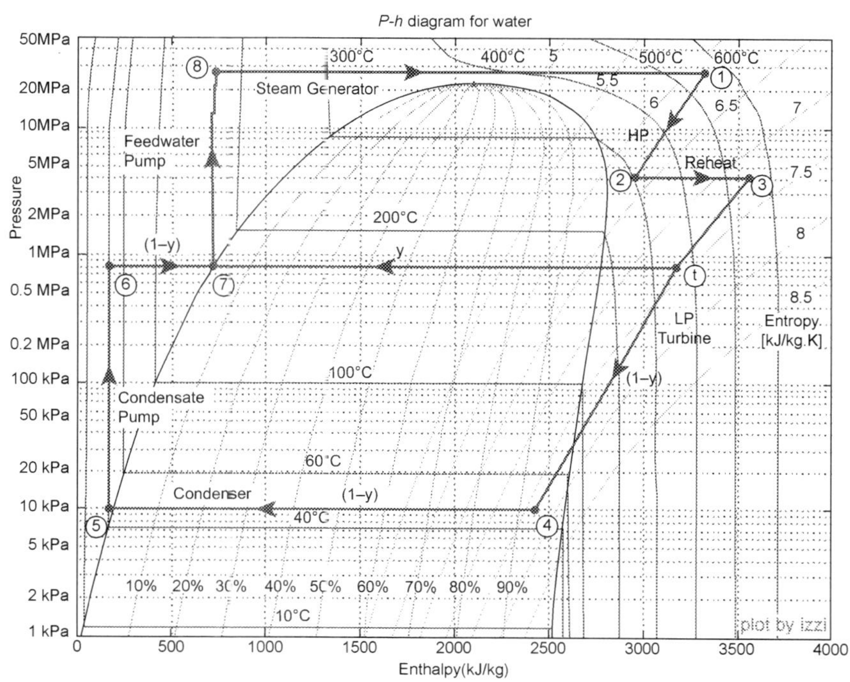

A considerable portion of the steam (y) can be extracted from the liquid petroleum (LP) turbine tap (t) to such an extent that merging it with $(1 - y)$ of the fluid water at the station (6) will bring the liquid to a soaked fluid state at the station (7). After that the feedwater pump directs the fluid to the station (8), and in this way, lot of heat is used from the steam generator in heating the liquid from the station (8) to the turbine gulf station (8). With a mass division of $(1 - y)$, there is less power yield because of a reduced mass stream rate in LP turbine from the tap (t) to the station (4), although the final net outcome is an expansion in heat energy productivity.

In this way, it can be seen that despite the versatility of the system, the P-h graph empowers an instinctive and subjective understanding of the system. By using the strategies portrayed previously for the investigation of every phase, using the steam table for assessing the enthalpy at different stations (appears in red), and releasing motor and potential energy impacts, it can be concluded that

1. The open feedwater pump is adiabatic, and the mass division of steam "y" is required to seep off the LP turbine, which will bring the liquid from station (6) to an immersed fluid state in the deaerator ($y = 0.18$).

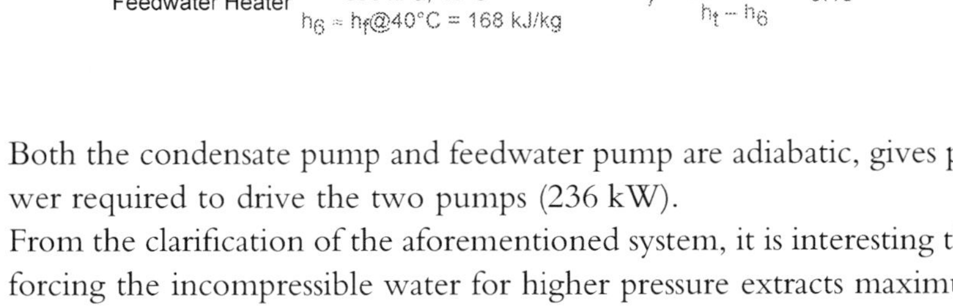

2. Both the condensate pump and feedwater pump are adiabatic, gives power required to drive the two pumps (236 kW).

   From the clarification of the aforementioned system, it is interesting that forcing the incompressible water for higher pressure extracts maximum heat from the geothermal magma. Thus it is necessary that the primary compressible fluid equation for measuring pump work shall assess the change in enthalpy from the compressed liquid water tables to capture the heat, and this can be explained by

T	160°C	175°C	180°C	
h	690	754	776	kJ/kg

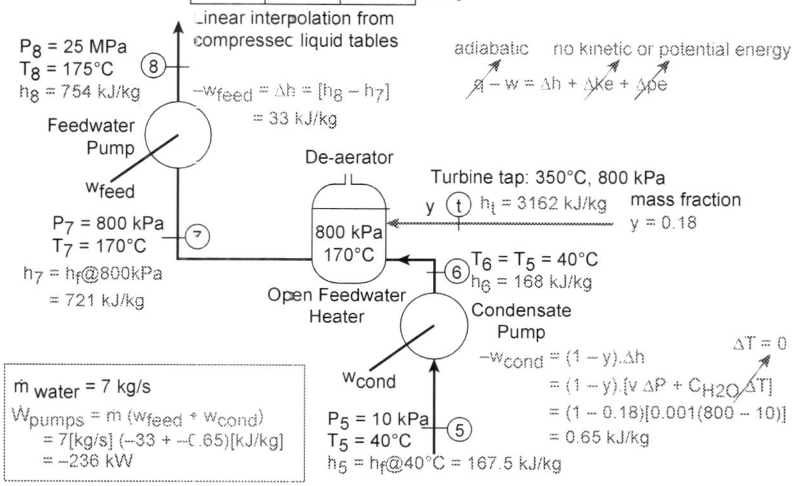

3. Two turbines are adiabatic, gives the newly consolidated power yield of the two turbines. Power yield of the turbines is around 10.6 MW if no steam seeps from the LP turbine (9.65 MW).

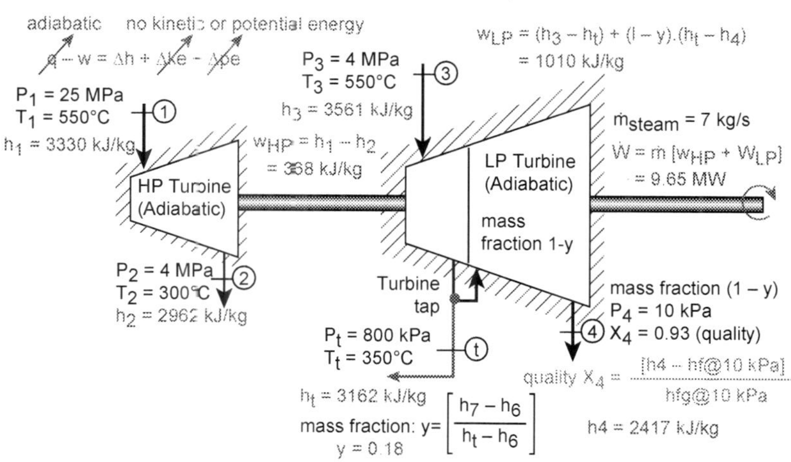

In this manner, it can be determined that the net power yield is equivalent to the efficiency of the thermal energy plant, where the speed of the feed-water pump must be controlled to locate a fluid water storage tank connected to deaerator and the water mass stream rate. It just needs to increase the stream rate from 7 kg/s to 8 kg/s, keeping in mind the end goal to recapture our unique power yield to give the aggregate heat exchange of 22.2 MW to the power plant, including the reheat system.

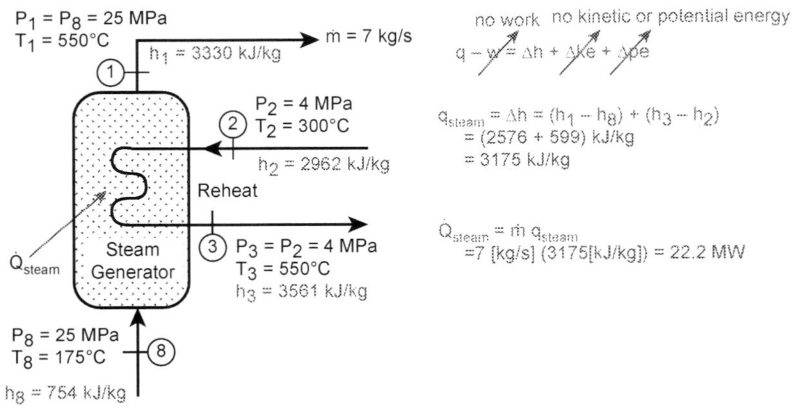

The thermal efficiency of the plant can be characterized as the net work done (turbines, pumps) divided by the aggregate heat provided to the plant which gives heat effectiveness of 42%.

$$\eta_{tth} = \left[\frac{W_{\text{turbines}} + W_{\text{pumps}}}{Q_{\text{steam}}}\right] = \left[\frac{(9.65 - 0.24)\ \text{MW}}{22.2\ \text{MW}}\right] = 42\% \qquad (3.71)$$

The net power (4.053 MW) can be supplied to the local building structures and industry to meet their total energy demand.

Depending on the financial capability, geothermal assets, having the potential of 3,800,000 MW, can be used to generate 30.8 billion MWh of power every year around the world. Besides, when we look at the complete life cycle of the plant, in situ geothermal power dominates other technologies. Geothermal plant neither has any fuel costs nor any secondary expenses. Once a plant is started, it can generate power for a long time. Therefore in situ geothermal plant is having a bright future as a prime energy source for home and business purposes.

3.2.3.2 Sustainability

Geothermal energy is heat stored in the earth. It is clean, pollution-free, and sustainable compared to the fossil fuel reserves such as coal, oil, and gas. The resources from which energy can repeatedly be extracted without getting extinct are classified as renewable resources. The term "renewable" describes a characteristic of the resource, which is the energy extracted from the resource is continuously replaced by more energy time to time. Consequently, geothermal exploitation is not a "mining" process where the production of geothermal energy simultaneously creates a heat sink to clear a path to generate clean energy. This leads to pressure and temperature gradients, which in turn, after completion of production, create fluid/heat inflow to reestablish the preproduction state. The regeneration of geothermal resources is a process, which occurs, depending on many factors such as the method of production, the rate of extraction, and on the attributes of the resource as the clean energy.

3.2.4 Biogas Energy

Biogas produced by the decay of organic matter in the absence of oxygen is considered as a renewable source of energy. It can be produced from materials such as agricultural waste, compost, metropolitan waste, plant material, or sewage. The fundamental structure of biogas consists of methane (CH_4) and carbon dioxide (CO_2) and might have the small percentage of hydrogen sulfide (H_2S), dampness, and siloxanes where methane and hydrogen can be combusted to generate energy for cooking and heating ventilating and air conditioning (HVAC) purpose. It can also be used in a gas engine to convert it into heat and electricity.

3.2.4.1 Energy Science and Technology

A study has been conducted to produce biogas from domestic waste including human excreta. Building solid wastewater can be collected into closed detention tank and separated into two chambers, one having domestic wastewater and another having excreta. After that, domestic wastewater is processed in five stages: (1) preliminary treatment, (2) primary treatment, (3) secondary treatment, (4) disinfection, and, finally, (5) sludge treatment in the plant. This process removes all pollutants from the wastewater, and the disinfected wastewater called effluent can be used for local gardening.

Sludge can be treated electrochemically by dynamic carbon nanotubes (CNTs), which can adsorb and oxidize chemicals in the anode successfully. A unique arrangement has been made to combine both adsorption and oxidation at the anode of CNT, and further oxidation is done by hydrogen peroxide (H_2O_2) at the cathode of CNT. Factors affecting treatment efficiency and oxidation mechanism are carefully studied. The result showed that H_2O_2 flow might be affected by the electrode material, cathode potential, pH, and flow rate. The maximum flow of H_2O_2 (1.38 mol/L m^2 was achieved by CNT with an applied cathodic potential $V = -0.4$ (vs. Ag/AgCl), a pH of 6.46, a rate flow of 1.5 mL/min, and a dissolved oxygen (DO) influent flow of 1.95 mol/L m^2 [68,69]. Also, phenol is used in the process as an aromatic compound for assessing the removal efficiency of the process, and its oxidation rate is directly related to the H_2O_2 flow. H_2O_2 was probably reacting with a phenol species and therefore cannot remove phenol efficiently. The formation of the electrochemical polymer through a chain reaction of phenolic radicals can also contribute to 13% of phenol removal. The electrochemical CNT filtering system developed with H_2O_2 generated in situ is a new application of filters and CNTs that can be used as an effective treatment for removing all organic pollutants. After that, it is stored in a closed chamber to allow the thermophilic anaerobic codigestion process to thicken.

Once it is thickened, slurry can be placed in free oxygen tanks called digesters to make it safe for the environment. It is heated to at least 95°F for 10–15 days in the digesters. Anaerobic bacteria such as *Desulfovibrio* and *Methanococcus* are produced, which consumes organic matter in the sludge. The digestion process stabilizes the thickened sludge by converting most of the material into the methane gas (Fig. 3.19).

Once methanogenesis starts, the organic loading rate on volumetric methane generation shown in Fig. 3.20 was examined by digital gas chromatograph, which can be collected and stored in biogas recovery unit. Later on, it is supplied for domestic fuel and HAVC equipment operation.

Water washing is a popular technology that can be implemented. In this method, heavyweight gas streams into a chamber where the carbon dioxide and other components are scrubbed by falling water, running opposite to the gas stream. This technique has an efficiency of 98% for extracting methane.

3.2.4.2 Sustainability

Bioenergy is produced by methanogenesis and is sustainable and completely pollution-free. It can make a considerable contribution to satisfy global energy demand or maybe even more than that along with providing environmental benefits. Bioenergy also contributes to the security of energy, improving trade balances, improving socioeconomic development, and helping with the management of wastes, thus improving resource management.

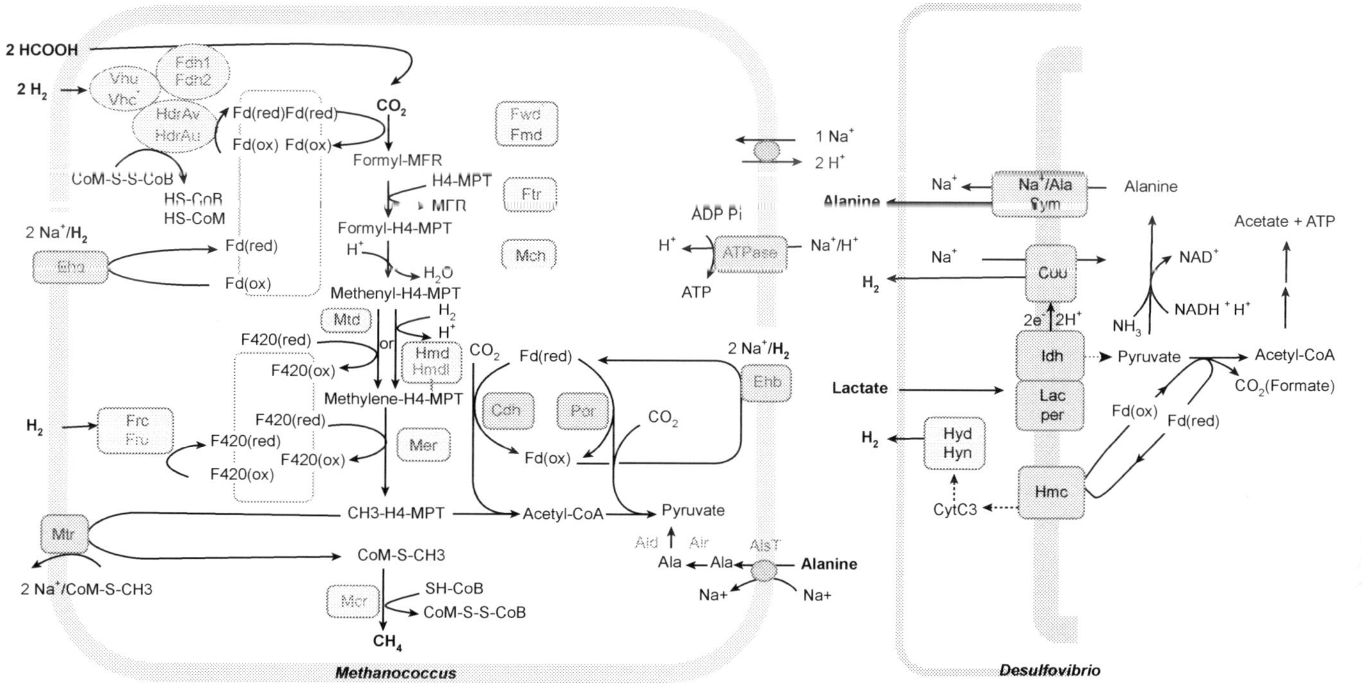

Figure 3.19 The methanogenesis process shows the detailed diagram of *M. maripaludis* and *D. vulgaris* bacterial syntrophic interaction, highlighting the central energy generation and consuming the pathways to produce methane.

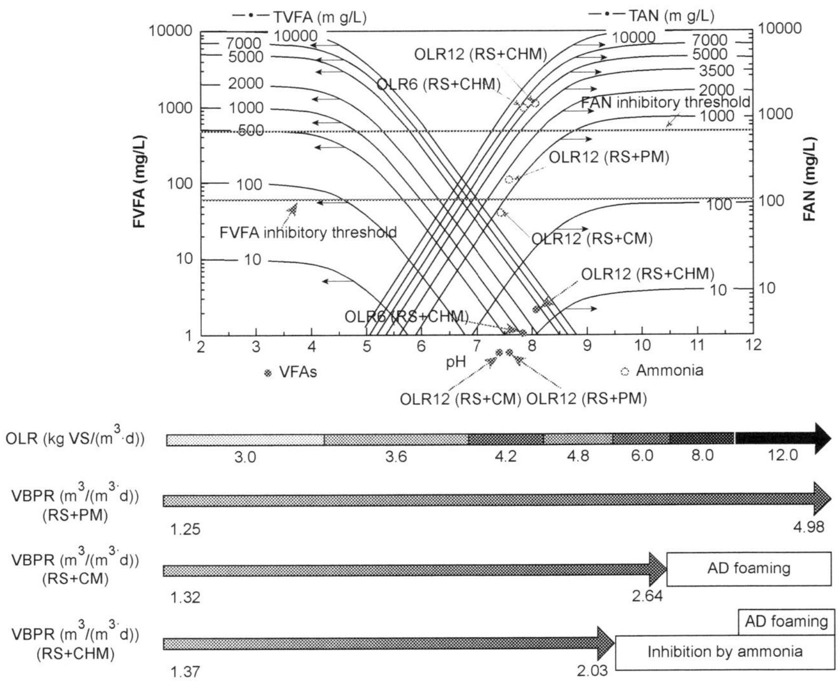

Figure 3.20 Anaerobic thermophilic codigestion of domestic perishable waste and human feces was conducted. The effects of organic loading rate on volumetric biogas production were examined. Anaerobic digestion foaming was observed. A diagram of the inhibitory conditions of thermophilic codigestion was plotted.

REFERENCES

[1] Al-Mashhadani A, Elseed MG, Alkhaldi R, Al-Sachit S, Al-Washahi W. Optimization of UK generation mix for 2030. Int J Energy Statistics 2015.

[2] Alammari R, Iqbal A. Modelling and simulation of single- and triple-junction solar cells using MATLAB/SIMULINK. Int J Ambient Energy 2016.

[3] Baur G, Hencken K, Trautmann D. Revisiting unitarity corrections for electromagnetic processes in collisions of relativistic nuclei. Phys Rep 2007;453(1).

[4] Debashisha J, Rajendran S. A review of estimation of effective wind speed based control of wind turbines. Renew Sustain Energy Rev 2015.

[5] Hossain F. Solar energy integration into advanced building design for meeting energy demand and environment problem. J Energy Res 2016;17:49−55.

[6] Howlader AM, Urasaki N, Yona A, Senjyu T, Saber AY. A review of output power smoothing methods for wind energy conversion systems. Renew Sustain Energy Rev 2013.

[7] Abdullah MA, Yatim AHM, Tan CW. A study of maximum power point tracking algorithms for wind energy system. 2011 IEEE Conference on Clean Energy and Technology (CET) 2011.

[8] Agger AK, Sørensen AH. Atomic and molecular structure and dynamics. Phys Rev A 1997;55:402.

[9] Arnold P. Photon emission from ultrarelativistic plasmas. J High Energy Phys 11/27/2001.
[10] Artemyev N, Jentschura UD, Serbo VG, Surzhykov A. Strong electromagnetic field effects in ultra-relativistic heavy-ion collisions. Eur Phys J C 2012;72:1935.
[11] Kuhn CCN. Phase diagrams of three- component attractive ultracold fermions in one dimension. N J Phys 01/06/2012.
[12] Yu G, Jung YS, Choi JY, Kim GS. A novel two-mode MPPT control algorithm based on comparative study of existing algorithms. Sol Energy 2004;(04).
[13] Baldev R, Chellapandi P, Rao P. Fast spectrum reactor vis-à-vis pressurized water reactors. Sodium Fast Reactors with Closed Fuel Cycle. 2015.
[14] Banos R. Optimization methods applied to renewable and sustainable energy: a review. Renew Sustain Energy Rev 2011.
[15] Baur G, Hencken K, Trautmann D, Sadovsky S, Kharlov Y. Dense laser-driven electron sheets as relativistic mirrors for coherent production of brilliant X-ray and γ-ray beams. Phys Rep 2002;364:359.
[16] Belkacem GH, Feinberg B, Bossingham R, Meyerhof WE. Semi classical dynamics and relaxation. Phys Rev Lett 1993;71:1514.
[17] Benavides ND, Chapman PL. Modeling the effect of voltage ripple on the power output of photovoltaic modules. IEEE Trans Ind Electron Jul. 2008;55(7):2638−43.
[18] Mahmoud Y, Xiao W, Zeineldin H. A parameterization approach for enhancing PV model accuracy. IEEE Trans Ind Electron 2012.
[19] Bhutto WA, Bazmi AA, Zahedi G. Greener energy: issues and challenges for Pakistan—wind power prospective. Renew Sustain Energy Rev 2013.
[20] Biswas W, Lund C. Investigation on wind power potential on Hong Kong islands-an analysis of wind power and wind turbine characteristics. Renew Energy September 2002.
[21] Boukhezzar B, Siguerdidjane H. Nonlinear control with wind estimation of a DFIG variable speed wind turbine for power capture optimization. Energy Convers Manag 2009;50(4):885−92.
[22] Elkhatib K, Aitouche A. Fuzzy controller for PV systems based on the fuzzy model reference. 2014 16th European Conference on Power Electronics and Applications 2014.
[23] Celik AN, Acikgoz N. Modelling and experimental verification of the operating current of mono-crystalline photovoltaic modules using four- and five-parameter models. Appl Energy January 2007;84(1):1−15.
[24] Sharma KG, Annapurna B, Gajrani K. Stability analysis of DFIG based wind turbines connected to electric grid. Int Rev Model Simul 2013.
[25] Connor A, Northrop R. Mitigations of human impacts through technology. Ecological Sustainability Understanding Complex Issues. 2013.
[26] Davis M, Ramos G. Combining digital signal processing, artificial intelligence and graphs for modeling power quality disturbances. Int Rev Model Simul 2015.
[27] De Soto W, Klein SA, Beckman WA. "Improvement and validation of a model for photovoltaic array performance. Sol Energy Jan. 2006;80(1):78−88.
[28] Faida H, Saadi J. Modelling, control strategy of DFIG in a wind energy system and feasibility study of a wind farm in Morocco. Int Rev Model Simul 2010;3(6):1350−62.
[29] Soedibyo P, Agung F, Ashari M. Grid quality hybrid power system control of microhydro, wind turbine and fuel cell using fuzzy logic. Int Rev Model Simul 2013.
[30] Eichler J, Stöhlker T. Radiative electron capture in relativistic ion-atom collisions and the photoelectric effect in hydrogen-like high-Z systems. Phys Rep 2007;439(1).

[31] Sivasankar G, Kumar VS. Improving low voltage ride through of wind generators using STATCOM under symmetric and asymmetric fault conditions. Int Rev Model Simul 2013.
[32] Walker CB, Redding-Johanson AM, Baidoo EE, Rajeev L, He Z, Hendrickson EL, Joachimiak MP, Stolyar S, Arkin AP, Leigh JA, Zhou J, Keasling JD, Mukhopadhyay A, Stahl DA. Functional responses of methanogenic archaea to syntrophic growth. ISME J 2012.
[33] Robyns B, Francois B, Degobert P, Hautier JP. Vector control of induction machines. London: Springer-Verlag; 2012.
[34] Valluri SR, Becker U, Grün N, Scheid W. Relativistic collisions of highly-charged ions. J Phys B At Mol Phys 1984;17:4359.
[35] Yang L, Wang S, Zeng Q, Zhang Z, Pei T, Li Y, Peng L. Efficient photovoltage multiplication in carbon nanotubes. Nat Photon 2011:672−6.
[36] Ghennam T, Berkouk EM, Francois B. A vector hysteresis current control applied on three-level inverter. Application to the active and reactive power control of doubly fed induction generator based wind turbine. Int Rev Electric Eng 2007;2(2): 250−9.
[37] Tan YT, Kirschen DS, Jenkins N. A model of PV generation suitable for stability analysis. IEEE Trans Energy Convers Dec. 2004;19(4):748−55.
[38] Shet MSK, Yaragatti UKR. Design of computer application for 3 phase vector control induction motor drive. In: IET-UK International Conference on Information and Communication Technology in Electrical Sciences (ICTES 2007); 2007.
[39] Widyan M. Operational performance of synchronous generator hybrid-excited by PMDC and PV generators. Int Rev Electric Eng 2014.
[40] Xiao W, Dunford WG, Capal A. A novel modeling method for photovoltaic cells. In: 35th Annual IEEE Power Electronics Specialists Conference, Aachen, Germany; 2004. p. 1950−6.
[41] Gopal C, Mohanraj M, Chandramohan P, Chandrasekar P. Renewable energy source water pumping systems-A literature review. Renew Sustain Energy Rev 2013. https://doi.org/10.1016/j.rser.2013.04.012.
[42] Gupta N, Singh SP, Dubey SP, Palwalia DK. Fuzzy logic controlled three-phase three-wired shunt active power filter for power quality improvement. Int Rev Electric Eng 2011;6(3):1118−29.
[43] Sandhu J, Serov V. Inverse backscattering Born approximation for the magnetic Schrödinger equation in three dimensions. Math Meth Appl Sci 2013.
[44] Harikrishnan T. Nuclear power. Green Energy Technol Econ Policy 2010.
[45] Ruiz A. Partial recovery of a potential from backscattering data. Commun Part Differ Equ 2014. Springer Tracts in Modern Physics.
[46] Gould RJ. Pair production in photon-photon collisions. Phys Rev 03/1967.
[47] Imam A, Musyafa A, Soeprijanto A. Design and implementation of active two axes solar tracking system using particle swarm optimization based fuzzy logic controller. Int Rev Model Simul 2015.
[48] Güçlüa MC, Lib J, Umarb AS, Ernstb DJ, Strayer MR. Electromagnetic lepton pair production in relativistic heavy-ion collisions. Ann Phys 1999;272:7.
[49] Hossain F, Fara N. Integration of wind into running vehicles to meet its total energy demand. Energy Ecol Environ 2016.
[50] Hossain F. Invisible transportation infrastructure technology to mitigate energy and environment. Energy Sustain Soc 2017.
[51] Najjari B, Voitkiv AB, Artemyev A, Surzhykov A. Simultaneous electron capture and bound-free pair production in relativistic collisions of heavy nuclei with atoms. Phys Rev A 2009;80(012701).

[52] Shannan A, Abd NM, Yahaya NZ, Singh B. Single-diode model and two-diode model of PV modules: a comparison. In: 2013 IEEE International Conference on Control System Computing and Engineering; 2013.
[53] Soon JJ, Low K. Optimizing photovoltaic model parameters for simulation. In: IEEE International Symposium on Industrial Electronics; 2012.
[54] Liu Y, Xie J, Ong CN, Vecitis CD, Zhou Z. Electrochemical wastewater treatment with carbon nanotube filters coupled with in situ generated H_2O_2. Environ Sci Water Res Technol 2015.
[55] Park J, Kim H, Cho Y, Shin C. Simple modeling and simulation of photovoltaic panels using Matlab/Simulink. Adv Sci Technol Lett 2014;73(FGCN 2014):147–55.
[56] Shouman ER, Khattab NM. Future economic of concentrating solar power (CSP) for electricity generation in Egypt. Renew Sustain Energy Rev 2015.
[57] Li G, Wu D, Hu J, Li Y, Hossain MS, Ghoneim A. HELOS: heterogeneous load scheduling for electric vehicle-integrated microgrids. IEEE Trans Veh Technol 2016.
[58] Nafeh AESA. A modified MPPT control loop for PV/battery-charging system using PI controller optimally tuned with GA. Int J Numer Model Electron Network Dev Field 2010.
[59] Hencken K. Transverse momentum distribution of vector mesons produced in ultraperipheral relativistic heavy ion collisions. Phys Rev Lett 01/2006.
[60] Hossain F. Solar energy integration into advanced building design for meeting energy demand and environment problem: climate change, photoenergy, solar panel, and clean energy. Int J Energy Res 2016.
[61] Lee AS. Mathematical modeling of hybrid renewable energy system: a review on small hydro- solar-wind power generation. Int J Precis Eng Manufact Green Technol 2014.
[62] Li L, Johnson B, Segawa R. Evaluation of weather conditions for methyl bromide ambient air monitoring in Monterey, Santa Cruz, and kern counties in the year 2000. J Air Waste Manag Assoc 2012.
[63] Hossain F. In situ geothermal energy technology: an approach for building cleaner and greener environment. J Ecol Eng 2016;17:49–55.
[64] Konstantin S, Gerasimov D, Solodusha S. Smart grid: algorithms for control of active-adaptive network components. In: 2015 IEEE Eindhoven PowerTech; 2015.
[65] Kalantar M, Mousavi SM. Dynamic behavior of a stand-alone hybrid power generation system of wind turbine, microturbine, solar array and battery storage. Appl Energy October 2010.
[66] Ionescu DC. Electron capture from pair creation in relativistic heavy-ion collisions as a charge transfer process. Phys Scripta 01/01/1997.
[67] Kłusek-Gawenda M, Szczurek A. Double scattering production of two positron–electron pairs in ultraperipheral heavy-ion collisions. Phys Lett B 2016.
[68] Kamal E, Koutb M, Sobaih AA, Abozalam B. An intelligent maximum power extraction algorithm for hybrid wind-diesel-storage system. Int J Electr Power Energy Syst 2010;32(3):170–7.
[69] Klein SA. Calculation of flat-plate collector loss coefficients. Sol Energy 1975;17: 79–80.

CHAPTER FOUR

Advanced Building Design

4.1 ENERGY-PRODUCING BUILDING

Owing to the harmful effects of burning fossil fuels on global warming, alternative energy sources will become indeed important in the future [1−3]. Fossil fuels shall run out by the early 22nd century given the present rate of consumption. Atmospheric concentration of greenhouse gases is trapping heat radiated from the earth's surface, thereby causing global warming and environmental problems such as greenhouses effect, stratospheric ozone depletion, acid precipitation, and flooding of coastal settlements. This implies that sooner or later humanity will rely heavily on renewable energy sources [4−6]. Here I have introduced light energy at an idealized large-scale application to produce solar energy, where exterior skin and roof of buildings shall be at least 25% blackbody—assisted photovoltaic (PV) to capture solar energy during the whole year. Simply it is a calculative reaction of solar irradiance on innovative building design to capture sunlight most efficiently, which would be the cutting-edge technology for the ultimate solution of global energy, environmental, and climate crisis

4.1.1 Background and Technology

The population growth and industrial development are seriously impacting environment because conventional energy generation processes are polluting the environment and causing the climate change [7,8]. Global demand for energy services is expected to increase by as much as one order of magnitude by 2050, whereas primary energy demands are expected to increase by nearly two times. As per Energy Information Administration (EIA), US government *International Energy Annual 2004*, the world marketed energy consumption (1980−2050) is 283 EJ/year at 1980, 347 EJ/year at 1990, 400 EJ/year at 2000, and 511 EJ/year at 2010 and would be 607 EJ/year at 2020, 702 EJ/year at 2030, 855 EJ/year at 2040, and 988 EJ/year at 2050 [7,9,10]. Simultaneously, energy-related environmental concerns such as acid precipitation, stratospheric ozone depletion, and global temperature change will also increase subsequently [11,12].

Sustainable Design and Build
ISBN: 978-0-12-816722-9
https://doi.org/10.1016/B978-0-12-816722-9.00004-5

Several definitions of sustainable development have been put forth, including the following common one: "development that meets the needs of the present without compromising the ability of future generations to meet their own needs." The notion of renewable energy engineering is defined as and the term renewable energy suggests an energy that can be renewed or, in other words, cannot be depleted such as solar energy. According to well-established measurements, the average power density of solar radiation just outside the atmosphere of the earth is 1366 W/m^2, widely known as the solar constant [13,14]. The definition of the meter is one over 10,000,000 of earth's meridian, from the North Pole to the equator [3,15,16].

The radius of earth is $(2/\pi) \times 10^7$ m. The total power of solar radiation reaching earth is $1366 \times (4/\pi) \times 10^{14} \cong 1.73 \times 10^{17}$ W. Each day has 86,400 s, and on average, each year has 365.2422 days [17,18]. The total energy of solar radiation reaching earth per year is $1.73 \times 10^{17} \times 86{,}400 \times 365.2422 \cong 5.46 \times 10^{24}$ J, which is equivalent to 5,460,000 EJ/year [2,19,20]. To have an idea of how much energy that is, let us compare it with annual global energy consumption; world energy consumption in the year 2015 was 5.59×10^{20} J = 559 EJ, out of which nearly 50% is consumed by residential and commercial buildings [19,21]. A mere 0.01% of the annual solar energy reaching earth can satisfy the energy need of the entire world [22,23]. Total proved fossil fuel reserve on earth is 36,600 EJ (where crude oil, 1.65×10^{11} tons or unit energy 4.2×10^{10} J/ton equivalent to 6930 EJ; natural gas, 1.81×10^{14} tons or unit energy 3.6×10^7 J/m^3 equivalent to 6500 EJ; high-quality coal, 4.90×10^{11} tons or unit energy 3.1×10^{10} J/ton equivalent to 15,000 EJ; low-quality coal, 4.3×10^{11} tons or unit energy 1.9×10^{10} J/ton equivalent to 8200 EJ) as per *BP Statistical Review of World Energy*, June 2007, British Petroleum [12,24]. It means that if the current level of consumption of fossil fuel continues, the entire fossil energy reserve will be finished in 65 or less years. Currently, the utilization of renewable energy is still a small percentage of total energy consumption, and the utilization of solar energy through PV technology only accounts for 0.07% of total energy consumption [25,26]. The approach of this research is the innovative design for all houses and buildings where at least 25% of exterior skin and 25% of the roof should have the capacity to receive PV energy by using innovative solar panel. Consequently, we can meet 100% global energy demand and climate change will come down at zero (0) for each year.

4.1.2 Methods and Materials

Because the photo-physics is related to photo-induced charge [15,27], I therefore analyzed the properties of sunlight by means electromagnetic wave and motion/flow of photons before it is introduced to solar panel [28,29]. The first point of view is essential for all solar thermal applications and the antireflection coatings for solar cells. The second point of view is essential regarding solar cells and solar photochemistry. The unification of the two points of view is represented by quantum electrodynamics (QED), one of the most fruitful and matured fields in modern physics [23,30]. In that paper, I have developed a set of equations explaining electromagnetic phenomena upon modification of *Maxwell's equations*. Based on these equations, I have clarified electromagnetic waves in free space with a speed that equals to the speed of light, which was then calculated as per following series of mathematics.

4.1.2.1 Electromagnetic Waves

In this section, the electromagnetic wave was studied considering mass–luminosity relation for considering atmospheric temperature changes. An electromagnetic wave with circular frequency ω in space can be defined as

$$A(x, y, z, t) = A(x, y, z)e^{-i\omega t} \tag{4.1}$$

To study the properties of electromagnetic waves, we consider the case that the wave propagates in one direction, say z. In this case the field intensities only depend on z and equation becomes

$$\frac{d^2A}{dz^2} + \frac{\omega^2}{c^2}A = 0 \tag{4.2}$$

The general solution is

$$A = A_0 e^{i(k_z z - \omega t)} \tag{4.3}$$

where A_0 is a constant, and the z-component of the *wave vector* k_z is defined as

$$k_z = \frac{w}{c} \tag{4.4}$$

4.1.2.2 Motion of an Electron in Electric and Magnetic Fields

In this section the interaction of the radiation electric field and magnetic fields varying with time with the electrons is studied using classical mechanics as a preparation for a quantum-mechanical treatment [31,32]. The standard method of developing the quantum mechanics of a dynamic system is first to cast the classical equation of motion into Hamiltonian format. The Hamiltonian **H** (**p**, **r**) of a dynamic system is a function of its coordinate **r** and corresponding momentum **p**, representing the total energy. For example, for an electron with charge q moving in an electric field with potential φ, the Hamiltonian format is

$$\mathbf{H} = \frac{1}{2\mathbf{me}}\mathbf{p}^2 + q\varphi(\mathbf{r}) \tag{4.5}$$

The equations of motion in the Hamiltonian format are a pair of first-order ordinary differential equations:

$$\dot{p}x = \frac{\partial H}{\partial x} \tag{4.6}$$

$$\dot{x} = \frac{\partial H}{\partial px} \tag{4.7}$$

Using Eqs. (4.6) and (4.7), the expression of momentum can be found to be identical to the usual definition as follows,

$$\mathbf{p} = m_e\mathbf{v} = m_e\dot{\mathbf{r}} \tag{4.8}$$

where a dot means taking a derivative, and by applying Eq. (4.8), the following equation can be derived,

$$m_e\ddot{\mathbf{r}} = -q\nabla\varphi(\mathbf{r}) = q\mathbf{E}, \tag{4.9}$$

which is Newton's equation of motion, where **E** is electric field intensity.

For the motion of an electron in both electric and magnetic fields, the standard method is to insert the vector potential **A** into the expression of momentum by simply substituting **p** with $\mathbf{p} - q\mathbf{A}$ in the Hamiltonian (**H**)

$$\mathbf{H} = \frac{1}{2m_e}(\mathbf{p} - q\mathbf{A})^2 + q\varphi(\mathbf{r}) \tag{4.10}$$

The Hamiltonian that describes the interaction of the atom and the electromagnetic waves can be made identical with the interaction of the atom with an assembly of particles moving with the velocity of light and satisfying the Einstein—Bose statistics, by a suitable choice to deliver interactive energy for electric magnetic field [32,33].

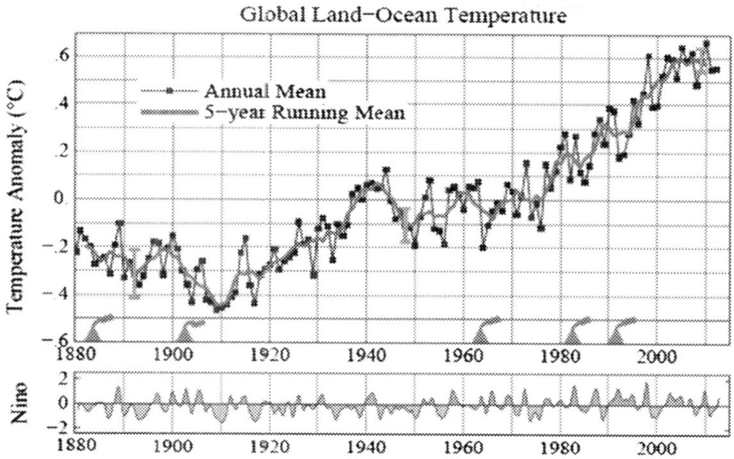

Figure 4.1 Global surface temperature anomalies relative to the period 1880 to 2015 (NASA Goddard Institute for Space Studies, 2012). The Nino index is based on the temperature in the Nino 3.4 area in the eastern tropical Pacific. *Dark green triangles* mark the times of volcanic eruptions that produced an extensive stratospheric aerosol layer.

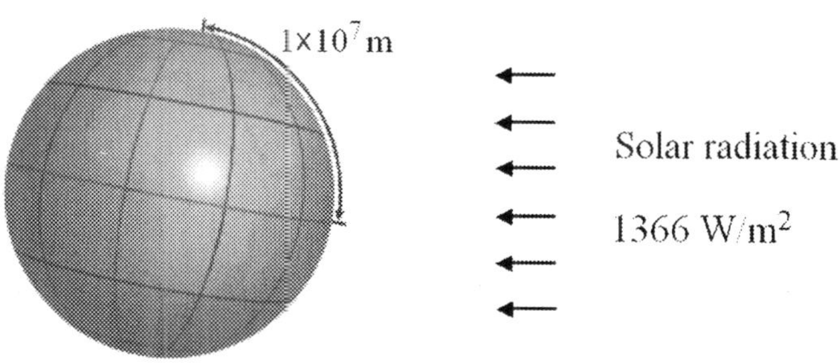

Figure 4.2 Annual solar energy arriving at the surface of the Earth. The average solar power on the Earth is 1366 W/m^2. The length of the meridian of Earth, according to the definition of the meter, is 10,000,000 m. The total solar energy that arrives at the surface of the Earth per year is 5,460,000 EJ.

4.1.2.3 Impact of Electromagnetic Waves and Motion of Electron on Blackbody Radiation

Electromagnetic wave and electron motion conducting can be analyzed by using IR photoconductive detector using classical physics, and it can be found that a hot body emits radiation maximum at around 700°C. Although all hot bodies emit radiation, blackbodies emit the maximum amount of radiation at a given temperature, but the body color changes to orange, yellow, white, and blue (Fig. 4.3).

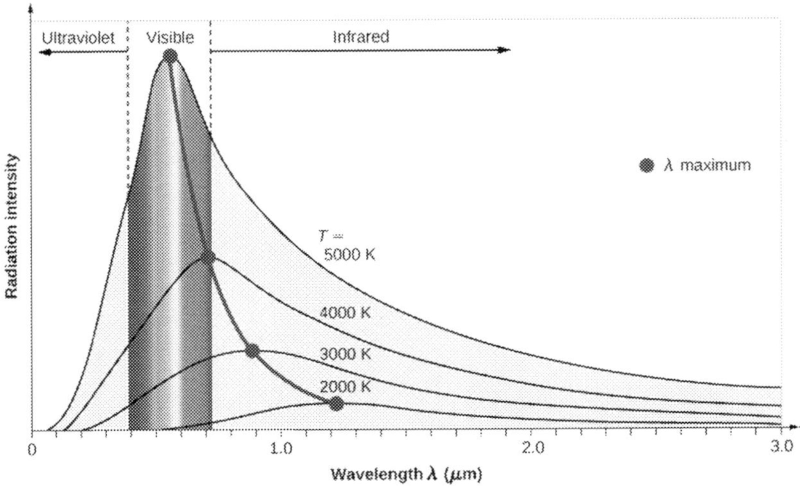

Figure 4.3 Light is an electromagnetic wave that is produced when an electric charge vibrates. In a hot object, electrons vibrate in random directions and produce light as a result. A hotter object means more energetic vibrations, and so light is emitted by a hotter object. The electrons in a hot object can vibrate with a range of frequencies, ranging from very few vibrations per second to a huge number of vibrations per second.

The energy density of radiation function of its frequency has been theorized using classical statistical physics. Consequently, standing electromagnetic waves can be treated as individual mode to follow the equal-partition law of Maxwell–Boltzmann statistics considering a closed cubic cavity with reflective inner flat surfaces of side (Fig. 4.4).

Thereafter, the blackbody can be experimented by opening a small hole on a large cavity, as shown in Fig. 4.5. The light ray passing through the hole with area A experiences multiple reflections on the internal surface of the cavity. Because the material is not shiny, after several impingements, the light will eventually be completely absorbed by the cavity. Therefore the small hole on the large cavity always shall look black, which is an excellent example of a blackbody [8,28]. On the walls of the cavity, electrical field intensity vanishes. Therefore the vector potential vanishes.

The following blackbody irradiance figure reveals the maximum of solar irradiance at 1.4 eV, with an energy value of 27.77 $MW/m^2 \cdot eV$. This estimate is as per year at an average 5-h peak levels daily, which is equivalent energy to 27,770 kW/year or 7.6 kW/day [33,34] (Fig. 4.6).

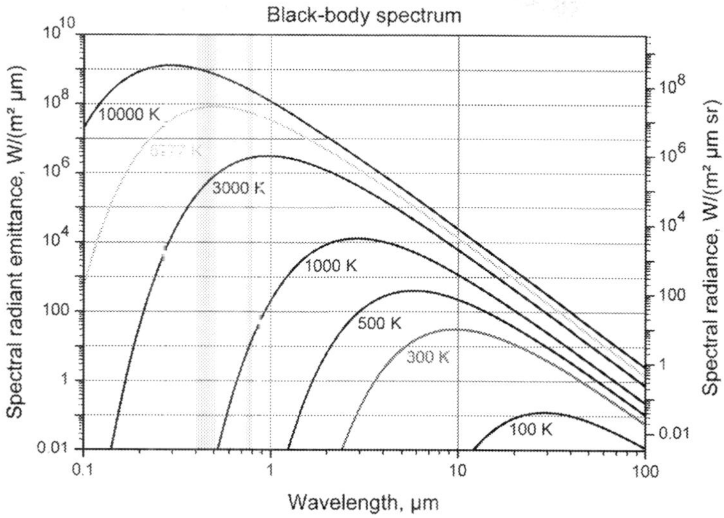

Figure 4.4 Blackbody radiation Planck, Wien, and Stefan–Boltzman laws calculated that the electromagnetic waves in a cavity of each as individual modes to follow the equal-partition radiation intensity once it is closed to cubic cavity with reflective inner surfaces of side.

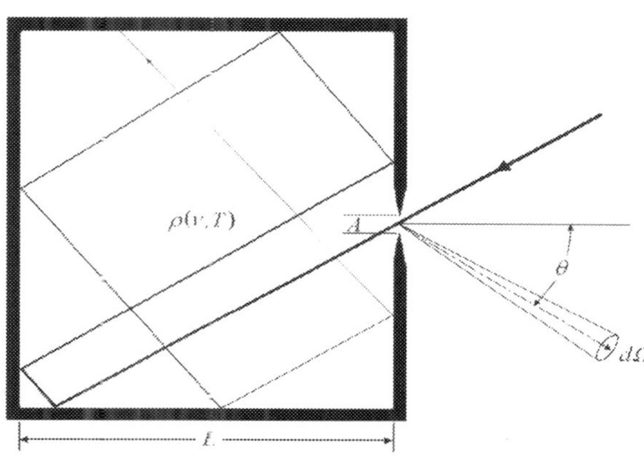

Figure 4.5 Blackbody radiation. A large cavity with a small hole is a good blackbody. The light that enters the hole will experience multiple reflections and all be absorbed and thus looks black. A blackbody emits maximum amount of radiation when heated.

4.1.2.4 Innovative Building Design and Solar Radiation Capture

The approach was to confirm to meet 100% energy consumption for a building. Thus the building can be designed in such a way that 25% exterior skin and 25% roof are constructed with solar panels. The panel stayed flat

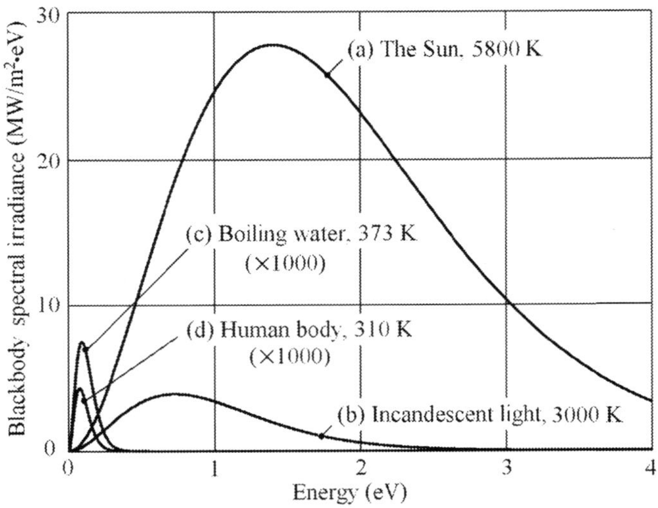

Figure 4.6 The blackbody spectral irradiance or the radiation power emitted per square meter per unit energy interval (here in electron volts) at an energy value (also in electron volts) at four different temperatures rise shown. The spectral irradiance from a blackbody at the boiling points of water and the human body are also shown just for information only, in units of kW/m$^2\cdot$eV. Blackbody radiation at different temperatures at 5800K power is 6.31 $\times$ 107(W/m^2); Peak E is 1.410 (eV); Peak λ is 0.88 (μm); Peak u is 2.81 $\times$ 107(W/m$^2\cdot$eV).

both horizontal (roof) and vertical (exterior skin), but the inner crystal structure can be designed as per following angle, latitude and longitude, and coordinate transformation in Cartesian coordinates consideration to make sure that the panel can capture maximum amount of radiation.

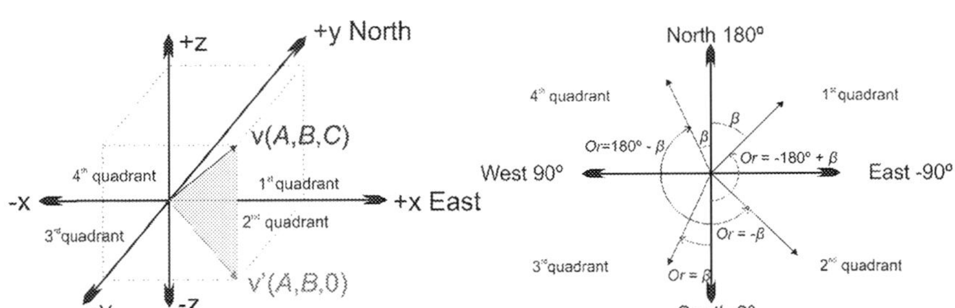

Figure 4.7 Solar panel angle consideration calculated upon effect of latitude and module tilt on the solar radiation received throughout the year. The module is to be facing south in the northern hemisphere and north in the southern hemisphere.

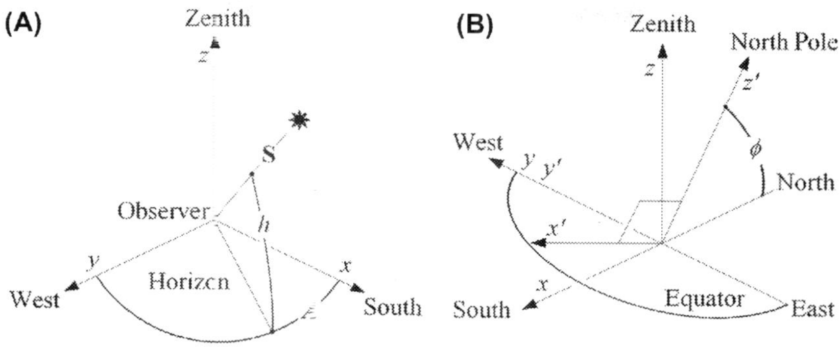

Figure 4.8 (A) Cartesian coordinates for the horizon system use the conventions that south is *x*, west is *y*, and the zenith is *z*. The position of a celestial body is determined by two angles, height h and azimuth angle *A*. (B) Cartesian coordinates for the equatorial system use the convention that *zt* axis points to the North Pole, whereas the east—west axis (*yt* axis) is identical to that in the horizon system. The *xt* axis is perpendicular to both. The position of a celestial body is determined by declination δ and hour angle ω.

In addition to the angle, the coordinate frequently in place of the hour angle is the right ascension, which takes the vernal equinox as the reference point, and was also considered very carefully [35,36] (Fig. 4.8).

Finally, the latitude and longitude are given attention. The zero point of latitude, the prime meridian, can be defined as the meridian passing through the Royal Greenwich Observatory. East of the prime meridian is the Eastern Hemisphere and in the west is the Western Hemisphere. Similarly, north of the equator is the Northern Hemisphere, and south of the equator is the Southern Hemisphere (Fig. 4.9).

Once the angle of the solar panel was confirmed as per the aforementioned description, the panel wiring diagram was focused as per the following circuit diagram to make sure that maximum solar energy can transport via photo electrical system. The following complex wiring system shows a simple three parallel circuit to increase direct current (DC) to PV module [24,37].

4.1.2.5 Energy Conversion by Solar Panel

The maximum of solar irradiance can be received at 1.4 eV, with an energy value of 27.77 $MW/m^2 \cdot eV$ as per the abovementioned energy diagram as estimated per year basis at an average 5-h peak levels daily, which is equivalent energy to 27,770 kW/year or 7.6 kW/day. Thus the solar-selective metal surface texturing is excellent in energy conversation because the net current conversion by solar panels is 125% increased and efficiency of the

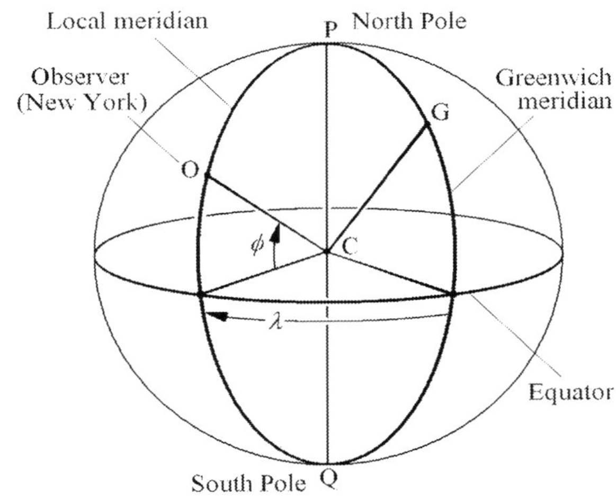

Figure 4.9 Latitude and longitude. The position of the observer is identified by its latitude φ and longitude λ, as marked on the map, and can be determined using GPS. The conversion of sign is eastward represents positive and westward is negative. For example, the latitude of New York City is $\varphi = 40°47$t N or $+0.712$ rad, and its longitude is $\lambda = 73°58$t W or -1.29 rad.

solar panel is 80% which means $(27{,}770 \times 1.25 \times 0.8) = 27{,}770$ kW/year or 7.6 kW/day. The energy remains same as solar energy, originally what it was before introducing to the solar panel (Fig. 4.10).

4.1.3 Results and Discussion

Because the fossil energy resources are finite, Marion King Hubbert at Shell (1956) predicted that crude oil production in the world would peak around 1970, then it will start to decline due to the shortage of fossil fuel. His theory started with the discovery that the plots of x, the cumulative production of crude oil Q, versus y, the ratio of production rate P over Q, in the whole world follows a straight line (Fig. 4.11). The two intersections of the straight line with the coordinate axes are defined as follows. The intersection with the x-axis, Q_0, is the total recoverable crude oil reserve. The value found from Fig. 1.5 is $Q_0 = 228$ billion barrels. The intersection with the y-axis, a, has a dimension of inversed time. The inverse of "a" is a measure of the duration of crude oil depletion.

It means that if peak production is 1970 from the last 100 years, then the least production (5% or less) would be during 2070 due to crude oil depletion. So, the only option is renewable energy that the sun offers to mankind, which is a virtually unlimited source of energy that can provide faster access

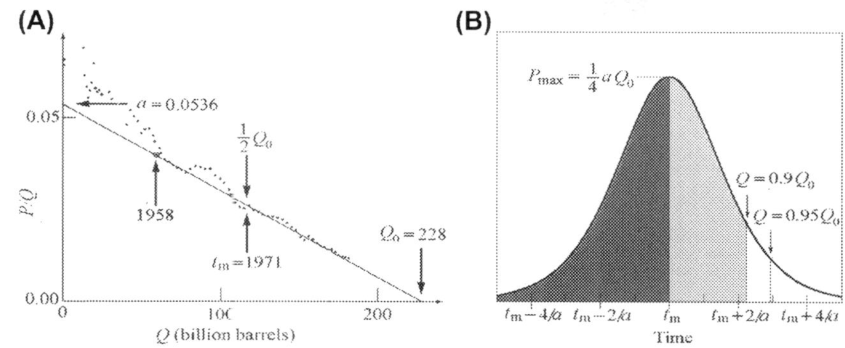

Figure 4.10 (A) In 1956 Merion King Hubbert of Shell Oil studied the data of the cumulative production of crude oil, Q, and the rate of production, P, in the whole world. He discovered a linear dependence between P/Q and Q. (B) A curve of Q versus time can be derived from the linear relation, Hubbert's curve (Eq. 1.9). The peak of production occurs at time tm when one-half of the crude oil is depleted. At time $t_m + 2.197/a$, 90% of the recoverable crude oil is depleted. At time $t_m + 2.944/a$, 95% of the crude oil is depleted. *(A) After Born M, Wolf E. Principles of optics. 7th ed. Cambridge: University Press Cambridge; 1999.*

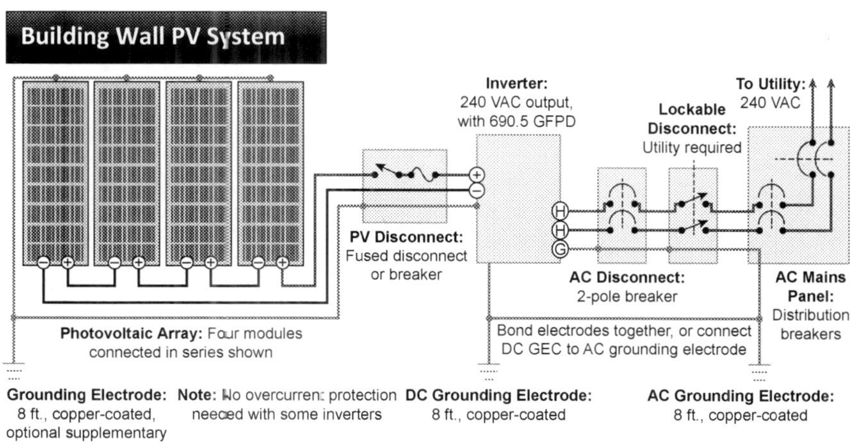

Figure 4.11 Building wall photovoltaic system is theorized to use at commercial building and offices and to be an additional power supply to the national grid or utility company. The four modules connected in a series of photovoltaic array supply power at inverter that has been used to meet building's energy demand. Additional power supply to the grid or utility company via 2 poles AC disconnect breaker has been used as grid/utility required lockable disconnector.

to modern energy services for any part of the world to meet the global energy demand.

According to the US EIA, the average monthly electricity consumption for a residential utility customer is about 903 kWh; for office and industrial consumer, it is about 450,000 kWh for a footprint of 32 m × 31 m with height 30 m (10 stories) building; and for transportation and others, it is about equivalent to industrial consumers energy. In average it should be about 30 kWh/day [37,38]. The estimate about 5 h per day to generate that is 30 kWh energy. So, the kWh divided by the hours of sun equals the kW needed. Or, 30 kWh/5 h of sun = 6 kW of alternating current (AC) output that is needed to cover 100% of energy usage. So now that we know we need to produce about 6 kW of AC output, we can work backward to figure out how many solar panels you need. Solar panels produce DC, and our home runs on AC. Because of physics, there are losses in converting the energy from the sun into DC power and turning the DC power into AC power. This ratio of AC to DC is called the "derate factor," and is typically about 0.8 [39,40]. This means we can convert about 80% of the DC power into AC power. So, we can take the AC amount we need, that is, 6 kW, and divide by 0.8 (6 kW/.8 = 7.5 kW DC). This means that we will need just one (1 × 1 m^2) solar panel to meet the 100% energy demand for the residential houses. Subsequently for office and industries, we will need average 500 (1 × 1 m^2) solar panels to meet their 100% energy demand. If we use 25% exterior skin of the tall building and 25% of the roof, it is quite possible to supply additional energy to the national grid.

The finding can be calculated as per the aforementioned graph and maximum solar irradiance can be found at 1.4 eV, with an energy value of 27.77 $mW/m^2 \cdot eV$ per year at an average 5-h peak levels daily for 365 days concerning per solar panel and blackbody modeling described previously. This energy is equivalent to 27,770 kW/year or 7.6 kW/day, which indeed will meet the energy demand for a residential house by using a 1 m^2 solar panel. Considering a tall building of an average footprint of 32 m × 31 m with a height of 30 m, the total panel case to be installed is 945 + 250 = 1195 unit, which can provide total energy of 7.6 kW × 1195 = 9,082 kW/day. If the building energy demand per day is 7.6 kW × 500 = 3800 kW, the additional energy that can be supplied to the utility grid is 5282 kW/day, which can be used for transportation sectors where energy is needed. Because transportation and other energy demands are approximately close to office and industry usages, this setup is more than enough for meeting global energy demand.

4.1.3.1 Savings on Energy Cost

In contrary the total cost for 30 years of energy consumption from a conventional source for a standard industry (100 people capacity) at 0.12/kWh of 4000 kWh per month is (30 × 12 × 4000 × 0.12) $172,800. This comparison between conventional energy use and solar panel energy production clearly indicates the cost saving of $68,400 when solar panel can be used as the energy source.

4.1.4 Conclusions

During the past several decades, there has been growing concern over the potential dangers associated with the accumulation of greenhouse gases (COx, CHx, NOx) in the atmosphere. Hazardous air pollutants are usually emitted in smaller quantities than those that are the focus of ambient air quality concerns. Many energy-related activities emit hazardous air pollutants, such as hydrocarbons (such as benzene) emitted from oil and gas extraction and processing industries; hydrocarbon and dioxin emissions are caused by the use and combustion of petrol and diesel oil for transport, and radionuclides are released during the combustion of coal and heavy fuel oil. Excessive concentrations of these pollutants that have demonstrated health, welfare, and ecological effects felt globally in addition to climate change. Because energy utilization is a major contributor to environment and climate change, deploying solar power at the scale is indeed to alleviate environmental and climate change; will pose serious challenges for today's electric power system along with finite nature of fossil fuel. In this research, it suggests ensuring an economic, reliable, and climate-friendly power system for future energy required for global energy demand. The demonstration of simple solar energy trapping and capturing it by innovative solar panel shall indeed fully meet the global energy demand. These results are not only promising but also very much calculative that it could encourage harvesting solar energy globally to its maximum utilization, which would be a sustainable energy source as well as environment friendly.

4.2 INTEGRATED BUILDING DESIGN

A modern building, either a residential or commercial one, is indeed connected with electricity, gas, and water to confirm a modern daily life. Adversely the impact of these conventional connections is indeed causing environmental pollution and climate change severely. In this article, a renewable technology has been proposed to meet total energy, water, and gas demand for a building without any outsource connection, which certainly would reduce 40% global warming [24,40]. For meeting the 100% energy demand of a building, a design theory has been proposed to use light irradiance as an ideal source to produce solar energy, by implementing construction of exterior skin and roof of buildings with at least 25% blackbody—assisted PV panel to capture solar energy and convert it into electricity energy. Domestic water supply is to be performed by PV panel—assisted in situ pump to extract groundwater and then treated with chlorination, ultraviolet application, and purification technology to meet all domestic water. Thereafter, the domestic wastewater can be subjected to treatment process applying all primary, secondary, and tertiary processes, and then the sludge, mainly humane feces, shall be collected into a closed anaerobic detention tank at cellar to introduce into a bioreactor, allowing to producing biogas by methenogenesis, storing the same, and using it for cooking. Finally, the separated treated wastewater can be used for gardening and irrigation for green roofing and landscaping. Combining all these technologies simply would be an innovative one where a building itself can produce electricity, water, and gas without outsource connection which certainly shall mitigate global energy, environment, and climate change crisis dramatically.

4.2.1 Background and Technology

Buildings have a significant impact on climate change because it uses almost 40% of energy consumption from burning fossil fuel [14]. To have an idea of how much energy that is, lets calculate it with annual global energy consumption; world energy consumption in the year 2015 was 5.59×10^{20} J = 559 EJ; where nearly 2.236×10^{20} EJ is consumed by residential and commercial buildings producing nearly 8.01×10^{11} ton CO_2 (218 gtC by building of global total 545 gtC; 1 gtC = 10^9 ton C = 3.67 gt CO_2). This amount of energy used by buildings continues to increase especially because the new buildings are built faster than old ones removed and the development is running fast in every corner of the world. Energy consumption by buildings will continue to increase until the buildings can be designed to produce enough renewable energy by a sophisticated technology. Interestingly, light emitted by the sun and falling on Earth is one of the most plentiful renewable energy resources [24,33]. In concept a zero-emission building (ZEB) shall greatly be an innovative one to balance its energy demand if the building's skin and roof are designed with at least 25% blackbody—assisted PV panel to capture solar energy to meet its total energy demand, which will reduce the 8.01×10^{11} ton CO_2 emission in a year [14,20]. It is estimated that currently 402 ppm CO_2 is present in the atmosphere causing global warming, which is required to cut down to 300 ppm CO_2 to confirm global cooling at comfortable stage [41,42]. Once blackbody—assisted PV panel is used at each building implemented by innovative building design, it will reduce 40% CO_2 per year. Interestingly, it will take only $\left\{ \int_{300}^{402} (1 - 0.4)dx \right\} = 61.2$ years to cool the atmosphere, resulting no more climate change after 62 years.

On the other hand, it is well known that the water reservoir plays a significant role on earth's ecosystem and may contribute to changes in earth's climate once reservoir get stratified due to water being extracted for domestic purposes. Subsequently reservoirs get warm and generate methane, a greenhouse gas, in which the bottom layers are anoxic (i.e., they lack oxygen), leading to degradation of biomass through anaerobic processes. In some cases, where flooded basins are wide and biomass volumes are high, the amount of biomass converted to methane results in pollution potential 3.5 times more than an oil-fired power plant would for the same generation capacity. The methane gas is one of the top significant players to global climate change. Besides, reservoir water treatment, supply, and maintaining city water lines are even more costly to implement in each building to meet our daily water demand. Because groundwater is available within 200-m energy line everywhere in the world which are mostly purified by natural process of aquifer, it can be pumped and further treated very cheaply in site to meet the building's water demand without harming the climate [35,43].

Eventually buildings' wastewater can be collected into the closed detention tank, and then wastewater and feces can be separated into two different chambers. In context with feces (solid waste), a person produces 0.4—0.5 kg/day, and 0.5 kg of human feces can produce 0.4 m^3 biogas/day, which can cook three meals for a family of 4 persons. On the other hand, treated wastewater can be supplied to the rooftop gardening and building's landscape irrigation. Indeed, it would be an independent building technology that can meet all its energy, water, and gas demand, and thus it shall play a key role to reduce the climate change.

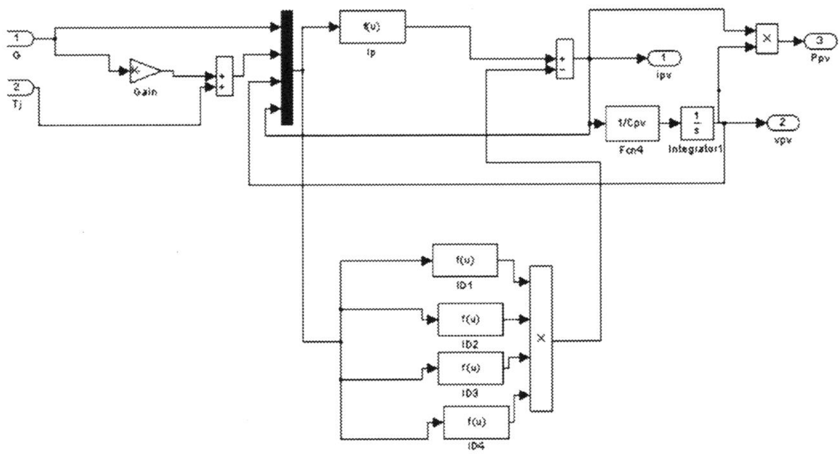

Figure 4.12 Diagram of PV system model.

4.2.2 Methods and Materials

4.2.2.1 PV Array Modeling

In this article a mathematical calculation can be analyzed that describes the operation and behavior of the photovoltaic generator in which the maximum power output can be calculated for a module once photovoltaic solar irradiance on the photovoltaic module and temperature. Interestingly, it can be introduced as a simplified model of the maximum power output which has a reciprocal relationship with the temperature module and logarithmic relationship with the solar radiation absorbed by the photovoltaic module [44,45]. These models in the calculation procedure, accuracy, and the number of parameters involved in the calculation of the current—voltage characteristic can be presented as follows (Fig. 4.12) and explained further in a single-diode equivalent circuit of a PV cell model module (Fig. 4.13).

With $I_{pv\text{-sim}}$ calculated from the model with a diode, using a wide range of variation of the illumination received by the photovoltaic panel, p contains various parameters to determine (P_1, P_2, P_3, P_4, A, R_s, R_{sh}), and thus it was used as the least control strategy to control to achieve simultaneously a green active and reactive power source with active filtering capability [46,47] (Fig. 4.14).

The power output of a PV array is based on solar irradiance and ambient temperature, and thus, the model can be calculated as

$$P_{pv} = \eta_{pvg} A_{pvg} G_t \tag{4.11}$$

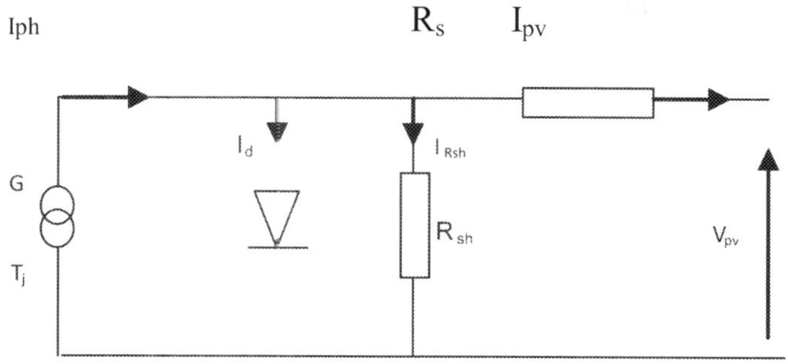

Figure 4.13 Single-diode equivalent circuit of a PV cell model module or array as a continuous function for a given set of operating conditions.

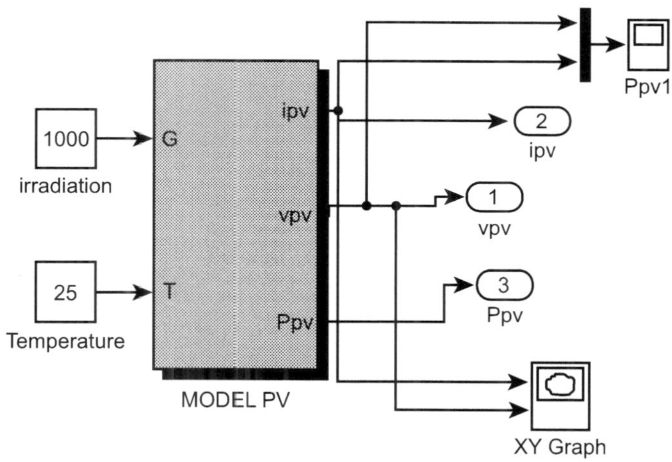

Figure 4.14 Activated Bloc diagram of the PV model.

where η_{pvg} is the PV-generation efficiency, A_{pvg} is the PV generator area (m^2), and G_t is the solar radiation in a tilted module plane (W/m^2). η_{pvg} can be further defined as

$$\eta_{pvg} = \eta_r \eta_{pc}[1 - \beta(T_c - T_{c\ \mathrm{ref}})] \tag{4.12}$$

where η_{pc} is the power conditioning efficiency, which is equal to one when MPPT is used; β is a temperature coefficient (0.004–0.006 per °C); η_r is the reference module efficiency; and $T_{c\mathrm{ref}}$ is the reference cell temperature in °C. The reference cell temperature ($T_{c\mathrm{ref}}$) can be obtained from the relation

$$T_c = T_a + \left(\frac{\mathrm{NOCT} - 20}{800}\right) G_t \tag{4.13}$$

where T_a is the ambient temperature in °C, NOCT is the nominal operating cell temperature in °C, and G_t is the solar radiation in a tilted module plane (W/m^2). The total radiation in the solar cell, considering both normal and diffuse solar radiation, can be estimated as

$$I_t = I_b R_b + I_d R_d + (I_b + I_d) R_r \tag{4.14}$$

The solar cell, which is the building block of the solar array, is essentially a P—N junction semiconductor capable of producing electricity via the PV effect. PV cells are interconnected in a series-parallel configuration to form a PV array, and graphene is integrated in the PV module to improve the efficiency of the resulting PV [47,48].

Using an ideal single diode, as shown in Fig. 4.2, for an array with N_s series-connected cells and N_p parallel-connected cells, the array current may be related to the array voltage as

$$I = N_p \left[I_{ph} - I_{rs} \left[\exp\left(\frac{q(V + IR_s)}{AKTN_s} - 1 \right) \right] \right] \tag{4.15}$$

where

$$I_{rs} = I_{rr} \left(\frac{T}{T_r} \right)^3 \exp\left[\frac{E_G}{AK} \left(\frac{1}{T_r} - \frac{1}{T} \right) \right] \tag{4.16}$$

and q is the electron charge (1.6×10^{-9} C), K is Boltzmann's constant, A is the diode ideality factor, and T is the cell temperature (K). I_{rs} is the cell reverse saturation current at T, T_r is the cell-referred temperature, I_{rr} is the reverse saturation current at T_r, and E_G is the bandgap energy of the semiconductor used in the cell. The photocurrent I_{ph} varies with the cell's temperature and radiation as follows:

$$I_{ph} = \left[I_{SCR} + k_i (T - T_r) \frac{S}{100} \right] \tag{4.17}$$

where I_{SCR} is the cell short-circuit current at the reference temperature and radiation, k_i is the short-circuit current temperature coefficient, and S is the solar radiation (mW/cm^2). The single-diode model includes an additional shunt resistance in parallel to the ideal shunt diode model. The $I-V$ characteristics of the PV cell can be derived using a single-diode model, as follows:

$$I = I_{ph} - I_D \tag{4.18}$$

$$I = I_{ph} - I_0\left[\exp\left(\frac{q(V + R_s I)}{AKT} - 1\right) - \frac{V + R_s I}{R_{sh}}\right] \tag{4.19}$$

where I_{ph} is the photocurrent (A), I_D is the diode current (A), I_0 is the inverse saturation current (A), A is the diode constant, q is the charge of the electron (1.6×10^{-9} C), K is Boltzmann's constant, T is the cell temperature (°C), R_s is the series resistance (ohm), R_{sh} is the shunt resistance (Ohm), I is the cell current (A), and V is the cell voltage (V). The output current of the PV cell using the diode model can be described as

$$I = I_{pv} - I_{D1} - I_{D2} - \left(\frac{V + IR_s}{R_{sh}}\right) \tag{4.20}$$

where

$$I_{D1} = I_{01}\left[\exp\left(\frac{V + IR_s}{a_1 V_{T1}}\right) - 1\right] \tag{4.21}$$

$$I_{D2} = I_{02}\left[\exp\left(\frac{V + IR_s}{a_2 V_{T2}}\right) - 1\right] \tag{4.22}$$

I_{01} and I_{02} are the reverse saturation currents of diode 1 and diode 2, respectively, and V_{T1} and V_{T2} are the thermal voltages of the respective diodes. The diode ideality constants are represented by a_1 and a_2. The simplified model of the PV system model can be presented as follows:

$$v_{oc} = \frac{V_{oc}}{cKT/q} \tag{4.23}$$

$$P_{\max} = \frac{\frac{V_{oc}}{cKT/q} - \ln\left(\frac{V_{oc}}{cKT/q} + 0.72\right)}{\left(1 + \frac{V_{oc}}{nKT/q}\right)}\left(1 - \frac{V_{oc}}{\frac{V_{oc}}{I_{sc}}}\right) \times \left(\frac{V_{oc0}}{1 + \beta \ln\frac{G_0}{G}}\right)\left(\frac{T_0}{T}\right)^{\gamma} I_{sc0}\left(\frac{G}{G_o}\right)^{a} \tag{4.24}$$

where v_{oc} is the normalized value of the open-circuit voltage V_{oc} with respect to the thermal voltage $V_t = nKT/q$, n is the ideality factor ($1 < n < 2$), K is Boltzmann's constant, T is the PV module temperature in kelvin, q is the electron charge, α is the factor responsible for all the nonlinear effects on which the photocurrent depends, β is a PV module technology-specific dimensionless coefficient, and γ is the factor representing all the nonlinear temperature–voltage effects. Eq. (4.24) represents the maximum power output of a single PV module [49,50]. A real system consists of a number of PV modules connected in series and in parallel. The total power output for an array with N_s cells connected in series and N_p cells connected in parallel with power P_M for each module will be

$$P_{\text{array}} = N_s N_p P_M \tag{4.25}$$

4.2.2.2 Design of Solar Panel

So far, photovoltaics, the conversion of sunlight into electricity, has been dominated by linking devices solid state, often made of silicon. But this domain is now being challenged by the emergence of a new generation of two layer photovoltaic cells, based, for example, on nanocrystalline materials and films of conductive polymers [49,51]. Thus a copper indium gallium selenide solar cell is a thin-film solar cell selected to convert sunlight into electricity because it has a high efficiency of conversion rate. To design long-lasting and such sophisticated PV panel considering load-resistant factor design, primarily uniformly distributed wind load is calculated to confirm that the PV panels are strong enough to protect itself even if heavy storm occurs and to deliver continuously without any interruption [52]. To confirm tornado-resistant PV panel (20,000 ibf $>$ 19, 375 ibf), it is needed that the PV panels should have the following wind load–resistant capacity if the wind velocity (F6 tornado level) is 379 mile/h at standard air density of 1.2 kg/m^3 and wind pressure and drag coefficient 1 considering per meter square (m^2). Because the wind stagnation pressure is half the density of the air times the square of the velocity, Eq. (4.26) can then be expressed as follows:

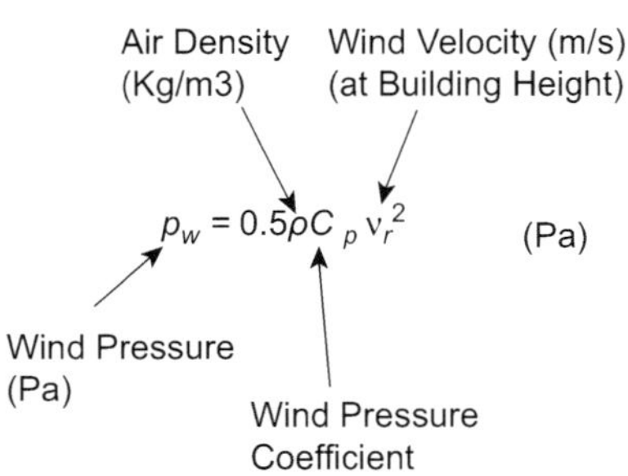

(4.26)

For which stagnation pressure can be determined from the following calculation: $P_w = 0.5 \times 1.2\ \text{kg/m}^3 \times 379^2\ \text{m/s} = 86{,}185\ \text{Pa}$. Finally the wind load can be calculated as $F = \text{area} \times \text{drag coefficient} \times \text{stagnation pressure}$.

$$F = 1\ \text{m}^2 \times 1.0 \times 86,185 = 86,185\ \text{N}(8,788\ \text{kgf} = 19,375\ \text{ibf}) \tag{4.27}$$

Once the sophisticated wind load resistance capacity has been determined, then the PV panel to be analyzed to capture maximum sunlight throughout the whole year, and the different directional angles, which can be defined a spherical coordinate system called horizontal coordinates, as shown in Fig. 4.15, where Cartesian coordinate system used for horizon conventions x is south west (y) and z is the zenith [52,53]. The position of a

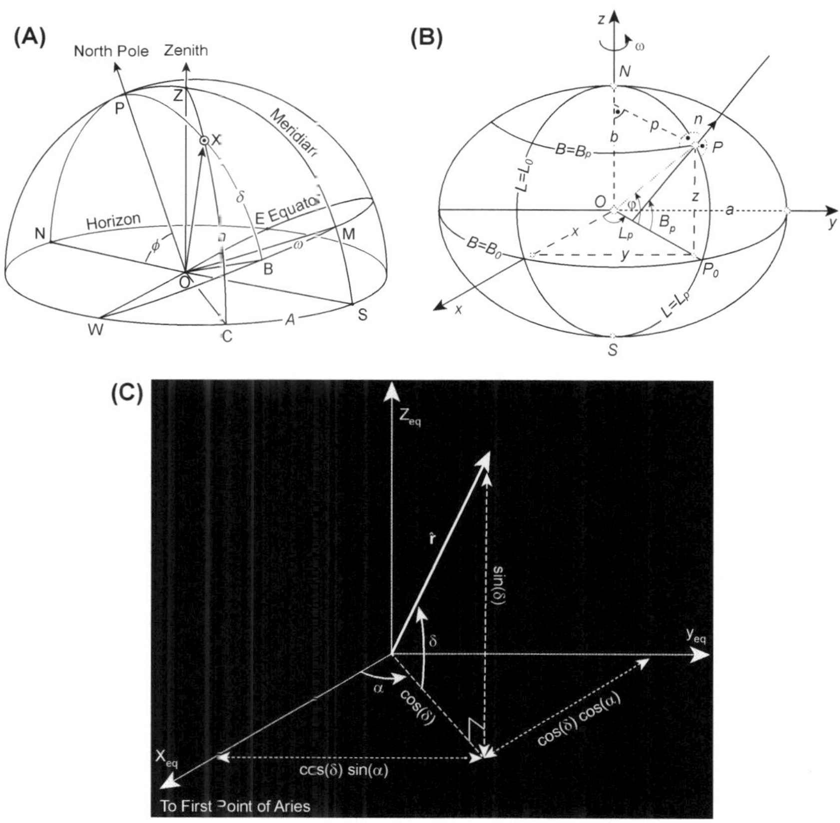

Figure 4.15 (A) Cartesian coordinate analysis shows the equatorial system, (B) position of angles, and (C) vector control.

celestial body is determined by two angles, the height h and the azimuth angle A. Cartesian coordinates for the equatorial system use the convention that the z axis points to the North Pole, whereas the y-axis, the east—west axis, is identical to the system horizon. The x-axis is perpendicular to both. The position of a celestial body is determined by δ decline and ω angle hours, and analysis naturally leads to the introduction of vector control strategies to control the active and reactive power based on the combination of different controls, which is reproducible and applicable to other complex systems [29,45].

Because the photo-physics relates to photo-induced charge, the properties of sunlight were considered by means of electromagnetic waves and movement/photon flux introduced to solar panel. The first view is essential for all applications of solar thermal energy and antireflective coatings for solar cells. The second view is essential with respect to solar cells and solar photochemistry [54,55]. The unification of the two views is represented by QED, one of the most fruitful and matured in modern physics fields. Despite all the hot bodies that emit radiation, blackbodies emit the maximum amount of radiation at a given approximately 700°C temperature, but changes in body color to orange, yellow, white, and blue and efficiencies in overlapping loads have been calculated as bandgap and the sun and cells that are supposed to blackbody temperatures of 6000 and 300K, respectively [56]. The energy density of radiation of frequency theorized the use of classical statistical physics and revealed the maximum solar radiation at 1.4 eV with an energy value of 27.77 $MW/m^2 \cdot eV$ [52,57]. This estimate is that each year an average of 5 levels peak hours of the day which is the equivalent of 27,770 kW/year or 7.6 kW/day energy (Fig. 4.16).

4.2.2.3 Energy Conversion by Solar Panel

To establish a connection between the number of light quanta by steady state and intensity of radiation, a finite volume enclosure containing radiation is considered. The number of stationary states of light quanta of a certain type of polarization whose frequency is in the range of ν_r to $\nu_r + d\nu_r$ and maximum solar radiation can be achieved at 1.4 eV with an energy value of 27. 77 MW/ m^2 eV according to the abovementioned scheme estimated by year based on an average of 5 h a day peak levels, which is the equivalent of 27,770 kW/ year or 7.6 kW/day energy. Thus the surface texture of selective solar metal is excellent in energy conversation [58] because the current net conversion by solar panels is 125% higher and an efficiency of 80% of solar panels means

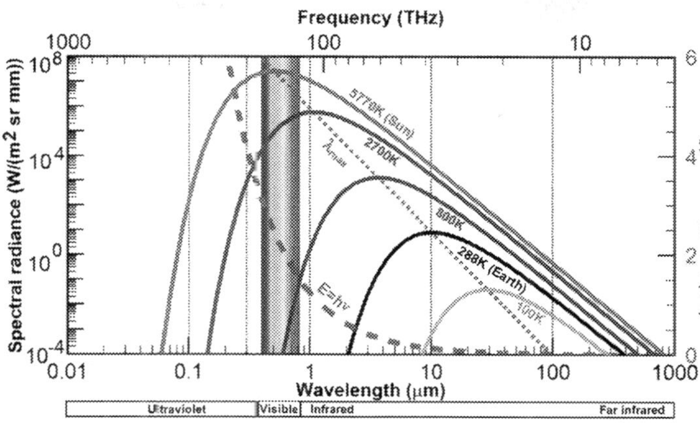

Figure 4.16 The figure depicted the blackbody radiation in various temperature at 5800K; power is 6.31×10^7(W/m^2); Peak E is 1.410 (eV); Peak λ is 0.88 (μm); Peak u is 2.81×10^7(W/m$^2 \cdot$eV).

$(27{,}770 \times 1.25 \times 0.8) = 27{,}770$ kW/year or 7.6 kW/day. Energy remains equal to what it was the initially before the introduction to the solar panel.

4.2.2.4 Groundwater by Solar Photovoltaic Pumping System

Groundwater is an extremely attractive source of water for construction and homeowners as it is available all year and exists in almost everywhere. It can be extracted by photovoltaic solar water pumping system (SMEs) consisting of solar photovoltaic panels, a motor, and a pump, which is dependent on system design and requires storage batteries and a charge controller. The engine is chosen according to the power demand and the type of output current system, which in case of change of frequency and voltage imbalance sinks backup representations showing promising successful transition. If the engine uses AC, it must be installed with a DC to AC converter. DC motors are not suitable for the high-power (above 7 kW) applications where an induction motor AC converter is required [54,59]. Using an inverter will result in additional costs and energy losses. For submersible DC water pumps, maintenance and replacement of brushes of DC motors require the pump to be removed from the deep well, which increases operating costs and maintenance and reduces their Solar Power Water Pump System (SPWPSs) without battery and reliability life, and they are inexpensive, which requires less maintenance compared to systems for the accumulator. However, the storage batteries have the advantage of providing consistent performance during lean times and off sun. The addition of a water storage tank in SPWP is cheaper than the backup battery storage. The use of photovoltaic solar energy is a key for countries in tropical regions where direct sunlight can reach up to 1000 W/m^2 resource [60,61]. A brief analysis of the studies that reported the performance, types of motors and pumps, the optimum sizing of photovoltaic panels, cooling photovoltaic solar panels, control SPWP, and economic and environmental considerations is discussed in this subsection. Then the Dupuit equation (Eq. 4.28) has been used to pump groundwater for domestic water supply by solar panel—assisted electricity supply, where h_1, H_w aquifer depth to the radial distance of R_1, and r_w are calculated to extract water without experiencing unsafe yield such as water quality, large pump compressed air, and water violation in factors affecting its surroundings (Figs. 4.17 and 4.18).

$$Q = \pi K \frac{h_1^2 - h_w^2}{\ln(r_1/r_w)} \tag{4.28}$$

Before the water is stored into the tank, it shall be treated in situ in water house by activated carbon method including disinfection, ozonation, and

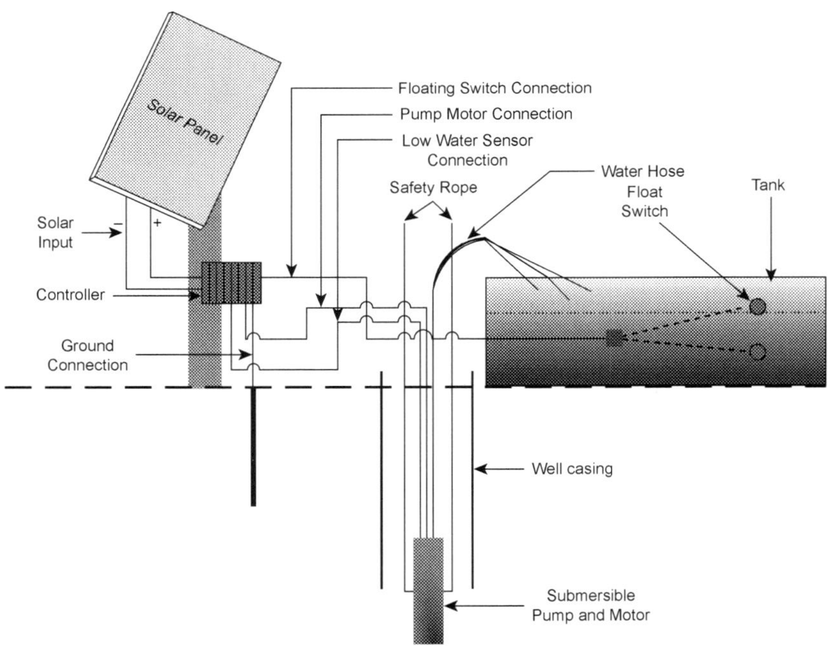

Figure 4.17 Layout of solar photovoltaic water pumping system using a submersible pump.

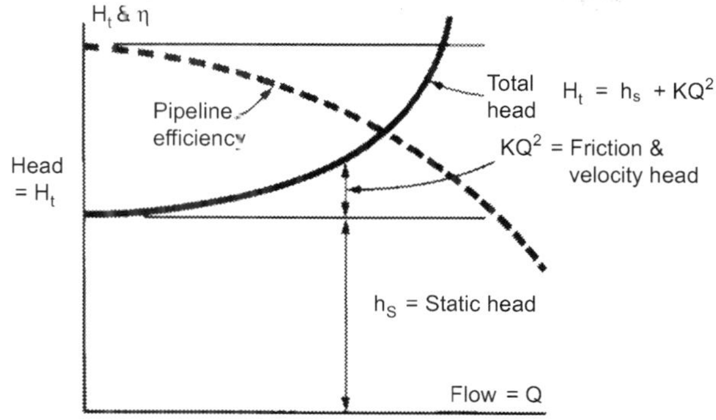

Figure 4.18 The water flow capacity against water energy line (head) once used solar PV power-assisted pump (considering hp and efficiency of the pump) to pump groundwater.

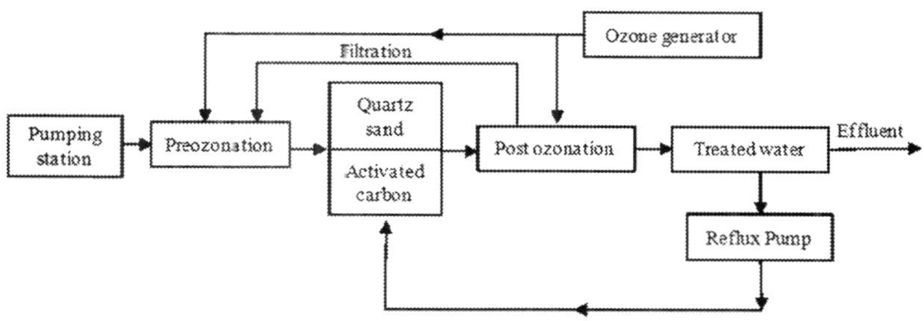

Figure 4.19 The process of in situ water treatment into the water house.

filtration to confirm that the water is free from any metal and microorganism and 100% potable as using the following diagram shown (Fig. 4.19).

4.2.2.5 Biogas Conversion from Human Feces and Domestic Waste

Interestingly, building solid wastewater can be collected into the closed detention tank and then separated into wastewater and feces into two different chambers. Thereafter, domestic wastewater shall undergo five major processes: (1) preliminary treatment, (2) primary treatment, (3) secondary treatment, (4) disinfection, and (5) sludge treatment in in situ plant. The whole treatment process will remove 100% pollutants from the wastewater and disinfect the wastewater called effluent which can be used for local gardening (Fig. 4.20).

Sludge, the product thereafter, must go beyond treatment process by performing filtration using electrochemically active carbon nanotubes

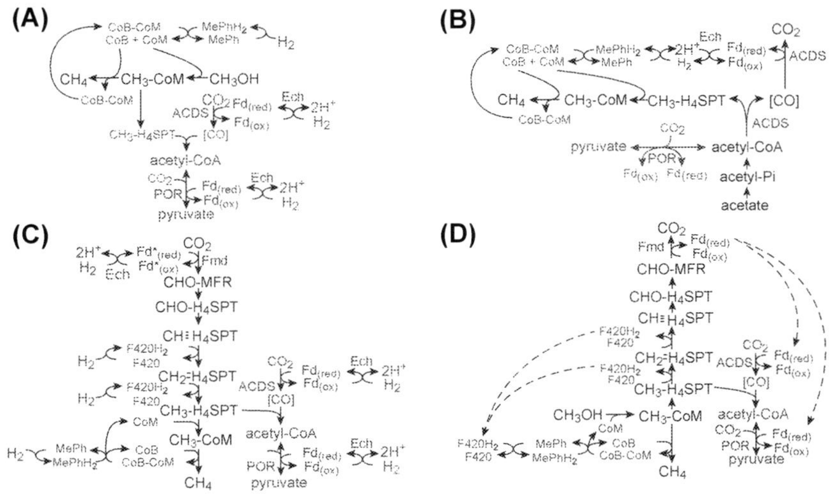

Figure 4.20 (A) Models for growth on methane plus H_2/CO_2, (B) acetate, (C) H_2/CO_2, and (D) methane are shown. The methanogenic pathway is shown in blue, the energy-conserving electron transport chain in red, and biosynthetic pathway in purple. The steps catalyzed by Ech and ferredoxin-dependent enzymes are labeled in green.

(CNTs), which can adsorb and oxidize chemicals in the anode effectively. In this study a novel system of wastewater treatment has been developed to combine both adsorption and oxidation at the anode CNT and further oxidation in situ generated hydrogen peroxide (H_2O_2) in the cathode CNT in a small-scale experiment [62,63]. Factors affecting the efficiency of the treatment and the oxidation mechanism of the system are studied systematically. The results demonstrated that H_2O_2 flow may be affected by the electrode material, the cathode potential, pH, flow rate, and oxygen dissolved. The maximum flow H_2O_2 of 1.38 mol/L m^2 C was achieved with CNT L^{-1} m^{-2} with an applied cathodic potential V of −0.4 (vs. Ag/AgCl), a pH of 6.46, a flow rate of 1.5 mL/min, and a DO influent flow of 1.95 mol/L m^2. Furthermore, phenol is used as aromatic compound model for assessing the removal efficiency of the system, and its oxidation rate correlated directly with H_2O_2 flow. H_2O_2 probably reacting with a phenol species is anodically activated to itself, and H_2O_2 radical form cannot remove phenol efficiently. Furthermore, the formation of electrochemical polymer through chain reactions of phenolic radicals can also contribute to 13% of phenol removal [63,64]. A stable removal efficiency of 87.0 ± phenol 1.8% to 4 h of continuous operation was achieved with an average rate of oxidation of 0.059 ± 0.001 mol/h m^2. The electrochemical

CNT filtering system developed with H_2O_2 generated in situ is a new application of filters, and CNTs can be used as an effective treatment for removing organic pollutants 100%. Then, store it in a closed chamber to allow the thermophilic anaerobic codigestion process to thicken. After thickening, the slurry is further processed to make it safer for the environment, and the sludge is placed in oxygen-free tanks, called digesters, and heated to at least 95 degrees Fahrenheit for 10–15 days [65]. This stimulates the growth of anaerobic bacteria of genus *Desulfovibrio* and *Methanecoccus*, which consume organic matter in the sludge. Unlike bacteria in the aeration tanks, these bacteria thrive in a free anaerobic environment. The digestion process stabilizes the thickened sludge by converting much of the material into methane gas.

Once methanogenesis started, the organic loading rate from the following diagram on volumetric methane (biogas) production can be examined by computerized gas chromatograph, which can be collected and stored into biogas recovery unit and then supplied for domestic cooking fuel and HVAC equipment operation.

4.2.3 Results and Discussion

Because the fossil fuel energy resources are finite, the only option is renewable energy, where the sun provides to humanity the energy virtually unlimited and that can provide faster access to modern energy services anywhere in the world to satisfy all household demand for energy. According to the US EIA, average monthly electricity energy consumption by a customer for residential use is about 903 kWh, office and commercial building consumption is about 450,000 kWh for a building of footprint 32 m × 31 m with height of 30 m (10 floors). On average, it should be approximately 30 kWh/day. It takes about 5 h a day to generate 30 kWh of energy. So, the kWh divided by the hours of sunshine equals the kW required. In other words, 30 kWh/5 h of sunshine = 6 kW AC output current required to cover 100% of energy consumption [66,67]. As now we know we need to produce 6 kW AC output, we can work backward to figure out how many solar panels are needed. Solar panels produce DC, and our house runs on AC. Owing to the physical reasons, there are losses in the conversion of solar energy into DC power and converting DC into AC. This ratio of AC to DC is called "derating factor" and is typically 0.8 [62,68]. This means we can convert about 80% of the DC power to AC So we can take the amount of CA we need, 6 kW, and divide by 0.8 (6 kW/0.8 = 7.5 kW DC). This means that we will need only one (1 × 1 m^2) solar panel to meet energy demand of 100% for residential houses. An office building of 10 standard floors and industry need an average of 500 (1 × 1 m^2) solar panels Standard Unit (SU) to meet 100% energy demand. By using only 25% of the tall buildings' exterior skin and 25% of the ceiling, it is quite possible to supply additional power to the national grid in addition to meeting the total energy demand.

The finding can be calculated according to the maximum solar irradiance found, that is, 1.4 eV, with an energy value of 27.77 mW/m^2 eV per year in an average of 5 h a day maximum levels for 365 days referents by solar panel and blackbody described previously (Fig. 4.21). This energy is equivalent to 27,770 kW/year or 7.6 kW/day, which in fact will meet the energy demand for a residential house using a solar panel of 1 m^2. In view of a tall building, with an average 32 m × 31 m footprint and a height of 30 m, total installed cased panel (945 + 250 = 1195) can provide total energy of 7.6 kW unit × 1195 = 9,082 kW/day. If the building energy demand per day is 7.6 kW × 500 = 3800 kW, additional energy can be supplied to the power grid, 5282 kW/day, which can be used for other industrial sectors where the energy is required.

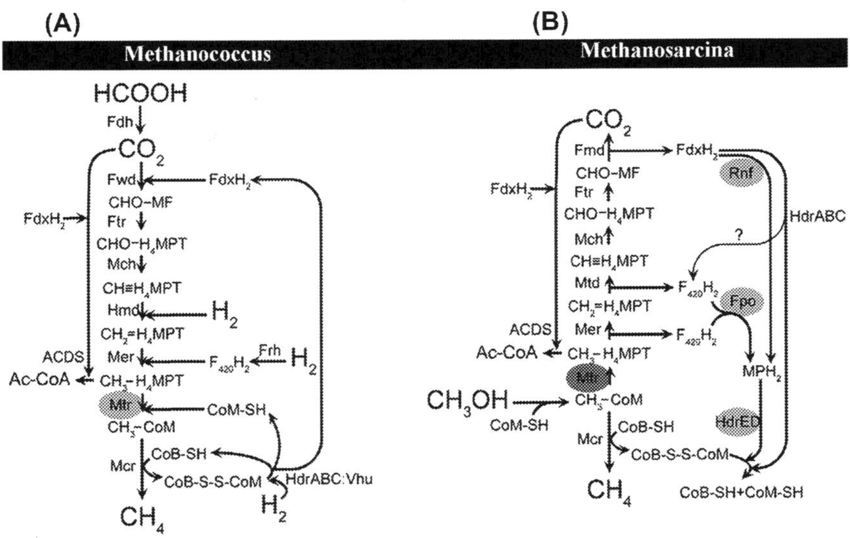

Figure 4.21 The methanogenesis process shows the detail diagram of (A) *Methonococcus* and (B) *Methanosarcina* bacterial syntrophic interaction, highlighting the central energy-generating and consuming pathways to produce methane.

Table 4.1 Design and construction cost of the solar panel for energy production

List of component	Materials cost	Labor cost	Equipment cost	GC and OH cost	Total cost
Solar panel	$10,000	$5000	$2500	$3500	$21,000
Instrumentation	$2000	$1000	$2000	$1000	$6000
Electrical and mechanical control	$2500	$1000	$1000	$900	$5400
Supply for 30 years cost at $0.05/kWh for monthly 4000 kWh for 100 people					$72,000
Total cost					$104,400

Note: This estimate was prepared by using current (January 2016) material cost from the top manufacturer and labor cost to install materials as per international union labor wages. The equipment rental was calculated as per current market rental cost in conjunction with production rate standard construction practice.

Table 4.2 Design and construction cost of the solar panel for energy production for a 10-story building

List of component	Materials cost	Labor cost	Equipment cost	GC and OH cost	Total cost
Solar panel	$10,000	$5000	$2500	$3500	$21,000
Instrumentation	$2000	$1000	$2000	$1000	$6000
Electrical and mechanical control	$2500	$1000	$1000	$900	$5400
Supply for 30 years cost at $0.05/kWh for monthly 4000 kWh for 100 people					$72,000
Total cost					$104,400

Note: This estimate was prepared by confirming recent (June 2016) cost of material for selective manufacturer and labor rate added in accordance with international union of specified trade workers. The equipment rental cost is calculated as current rental market in conjunction with the standard practice of construction of the production rate.

4.2.3.1 Savings on Energy Cost

In contrary, the total cost for 30 years energy consumption from a conventional source for a standard industry (100 people capacity) at 0.12 of

4000 kWh per month is (30 × 12 × 4000 × 0.12) $172,800. This comparison between conventional energy use and solar panel energy production clearly indicate the cost saving of $68,400 when solar panel is used as the energy source.

4.2.3.2 Groundwater

Global estimates show that groundwater comprises one-sixth of the total freshwater resources available in the world and is a flexible resource that can be used whenever required, which also has the added benefit of incurring no evaporation losses [43,68]. Besides, reservoir water treatment process is very much complex and costly and maintenance requires tremendous technical expertise and various agencies' coordination. Therefore groundwater has emerged as an exceedingly important freshwater resource for domestic uses and the top ranked resource as it is of strategic importance because of easy collection, which also can play notable role to mitigate climate change. An average of 100 gallons of water is required per person per day, which is approximately (50 people/each floor) 5000 gallons required for a 10-story building Table 4.3.

Table 4.3 Design and construction cost of the groundwater for a 10-story building

List of component	Materials cost	Labor cost	Equipment cost	GC and OH cost	Total cost
Shallow tube	$5000	$3000	$2500	$2100	$12,600
Filters	$2000	$1000	$2000	$1000	$6000
Solar panel—assisted pump	$10,500	$5000	$1000	$3300	$19,800
In situ treatment	$100,000	$50,000	$100,000	$50,000	$300,000
30 years of maintenance cost at $1500/year					$45,000
Total cost					$383,400

Note: This estimate was prepared by confirming recent (June 2016) cost of material for selective manufacturer and labor rate added in accordance with international union of specified trade workers. The equipment rental cost is calculated as current rental market in conjunction with the standard practice of construction of the production rate.

4.2.3.3 Savings on Energy Cost

In contrary the total cost for water consumption from a conventional city and/or pubic department source for a standard [6] 10-story building at

200 gallon/day person for 50 persons/floor for 30 years requires (200 × 50 × 10 × 30 × 365) 1.09×10^{10} gallon water. The average cost was 0.00,015/gallon water supply, which is equivalent to $1,642,500. This comparison between conventional water use and in situ solar panel—assisted water pumping production and treating clearly indicate the cost saving of $1,259,100.

4.2.3.4 Biogas

Biogas is indeed the most significant renewable energy source because it offers an economically attractive alternative to fossil fuels. The success of biogas production will come from the availability at low costs and the broad variety of usable forms of biogas for the production of heat, steam, electricity, and hydrogen and for the utilization as a household kitchen gas.

This estimate shown in Table 4.4 was prepared by confirming recent (June 2016) cost of material for selective manufacturer and labor rate added in accordance with international union of specified trade workers. The equipment rental cost is calculated as per current rental market in conjunction with the standard practice of construction of the production rate.

4.2.3.5 Savings on Energy Cost

Similarly, the total cost of gas supply from a conventional source of utility company and/or agency for a standard [6] 10-story building at $100/floors/month is (1000 × 50 × 12 × 30) $1,800,000 for 30 years of consumption. This comparison between conventional utility company and/or agency gas use and on-site biogas production and supply to the domestic

Table 4.4 Design and construction cost of the biogas production for a 10-story building

List of component	Materials cost	Labor cost	Equipment cost	GC and OH cost	Total cost
Detention chamber	$10,000	$5000	$2500	$3500	$21,000
Bioreactor	$100,000	$75,000	$50,000	$45,000	$270,000
Electrical and mechanical control	$20,500	$10,000	$10,000	$8100	$48,600
Gas supply and maintenance for 30 years at $2000/year					$60,000
Total cost					$399,600

use revealed that the cost saving of $1,400,400, which is also environmentally benign.

4.2.4 Conclusions

Owing to the population growth, the development of residential and commercial building in cities and suburban regions including rural area is accelerating exponentially for the past several decades. Subsequently, renewable energy resources are getting more importance in recent years due to their tremendous contributions to the power-generation industry conveniently comparing from traditional fossil energy productions. Naturally there has been growing concern over the potential dangers associated with the accumulation of greenhouse gases in the atmosphere due to rapidly increasing global conventional energy usages in the building and houses traditionally [5,18]. Besides, meeting domestic water supply from reservoir water extraction also causing severe global warming and the management of city water is costly and highly technical oriented. Similarly, traditional way of wastewater management is costly and creates serious environmental pollution, causing harm to human health and damaging fauna in aquatic environment. Simply, conventional sources of energy, water, and gas supply for any building and house are indeed creating serious threat to the environment and climate. Therefore deploying sustainable technology is indeed to alleviate environmental and climate change. Hence a better technology is required for building and house development to confirm a cleaner, greener environment. Interestingly, Green Science, the combined technology described previously, could be a cutting-edge technology to solve the total demand of energy, water, and gas for a building or house without any outsource connection because it can produce itself all energy, water, and gas with this sustainable technology which involves an innovative design for all houses and buildings with at least 25% of exterior skin and 25% of the roof having the capacity to receive PV energy by sophisticated solar panels to meet 100% of its energy demand. Naturally in situ groundwater pumping and treating for drinking water shall be the most promising application of green science for building which can play a significant role to mitigate climate change. Subsequently, utilization of environmental pollutants (human feces and domestic waste) which shall be transformed into biogas is indeed an innovative technology to meet a buildings' total gas needed for cooking and gas-fired HVAC equipment. Finally, a detailed cost comparison for these two scenarios (conventional and independent building technology) and the results show that the application of independent building technology is much cheaper and offers better technical advantages. These results are not only promising but also very much calculative that it could encourage harvesting solar energy, in situ groundwater pumping, and biogas producing to its maximum utilization globally to confirm an independent building technology that meets all these three vital needs by self-generating technology, which indeed would be the most interesting technology in the whole world and also an environmentally friendly one.

4.3 ENERGY MODELING TO COOL AND HEAT THE BUILDING NATURALLY

Photon particle can be modeled to be decoded by the curtain wall of the building to create a natural cooling and heating system for a building. For this, Bose—Einstein *(B-E)* photon distribution mechanism into helium-assisted building curtain wall can be implemented where the photonic bandgap (PBG)-state photon will be locally induced to cool the building naturally by the cooling state of the photons. This cooling-state photon is denominated as *Hossain Cooling Photon* (HcP^-). Once needed, this (HcP^-) can be transformed into thermal state photon by implementing the quantum Higgs Boson BR ($H \rightarrow \gamma\gamma^-$) to create the electromagnetic field (EmF) using two-diode thermal semiconductors. Just because Higgs boson BR ($H \rightarrow \gamma\gamma$) quantum field has extremely short range of weak force that initiates Higgs Boson BR ($H \rightarrow \gamma\gamma$), the quantum gets excited. Therefore electrically charged cooling-state photon (HcP^-) will be transformed into a thermal stage photon, referred to here as the *Hossain Thermal Photon* (HtP^-). To confirm this HcP^- formation and its transformation into HtP^-, a series of mathematical tests has been performed which revealed that formation and transformation of HcP^- and HtP^- are indeed doable to decode the photons into the curtain wall to cool and heat the building naturally.

4.3.1 Background and Technology

Conventional heating and cooling systems installed in buildings are causing serious environmental and atmospheric problems. Traditional heating technology consumes fossil fuels and releases CO_2, a major contributor to climate change. Excessive CO_2 emissions threaten the environment and can potentially trigger catastrophic natural disasters. Conventional cooling technologies also release chlorofluorocarbons, which create holes in the ozone layer. The ozone layer, which lies between 9.3 and 18.6 miles (between 15 and 30 km) above the Earth's surface, is a protective blanket layer that blocks most of the sun's high-frequency ultraviolet rays that induce skin cancer in humans and serious reproductive problems in all mammals [68]. Although clean-energy technology, climate change, and conventional heating/cooling systems have been extensively researched, natural cooling and heating systems in the building sector are rarely reported. Therefore in this study, I propose a natural cooling and heating technology based on the Bose—Einstein photon distribution mechanism and the quantum activation of Higgs bosons. The proposed system decodes photons (solar energy) into cooling and heating states. Cooling photon emission panels consisting of nano-breakpoint and waveguides, which install helium in a portion of the exterior curtain wall, have been already proposed. QED waveguides naturally cool the photons emitted by the sun using photon band edges (PBEs). Mediated by two-diode semiconductors, the cooling-state photon can then be transformed into a heating-state photon through the Bremsstrahlung radiation (BR) emitted by quantum Higgs bosons ($H \rightarrow \gamma\gamma^{-}$), creating an EmF that naturally heats the building. This cooling and heating transformation process is a novel approach that mitigates unsustainable energy usage, damage to the environment, and depletion of the ozone layer.

4.3.2 Methods and Simulation

4.3.2.1 Cooling Mechanism

Activated photons can be decoded into the cooling state in photon emission networks of nano-breakpoint, waveguides, and helium-assisted curtain walls, which create point defects in the photon emission panel. In the same way, arrays of PBG waveguide defects can be incorporated into curtain walls [69]. Such point defects and PBG waveguides decode the quantum dynamics of photons under helium-cooling conditions, providing a means of cooling solar photons. The present study calculates the formation of the cooling state by the conversion of solar photons in MATLAB (v. 9.0) software [66,70]. In these calculations, helium waveguides embedded in the curtain wall are treated as photon reservoirs. The electrodynamics of the cooling-state photons can be expressed by the following Hamiltonian:

$$H = \sum \omega_{ci} a_i^{\dagger} a_i + \sum_k \omega_k \, b_k^{\dagger} b_k + \sum_{ik} V_{ik} a^{\dagger} b_k + V_{ik}^{*} b_k^{\dagger} a_i, \tag{4.29}$$

where $a_i\left(a_i^{\dagger}\right)$ and $b_k\left(b_k^{\dagger}\right)$ represent the drivers of the nano-breakpoint modes and the photodynamic modes of the photon nanostructure,

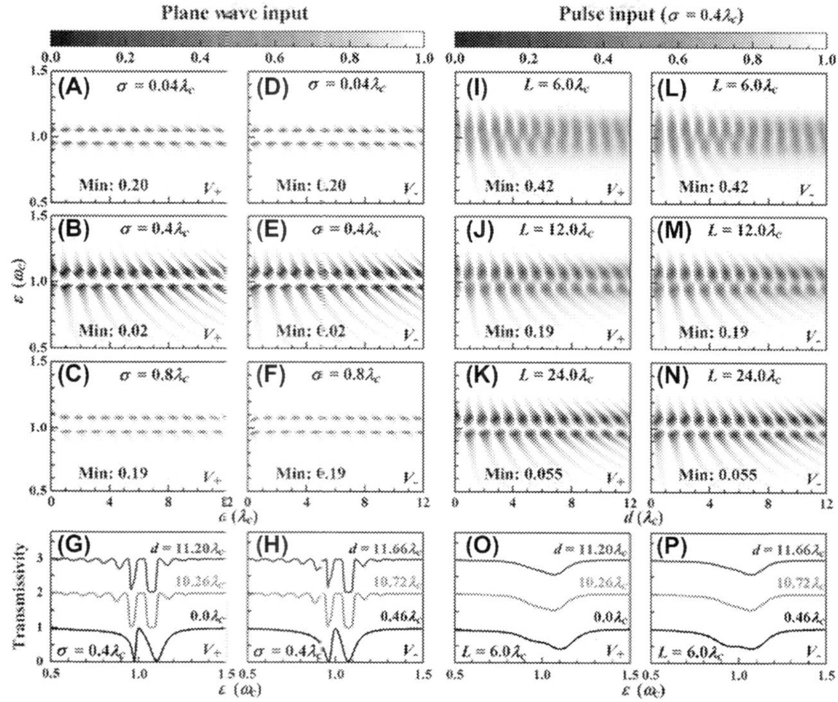

Figure 4.22 Contour maps of the transmissivity of a single photon (A–F) plane wave and (I–N) pulse as functions of *d* and ε. Transmission spectra of the (G and H) plane wave and (O and P) pulse for different values of *d* (offset in steps of 1).

respectively, and the coefficients V_{ik} represent the magnitudes of the photonic modes among the nano-breakpoints and photon nanostructures. Fig. 4.1 displays the transmissivity contours and spectra of photon plane waves and pulses (Fig. 4.22).

The proposed photon module, which generates HcP^{-} to cool a building, comprises helium-assisted point breaks, two diodes, and two resistors (Fig 4.23). The current in the module can be converted from the photon energy.

The current–voltage (I–V) characteristics of the photon cells in the single-diode mode can be given by

$$I = I_L - I_O\left\{\exp\left[\frac{q(V+I_{rs})}{AkT_c}\right] - 1\right\} - \frac{(V+I_{rs})}{R_s}, \qquad (4.30)$$

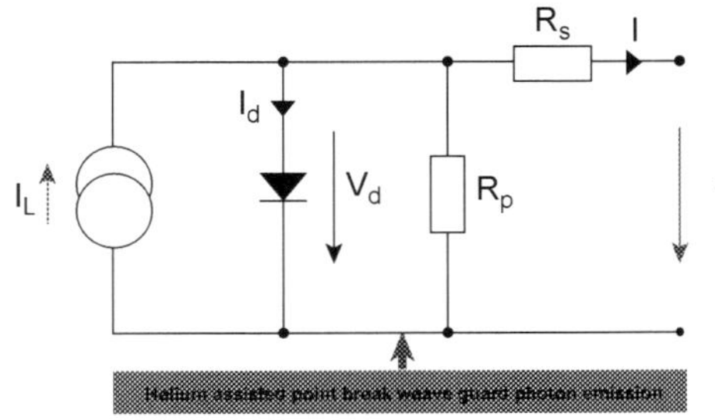

Figure 4.23 Diagram of the two-diode model of the solar irradiance receptor. Electrons are cooled in the curtain wall skin by photon induction aided by helium-assisted point breaks.

where I_L represents the photon generating current, I_O represents the saturated current in the diode, and R_s is the series resistance. A represents the passive function of the diode, k ($=1.38 \times 10^{-23}$ W/m^{2}K) is the Boltzmann's constant, q ($=1.6 \times 10^{-19}$C) is the magnitude of the charge of an electron, and T_c represents the functional cell temperature. Subsequently, the $I-q$ relationship in the photon cells varies with the diode and/or saturation current, which can be given by

$$I_O = I_{rs}\left(\frac{T_c}{T_{\text{ref}}}\right)^3 \exp\left[\frac{qEG\left(\frac{1}{T_{\text{ref}}} - \frac{1}{T_c}\right)}{KA}\right] \tag{4.31}$$

In Eq. (4.31), I_{rs} represents the saturation current, which depends on the functional temperature and solar irradiance speed, and qEG represents the bandgap energy of the electrons per unit area of the photon cell. The $I-V$ characteristics of various two-diode models are shown in Fig. 4.24.

In the photon module, the $I-V$ equation integrates the $I-V$ curves of all cells in the photon emission panel. The $V-R$ relationship in the module is given by

$$V = -IR_s + K\log\left[\frac{I_L - I + I_O}{I_O}\right], \tag{4.32}$$

where K is a constant $\left(=\frac{AkT}{q}\right)$, and I_{mo} and V_{mo} are the current and voltage in the PV panel, respectively. Therefore the relationship between I_{mo} and V_{mo} is the same as the $I-V$ relationship in the PV cell:

$$V_{mo} = -I_{mo}R_{smo} + K_{mo}\log\left(\frac{I_{Lmo} - I_{mo} + I_{Omo}}{I_{Omo}}\right), \tag{4.33}$$

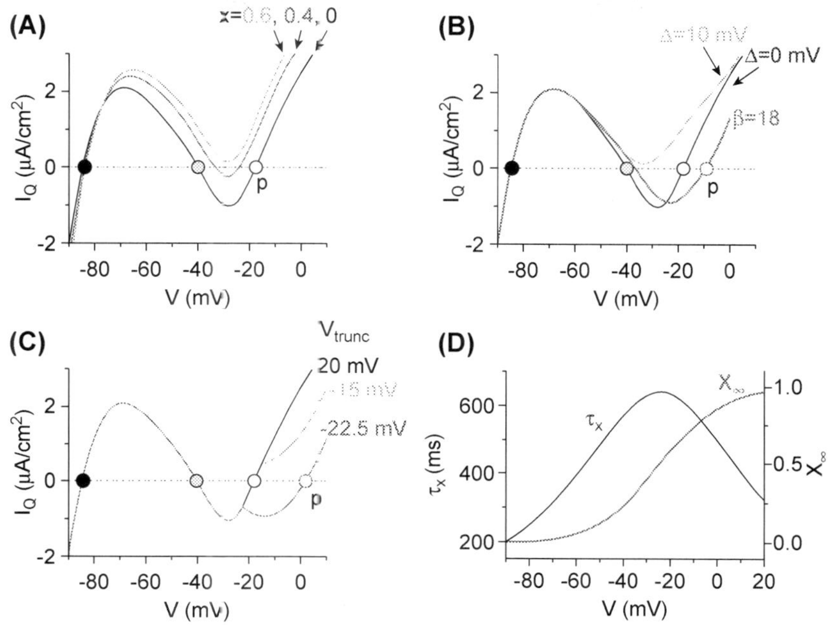

Figure 4.24 I—V characteristics of the two-diode cooling mechanism: (A) curtain wall skin cell in the normal state, (B) curtain wall skin cell under the normalized condition, (C) curtain wall skin cell module in the normal state, and (D) curtain wall skin cell module under the normalized condition.

where I_{Lmo} represents the photon-generated current, I_{Omo} represents the saturated current in the diode, R_{smo} is the series resistance, and K_{mo} is a constant. When the resistances of all nonseries (NS) cells are connected in series, the total resistance is $R_{smo} = N_s \times R_s$, and the constant is $K_{mo} = N_s \times K$. The current flowing into the series of connected cells is the same in each component, i.e., $I_{Omo} = I_O$ and $I_{Lmo} = I_L$. Thus, the $I_{mo}-V_{mo}$ relationship in the N_s connected cells is given by

$$V_{mo} = -I_{mo}N_sR_s + N_sK\log\left(\frac{I_L - I_{mo} + I_O}{I_O}\right). \tag{4.34}$$

Similarly, when all N_p cells are connected in parallel, the $I_{mo}-V_{mo}$ relationship is given by

$$V_{mo} = -I_{mo}\frac{R_s}{N_p} + K\log\left(\frac{N_{sh}I_L - I_{mo} + N_pI_O}{N_pI_O}\right), \tag{4.35}$$

Because the photon-generated current depends primarily on the solar irradiance and relativistic temperature conditions of the photon emission panel, the current can be calculated as follows:

$$I_L = G[I_{sc} + K_I(T_{\text{cool}})]V_{mo} \tag{4.36}$$

$$T_{\text{cool}} = \left(\frac{I_L}{(G^* V_{mo})x(I_{sc} + K_I)}\right),$$

where I_{sc} is the photon current per unit area at 25°C, K_I denotes the relativistic photon panel coefficient, T_{cool} represents the cooling temperature of the photon cell, and G represents the solar energy per unit area.

4.3.2.2 Heating Mechanism

To convert cooling photons to heating photons, researchers have exploited the Higgs boson EmF and determined the accuracy and parameters of the photon-heating relationship. Thus to create a local Higgs quantum field in the curtain wall skin, I simulated Abelian local symmetries in MATLAB 9.0 software. The penetration of solar irradiance breaks the gauge field symmetry, and the Goldstone scalar particle becomes the longitudinal mode of the vector boson. In the Abelian case the local symmetry of each spontaneously broken particle T^α is the corresponding gauge field of $A^\alpha_\mu(x)$. The Higgs quantum field begins to operate in local U(1) phase symmetry. Thus the model can comprise a complex scalar field $\Phi(x)$ of electric charge q coupled to the EM field $A^\mu(x)$ which can be expressed by the following Lagrangian function:

$$\mathrm{L} = -\frac{1}{4}F_{\mu\nu}F^{\mu\nu} + D_\mu\Phi^* D^\mu\Phi - V(\Phi^*\Phi), \tag{4.37}$$

where

$$D_\mu\Phi(x) = \partial_\mu\Phi(x) + iqA_\mu(x)\Phi(x)$$

$$D_\mu\Phi^*(x) = \partial_\mu\Phi^*(x) - iqA_\mu(x)\Phi^*(x), \tag{4.38}$$

and

$$V(\Phi^*\Phi) = \frac{\lambda}{2}(\Phi^*\Phi)^2 + m^2(\Phi^*\Phi). \tag{4.39}$$

Suppose that $\lambda > 0$ but $m^2 < 0$ so that $\Phi = 0$ is a local maximum of the scalar potential, the minima form a degenerate circle $\Phi = \frac{v}{\sqrt{2}} * e^{i\theta}$ with

$$v = \sqrt{\frac{-2m^2}{\lambda}} \quad \text{for any real } \theta \tag{4.40}$$

Consequently, the scalar field Φ develops a nonzero vacuum expectation value $\Phi \neq 0$, which spontaneously creates the U(1) symmetry of the magnetic field. Breakdown of this symmetry creates a massless Goldstone scalar from the phase of the complex field $\Phi(x)$. However, in local U(1) symmetry, the phase of $\Phi(x)$ is the x-dependent phase of the dynamic $\Phi(x)$ field rather than the phase of the expectation value Φ.

To confirm this mechanism, I express the scalar field space in polar coordinates:

$$\Phi(x) = \frac{1}{\sqrt{2}}\Phi_r(x)^* e^{i\Theta(x)}, \quad \text{real } \Phi_r(x) > 0, \quad \text{real } \Phi(x) \tag{4.41}$$

As the field in this formulation is singular at $\Phi(x) = 0$, it is inapplicable to theories with $\Phi \neq 0$ but is adequate for spontaneously broken theories where $\Phi x \neq 0$ is expected almost everywhere. In terms of the real fields $\phi_r(x)$ and $\Theta(x)$, the scalar potential depends only on the radial field ϕ_r,

$$V(\phi) = \frac{\lambda}{8}\left(\phi_r^2 - v^2\right)^2 + \text{const.} \tag{4.42}$$

If the radial field shifted by a variable scalar, $\Phi_r(x) = v + \sigma(x)$, we have

$$\phi_r^2 - v^2 = (v+\sigma)^2 - v^2 = 2v\sigma + \sigma^2 \tag{4.43}$$

$$V = \frac{\lambda}{8}\left(2v\sigma - \sigma^2\right)^2 = \frac{\lambda v^2}{2} * \sigma^2 + \frac{\lambda v}{2} * \sigma^3 + \frac{\lambda}{8} * \sigma^4. \tag{4.44}$$

Meanwhile, the covariant derivative $D_\mu\phi$ becomes

$$\begin{aligned} D_\mu\phi &= \frac{1}{\sqrt{2}}\left(\partial_\mu\left(\varphi_r e^{i\Theta}\right) + iqA_\mu * \phi_r e^{i\Theta}\right) \\ &= \frac{e^{i\Theta}}{\sqrt{2}}\left(\partial_\mu\phi_r + \phi_r * i\partial_\mu\Theta + \phi_r * iqA_\mu\right) \end{aligned} \tag{4.45}$$

$$\left|D_\mu\phi\right|^2 = \frac{1}{2}\left|\partial_\mu\phi_r + \phi_r * i\partial_\mu\Theta + \phi_r * iqA_\mu\right|^2 \tag{4.46}$$

$$\begin{aligned} &= \frac{1}{2}\left(\partial_\mu\phi_r\right) + \frac{\phi_r^2}{2} * \left(\partial_\mu\Theta qA_\mu\right)^2 \\ &= \frac{1}{2}\left(\partial_\mu\sigma\right)^2 + \frac{(v+\sigma)^2}{2} * \left(\partial_\mu\Theta + qA_\mu\right)^2 \end{aligned} \tag{4.47}$$

The Lagrangian can then be given by

$$\mathrm{L}_{\text{heat}} = \frac{1}{2}\left(\partial_\mu \sigma\right)^2 - v(\sigma) - \frac{1}{4}F_{\mu\nu}F^{\mu\nu} + \frac{(v+\sigma)^2}{2} * \left(\partial_\mu \Theta + qA_\mu\right)^2. \tag{4.48}$$

To incorporate the heating (L_{heat}) into the magnetic field properties of this Lagrangian, L_{heat} can be expanded as a power series of the fields (and their derivatives) and extract the quadratic part describing the free particles:

$$\mathrm{L}_{\text{heat}} = \frac{1}{2}\left(\partial_\mu \sigma\right)^2 - \frac{\lambda v^2}{2} * \sigma^2 - \frac{1}{4}F_{\mu\nu}F^{\mu\nu} + \frac{v^2}{2} * \left(qA_\mu + \partial_\mu \Theta\right)^2. \tag{4.49}$$

Obviously, to initiate high heating within the quantum field of the curtain wall, the free particles (with Lagrangian L_{free}) must be real scalar particles with positive $m^2 = \lambda v^2$ (where m denotes the particle mass; Fig. 4.25).

4.3.3 Results and Discussion

4.3.3.1 Cooling Mechanism

To mathematically demonstrate the formation of cooling photons by the helium-assisted curtain wall skin, I determined the dynamic photon proliferation by integrating Eqs. (4.43) and (4.44). Owing to the cooling unit areal

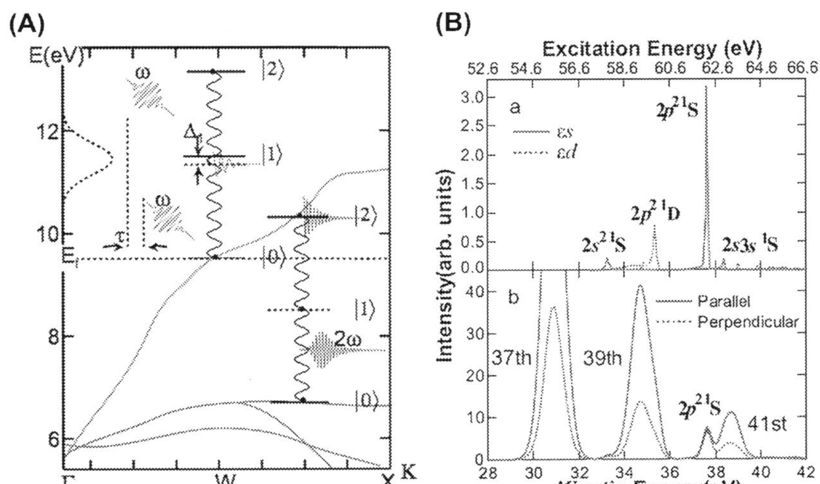

Figure 4.25 Transformation mechanism of a photon from energy level ω to 2ω (eV) in the two-diode feed semiconductors (A) and quantum field intensity spectra of the electrons (B). The photon excitation energy (eV) is transformed into kinetic energy (eV) by conversion of the heating state of photons.

condition $J(\omega)$ and the persistent weak-coupling limit, the curtain wall skin is expected to proliferate photons. Here $J(\omega)$ is the quantum field area defining the density of states (DOS) field produced in the PV cell by the fine cooling photonic magnitude $V(\omega)$ within the photonic band (PB) and the PV cell. Moreover, photon production should follow the Weisskopf–Winger approximation and/or the Markovian master equation. Consequently, all of the proliferated *HcPs* will possess a dynamic state mode (1D, 2D, and 3D) in the curtain wall skin, as described in Table 4.5.

In the 3D curtain wall skin, Ω_C is a fine frequency cutoff that avoids bifurcation of the DOS. Similarly, the 1D and 2D curtain wall skins require a sharp frequency cutoff at Ω_d to avoid negative DOSs (Fig. 4.26). Hence $Li_2(x)$ and $e_{rfc}(x)$ are di-logarithmic and additive variables, respectively. Subsequently, the DOS of various curtain wall skins, denoted as $\varrho_{PC}(\omega)$, is determined by calculating the photon eigenfrequencies and eigenfunctions of Maxwell's rules in the nanostructures. In a 1D curtain wall skin, the DOS is given by $\varrho_{PC}(\omega) \propto \frac{1}{\sqrt{\omega-\omega_e}}\Theta(\omega - \omega_e)$, where $\Theta(\omega - \omega_e)$ is the Heaviside step function and ω_e represents the frequency of the PBE at the given DOS.

The DOS is required for accurately predicting the qualitative state of the non–Weisskopf–Winger mode and the photon-cooling state of the photon cell in a 3D isotropic analysis in the curtain wall skin. The DOS and

Table 4.5 Photonic structures of density of states (DOS) in different dimensional modes of the curtain wall skin

Photon	Unit area $J(\omega)$ for different DOS[a]	Reservoir-induced self-energy correction $\Sigma(\omega)$[a]
1D	$\frac{C}{\pi}\frac{1}{\sqrt{\omega-\omega_e}}\Theta(\omega-\omega_e)$	$-\frac{C}{\sqrt{\omega_e-\omega}}$
2D	$-\eta\left[\ln\left\lvert\frac{\omega-\omega_0}{\omega_0}\right\rvert-1\right]\Theta(\omega-\omega_e)\Theta(\Omega_d-\omega)$	$\eta\left[Li_2\left(\frac{\Omega_d-\omega_0}{\omega-\omega_0}\right)-Li_2\left(\frac{\omega_0-\omega_e}{\omega_0-\omega}\right)-\ln\frac{\omega_0-\omega_e}{\Omega_d-\omega_0}\ln\frac{\omega_e-\omega}{\omega_0-\omega}\right]$
3D	$\chi\sqrt{\frac{\omega-\omega_e}{\Omega_C}}\exp\left(-\frac{\omega-\omega_e}{\Omega_C}\right)\Theta(\omega-\omega_e)$	$\chi\left[\pi\sqrt{\frac{\omega_e-\omega}{\Omega_C}}\exp\left(-\frac{\omega-\omega_e}{\Omega_C}\right)\operatorname{erfc}\sqrt{\frac{\omega_e-\omega}{\Omega_C}}-\sqrt{\pi}\right]$

[a]The unit area $J(\omega)$ and self-energy induction in the reservoir $\Sigma(\omega)$, determined by the photon dynamics into the extreme relativistic curtain wall skin, differ among the structures. The variables C, η, and χ function as coupled forces between the point break and PV in the curtain wall skin in 1, 2, and 3 dimensions.

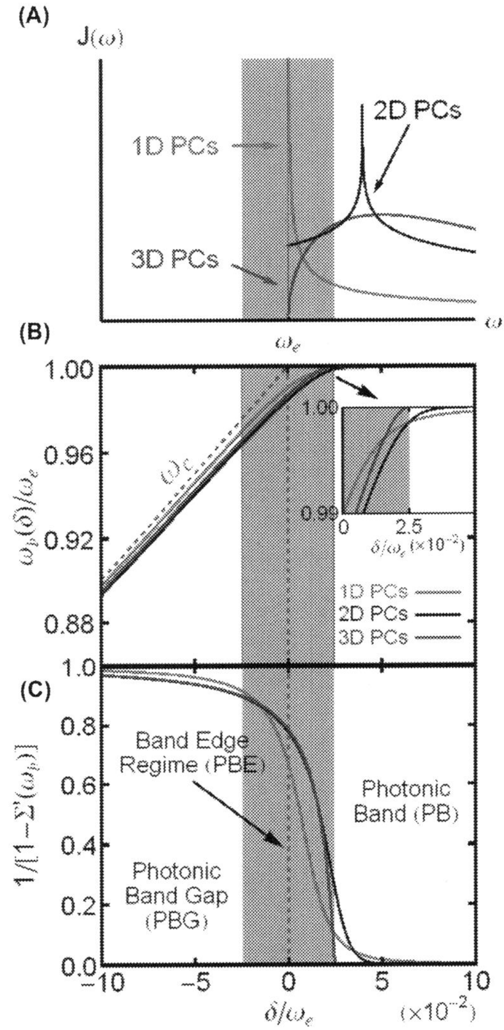

Figure 4.26 Photonic band structure and energy conversion modes. (A) Unit area versus frequency at various DOSs in 1D, 2D, and 3D curtain wall skins. (B) Photonic modes of frequencies for functional tuning. (C) Photonic modes of magnitudes to release the energy calculated by Eq. (4.30). The photonic modes depict the crossover into the curtain wall skin in 1D and 2D. In the complex 3D transitional state, they depict the crossover into the PV cell once the point break frequency "vc" transforms from a PBG area to a photonic band (PB) area [34,70].

projected DOS (PDOS) are displayed in Fig. 4.27. In a 3D curtain wall skin the DOS close to the PBE is anisotropic and given by $\varrho_{PC}(\omega) \propto \frac{1}{\sqrt{\omega-\omega_e}} \Theta(\omega - \omega_e)$. This DOS is then clarified with respect to the EmF vector [2,17,26,30]. In 2D and 1D curtain wall skins, the cooling photon DOS exhibits a pure logarithmic divergence close to the PBE, which is approximated as $\varrho_{PC}(\omega) \propto -[\ln|(\omega - \omega_0)/\omega_0| - 1]\Theta(\omega - \omega_e)$, where ω_e represents the central point of the peak in the DOS distribution.

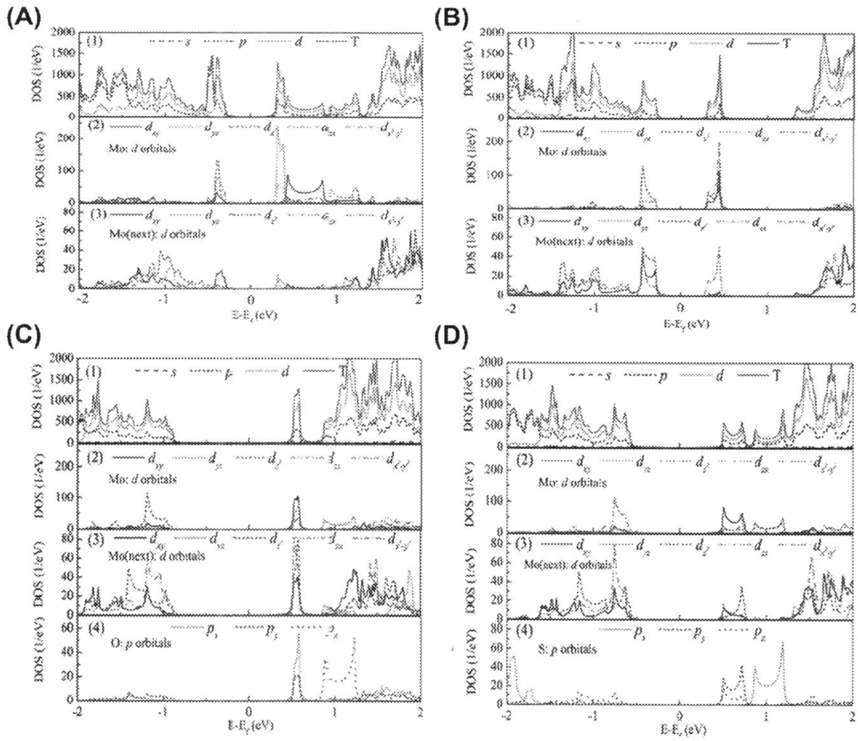

Figure 4.27 Total density of states (DOS) and projected density of states (PDOS) of decoded photons for transformation into the cooling state: (Panel A) [4] total DOS (T) and DOS projected onto the *s*, *p*, and d orbitals [5], PDOS of d orbitals on the fourth level of Mo atoms and [1] PDOS of d orbitals on the third level of Mo atoms; (Panel B) as in Panel (A) but for projected DOS of Mo atoms; (Panel C) [1,4,5] as in Panel (A) and [7] PDOS of *p* orbitals of O atoms. (Panel D) [1,4,5] as in Panel (B) and [7] PDOS of *p* orbitals of external S atoms.

As mentioned previously, $J(\omega)$ defines the DOS field produced in the PV cell by the fine cooling photonic magnitude $V(\omega)$ within the PB and the PV cell:

$$J(\omega) = \varrho(\omega)|V(\omega)|^2. \tag{4.50}$$

Hereafter, I consider the PB frequency ω_c and the proliferative photon dynamics $\langle a(t)\rangle = u(t, t_0)\langle a(t_0)\rangle$, where the function $u(t, t_0)$ describes the photon structure. $u(t, t_0)$ can be calculated using the dissipative integral–differential equation given in Eq. (4.47), that is,

$$u(t, t_0) = \frac{1}{1 - \Sigma'(\omega_b)} e^{-i\omega(t-t_0)} + \int_{\omega_e}^{\infty} d\omega \frac{J(\omega)e^{-i\omega(t-t_0)}}{[\omega - \omega_c - \Delta(\omega)]^2 + \pi^2 J^2(\omega)}, \tag{4.51}$$

where $\Sigma'(\omega_b) = [\partial\Sigma(\omega)/\partial\omega]_{\omega=\omega_b}$ and $\Sigma(\omega)$ represents the PB photon self-energy correction induced in the reservoir,

$$\Sigma(\omega) = \int_{\omega_e}^{\infty} d\omega' \frac{J(\omega')}{\omega - \omega'}. \tag{4.52}$$

Here the frequency ω_b in Eq. (4.30) represents the cooling photonic frequency mode in the PBG ($0 < \omega_b < \omega_e$), calculated under the pole condition $\omega_b - \omega_c - n(\omega_b) = 0$, where $\lesssim\Delta(\omega) = \mathrm{P}\left[\int d\omega' \frac{J(\omega')}{\omega - \omega'}\right]$ is a principal-value integral.

Fig. 4.28A plots the cooling photonic dynamics of the proliferation magnitude $|u(t, t_0)|$, calculated in 1D, 2D, and 3D photon cells for various values of the detuning parameter δ and integrated from the PBG area to the PB area. The cooling photonic dynamic rates $\kappa(t)$ are plotted in Fig. 4.28B. The results indicate that once ω_c has crossed from the PBG to the PB area, dynamic photons are produced at a high rate. Because the $u(t, t_0)$ range is $1 \geq |u(t, t_0)| \geq 0$, I have defined the crossover area to satisfy $0.9 \gtrsim |u(t \rightarrow \infty, t_0)| \geq 0$. This corresponds to $-0.025\omega_e \lesssim \delta \lesssim 0.025\omega_e$, with a cooling photon induction rate $\kappa(t)$ within the PBG ($\delta < -0.025\omega_e$) and near the PBE($-0.025\omega_e \lesssim \delta \lesssim 0.025\omega_e$).

More specifically, the PB can be considered as the Fock cooling determination n_0, *i.e.*, $\rho(t_0) = |n_0 n_0|$, which is obtained theoretically through real-time quantum feedback control and then by solving Eq. (4.1), considering the state of cooling photon induction at time t:

$$\rho(t) = \sum_{n=0}^{\infty} \mathrm{P}_n^{(n_0)}(t)|n_0 n_0| \tag{4.53}$$

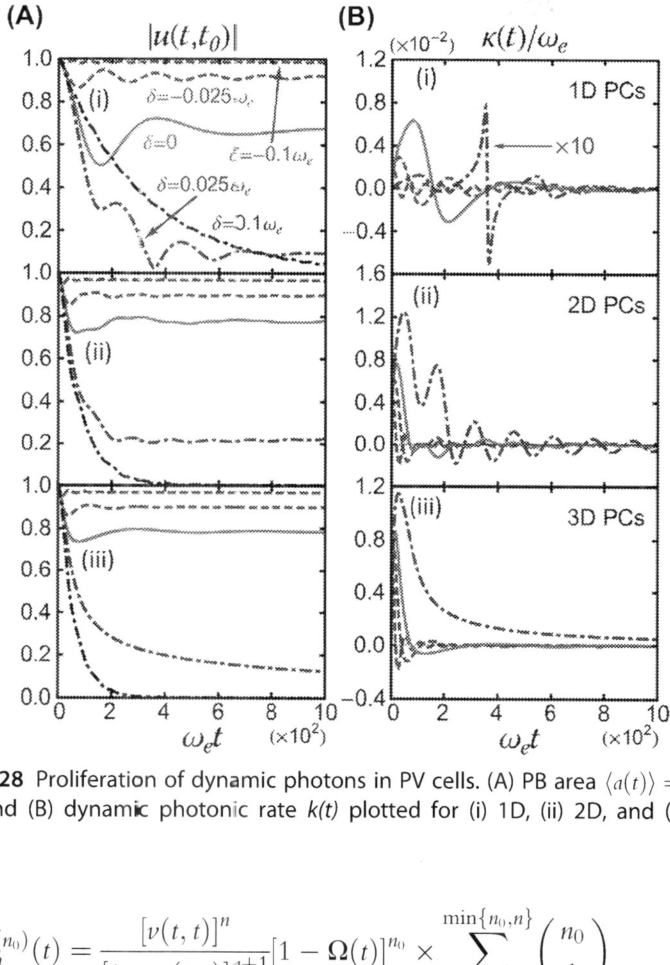

Figure 4.28 Proliferation of dynamic photons in PV cells. (A) PB area $\langle a(t)\rangle = 5\, u(t, t_0)$ $\langle a(t_0)\rangle$ and (B) dynamic photonic rate *k(t)* plotted for (i) 1D, (ii) 2D, and (iii) 3D PV cells [17].

$$P_n^{(n_0)}(t) = \frac{[v(t,t)]^n}{[1+v(t,t)]^{n+1}}[1-\Omega(t)]^{n_0} \times \sum_{k=0}^{\min\{n_0,n\}} \binom{n_0}{k} \times \binom{n}{k}\left[\frac{1}{v(t,t)}\frac{\Omega(t)}{1-\Omega(t)}\right]^k, \tag{4.54}$$

where $\Omega(t) = \frac{|u(t,t_0)|^2}{1+v(t,t)}$. This result suggests that a Fock state cooling photon is induced into dynamic states $P_n^{(n_0)}(t)$of$|n_0\rangle$. In fact, Fig. 4.7 plots the proliferation of photon dissipation $P_n^{(n_0)}(t)$ in the primary state $|n_0 = 5\rangle$ and in the steady-state limit $P_n^{(n_0)}(t\rightarrow\infty)$. Therefore the proliferation of the produced cooling photons will ultimately reach a nonequilibrium cooling state that cools the building.

4.3.3.2 Heating Mechanism

In the proposed system, two semiconductors use the EmF created by the Higgs boson quantum field. Therefore local U(1) gauge-invariant QED allows an additional mass term for the gauge particle under $\varnothing' \rightarrow e^{i\alpha(x)}\varnothing$; that is, the cooling photons can be transformed into heating photons. This mechanism can be explained by a covariant derivative with a special transformation rule for the scalar field, which can be given by

$$\partial_\mu \rightarrow D_\mu = \partial_\mu = ieA_\mu \quad \text{[covariant derivatives]} \tag{4.55}$$

$$A'_\mu = A_\mu + \frac{1}{e}\ \partial_\mu\alpha \quad \left[A_\mu\ \text{derivatives}\right], \tag{4.56}$$

where the local U(1) gauge-invariant Lagrangian for a complex scalar field can be given by

$$\mathrm{L} = (D^\mu)^\dagger\left(D_\mu\varnothing\right) - \frac{1}{4}F_{\mu\nu}F^{\mu\nu} - V(\varnothing) \tag{4.57}$$

The term $\frac{1}{4}F_{\mu\nu}F^{\mu\nu}$ is the kinetic term in the gauge field (heating photons) and $V(\varnothing)$ is an additional term expressed as

$$V(\varnothing^*\varnothing) = \mu^2(\varnothing^*\varnothing) + \lambda(\varnothing^*\varnothing)^2. \tag{4.58}$$

According to the Lagrangian L, perturbations in the quantum field initiate the production of massive scalar particles ϕ_1 and ϕ_2 and a mass μ. In this situation, $\mu^2 < 0$ admits an infinite number of quanta, each satisfying $\phi_1^2 + \phi_2^2 = -\mu^2/\lambda = v^2$.In terms of the shifted fields η and ξ, the quantum field can be defined as $\phi_0 = \frac{1}{\sqrt{2}}[(v + \eta) + i\xi]$, and the covariant derivatives of the Lagrangian become

$$\text{Kinetic term: } \mathrm{L}_{\text{kin}}(\eta, \xi) = (D^\mu\phi)^\dagger(D^\mu\phi) \tag{4.59}$$

$$= (\partial^\mu + ieA^\mu)\phi^*\ \left(\partial_\mu - ieA_\mu\right)\phi \tag{4.60}$$

Potential term (to second order) $V(\eta, \xi) = \lambda v^2\eta^2$. Thus the full Lagrangian can be written as

$$\mathrm{L}_{\text{kin}}(\eta, \xi) = \frac{1}{2}\left(\partial_\mu\eta\right)^2 - \lambda v^2\eta^2 + \frac{1}{2}\left(\partial_\mu\xi\right)^2 - \frac{1}{4}F_{\mu\nu}F^{\mu\nu} + \frac{1}{2}e^2v^2A_\mu^2 - evA_\mu(\partial^\mu\xi) + \text{int.terms.} \tag{4.61}$$

Here, η is massive, ξ is massless (as before), μ is the mass term for the quantum, and A_μ is fixed up to a term $\partial_\mu\alpha$, as is evident in Eq. (4.57). In general, A_μ and ϕ change simultaneously, so Eq. (4.60) can be redefined to accommodate the heating photon particle spectrum within the quantum field:

$$L_{\text{scalar}} = (D^{\mu}\phi)^{\dagger}(D^{\mu}\phi) - V(\phi^{\dagger}\phi)$$

$$= (\partial^{\mu} + ieA^{\mu})\frac{1}{\sqrt{2}}(v+h)(\partial_{\mu} - ieA_{\mu})\frac{1}{\sqrt{2}}(v+h) - V(\phi^{\dagger}\phi) \tag{4.62}$$

$$= \frac{1}{2}(\partial_{\mu}h)^2 + \frac{1}{2}e^2A_{\mu}^2(v+h)^2 - \lambda v^2h^2 - \lambda vh^3 - \frac{1}{4}\lambda h^4 + \frac{1}{4}\lambda h^4 \tag{4.63}$$

The expanded term in the Lagrangian of the scalar field suggests that the Higgs boson quantum field can initiate heating photons.

To confirm this heating photon transformation, I calculated the isotropic distribution of movement on the differential cone with respect to the angle θ from the vertical axis. The differential between θ and $\theta + d\theta$ is ½sin$\theta d\theta$. The differential photon density at energy $\in$ and angle θ can be then given by

$$dn = \frac{1}{2}n(\in)sin\theta d\in d\theta. \tag{4.64}$$

Consequently, the functional speeds of the high-energy photons were calculated as $c\ (1 - cos\theta)$, and the absorption per unit path length is

$$\frac{d\tau_{\text{abs}}}{dx} = \int\int \frac{1}{2}\sigma n(\in)(1 - cos\theta)sin\theta d\in d\theta. \tag{4.65}$$

Reexpressing these functions as integrals over s instead of θ, by Eqs. (63) and (65), the following can be obtained,

$$\frac{d\tau_{\text{abs}}}{dx} = \pi r_0^2\left(\frac{m^2c^4}{E}\right)^2\int_{\frac{m^2c^4}{E}}^{\infty} \in^{-2} n(\in)\overline{\varphi}[s_0(\in)]de, \tag{4.66}$$

where

$$\overline{\varphi}[s_0(\in)] = \int_1^{s_0(\in)} s\overline{\sigma}(s)ds, \overline{\sigma}(s) = \frac{2\sigma(s)}{\pi r_0^2}. \tag{4.67}$$

This result defines the dimensional variable $\overline{\phi}$ and dimensionless cross section $\overline{\sigma}$. The variable $\overline{\phi}[s_0]$ can be calculated based on a detailed graphical frame for $1 < s_0 < 10$. $\overline{\phi}$ can be calculated by a functional asymptotic calculation

$$\overline{\varphi}[s_0] = \frac{1+\beta_0^2}{1-\beta_0^2}\ln\omega_0 - \beta_0^2\ln\omega_0 - \ln^2\omega_0 - \frac{4\beta_0}{1-\beta_0^2} + 2\beta_0$$
$$+ 4\ln\omega_0\ln(\omega_0+1) - \mathrm{L}(\omega_0), \tag{4.68}$$

where $s_0 - 1 \ll 1$ or $s_0 \gg 1$,

$$\beta_0^2 = \frac{1-1}{s_0}, \quad \omega_0 = \frac{(1+\beta_0)}{(1-\beta_0)}, \tag{4.69}$$

and

$$L(\omega_0) = \int_1^{\omega_0} \omega^{-1}\ln(\omega+1)d\omega.$$

The last integral can be written as

$$(\omega+1) = \omega\left(\frac{1+1}{\omega}\right), \ L(\omega_0) = \frac{1}{2}\ln^2\omega_0 + L'(\omega_0), \tag{4.70}$$

where

$$L'(\omega_0) = \int_1^{\omega_0} \omega^{-1}\ln\left(1+\frac{1}{\omega}\right)d\omega, \tag{4.71}$$

$$= \frac{\pi^2}{12} - \sum_{n=1}^{\infty}(-1)n^{-1}n^{-2}\omega_0^{-n}. \tag{4.72}$$

This accurate representation of the heating photons readily allows the calculation of $\overline{\phi}[s_0]$ to the desired accuracy for the expected value of s_0. Thus the corrective functional asymptotic formulas may be expressed as follows:

$$\overline{\varphi}[s_0] = 2s_0(\ln 4s_0 - 2) + \ln 4s_0\,(\ln 4s_0 - 2)$$
$$- \frac{(\pi^2-9)}{3} + s_0^{-1}\left(\ln 4s_0 + \frac{9}{8}\right) + \cdots (s_0 >> 1); \tag{4.73}$$

$$\overline{\varphi}[s_0] = \left(\frac{2}{3}\right)(s_0-1)^{\frac{3}{2}} + \left(\frac{5}{3}\right)(s_0-1)^{\frac{5}{2}} - \left(\frac{1507}{420}\right)(s_0-1)^{\frac{7}{2}}$$
$$+ \cdots (s_0 - 1 << 1). \tag{4.74}$$

The function $\frac{\overline{\phi}[s_0]}{(s_0-1)}$ is shown in Fig. 4.28 for $1 < s_0 < 10$; at larger s_0, it becomes a natural logarithmic function of s_0. The power law spectrum of the heating photons is expressed in the form $n(\in) \propto \in^{m}$ for two systems in a pristine state and for a system with a graphene sheet.

Thus the light-absorption spectrum should feature a high-energy cutoff with $m > 0$. We now derive the heating photon spectrum with a high-energy cutoff.

Consider a spectrum of the form

$$n(\in) = D\in^{\beta}, \in < \in_{m}, \beta \leq 0 \tag{4.75}$$

$$= 0, \in > \in_{m}. \tag{4.76}$$

For this spectrum, following can be obtained

$$\frac{d\tau_{\mathrm{abs}}}{dx} = \pi r_0^2 D\left(\frac{m^2c^4}{E}\right)^{1+\beta} \times \begin{cases} 0, E < E_m \\ F_\beta(\sigma_m), E > E_m \end{cases}, \tag{4.77}$$

where

$$\sigma_m = \frac{E}{E_m} = \frac{\in_m E}{m^2c^4}, \tag{4.78}$$

$$F_\beta(\sigma_m) = \int_1^{\sigma m} s_0^{\beta-2}\overline{\varphi}[s_0]ds_0. \tag{4.79}$$

Again, by Eqs. (4.75) and (4.76), we can obtain the asymptotic forms

$$\beta = 0: F_\beta(\sigma_m) \to A_\beta + \ln^2\sigma_m - 4\ln\sigma_m + \cdots,$$
$$\beta \neq 0: F_\beta(\sigma_m) \to A_\beta + 2\beta^{-1}\sigma_m^{\beta}(\ln 4\sigma_m - \beta^{-1} - 2) + \cdots, \sigma_m > 10 \tag{4.80}$$

$$\beta: F_\beta(\sigma_m) \to \left(\frac{4}{15}\right)(\sigma_m - 1)^{\frac{5}{2}}\left[\frac{2(2\beta+1)}{21}\right](\sigma_m - 1)^{\frac{7}{2}} + \cdots, \sigma_m - 1 << 1. \tag{4.81}$$

Fig. 4.29 plots $\sigma_m^{-\beta}F_\beta(\sigma_m)$ for $\beta = 0{-}3.0\ A_\beta$ in $0.5 - A_\beta$ intervals, which contributes to the integral in the region [1, 10]. The values were calculated as $A_\beta = 8.111$ ($\beta = 0$), 13.53 ($\beta = 0.5$), 9.489 ($\beta = 1.0$), 15.675 ($\beta = 1.5$), 34.54 ($\beta = 2.0$), 85.29 ($\beta = 2.5$), and 222.9 ($\beta = 3.0$). Subsequently, the heating photon terms can be calculated corresponding to the spectra for both negative and positive indexes:

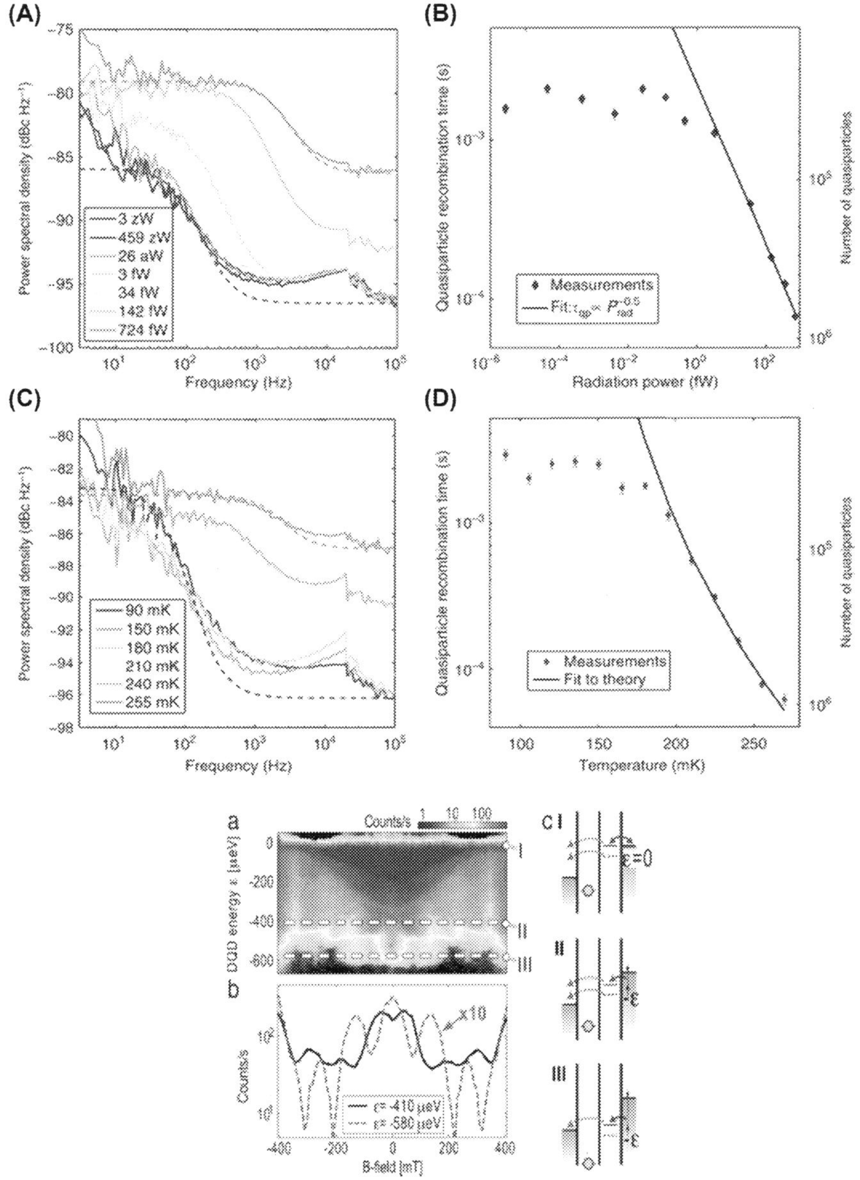

Figure 4.29 (A) Thermal photon power spectral density of the resonator amplitude as a function of frequency. (B) Quasiparticle recombination time as a function of thermal photon power obtained from the roll-off frequency in the measured spectra. (C) Thermal photon power spectral density of the resonator amplitude as a function of frequency for different bath temperatures. (D) Average heating photon energy measurements of (A, B, and C), incorporating photon excitation into the Higgs boson quantum B-field.

$$\begin{aligned} a(\in) &= 0, \in < \in_0 \\ &= C_{\in}^{-\alpha} \text{ or } D_{\in}^{\beta}, \in_0 < \in < \in_m \\ &= 0, \in > \in_m. \end{aligned} \tag{4.82}$$

then,

$$\left(\frac{d\tau_{abs}}{dx}\right)_{\alpha} = \pi r_0^2 C\left(\frac{m^2c^4}{E}\right)^{1-\alpha}$$

$$X\left\{\begin{array}{ll} 0, & E < E_m \\ (F_{\alpha}(\sigma) - F_{\alpha}(\sigma_m)), & E_m < E_0 \\ (F_{\alpha}(\sigma_0) - F_{\alpha}(\sigma_m)), & E > E_{0,} \end{array}\right\} \tag{4.83}$$

$$\left(\frac{d\tau_{abs}}{dx}\right)_{\beta} = \pi r_0^2 D\left(\frac{m^2c^4}{E}\right)^{1+\beta}$$

$$\left\{\begin{array}{ll} 0, & E < E_m \\ \left(F_{\beta}(\sigma_m)\right), & E_m < E < E_0 \\ \left(F_{\beta}(\sigma_m) - F_{\beta}(\sigma_0)\right), & E > E_{0,} \end{array}\right. \tag{4.84}$$

In this case the heating photon spectrum can be adequately described by asymptotic formulas. The term Γ_{γ}^{LPM} denotes the photon's contribution to the emitted irradiance per unit volume due to bremsstrahlung processes:

$$\Gamma_{\gamma} \equiv \frac{dn_{\gamma}}{dVdt}. \tag{4.85}$$

After summing the contributions Γ_{γ}^{LPM}, the rate of [energy] was confirmed as $O(\alpha_{EM}\ \alpha_s)$. Here the following expression can be assumed for the polarized emission rate Γ_{γ}^{LPM} at the thermodynamically controlled equilibrium of the plasma surface at temperature T and photophysical reaction μ:

$$\frac{d\Gamma_{\gamma}^{LPM}}{d^3k} = \frac{d_F q_s^2 \alpha_{EM}}{4\pi^2 k}\int_{-\infty}^{\infty}\frac{dp_{\parallel}}{2\pi}\int\frac{d^2\mathrm{p}_{\perp}}{(2\pi)^2}A\left(p_{\parallel},k\right)\mathrm{Re}\left\{2P_{\perp}\cdot f\left(\mathrm{p}_{\perp};p_{\parallel},k\right)\right\}, \tag{4.86}$$

where d_F represents the functional strategy of a quark particle [N_c in SU(N_c)], q_s represents the Abelian charge of a quark, $k \equiv |k|$, and the kinetic function $A(p_{||}, k)$ of the emitted particle can be given by

$$A\left(p_{||}, k\right) \equiv \begin{cases} \dfrac{n_b\left(k + p_{||}\right)\left[1 + n_b\left(p_{||}\right)\right]}{2p_{||}\left(p_{||} + k\right)}, & \text{scalars} \\ \dfrac{n_f\left(k + p_{||}\right)\left[1 - n_f\left(p_{||}\right)\right]}{2\left[p_{||}\left(p_{||} + k\right)\right]^2}\left[p_{||}^2 + \left(p_{||} + k\right)^2\right], & \text{fermions} \end{cases} \tag{4.87}$$

with

$$n_b(p) \equiv \frac{1}{\exp[\beta(p - \mu)] - 1}, \quad n_f(p) \equiv \frac{1}{\exp[\beta(p - \mu)] + 1}. \tag{4.88}$$

The function $f(p\perp; p||, k)$ in Eq. (4.86) may be introduced to solve the following linear integral equation, which confirms multiple photon production by three-diode scattering:

$$p_\perp = i\delta E f\left(p_\perp; p_{||}, k\right) + \frac{\pi}{2} C_F g_s^2 m_D^2 \int \frac{d^2 q_\perp}{(2\pi)^2} \frac{dq_{||}}{2\pi} \frac{dq^0}{2\pi} 2\pi\delta\left(q^0 - q_{||}\right)$$
$$\times \frac{T}{|q|}\left[\frac{2}{\left|q^2 - \Pi_L(Q)\right|^2} + \frac{\left[1 - \left(q^0/\left|q_{||}\right|\right)^2\right]^2}{\left|(q^0)^2 - q^2 - \Pi_T(Q)\right|^2}\right]\left[f\left(p_\perp; p_{||}, k\right)\right.$$
$$\left. - f\left(q + p_\perp; p_{||}, k\right)\right]. \tag{4.89}$$

In Eq. (4.89), C_F is a quadratic quark [$C_F = (N_c^2 - 1)/2N_c = 4/3$ in QCD], m_D is the leading-order Debye mass, and δE is the energy difference between quasiparticles, which considers the photon emission and the state of the thermodynamic temperature equilibrium:

$$\delta E \equiv k^0 + E_p \text{sign}\left(p_{||}\right) - E_{p+k} \text{sign}\left(p_{||} + k\right). \tag{4.90}$$

For an SU(N) gauge theory with N_s complex scalars and N_f Dirac fermions, the Debye mass in the fundamental representation can be given by

$$m_D^2 = \frac{1}{6}(2N + N_s + N_f)g^2T^2 + \frac{N_f}{2\pi^2}g^2\mu^2. \tag{4.91}$$

To accurately determine the photon energy emission rate in the region $p|| > 0$, the distribution of $n(k + p||)[1 \pm n(p||)]$ can be calculated in the integral containing $A(p||, k)$ in Eq. (4.86), which determines the distribution of pair annihilations, using the following equation:

$$n_b(-p) = -[1 + \overline{n}_b(p)], \quad n_f(-p) = [1 - \overline{n}_f(p)], \tag{4.92}$$

where $n(p) \equiv 1/[e\beta(p + \mu) \mp 1]$ is the appropriate antiparticle distribution function. Therefore the factor $A(p, k)$ in this interval may be rewritten as

$$A\left(p_{\|}, k\right) \equiv \begin{cases} \dfrac{n_b\left(k - \left|p_{\|}\right|\right)\overline{n}_b\left(\left|p_{\|}\right|\right)}{2\left|p_{\|}\right|\left(k - \left|p_{\|}\right|\right)}, & \text{scalars} \\ \dfrac{n_f\left(k - \left|p_{\|}\right|\right)\overline{n}_f\left(\left|p_{\|}\right|\right)}{2\left[\left|p_{\|}\right|\left(k - \left|p_{\|}\right|\right)\right]^2}\left[p_{\|}^2 + \left(k - \left|p_{\|}\right|\right)^2\right], & \text{fermions} \end{cases}. \tag{4.93}$$

Thus the energy E_p of a hard quark with momentum $|p|$ is explicitly given by

$$E_p = \sqrt{p^2 + m_\infty^2} \simeq |\mathrm{p}| + \frac{m_\infty^2}{2|p|} \simeq \left|p_{\|}\right| + \frac{\mathrm{p}_\perp^2 + m_\infty^2}{2\left|p_{\|}\right|}, \tag{4.94}$$

where the asymptotic thermal "mass" is

$$m_\infty^2 = \frac{C_f g^2 T^2}{4}. \tag{4.95}$$

Substituting the explicit form of E_p into definition (95), we can get

$$\delta E = \left[\frac{\mathrm{p}_\perp^2 + m_\infty^2}{2}\right]\left[\frac{k}{p_{\|}\left(k + p_{\|}\right)}\right]. \tag{4.96}$$

Previously, I have derived the explicit forms of Eqs. (4.87) and (4.90). Fig. 4.29 plots the leading-order heating photoemission rates that maximize the power given the electron time of flight of the photons into the curtain wall skin.

4.3.4 Conclusions

Traditional cooling and heating systems adopted by the building sector are problematic as they contribute to climate change and threaten the ozone layer. To mitigate these potentially catastrophic effects, the present paper has proposed transforming solar irradiation into cooled photons by implementing the Bose—Einstein photon distribution mechanism on helium-assisted curtain walls, which can be installed on the external surfaces of buildings. The cooled photons are then available for cooling the building. The feasibility of the approach was demonstrated in mathematical tests implemented in MATLAB 9.0. Moreover, the cooling photons can be converted into the heating state by the Higgs boson [BR $(H \rightarrow \gamma\gamma^{-})$] quantum field, which can be created by two thermal semiconductor diodes installed in the helium-assisted curtain wall, providing a natural heat source for buildings. Simply, we conclude that cooling-state photons (HcP^{-}s) can be extracted from the solar irradiance and transformed into heating-state photons (HtP^{-}s), providing a natural cooling—heating system that reduces global energy consumption, environmental damage, and ozone layer depletion.

4.4 BUILDING INFORMATION MODELING

Building Information Modeling (BIM) can be comprehended as a cardinal representation of the natural and working features of any given facility. A BIM is an alloted source of data to get any information regarding the given facility and thus form a trustworthy base to make choices throughout its life cycle, i.e., since its primary concept to its destruction. Subsequently, BIM represents a process that involves generating and managing digital representations of natural and working features of places. BIMs are documents (not necessarily in protective formats which contain protective data) that can be extracted, replaced, or exchanged and thus assist in arriving at a conclusion about a building or any other property. The existing BIM software can be operated by anyone (people, industries, or government organizations) who intend to plan, design, build, manage, and sustain various physical infrastructures such as roads, bridges, ports, tunnels, communication utilities, water, gas, electricity, refuse, and so forth.

4.4.1 BIM Origins and Elements

The term "building model" (i.e., BIM as used today) was used for the first time in the papers published in the 1980s; a 1985 paper by Simon Ruffle which was finally printed in 1986, followed by a paper published in 1986 by Robert Aish, later at GMW Computers Ltd, who developed the RUCAPS software, mentioning the use of the software at the Heathrow Airport, London. "Building Information Model" as a phrase was first published in a paper published in the year 1992 by F. P. Tolman and G.A. van Nederveen.

Nevertheless, the phrases "Building Information Modeling" and "Building Information Model" (together with the abbreviation "BIM") were unpopular even a decade after their first usages. In 2002 a white paper was released by Autodesk named "Building Information Modeling", which encouraged other software vendors to start participating in this area. Jerry Laiserin, in 2003, popularized and standardized the term BIM as a common name for the cardinal depiction of the building procedure. He achieved this by hosting assistances from Bentley Systems, Graphisoft, Autodesk, and even other industry observers. Simplifying exchange and usage of information in digital format were previously presented underneath different terminologies by Graphisoft as "Virtual Building," by Bentley Systems as "Integrated Project Models," in addition to "Building Information Modeling" by Autodesk or Vectorworks. Laiserin and the UK's Royal Academy of Engineering have recognized the pioneering role of applications such as Sonata, Reflex, and RUCAPS. Laiserin regarded ArchiCAD by Graphisoft as "one of the advanced BIM solutions available in the market." This is because Graphisoft had been creating such solutions a long time before its rivals. After its launch in 1987, some people regarded ArchiCAD as the first implementation of BIM because it was the first Computer Aided Design (CAD) product on a personal computer which was able to create both 2D and 3D geometry, in addition to being the first commercial BIM product for personal computers [71—73].

Conventional building design was heavily dependent on 2D technical drawings which included plans, sections, elevations, and so forth. BIM takes this to a new level, even past 3D, supplementing the three main spatial dimensions (width, height, and depth) with the additional fourth dimension as time (4D) and fifth one as cost (5D). Thus BIM goes beyond geometry. In addition to this, spatial relationships are also covered along with evaluation of light, geographic information, and properties and quantities of components of the building (for instance, details of the manufacturers). BIM defines a design as mixture of "objects"—unclear and indefinite, general or product-specific, solid shapes or void-space adjusted (such as the shape of a room), or which show their geometry, relationships, and qualities. BIM design tools offer drawing construction and other uses by extracting various views from a building model. These various views are reliable without any doubt because they have a single definition of each object case as their basis. In addition, BIM software defines the objects as parameters, and their relationships with different objects are also mentioned, so that if any linked object is edited, related ones will get changed too. Every model element can include features for choosing and organizing them on their own, which provides cost estimations along with material tracing and buying. The design team of BIM provides a simulated info model that can be used by the professionals included in a project, which include architects, landscape architects, surveyors, civil, structural and building services engineers, and so forth, along with the main contractor and subcontractors and even the owner/operator; every professional improves the single shared model by adding discipline-specific data to it. The problem of information losses that occurred in traditional methods when a fresh team which procures the "ownership" of the project passes widespread information to the owners of complex structures is drastically reduced here.

4.4.1.1 BIM Throughout the Project Life cycle

BIM usage is not only limited to the planning and design stage of the project but spans all through the building life cycle. It also supports practices involving cost management, construction management, project management, and facility operation.

4.4.1.2 Management of Building Information Models

Building information models extend throughout the idea-to-occupancy time span. A BIM manager might be assigned to achieve an effective organization of data processes all through this span. The BIM manager is reserved by a design development team on behalf of the client starting from the pre-design phase to the development phase and to trace the object-focused BIM against the foreseen and measured performance objectives, which will support multidisciplinary building information models that drive scrutiny, agendas, take-off, and logistics. Currently, the companies are also bearing in mind creating BIMs in various amounts of detail because relying on the application of BIM, some detail is needed, and there is differing modeling attempt associated with creating building information models at various levels of detail.

4.4.1.3 BIM in Construction Management

Members of the building procedure are tested to provide effective projects on a constant basis in spite of stringent budgets, restrained manpower, enhanced schedules, and incomplete or contradictory information. The important fields such as architectural, structural, and MEP designs should have good harmonization, as it is not possible to conduct two things simultaneously. BIM helps in recognizing crash at the preliminary stage and thus identifies the precise location of inconsistencies. The BIM model foresees creation of a facility before it is physically built. This is done for decreasing ambiguity, increasing safety, solving problems, and replicating and scrutinizing possible impacts. Data can be put by the subcontractors from each trade into the model before starting the construction, with chances to preconstruct or preassemble some systems off-site. Waste can be reduced on-site, and by-products could be transported and received on a just-in-time basis instead of keeping the stocks as a pile on the site. Easy extraction of magnitudes and shared characteristics of materials is possible. Scopes of work can be isolated and identified. A relative scale can be used to show systems, assemblies, and sequences with the complete facility or a group of facilities. Easy detection and prevention of errors is possible as conflict or "clash detection" can be enabled in BIM in which the computer model visually focuses on areas of the building (e.g., structural frame and building services pipes or ducts) which may wrongly intersect.

4.4.1.4 BIM in Facility Operation

BIM acts as a link and thus reduces the data loss related with handling a project from design team, further the construction team, and lastly the building owner/operator, by letting every party to improve and look back to all the data they have received throughout the cycle when they are contributing to the BIM model. This results in added advantages to the building operator/owner. For instance, there might be signs of a leakage in the premises which the owner may find Instead of exploring the physical building, the model could be used to check any presence of a water valve in the suspicious area [48,59]. The model can have the details about the specific size of the valve, manufacturer, part number, and several other data that have been ever explored in the past. These problems were first tackled by Leite and Akinci when they were trying to make a susceptible illustration of the contents of the facility and risks to support the classification of weaknesses in building emergencies. The BIM software could also include vigorous facts about the building, such as measurements of sensors along with control signals from the building systems, which could allow assessment of building operation and maintenance. Various attempts have been made to create statistical models for the facilities already existing. These attempts involve providing important variables such as the facility condition index, or use of a 3D laser-scanning studies and photogrammetry methods (either individually or together) for capturing exact quantities of the facility which can act as a basis for a model. Modeling a facility built in, about 1927, is more complicated than creating a model during design as it needs various postulations regarding design standards, building codes, methods of construction, materials, and so forth and is hence more complicated than creating a model during design.

4.4.1.5 BIM in Land Administration and Cadastre

BIM could possibly provide certain advantages for dealing with bedded registered areas in urban developed ecosystems. The first advantage will be improving graphic interaction of intertwined, piled, and complicated registered areas for nonspecialists. The high quantity of 3D and well-formed data regarding physical structures in models can help in understanding of registered boundaries and thus provide a clear description of ownership, rights, responsibilities, and restrictions. In addition to this, the use of BIM to handle registered data can transform the existing land management systems from a 2D-based and similar factual setting into a 3D digital, intelligent, communicating, and lively one. In addition, BIM can create a bridge between the cadastral informational and the communicating life cycle and administration of buildings by unlocking the value in the cadastral information.

4.4.1.6 BIM Software

The first building modeling software tools were developed during the 1970s and 1980s and involved workstation products such as Chuck Eastman's Building Description System and GLIDE, RUCAPS, Sonata, and Reflex. The primary products along with the hardware required for running them were costly and restricted their extensive acceptance. The primary modeling software that was accessible on a personal computer was ArchiCAD's Radar CH, released in 1984. Certain firms have created software intended particularly to work in a BIM framework because of the complex process of collecting all the related data while working with BIM on a project. The basic difference between these packages and architectural drafting tools like AutoCAD is that these packages allow the input of extra data (time, cost, details of the manufacturers, sustainability, information regarding maintenance, and so forth) into the building model.

4.4.1.7 Nonproprietary or Open BIM Standards

BIM is frequently linked with Industry Foundation Classes (IFCs) and aecXML—data structures for signifying data. For the easy sharing of BIM information among various software applications (some private data structures are created developed by sellers of CAD integrating BIM with their software), buildingSMART (the earlier International Alliance for Interoperability) has created an unbiased, nonprivate, or open IFCs. Poor software exchange has been considered as a barrier to an effective industry in general and particularly to adopting BIM. A report by the US National Institute of Standards and Technology, in August 2004, conventionally projected an annual loss of $15.8 billion by the US investment agency industry because of absence of exchange resulting from "the extremely uneven characteristics of the industry, the industry's prolonged printed business exercises, absence of regulation, and unpredictable technology approval among participants." The American Institute of Steel Construction has accepted CIS/2 standard, a nonprivate standard with its base in the UK, which can be considered as a primary instance of a nationally accepted BIM standard.

4.4.1.8 Background and Technology

With the increase in knowledge of energy utilization in addition to the environmental effect of the operations of the building, it is required that more considerations are made by the architects, designers, and planners on sustainability and energy functioning of the building. Crucial design choices must be made by important parties initially in the design-development phase to ensure most of these points are shown in the building functioning [74,75]. The use of BIM in the building energy models has extremely enhanced the scrutiny of energy, and thus this method has achieved impetus. In spite of fast developments in BIM-based procedures, it continues to be unanswered that how conventional building parties could conduct energy functioning evaluation, which have earlier been done mainly by professionals, to exploit energy-effective building performance. To tackle this problem, two foremost building performance evaluation software programs have been recognized, namely, EnergyPlus and IES Virtual Environment. Their efficiency and appropriateness were compared as BIM-based energy simulation tools. For simplifying this study, a case study on Building Performance Model (BPM) of a building having just a single story and a door, each wall having many windows, a slab, and a roof were examined. The main focus was specifically on building energy functioning by varying alignment of the building and dimensions of the windows and a comparison of how efficient these two software programs evaluated the functioning was made. Standard decision-making procedures executing building energy simulation process in the early design phases in the United States are also considered. Inferences can be made as to how to show BIM-based building energy functioning calculations more effectively.

As per the US Environmental Protection Agency (EPA), 39% of the total energy usage and 68% of the electricity usage in the United States is used by buildings. In addition, information from the US EIA shows that buildings account for approximately half of all greenhouse gas emissions that occur in a year. Hence by adopting strict building codes, industry norms, and the green building rating systems such as Leadership in Energy & Environmental Design, attempts have been constantly going on for considerably reducing the usage of fossil fuel—dependent energy in the next two decades, such that net energy usage by the year 2030 is zero. Though proposing buildings and systems where in all the sections of the building, such as, architecture, HVAC and electrical systems, and so forth, are planned individually, they would never help in reaching the goals of energy saving. Existing design, construction, and operating methods are very uneven for permitting the judicious and efficient application and incorporation of energy-effective procedures and technologies on constructing a new building or restoration projects.

Yet there exist certain technologies and knowledge that could be of help in creating high-performance buildings. In fact if energy-effective methods, merged with an improved systems integration, are integrated at the design phase, the potential for energy saving is significant. It is clear that a building constructed with the constant association of all parties has a higher possibility of attaining optimum energy efficiency than one created on a series of handoffs.

4.4.1.9 Apply Whole Building Design Approach

For achieving high-functioning buildings, two major components of Whole Building Design are used, namely, an integrated design approach and an integrated team process. It comprises of all the parties taking part all through the life process of the building, beginning from the design stage to controlling and maintaining the building. An integrated team process is a must for Whole Building Design. This process includes the design team and all involved parties working simultaneously during all the phases of the project and evaluating the design for cost, efficiency, elasticity for future modifications, and complete environmental effect, along with other stuffs. The identification of project goals is made early on during the project. Evaluation, appropriate application, and coordination of all the interrelated building systems from the designing phase and scheduling stage are worked by the team. To achieve the targeted energy consumption goals, rigorous coordination among the design team members and use of an energy modeling software program BIM are very crucial. This is not only important for the most basic projects but also for energy modifications of current buildings.

4.4.2 Use EPA Energy Design Guidance

The Energy Design Guidance for the design of commercial facilities to attain an ENERGY STAR certification has been provided by the US EPA. It is mainly an organizational tactic with a series of recommended procedures for the facility owners and design experts to set energy competence aims and to make sure that energy usage is tackled at all levels of the project.

As briefed in Fig. 4.30, to achieve the ENERGY STAR certification, EPA encourages to pursue finest methods for energy plan as an element of the complete design, construction, and operation procedure for transforming design objective into buildings which accomplish the target. Among all the processes outlined in Fig. 4.1, initial computations and/or models for assessing the energy usage of different design policies when the design concepts take place happen at the schematic design phase. Suitable design tools such as BIM integrated with different energy imitation models could be used for energy analysis of design concepts [61,76]. Evaluation of comparative effectiveness of energy strategies could be used to improve design constantly. Whenever attainable energy performance targets, i.e., energy usage and CO_2 targets, are established and incorporated in descriptive language, design policies could be improved in the design-development phase.

4.4.3 Energy Simulations

As mentioned in the previous section, making decisions associated with sustainable design of the building during the primary design and

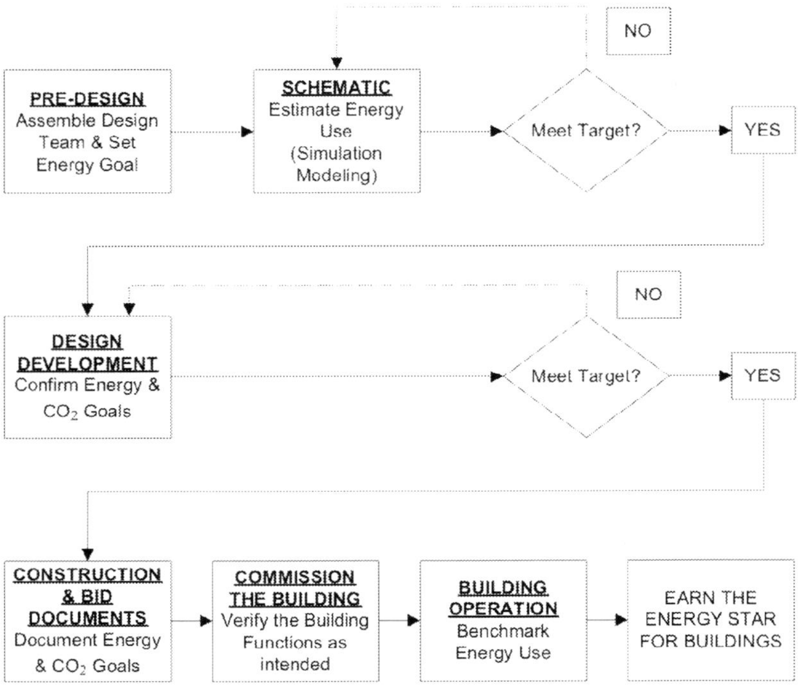

Figure 4.30 Integrating energy targets, evaluation, and tracing all through the design and operating stages of the facility. *Modified from the EPA Energy Design Guidance.*

preconstruction phases is considered to be the most effective method. For accurately measuring the functioning of the building in the initial design and preconstruction stages, an approach to a complete set of information related to the building's form, context, materials, and technical systems is necessary. An opportunity is created for performing sustainability methods and functioning evaluation during the design procedure as BIM permits multidisciplinary data to be covered in a single model.

For this study, energy replications and evaluation of energy functioning of a BPM are performed using two frequently used energy replication software packages, EnergyPlus and IES VE.

4.4.3.1 Energy Simulation Solutions

EnergyPlus, complete building recreation program, is created by the Department of Energy. An incorporated (loads and systems) imitation for temperature and relief forecasting is delivered at a customer-identified period. Accurate control of the systems, moisture adsorption and desorption in elements of the facility, interzone air flow, and radiant cooling and heating systems can also be calculated by the software.

Integrated Environmental Solution developed the software IES VE. Evaluation of building mobilization and thermal mass, building arrangement and alignment, shading, glazing, solar penetration, thermal insulation (type and placement), air-tightness, climate response, casual gains, motorized ventilation, multimode systems, natural ventilation, and HVAC systems is also possible using this software.

4.4.3.2 The Building Performance Model

The BPM calculated shown in Fig. 4.31 is of a building having one storey with a single door, every wall having numerous windows, a slab foundation, and a roof.

For Building Energy Simulation Processes, the primary stage was conducting energy simulations using the EnergyPlus software. Google SketchUp was used to create BPM. Open Studio, an ENergyPlus Plug-in for SketchUp was then used for simulating energy performance. The results of simulation are shown in Table 4.7.

Evaluation of energy performance of the BPM is then done using IES VE. BPM was first created using Revit Architecture. Energy simulation was then conducted using the IES VE plug-in for Revit. Results of the simulation process are summarized in Table 4.6.

Building energy performance is affected by various design parameters. A detailed study of all the design factors was not required in this study as the goal was to find the different ways in which energy effectiveness influences the building design decision-making procedure. Orientation and window sizes were the two design factors that were examined in the energy simulations.

The default HVAC system that was chosen for the energy simulations was VAV Single-Duct system.

4.4.3.3 Energy Simulation Results and Discussions

The simulation results are shown in the following tables.

Table 4.6 Shows that the total energy utilization, comprising cooling and heating, was minimum when the BPM was at 0 degree and 180 degrees angle. Hence the recommended angle is either the South or North. It was also observed that the total energy utilization increased because of larger windows.

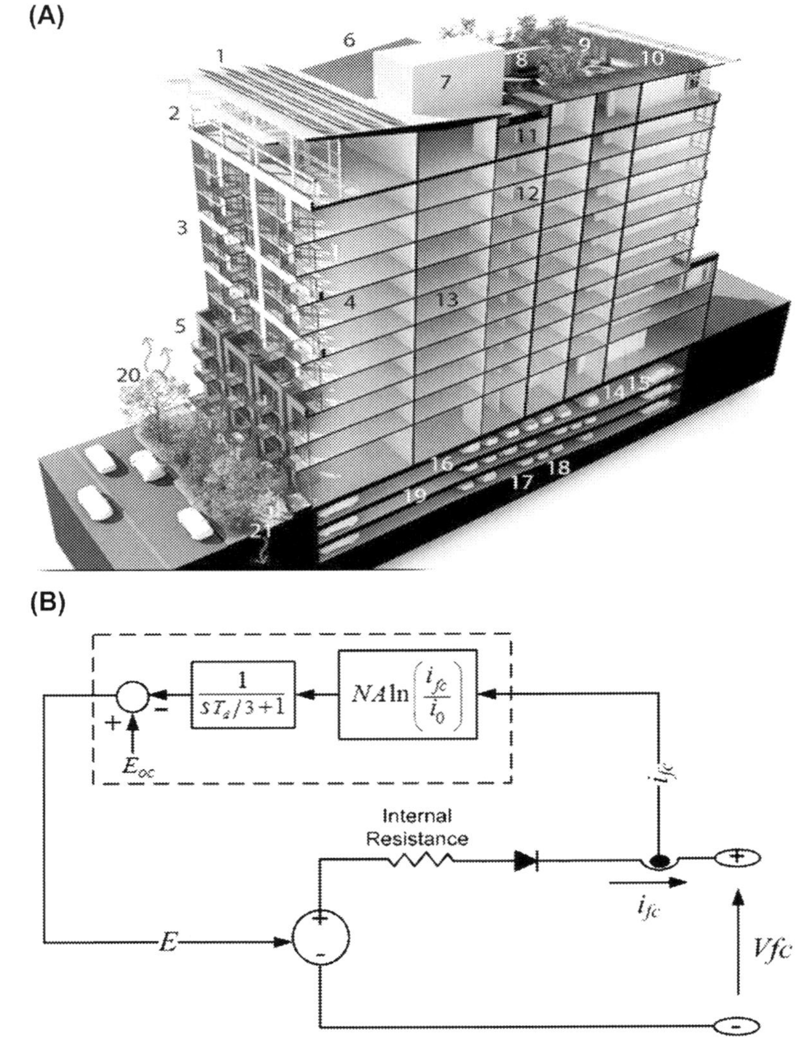

Figure 4.31 (A) Building Performance Model (BPM) in relation to various sections of building (1 through 19) has been modeled considering (B) BIM software's energy simulation process analysis.

Even though by adding larger windows, considerable improvement in the lighting conditions can be achieved, the HVAC system will consume higher heating energy for heating the building and similarly will consume higher cooling energy for cooling the building. Therefore a better design is not ensured because of larger windows.

Table 4.6 Energy simulation results using EnergyPlus

EnergyPlus simulation results (VAV single-duct system)			
Parameters		**Heating (GJ)**	**Cooling (GJ)**
Orientation (window size 48″ × 36″)	0 degree	417,764.1	275,803.1
	45 degrees	417,765.1	275,805.2
	90 degrees	417,765.8	275,805.2
	135 degrees	417,766.1	275,804.6
	180 degrees	417,765.2	275,802.3
Window (building orientation 0°)	24 × 36	417,755.5	275,803.1
	48 × 36	417,764.1	275,803.1
	72 × 36	417,772.9	275,803.1

Table 4.7 shows that majority of heating and cooling energy consumption in a building are a result of the HVAC system. No system other than the default HVAC system of VAV Single-Duct System was chosen during the energy simulations in this study for changing design parameters. This is the reason that energy differences in both the simulation models, as can be seen in Tables 4.1 and 4.2, were not noteworthy.

4.4.4 Conclusions

A stable increase in usage of energy and electricity in the US building sector can be seen in the recent statistics. However, there exist technologies and information that can be exploited for creating improved, high-performance buildings. The energy saving prospective is noteworthy only when (1) linking of energy-effective actions with improved systems incorporation is performed during the design phase, combining it with improved

Table 4.7 Results of energy simulation using IES VE

IES VE simulation results (VAV single-duct system)			
Parameters		**Heating (Mbtu)**	**Cooling (Mbtu)**
Orientation (window size 48″ × 36″)	0 degrees	77.8	6.4
	45 degrees	78.0	6.5
	90 degrees	78.3	6.5
	135 degrees	78.3	6.5
	180 degrees	77.9	6.3
Window (building orientation 0°)	24″ × 36″	77.8	5.9
	48″ × 36″	77.8	6.4
	72″ × 36″	77.9	6.9

systems incorporation; (2) the buildings are not just built on a sequence of hand-offs but are rather built through ongoing collaboration of all stakeholders; (3) the utmost efficient choices concerning energy effectiveness in a facility are made primarily during the planning and preconstruction phases; (4) simultaneous measurement of scope expansion takes place with the evaluation of energy effectiveness. Simultaneous constructability analysis could be done by including the evaluation of energy effectiveness as portion of the building/project design factors evaluation; and (5) a rigorous cooperation among the design members and an energy modeling like the BIM are ready.

Understanding results from the energy simulations are required for designing energy-efficient buildings. To fulfill this need, we have a number of energy analysis solutions available. Every software answer suggests energy utilizations depending on a number of design factors for helping the concerned parties in completing the design by concentrating on opposing features of energy functioning evaluation.

The most frequently used energy simulation solutions throughout the world are IES VE and EnergyPlus. IES VE has better interface and reports and thus is more user-friendly. Design parameters are easier to define in IES VE. Computational fluid dynamics analysis and life-cycle cost analysis can also be calculated using IES VE. Great flexibility in defining HVAC systems is offered by EnergyPlus. It is a powerful energy simulation tool and permits simulation of many HVAC systems aimed at any specific zone. Yet, to obtain accurate simulation results, extensive knowledge of HVAC systems is necessary.

4.5 SMART BUILDING TECHNOLOGY

Smart building technology, through a building automation system (BAS), is the concentrated controller of the facility's heating, lighting, air conditioning and ventilation, and energy supply via automation. Improvement in the comfort of the occupants, effective operation of building systems, reducing the operating costs and energy consumption, and improvement in the life cycle of utilities are the objectives of building automation. Building automation is a case of a dispersed regulatory procedure that involves the electrical products being devised such that they check and regulate the motorized equipments, fire and food safety, HVAC and humidity control, security, and lighting (specifically emergency lighting) along with ventilation systems of a building through a computer networking.

Keeping the building climate within a specified range, using the occupancy schedule to provide light to rooms (lack of obvious switches to be different), keeping a check on implementation and device crashes in all systems and providing breakdown alerts to the maintenance team of the building are some of the things possible because of BAS core functionality. An intelligent or a smart house is the one which is controlled by a BAS. Homes used proprietary protocols (such as X-10) in contrary to the robust proven protocols (such as BACnet), which are used by commercial and industrial buildings [77]. Accommodation of a BAS for the energy, air, and water conservation features is provided in almost all multistory green buildings. As and when more sophisticated ventilation and humidity inspection is required for tightly insulated buildings, the typical function of the BAS, electrical device demand response comes in to the picture. Maximum possible number of low-power DC devices are used by most green buildings. These are basically integrated with power over Ethernet wiring. Hence by using an Ethernet connectivity, they are always accessible to a BAS. A BAS will be required for managing shading and venting, scheduling device use, and managing net capture even for a passive design which is targeted to consume zero net energy (Fig. 4.32).

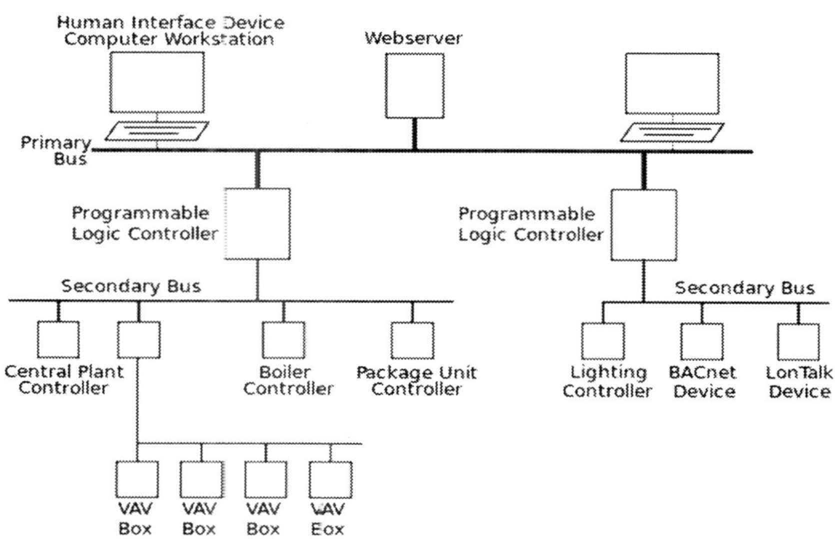

Figure 4.32 Monitoring of the performance of a building immediately using a Smart building technology for controlling the central plant, boiler, VAV box, lighting, energy BACnet, and LonTalk.

4.5.1 Automation System

Any electrical control system which controls a facility's heating, ventilation, and air conditioning (HVAS) system is termed as BAS. In addition to controlling indoor and outdoor lighting, modern BAS can also help in controlling fire alarm, security, and almost everything else that uses electricity in the building. The biggest advantages of modern HVAC system are flexibility and integration, which the former HVAC control systems, such as 24 V DC wired thermostats or pneumatic controls, lacked.

4.5.1.1 Protocols

Majority of building automation networks comprise of a main and subordinate bus. These help in integrating input/output appliances with a consumer interface (called a human interface tool too) and also low-level controllers with high-level controllers (usually dedicated to automation of the structure, however, might also be a programmable controller that is based on logic circuits in general). The way in which these devices work together even without being configured to do so is specified in ASHRAE's free procedure BACnet and the free procedure LonTalk. SNMP-based protocols have decades of history in the computer networking world [77,78]. Modern systems track events by exploiting such SNMP-based protocols. Ethernet, RS-232, RS-485, dedicated optical fiber, ARCNET, or a low-bandwidth specific target wireless network are traditionally used for providing physical connectivity among various devices. IEEE 1905.1 standard, which is confirmed by the nVoy reviewing indicator, specifies certain standard-based multiprotocol diverse networking on which the modern systems can rely on. Only Ip-based networking can be accommodated in these systems, although any existing wiring can still be used by them. Incorporation of powerline networking instead of AC circuits, power instead of Ethernet low-power DC circuits, along with high-bandwidth wireless networks such as LTE and IEEE 802.11n and 802.11ac is also possible. The building-specific wireless mesh open standard ZigBee is often used for incorporating these.

The controller market is dominated by the proprietary hardware. Each company holds controllers for certain precise applications. Some applications such as the simple packaged rooftop units for HVAC are designed with restricted controls and no interoperability. Packages from other vendors will generally not allow integration with this software. Coordination is possible only at the Zigbee/BACnet/LonTalk level. Interoperability in the existing systems during the functioning level lets the consumers to intermingle devices from various producers in addition to providing incorporation with various similar building control systems. Such systems depend on SNMP, which has been in use since a long time for integrating diverse computer networking devices into a single logical network.

4.5.1.2 Sensors

A variable measurement can be read by using analog inputs. Humidity, temperature, and pressure sensors such as the thermistor, 4–20 mA, 0–10 V, or a platinum resistance thermometer (to detect resistance temperature) or a wireless sensor can be considered as some examples.

In whichever way the device may be detected, a cardinal key shows whether a device has been started or not. An air flow switch, a 24 V DC/AC signal, a volta-free relay contact (dry contact), or a current switch can be considered as some of the examples of a fundamentally digital input. Pulse-type inputs that count the rate of pulses for certain duration could also be considered as types of digital inputs. A turbine flow meter that transmits revolution information as a rate of pulses to an input is a well known of example of pulse-type input.

For the discovery of device or other loads from electronic or magnetic characteristics of the circuit, discreet load supervising software relies on cardinal devices and systems. However, the detection takes place by analog means. These are not only highly useful for identification but their operation is also very cost-effective. Moreover, they also help in detecting line or equipment faults, start-up transients, and so forth.

4.5.1.3 Controls and Controller

Speed or location of any device such as an inconsistent frequency drive, a valve, an I–P (pneumatics from a current), or a damper actuator can be regulated using an analog output. A hot water valve being opened one fourth of total to maintain a setpoint is an example. A motor being ramped up slowly by a variable frequency drive to avoid a hard start is another example. Opening and closing of relays as well as switches in addition to driving a load upon command are made possible because of digital outputs [77,79]. Turning on a parking lot light whenever a photocell shows that it has become darker is an instance. Another example is passing a 24-V DC/AC through the output powering through a valve when the valve is opened. Emission of regular pulses in a given duration is also possible by pulse-type digital outputs. Emission of a frequency of pulses when an energy meter calculates kWh is an example.

Controllers are basically minute, aim-built computers that are equipped with input and output potentials. To control various devices that are usually found in a building in addition to controlling other subnetworks of controllers, such controllers are available in a variety of sizes and potentials.

Temperature, pressure, air flow, humidity, current flow, and various other essential factors can be measured using inputs. The controller sends command regulatory indicators to working device and various portions of the system with the help of outputs. Inputs and outputs are of two types: digital or analog. Depending on the manufacturer, digital outputs can also be called discrete.

There are three categories in which controllers for building automation could be classified: (1) programmable logic controllers, (2) system/network controllers, and (3) terminal unit controllers. Though, to achieve integration of third-party systems (such as a single AC system) with a centralized BAS, an additional device can also be added. Controlling lighting and/or simpler devices such as a heat pump, fan coil, a package rooftop unit, VAV box, and so forth is usually made possible with the use of terminal unit controllers. A new control logic need not be created as it is possible for the installer to select any accessible preprogrammed personalities that perfectly suits to the appliance that is to be regulated.

4.5.1.4 Occupancy

Occupancy can be considered as a functioning approach for a BAS. The other common approaches are, namely, unoccupied, morning warmup (MWU), and night-time setback. Occupancy is generally built on time of day schedules. Zone-based control system allows the users on any side of a building to have a dissimilar thermostat (or a separate system or a subsystem). This is possible in occupancy mode because the BAS targets to deliver a calm climate with sufficient lighting.

A feedback is provided by the temperature sensor in the area to the controller so that provision of heating or cooling is possible whenever required [80,81]. MWU model occurs before the occupancy on being permitted. The building is brought to setpoint at the proper time for occupancy by the BAS during the MWU. The optimization of MWU is made possible as the BAS frequently features in outdoor situations and historical incident. This is called optimized start too. An override is defined as a physically started instruction to the BAS. For instance, there are push-buttons on some wall-mounted temperature sensors which force the system into occupancy mode for a predefined amount of time. Users are allowed remote initiation of an override on the BAS by the Web boundaries wherever present.

Lighting or climate conditioning in some buildings can be activated by occupancy sensors. An occupancy sensor does not directly start climate conditioning, as it has the potential of extensive start times by the time any space turns enough cool or warm.

4.5.1.5 Lighting

Depending on the moment of the day or on occupancy sensor, photosensors, and timers, lighting could be switched on or off or even the lighting can be reduced by using building automation or lighting regulation system. Keeping lighting in a given area on for half an hour after the last movement in that area was detected as a good example of such lighting. Lights can be modulated in parking lots and outer offices by placing photocells outside the building which can detect darkness and the time of the day. Many control systems offer the facility of dimming (or turning off) lights so that the benefit of DC incentive and savings can be taken, which makes lighting a good candidate for demand response. The lighting control in new buildings could be established over the field bus Digital Addressable Lighting Interface (DALI). DALI ballasts lights are completely dimmable. DALI luminaires and signal failures having lamp and ballast failures can also be detected.

4.5.1.6 Air Handlers

Humidity/temperature conditioning is not much required as majority air handlers combine the return and the outside air. With the usage of less refrigerated or warmed water (however, all AHU's do not use chilled or hot water circuits), it could become a lot more economic. For maintaining the building's air healthy, it is required to have some external air. Demand control (or controlled) ventilation regulates the quantity of outside air depending on the computed levels of residence. This is performed to improve energy effectiveness at the same time retaining the healthy indoor air quality. Spaces or rooms, the ducts of return or supply air, and occasionally even the external air may be equipped with analog or digital temperature sensors. The dampers of outside and return air and the warm as well as refrigerated water valves have actuators placed over them. Depending on the time of the day, building pressures, temperatures, or both, as well as starting and stopping of the source fan (and return if applicable), is decided.

4.5.1.7 Constant Volume Air-Handling Units

The "constant volume air-handling unit (AHU)" or a CAV is a less efficient kind of air handler. The variable-speed controls are not present in the fans in CAVs. On the other hand, for maintaining temperatures in the areas of the building, CAVs lock and unlock dampers and valves of water supply valves. By locking or unlocking hot or chilled water valves, the ones which supply the internal heat exchangers of the CAVs, they warm or cool the areas. Usually a single CAV is enough to serve several spaces.

4.5.1.8 Variable Volume AHUs

The "variable air volume (VAV) AHU," or a VAV is a more effective kind of air handler. Usually a single VAV box is kept per room or area, which is further supplied with pressurized air from the VAVs. By altering the velocity of the fan or a blower by an adjustable regulatory drive or (less effectively) by shifting inlet controller blades to a set-speed fan, pressure supplied to the VAV boxes can be changed by a VAV air handler. The VAV boxes that serve the particular spaces decide the quantity of air. The air is then supplied from every VAV box to a tiny area, such as an orifice. Depending on the heating or cooling needed in a space, each box is equipped with a damper which can be opened or closed. The higher the number of boxes that are open, the more is the quantity of air required and thus the higher is the quantity of air that is provided by the VAV AHU. There are warm water valves and an inner heat exchanger provided in some VAV boxes. Depending on the requirement of the spaces, the valves for warm and cold water can be locked or unlocked accordingly. Sometimes the interior zones are cooling while its only the parameter that uses these heated VAV boxes. For achieving sufficient amount of ventilation and proper air balance, the VAV boxes must be set with a minimum and maximum CFM.

4.5.1.9 AHU Discharge Air Temperature Control

The discharge air temperature set point value should accordingly change in the span ranging from 55 to 70 F for the AHU and Roof Top units (RTU) which are used for serving multiple zones. This adjustment helps in reducing the energy consumption for cooling, heating, and fan. The use of reheat at the zone level is reduced for the zones having very low cooling loads when the temperature is below 70 F outside.

4.5.1.10 VAV Hybrid Systems

There is also an availability of a mixture of VAV and CAV systems. The interior zones in this hybrid operate in the manner like a VAV system. There is generally a heating fan located in the innermost location equipped with a heating coil that is supplied from the building boiler. This is how the heating in the outer zones differ from VAV. The warm air is sent to the external twin mingling boxes and dampers. The thermostat of the zone while controlling these boxes and dampers requests for the cooled as well as the heated air, however, required.

4.5.1.11 Central Plant

For supplying water to the AHUs, a central plant is required. A hot or a chilled water system or a condenser water system may be supplied to the central plant. In addition to this, supply of transformers and secondary power unit for emergency power is also possible. Managing these well would prove helpful to each other. Generating electric power at a plant during peak demand period by means of a gas turbine and further using the hot exhaust from this turbine to operate an absorptive cooler or a to heat water are some examples.

4.5.1.12 Chilled Water System

A building's air and equipments are often cooled with chilled water. Cooler(s) and pumps are parts of the chilled water system. The cold-water supply and return lines are measured using analog temperature sensors. For chilling the cold-water supply, the cooler(s) are arranged and scheduled on and off. For the purpose of space cooling, cool (chilled) water is produced using a cooler as a refrigeration unit. There are coils present in the AHUs, fan coils, or induction units where the cold water gets circulated. Chilled water systems are used in bigger buildings as the distribution of cold water is unrestricted by the 100-foot division boundary which is applied to DX systems. Modulation of water flow through the coils is generally achieved by the capacity control present in the chilled water. Hence a single cooler can serve multiple coils without even negotiating command of any single unit. Coolers might function on two theories, namely, the vapor compression theory and the absorption theory. Coolers functioning on vapor compression might use responding, eccentric, spiral, or rotational compressor conformations. For capacities below 200 tons, reciprocating coolers are used; for higher capacities, centrifugal coolers are generally used. The less commonly used are the rotary and screw coolers; however, their use is not rare. An air-cooled condenser or a cooling tower may be used for rejecting heat from a cooler. For providing a packaged cooler, an air-cooled condenser may be pushed into vapor compression coolers. This would further be fitted outside of the building envelope. For installing a cooler in a bounded central plant space, vapor compression coolers may be designed in such a way that they are installed separately from the condensing unit. Absorption coolers are designed in such a way that they are installed separately from the condensing unit.

4.5.1.13 Condenser Water System

Supply of cool condenser water to the coolers is made using the cooling and pumps. Flexible velocity drives are generally equipped at the cooling tower fans for controlling temperature as there has to be a steady supply of condensed water to the coolers. Appropriate temperature of the cooling tower makes sure that the appropriate refrigerant head force is achieved in the cooler. The refrigerant that is used decides the cooling tower set point that is being used. The condensed water supply lines and the return lines are measured by the analog temperature sensors.

4.5.1.14 Hot Water System

Together with the AHU of the building or VAV box heating coils, the internal warm water heating coils (Calorifer) also feed the supply of heat from the warm water system. The warm water system consists of a boiler(s) and pumps. Analog temperature sensors are positioned inside the warm water supply and return lines. For controlling the hot water loop temperature, a particular type of mixing valve is generally used. For maintaining the supply, the boiler(s) and pumps are arranged and scheduled on and off. Lowering of the energy utilization of the circulation pumps of a building to about 15% of what was already being consumed before can be a result of installing and incorporation of varying frequencies. An adjustable regulatory drive operates by moderating the rate of the electricity supplied to the motor that it has to power. The electrical grid in the United States consumes a frequency of 60 Hz or 60 cycles per second. By reducing this rate of the electricity that is supplied to the motor, varying frequency drives are capable of reducing the output and energy utilization of motors. Yet, the relationship between the motor output and the energy utilization is not straight. The output of the motor when a variable frequency drive supplies electricity to the motor at 30 Hz will be 50% as 30 Hz when divided by 60 Hz give 0.5% or 50%. As the relation between motor output and energy utilization is not straight, the energy consumption will be around 18% for a motor that runs at 50% or 30 Hz and not 50%. The effectiveness of the variable frequency drive decides the correct fractions of motor output or Hertz that is supplied to the motor (that are in fact similar) and the real energy utilization of the varying frequency drive/motor combination. For instance, the varying frequency drive requires power to connect with the BAS, run its cooling fan, and so forth. If the motor is continuously operated at 100% with the varying frequency drive connected, the operating cost or electricity utilization will in fact increase with the new varying frequency drive connected. The varying frequency drives consume negligible amount of energy, and such an amount is barely a factor when calculating savings. Nevertheless, it should be kept in mind that VFDs in fact use energy themselves. The energy consumption of the pumps has been reduced by the varying frequency drives to 15% of what they were using before. This is done as the varying frequency drives hardly operate at 100% and spend majority of the time in the 40% output range, in addition to the pumps being totally shut down whenever not in use.

4.5.1.15 Alarms and Security

Alarm systems are installed in the automation systems of all modern buildings. Detecting a possibly dangerous or an expensive situation is not of any help if there is no notification of the problem to its solver. There can be many types of notifications, namely, via a computer (an e-mail or a sms), pager, voice call on mobile, loud alarm, or a combination of all. All systems keep logs of who was notified, when, and how, for the purpose of insurance and accountability. Alarms may be designed to either notify someone at once or notify when some level of seriousness or urgency develops. Owing to momentary power failures at the site due to the presence of many buildings, hundreds or even thousands of alarms can be triggered from various equipments that have faced power failure. They must be repressed as well as acknowledged as signs of a bigger malfunction. At certain sites, there are crucial alarms that get sent again automatically at changing time durations. For instance, the intervals for a recurring crucial alarm (of an undisruptable supply of power in "bypass") can be at 10 min, half an hour, and even every 2–4 h subsequently till the alarms are answered.

Security systems and a BAS can be connected. Occupancy sensors can also be used as burglar alarms if present. There should be various sensors or cameras having battery standby and wireless network, and the capability to activate alarms even after being detached as security systems are frequently intentionally damaged. Ethernet (which is capable of operating a pan-tilt-zoom camera and supplementary tools fit for 30–90 W) is generally used in the modern systems. This has the capacity to charge such batteries as well as keep wireless connectivity unrestricted for actual wireless usages, like a standby communication in blackout. For overriding the building automation, fire alarm boards and the linked smoke alarm systems are generally hard-wired. For instance, when the smoke alarm is triggered, the complete outdoor air dampers shut, preventing air from arriving inside the premises, as well as an ejection system can separate the fire. Likewise, an electronic defect identification system could switch off the complete circuit, irrespective of the amount of alarms that will be triggered or persons who will be distressed. Fossil fuel ignition appliances too have their interrupters, such as a natural gas feed line which will be turned off whenever slow pressure falls are discovered (showing a leakage) or whenever extra methane is found in the supply of air of the building. Good BAS stays alert of such interrupters and identify complicated breakdown situations. Neither backup power is wasted in turning on the appliances which these safety interrupters have switched off, nor are excessive alerts sent by good BAS. As understood by its definition, a poor BAS emits an alarm for each alert and does not identify any physical, fire or electrical, or fuel safety interruption. For that reason, good BAS is generally based on safety and fire systems.

4.5.1.16 Information Security

As a result of increasing range of abilities and their association by the means of the Internet, BASs were constantly stated to be susceptible to hackers and cybercriminals who can infect their parts. Hackers can manipulate a building by measuring or changing its setting; scrutiny is possible because of sensors (such as activities of employees or behaviors of occupants can be monitored), whereas actions in a building are possible because of actuators (such as unlocking doors or windows for invaders). In addition, KNX, ZigBee, BACnet, and various sellers and groups have begun improving the safety aspects in their products and criteria (view atest norms or standard drafts). Yet, there are stories of many existing troubles in building automation safety.

4.5.1.17 Room Automation

Room automation stands as a part of building automation having the same principle, which is the association of multiple systems under a central control, although for this situation, in a single room. All places where the manual operation of the room would be made very complex because of presence of a great quantity of appliances that describe the room purpose have to be operated can be regarded as some of the examples of room automation. Such places include presentation suites, lecture halls, and corporate boardroom. A touchscreen is usually used for room automation systems to serve as a principle way of controlling each operation.

4.6 RESILIENCE IN URBAN AND RURAL DEVELOPMENT

For the effective management of assets and resources, if an urban zone used various kinds of electrical information gathering sensors for supplying data, then it is called a smart city. These data comprise of information that is gathered from civilians, tools, and assets, which are managed and examined for monitoring as well as managing traffic and transport systems, power plants, law enforcement, water supply connectivity, waste management, schools, libraries, hospitals, information systems, and other community services. For maximizing the efficiency of operations and services of a city connecting them to the citizens, the concept of smart city combines information and communication technology (ICT), as well as several material tools which are linked to the network (the Internet of things [IoT]). Interaction of officials directly with the community as well as the city infrastructure is possible because of smart city. It also helps in monitoring whatever is going on in the town and also examines how the town is developing.

ICT serves various purposes: (1) enhances superiority functioning as well as improved ability and working of urban facilities, (2) reduces amount and resource utilization, and (3) increases connection between the citizens and the government [12,47]. For managing urban movements and also allowing actual time reactions, there are smart city applications that are developed. Hence a city with the simple, traditional relationship with its citizens will not be as prepared as a smart city to counter challenges. Nevertheless, the term is still open to several interpretations as it is itself unclear. Digital city, flexicity, information city, knowledge-based city, MESH city, teletopia, wired city, telecity, cyberville, intelligent city, Ubiquitous city, and electronic communities are some of the terms which are used to describe analogous ideas.

Interests in smart cities, in addition to economic rearrangement, the transition toward online retails and entertainment, climate change, aging populations, pressures on public finances, and urban population growth are the results of key economic, environmental, and technological changes. For accomplishing "smart" urban development for the cosmopolitan cities, the

European Union (EU) has dedicated continuous efforts in planning a strategy for the same. There have been a variety of plans developed in "Europe's Digital Agenda" by the EU. In the year 2010 the EU stated its idea of developing community amenities and the condition of life by focusing on supporting innovation and investment in ICT services. According to an estimate, by the year 2020 the international market aimed at smart urban amenities will be around $400 billion yearly. Analyzing the topic through a particular framework is a must to accomplish an exact depiction and justification of the idea of smart city. There are 4 main dimensions in the framework:

4.6.1 Technology Framework

Technology is the basis for many concepts of the smart city; a technological smart city is not just a single idea. It includes various patterns of technical infrastructure which are used to create the idea of smart city.

Digital city: It includes combination of innovation services, communication infrastructure, and service-oriented infrastructure; a digital city is "a connected community that combines broadband communications infrastructure; a flexible, service-oriented computing infrastructure based on open industry standards; and comprises innovative services to meet the needs of governments and their employees, citizens, and businesses."

Creating an environment where the citizens are interconnected and they can freely share information anywhere in the city is the key purpose of a digital city.

Virtual city: Cyberspace executes functions in such cities; this contains the idea of a hybrid city, wherein there exists an actuality with actual civilians and objects along with a simulated city of actual objects and individuals simultaneously. The coexistence between these two realities can be found in a virtual smart city, yet it is difficult to manage the matter of physical distance and location. Having a world with no distance is yet not met in many ways. In reality, such an idea is delayed because of real IT infrastructure of cables, exchanges, and information centers.

Information city: The regional data are collected and delivered to the public portal; In such a city, living and working on the Internet is possible as any information could be found through various IT infrastructures. This is only possible because of the method of sharing information among the citizens themselves. The information of the city can be an urban hub generally addressing as well as economically with the use of this approach, with the linkage among the various civil services, government institutions, and people interactions being the most important thing.

Intelligent city: For supporting innovation procedure and learning, technological innovation or research is involved as a function. The most important resource in such a technological city is the human capital. There lies great importance in the learning process, knowledge, and creativity where the idea develops in a social context. One of the key characteristics of an intelligent city is having each infrastructure of the city latest thus having the most advanced machinery in telecommunications, electrical, and motorized technology. Attempting to build an "intelligent" smart city can be considered as more of a fundamental innovation instead of an incremental innovation because of large quantity of efforts for using IT in transforming the everyday life.

Ubiquitous city (U-city): An ecosystem is created by this city, such that any device can be connected to the citizens. Because of the facilities it provides in terms of accessibility to every infrastructure, U-city can be counted as a further extension of digital city concept [49]. Thus it makes it easier for the citizen to use any available devices and further connect with them parallely. Providing all the services to a citizen which he can avail anywhere and anytime with the use of various kinds of devices is the ultimate goal of such a city. However, it should be made clear that a ubiquitous city and a virtual city are different in concepts. The U-city inserts computer chips in the urban elements, whereas a virtual city visualizes real urban elements inside the virtual space by creating an additional space.

4.6.2 Human Framework

Human infrastructure (i.e., knowledge networks, voluntary organizations, creative occupations, and workforce) is a vital axis for the development of the city.

Creative city: For a smart city, creativity can be considered as a primary driver. In addition to this, it also represents a version of it. For building a city which is smart as per the humanoid standards, social infrastructures such as intellectual as well as social capital are very crucial factors. People and their relationships get concerned because of these infrastructures. Creating a smart city concept with a mixture of teaching and guidance, arts, music and literature, and trade and industry would be a lot easier as a smart city benefits from social capital.

Learning city: A learning city is occupied in preparing a skilled workforce. Competitiveness in the international knowledge economy is enhanced with such a type of city, and an analysis of cities that are trying to become smart was recognized by Campbell: (1) independently active city, (2) group of cities, (3) direct link among cities, and (4) city system. This taught the city that it ought to be possible and realistic by being smart from the learning process which is supported by a city staff.

Humane city: It uses the human capability, the informatics staffs. Such a tactic will help in focusing on learning as well as developing a hub for higher education, which can be the city, acquiring learned personalities. It is stated that this approach will transform a smart city into a city that is full of skilled workforces. High-tech knowledge-sensitive industries who desire to transfer into dynamic and a proactive community can be covered under the same reasoning. As a result of the abovementioned transfer, the diversity between a smart city and other cities is widening. This is because smart cities prove to be a magnet for creative people and workers and thus becoming smarter day by day while the other cities are becoming less smart.

Knowledge city: Such a city is based on knowledge wealth and invention procedure. Such a smart city is analogous to a learning city; the lone dissimilarity is that "a knowledge city is heavily related to knowledge economy, and its distinction is stressed on innovation."

The intelligent and educating cities are the other smart cities that can be linked to a knowledge based on similar evolving concepts. Development of knowledge cities can be related to one of the key characteristics of these cities, which is the basic concept of knowledge-dependent urban development. This has turned out to be a vital as well as a well-known mechanism for its development.

4.6.3 Institutional Framework

The Smart Communities movement transformed into an approach for widening the base of users who are occupied in IT [12], for motivating the usage of IT in improving the value of everyday life as a result of diverse deterioration in regular activities. It has to be made clear the resulting improvement because of the mentioned usage of IT can be very challenging without the help of institutions; certainly, for the success of smart city initiatives, institutional help is very important. Simply "a smart growth is a desire to build and plan a smart community"; a smart development is important because whatever the cooperation among the civilians and formal associations attempts at doing is just a response deteriorating trends in regular activities, such as traffic jams, school crowding, and air pollution. Yet, it should be noted that technical spread is just a medium to reinvent cities for a newer market and culture and is not an end in itself. To conclude, it could be stated that any smart city initiative would require government's support to achieve success.

4.6.4 Energy Framework

For enhancing the quality of life factors living and working in the city, smart cities make use of various data and technology in creating efficiencies, improving sustainability, and creating economic development. This even shows that the city is equipped with smarter energy infrastructure. An appropriate definition is that "an urban area that has strongly incorporated technology across the information and IoT sectors to better manage a city's assets."

"Smart connections" power several items in a smart city. These items include smart buildings, data analytics, street lighting, distributed energy resources, and smart transportation. Among the listed things, energy is the principal item. Hence the utility companies play a vital role in creating smart cities. Acceleration of the development of the smart cities in the entire world is possible because of the collaboration of electric companies with city officials, technology companies, and various other institutions.

4.6.5 Strategy of the Development of Smart City

The increase in the urban population is a vital driving factor behind urban expansion. However, this explanation is not sufficient for Europe. Cities in Europe are becoming much less dense. Since the mid-1950s population in the European cities has risen by only 33% as compared with the expansion of cities, which on an average is 78%. In the PLUREL project, which is a united project within the EU's sixth Research Framework Programme, a group of more than 100 researchers, all of them from 15 countries, examined the effects of urban land utilization at a pan-European level [35,47,70]. Also, six European and one Chinese case studies where thoroughly examined. These studies helped acknowledge how the land use conflicts as well as the pressure toward periurban areas can be skillfully managed under various development and regulatory circumstances. In conclusion the resulting strategies were pinpointed for being the key steps that would help in achieving better and efficient sustainable urban-rural futures: (1) a better organized and managed transport, land use, and open space planning; (2) urban control and densification for the advancement of a green compact city; (3) protecting the blue and green infrastructure; and (4) safeguarding the agricultural land and the encouragement of local production. The governance at the regional level should be strengthened. On the other hand, at the pan-European level, there is a necessity for better policy attention for the urban-rural linkages.

Periurbanization might become the main challenge of the 21st century, especially for the planning and designing at regional level as well as city as a whole. The zone of social and economic changes and restructuring is known as periurban. Because of the changes incorporated, it becomes a zone of rigorous chaotic development. It is a multipurpose landscape where urban renewal and development can take place rather than being just an in-between or edge space at the urban periphery.

This chapter summarizes the agenda of the policy, and various study methods for periurbanization are discussed. This summary suggests a number of paths for periurban development which are extending from urban shrinkage to regulated polycentric growth with some importance given to urban sprawl.

Urban sprawl is an easily noticeable trend with respect to the present changing aspects of cities and urban regions. Any low-density expansion or "Leapfrog development" of bigger urban areas into the neighboring rural landscape is called urban sprawl. According to other definitions, it is a development accompanied by weak regional land use planning and control. In the years 1990—2006, the territories of urban area and the related infrastructure all over Europe added area approximately at the rate of 1000 km^2 per year, which is the same as that of the city of Berlin.

Population growth, which is caused due to migration (for Europe), is considered as a vital driving force behind urban expansion. However, it is not the complete clarification. Since the mid-1950s population in the European cities has risen by only 33% as compared with the expansion of cities, which on an average is 78% [24,40].

The crucial drivers support this kind of development where a city trend shows increasing economic welfare as well as the living standards in those areas, along with smaller household sizes; however, there are increasing numbers of these households.

How will the future be developed? In the PLUREL project, established on the global situations of the Intergovernmental Panel on Climate Change,

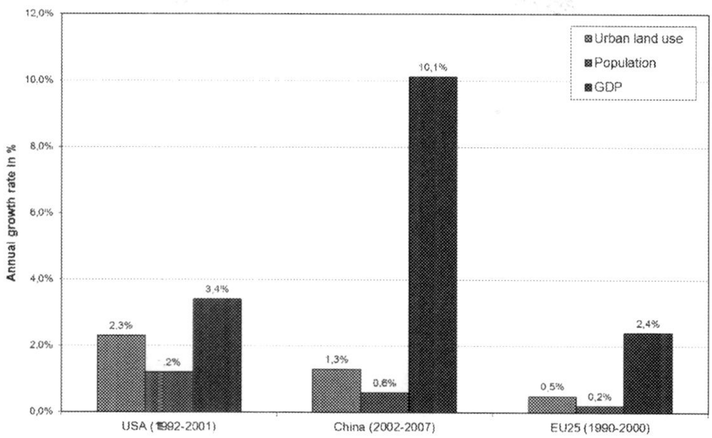

Figure 4.33 Progression rates for urban land use, population, and GDP in the EU, the United States, and China. *Zhu, et al., 2004; Ministry of Land and Resources P.R.C. 2007; U.S. Department of Agriculture, 2003; European Environment Agency 2005; and United Nations databases, 2009. Accessed at: http://data.un.org/.*

four probable situations aimed at achieving the future development of urban areas in Europe, as described in the "Special Report on Emissions Scenarios." The description of these PLUREL scenarios were based on two theoretical axes: (1) an axis where international and local are plotted against local dynamics, and (2) an axis where an emphasis on public, social, and environmental values is plotted against private enterprise/economic values (Fig. 4.33). The situations formed the basic foundation for successive "top-down" modeling work at the pan-European level on demographic, land use, economic, and environmental changes. They are also used as the basis for examination of regional "bottom-up" situations within each case study (Fig. 4.34).

A1 "Hypertech" is defined as a future world, wherein there exists rapid economic growth, rapid spread of new technologies, and falling energy prices. A2 "Self-reliance" is defined as a rural world where the economic development is mainly oriented regionally; there are divisions in per capita economic growth and technological change, and these are slow too. B1 "Peak oil" is described as a future resistant society of environmental and social awareness—a global methodology of achieving sustainable development and a dramatic rise in fossil fuel energy prices. B2 "Fragmentation" is defined as a world of slow economic growth and a segregated society in terms of age, ethnicity, and increased international distrust.

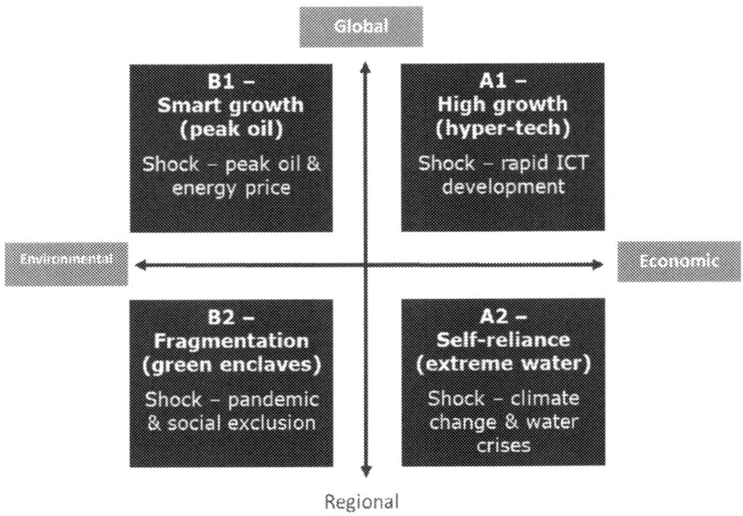

Figure 4.34 PLUREL situation framework and images (Ravetz and Rounsevell, 2008).

4.6.5.1 Strategies for Growth Management

Conflicts among the urban, agricultural, and nature-based values are triggered because of the development in the periurban areas. The disparities of the urban and rural areas are generally used in policy and regulatory terms. The six European and one Chinese PLUREL case study regions signify exceptional showcase instances of how these disagreements and the pressures with respect to periurban areas can be cleverly handled in numerous development and regulatory frameworks [37].

The inconsistent economic, geographic, and social settings that are existing in Europe can be examined from the six European case studies; however, these are different in terms of the dissimilar types of governance for each case. There is also a variation observed in the population patterns in the case study regions, extending from growing monocentric or polycentric areas [38,47] to regions with constant decline in population, which results in diverse rural—urban development patterns [43,51]. In economic terms, these include regions that are performing strongly (66.88), regions that are defined by economic regeneration [24,43], and those that are defined by a shifting economy. Ultimately, a Chinese reference study was chosen for comparing how much the results are relevant to the very rapidly urbanizing areas in Asia, referring to a cellular robotic model that advocates urban development as a growth process. The possibility of land use change is determined by the position of each grid cell and the appropriateness of the land

and zoning regulations, along with the status of neighboring cells. Regional storylines were expanded for modeling situations of land use change in the case study regions up to 2025.

The sustainability indicators help to measure sustainability. Also, a comprehensive response function approach, e.g., modifications in land use along the rural—urban band can be used for the measurement of sustainability [47]. Without the practitioners and stakeholders from the urban regions with expertise in measuring sustainability, the outcomes of peri-urbanization cannot be completely surveyed and/or understood. These practitioners can evaluate the sustainability indicators with anticipated outcomes and consequently modify their strategies to aim for required outcomes in a future that are doubtful but may be tackled by means of evaluating the situations.

The following subsections provide instances of probable approaches to manage peri-urban growth as it is recognized in the case study regions. These approaches were examined based on a combined analytical and assessment method of governance and spatial planning in the regions, in addition, to the defined situation.

The following are the four key land use problems that were used as benchmarks for assessing the approaches: (1) reduction in the pressure on housing in urban periphery, (2) strengthening agriculture in the urban fringe, (3) sustainable integration of tourism and recreation with regional development, and (4) preservation of high-value nature areas in the urban frontier.

4.6.5.2 Territorial Cohesion

The examples of modern advanced strategies for achieving better synchronization of transport and land use planning are The Regional Structure Plan of The Hague Region and the Scheme of Territorial Coherence of Montpellier Agglomération. The Regional Structure Plan of The Hague Region gives the key tactics of the region with respect to bracing the connection between spatial planning and transportation/traffic, managing sectoral objectives, and forming a base for having long-term politically sanctioned agreements between the participating municipalities [23,50]. In the Scheme of Territorial Coherence of Montpellier Agglomeration (SCOT: Schéma de Cohérence Territoriale), where in a highly attractive region is involved, there is obviously a level of political motivation for developing the region via a sustainable distribution of open green lands and urban development which is established on the grounds of quality of life and attractiveness measures, in addition to protecting the agriculture and green space in the urban periphery.

There are a number of divisions in the local municipalities which have certain degree of independence in planning, and this is an important issue. On the other hand, for managing urban development that will benefit the wider region, a combined territorial policy that will approach the local level is much needed. In majority of European countries, official power of the government and the capability of controlling and coordinating the complete urban region regarding the spatial development are very limited.

To accomplish sustainable development, effective governance, in other words, the scope of public debate, collaborations, communication, and dialog between the citizens, organizations, and local governments is a prerequisite. The Regional Structure Plan of The Hague Region and the Scheme of Territorial Cohesion of Montpellier Agglomération have been previously stated as good examples of effective governance for launching policies for guarding the agriculture and green space in the urban peripheral region. On the other hand, the condition in Warsaw is characterized by strong pressure on high-value nature and agricultural areas owing to excessive population and economic growth rates and unrestrained peri-urbanization in the areas of Warsaw Metropolitan Area, which surround the city as well as the broader Mazovian region. There have been several efforts by the regional authorities to work out a measure of control over this, which can be found in the prevailing strategies and plans, but they have not yet provided a satisfactory basis for sustainable development to take hold. The procedures and regulations that have been worked out emphasize only on certain issues, primarily of an economic nature, with the local interests in elements which have a priority over overall coordination and cooperation with neighboring communities [23,27].

4.6.5.3 Urban Containment and Densification—the Development of a Green Compact City

For controlling urban growth and achieving sustainable development that concentrates on effective land use and the conservation of rural functions such as ecosystem services, urban containment is an approach that is handled by definite boundary zone between urban and rural land use. For the accomplishment of such an approach, strong planning legislature (government) along with an open discussion with different stakeholder groups (governance) are the vital factors. Although the Green Belt Policy of the UK and the Green Heart in the Netherlands are policies that have shown success in quantitative terms, there exist negative consequences as can be seen in the Greater Manchester Area. A particular amount of the development, that could take place in the Green Belt of Manchester, is just being opposed, bypassing the belt itself, and this could be leading to larger influential effect on the countryside, as well as the commuters could face longer travel distances. The issue imposed here could become more serious as the commuters gradually prepare to accept longer traveling distances, which is observed by the growth in tele-working as stated previously.

Another popular strategy to avoid urban sprawl is smart urban densification which is designed into a pattern of multifunctional neighborhoods. The EC Thematic Strategy on the Urban Environment (CEC 2006) recommends enhanced coordination between urban transport and proper planning for land use for denser settlements. The Hague Region has considered these recommendations where the goal is to have 80% of all urban development taking place within the existing urban fabric. However, one of the limiting factors for urban densification can be the quality of life. For instance, although urban densification at the cost of decreasing area of parks and other public open spaces may perhaps seem to be a good approach at the regional level, it might have some serious negative effects at the local level. Hence the Regional Structure Plan of The Hague Region identifies numerous functions and better quality public space, in addition to a limited sprawl in a planned way which leads to integrated polycentric development as useful strategies to maintain quality of life in dense areas.

The matter of urban density versus green space is an instance of the capable clash between climate change improvement and adaptation concerns. For the reduction of energy by decreasing demands in travel and heating requirements, if multiplying density indicates the loss of green space, one effect would be the loss of an important adaptation resource. For instance, if the number of heavy rainstorms is increased, it can cause repeated issues with flooding that will be risky in areas having insufficient open spaces allowing infiltration of water. In addition to this, the urban heat island effect is likely to strengthen more common temperature extremes in urban environments if the average temperatures change in the future. According to the results obtained for the city of Manchester, even a 10% increment in the green space would balance the worst-case temperature scenario by 2080.

A judicious discussion of the view encouraged by the European Commission policy on urban planning and sustainability is thus required. New concepts need to be brought in for making compact cities more attractive so that residents who leave the compact cities for living in better environments in the urban fringe can be stopped. The main plan is to deal with suburbanization processes by improving the city and quality of life to keep the residents in the city center.

4.6.6 DISCUSSION AND FUTURE PERSPECTIVES

Eventually it needs to be thought about how the cities have a bottom to top approach when it comes to growth, wherein the main elements are economic and transportation behaviors, whereas the subordinate role is played by spatial planning and urban design. Some excellent models regarding good governance such as the regional structure plan and the scheme of territorial coherence of montpellier agglomération are also found. It can be noted from these examples how a robust form of government with a defined jurisdiction area for the particular urban unit and also an efficient spatial planning system is most important and an inevitable requirement. At strategic level, the local regional authority needs to have their input/viewpoint of the rural—urban region. This will eliminate a lot of differences that can be expected in various parts such as north/south, east/west, rich/poor, and, finally even, and urban/rural to grow broader. This will result in a state where a sustainable and complete world will become ever difficult to visualize.

There are two factors on which the capability that the public authorities have if they could oppose the thrust of those several market players which form an even more urban sprawl is based on, the power of either the administrative or governance-based planning instruments and the governmental level at which land use decisions are taken. The guiding potential for favoring the wider urban group weakens with increasing decentralization in the system and relaxations in the policy. Direct and indirect incentives can be created for the urban stretch by integrating spatial planning and the interconnection of the territories polices with the mechanisms of environmental stewardship (Table 4.8).

In spite of this, recent trends show that flow of power is in the opposite direction, i.e., moving toward achieving more decentralized growth management and deregulation. Considering control of spatial development, the regional level has grown weaker in countries such as France, the UK, and the Netherlands. According to all six European case studies, during the period of 2000—06, the urban area rose, even in Leipzig-Halle, which saw a significant reduction in its population, even though the utilization of urban area per new resident was very unusual. Manchester and

Table 4.8 Strength of the public sector in controlling urban development (Tosics and Gertheis, 2010)

	Haaglanden	Manchester	Montpellier	Leipzig	Koper	Warsaw
Financial transfer system	0.67	1.00	0.67	0.67	0.00	0.33
The local taxation system	0.67	0.67	0.67	0.33	1.00	0.67
Local government financing systems	0.67	0.83	0.67	0.50	0.50	0.50
Economic development and infrastructure	1.00	0.50	0.50	1.00	0.50	1.00
Transport	0.83	0.67	0.50	0.50	0.17	0.33
Housing	1.00	0.67	0.67	0.33	0.33	0.33
Sectoral policies	0.94	0.61	0.56	0.61	0.33	0.56
Tools to steer development	0.75	0.50	0.50	0.75	1.00	0.25
Summary	6.53	5.45	4.74	4.69	3.83	3.97

Montpellier became denser, whereas The Hague Region, Warsaw, Leipzig, and Koper became sparse in 2006 than in 2000. Surprisingly, although The Hague Region was anticipated to have the highest capability of controlling urban growth, the region saw the second largest additional urban area per new resident that measured 786 square meters, which succeeded Warsaw having 904 square meters. The Hague Region had the lowest urban area per resident of the regions in the year 2000. During this period, the large planned expansions in the urban area were achieved in The Hague.

There is a higher requirement of attention to be given to policies for urban–rural linkages at the regional level for having a better balanced, sustainably developed urban growth and reduction. However, there is also a clear requirement of new plans with respect to the EU policies and the resources that are linked with spatial development. The time has come when the ancient differences between the policy-making for urban and rural issues should be questioned. As an alternative, perceptions that are more general and are territory oriented should be designed to influence future EU agricultural and structural policies. For receiving any EU support from the Structural and Cohesion Funds or even from the Common Agricultural Policy, a prerequisite policy of a united urban–rural development should be designed.

For supporting sustainable urban—rural relationships, research plays a very important role. A number of vital knowledge breaches have been recognized with respect to the present research effort. To begin with, more attention is to be given to the type and the consequences of the various decision-making mechanisms. Land use management and change invites several uncertainties because different types of steering mechanisms, such as planning of the legislation, rules of taxation, housing policies, multilevel governance, and so forth, would be effective or not and to what extent. Also there is not much satisfaction achieved in our knowledge of effectiveness of various transport systems in relation to the shape and morphology of urban structures, i.e., whether they are compact or dispersed or they are monocentric or polycentric. At last, the idea of sustainable city will require an enhanced knowledge that will help to achieve an equalized balance of the urban densification and the preservation of the green infrastructure.

ACKNOWLEDGMENTS

This research was supported by Green Globe Technology under the grant RD-02016-09. Any findings, conclusions, and recommendations expressed in this chapter are solely those of the author and do not necessarily reflect those of Green Globe Technology.

REFERENCES

[1] Agger AK, Sørensen AH. Atomic and molecular structure and dynamics. Phys Rev A 1997;55:402.
[2] Born M, Wolf E. Principles of optics. 7th ed. Cambridge: Cambridge University Press; 1999.
[3] Boukhezzar B, Siguerdidjane H. Nonlinear control with wind estimation of a DFIG variable speed wind turbine for power capture optimization. Energy Convers Manag 2009;50:885—92.
[4] Reinhard A, Volz T, Winger M, Badolato A, Hennessy KJ, Hu EL, Imamoğlu A. Strongly correlated photons on a chip. Nat. Photonics 2012;6:93—6.
[5] Agarwal V, Aggarwal RK, Patidar P, Patki C. A novel scheme for rapid tracking of maximum power point in wind energy generation systems. IEEE Trans Energy Convers 2010;25(1):228—36.
[6] Baur G, Hencken K, Trautmann D, Sadovsky S, Kharlov Y. Dense laser-driven electron sheets as relativistic mirrors for coherent production of brilliant X-ray and γ-ray beams. Phys Rep 2002;364:359.
[7] Alsema EA, Wild-Scholten MJD, Fthenakis VM. Environmental im- pacts of PV electricity generation - a critical comparison of energy supply options. In: Proceedings of 21st European Photovoltaic Solar Energy Conference, vol. 50; 2006. p. 97—147.
[8] Faida H, Saadi J. Modelling, control strategy of DFIG in a wind energy system and feasibility study of a wind farm in Morocco. Int Rev Model Simul 2010;3(6):1350—62.
[9] Arnold P, Moore GD, Yaffe LG. Photon emission from ultrarelativistic plasmas. J High Energy Phys 2001;11:057.

[10] Douglas JS, Habibian H, Hung CL, Gorshkov AV, Kimble HJ, Chang DE. Quantum many-body models with cold atoms coupled to photonic crystals. Nat Photonics 2015; 9:326–31.

[11] Diaf S, Notton G, Belhamel M, Haddadi M, Louche A. Design and techno-economical optimization for hybrid PV/Wind system under various meteorological conditions. Appl Energy 2008;85:968–87. https://doi.org/10.1016/j.apenergy.2008.02.012.

[12] Dirac PAM. The quantum theory of the emission and absorption of radiation. Proc R Soc London 1927;A114:243–65.

[13] Artemyev N, Jentschura UD, Serbo VG, Surzhykov A. Strong electromagnetic field effects in ultra-relativistic heavy-ion collisions. Eur Phys J 2012;72:1935.

[14] De Soto W, Klein SA, Beckman WA. Improvement and validation of a model for photovoltaic array performance. Sol Energy 2006;80:78–88.

[15] Najjari B, Voitkiv AB, Artemyev A, Surzhykov A. Simultaneous electron capture and bound-free pair production in relativistic collisions of heavy nuclei with atoms. Phys Rev A 2009;80:012701.

[16] Baur G, Hencken K, Trautmann D. Revisiting unitarity corrections for electromagnetic processes in collisions of relativistic nuclei. Phys Rep 2007;453:1.

[17] Boyle G. Renewable energy. 2nd ed. Oxford (UK): Oxford University Press; 2004.

[18] Celik AN, Acikgoz N. Modelling and experimental verification of the operating current of mono-crystalline photovoltaic modules using four- and five-parameter models. Appl Energy 2007;84:1–15.

[19] Beatty JK, Peterson CC, Chaokin A. The solar system. Cambridge (UK): Cambridge University Press; 1999.

[20] Deffeyes KS. Beyond oil. New York: Hill and Wang; 2005.

[21] Bhandari B, Poudel SR, Lee KT, Ahn SH. Mathematical modeling of hybrid renewable energy system: a review on small hydro-solar-wind power generation. Int J Precis Eng Manuf Green Technol April 2014;1(2):157–73.

[22] Becker U, Grün N, Scheid W. K-shell ionisation in relativistic heavy-ion collisions. J Phys B At Mol Phys 1987;20:2075.

[23] Bethe HA. Energy production in stars. Phys Rev 1939;55:434–56.

[24] Energy Information Administration (EIA). Official energy statistics from U.S. government. History: international Energy Annual 2004. May–July 2006.

[25] Belkacem H, Gould B, Feinberg R, Bossingham WE, Meyerhof. Semiclassical dynamics and relaxation. Phys Rev Lett 1993;71:1514.

[26] Beltran B, Benbouzid M, Ahmed-Ali T, Mangel H. DFIG-based wind turbine robust control using high-order sliding modes and a high gain observer. Int Rev Model Simul 2011;4(3):1148–55.

[27] Benavides ND, Chapman PL. Modeling the effect of voltage ripple on the power output of photovoltaic modules. IEEE Trans Ind Electron 2008;55:2638–43.

[28] Gratzel. Photoelectrochemical cells. Nature 2001;414:338–44.

[29] Green MA, Zhao J, Wang A, Wenham SR. Progress and outlook for high-efficiency crystalline silicon solar cells. Sol Energy Mater Sol Cell 2001;65:9–16.

[30] Brabec CJ, Sariciftci NS, Hummelen JC. Plastic solar cells. Adv Funct Mater 2001;11: 15–26.

[31] Djamel L, Abdallah B. Power quality control strategy for grid-connected renewable energy sources using PV array, wind turbine and battery. In: Proceedings. Of the fourth international Conference on power engineering, energy and electrical drives. IEEE; 2013. p. 1671–5.

[32] Duan L, Sun S, Yue L, Qu W, Bian J. Study on different zero CO_2 emission IGCC systems with CO_2 capture by integrating OTM Study on different zero CO_2 emission

IGCC systems with CO_2 capture by integrating OTM. Int J Energy Res April 1, 2016. https://doi.org/10.1002/er.3533.

[33] Duffie A, Beckman WA. Solar engineering of thermal processes. 3rd ed. Hoboken (NJ): John Wiley and Sons; 2006. Ann Rev Energy 1979;4:353–401.

[34] Liu BYH, Jordan RC. The interrelationship and characteristic distribution of direct, diffuse, and total solar radiation. Sol Energy 1960;4(3):1–19.

[35] Kennedy CE. Reiew of mid- to high-temperature absorber materials. Natl Renew Energy Lab Rep 2002;520:31267.

[36] Kjell Nilsson K, Thomas Sick Nielsen TS, Aalbers C, Simon Bell S, Boitier B, Chery JP, Fertner C, Groschowski M, Haase D, Loibl W, Pauleit S, Pintar M, Piorr A, Ravetz J, Ristimäki M, Rounsevell M, Tosics I, Westerink J, Zasada I. Strategies for sustainable urban development and urban-rural linkages 2011.

[37] Duffie A, Beckman WA. Solar energy thermal processes. New York: John Wiley and Sons; 1974.

[38] Dürr M, Cruden A, Gair S, McDonald JR. Dynamic model of a lead acid battery for use in a domestic fuel cell system. J Power Sources 2006;161(2):1400–11.

[39] Tan YT, Kirschen DS, Jenkins N. A model of PV generation suitable for stability analysis. IEEE Trans Energy Convers 2004;19:748–55.

[40] Eichler J, Stöhlker T. Radiative electron capture in relativistic ion-atom collisions and the photoelectric effect in hydrogen-like high-Z systems. Phys Rep 2007;439:1–99.

[41] Valluri SR, Becker U, Grün N, Scheid W. Relativistic collisions of highly-charged ions. J Phys B At Mol Phys 1984;17:4359.

[42] Xiao YF, Li M, Liu YC, Li Y, Sun X, Gong Q. Asymmetric Fano resonance analysis in indirectly coupled microresonators. Phys Rev A 2010;82:065804.

[43] Kennedy CE, Price H. Progress in development of high-temperature solar selective coating. Proc ISEC 2005;520(36997):2005.

[44] Mamdani EH, Assilina S. An experiment in linguistic synthesises with a fuzzy logic controller. Int J Man Mach Stud 1975;7:1–13.

[45] Maxwell JC. A dynamic theory of the electromagnetic field. Reprinted by Wipf and Stock Publishers; 1996. p. 1864.

[46] Gaillard A, Poure P, Saadate S, Machmoum M. Variable speed DFIG wind energy system for power generation and harmonic current mitigation. Renew Energy 2009;34: 1545–53.

[47] Gopal C, Mohanraj M, Chandramohan P, Chandrasekar P. Renewable energy source water pumping systems—a literature review. Renew Sustain Energy Rev 2013. https://doi.org/10.1016/j.rser.2013.04.012.

[48] Park J, Kim H, Cho Y, Shin C. Simple modeling and simulation of photovoltaic panels using Matlab/Simulink. Adv Sci Technol Lett 2014;73:147–55.

[49] Garg HP, Mullik SC, Bhargava AK. Solar thermal energy storage. Dordricht: D. Reider Publishing Company; 1985.

[50] Ghennam T, Berkouk EM, Francois B. A vector hysteresis current control applied on three-level inverter. Application to the active and reactive power control of doubly fed induction generator based wind turbine. Int Rev Electr Eng 2007;2(2):250–9.

[51] Gould RJ, Schréder GP. Pair production in photon-photon collisions. Phys Rev 1967; 155:1404.

[52] Gupta N, Singh SP, Dubey SP, Palwalia DK. Fuzzy logic controlled three-phase three-wired shunt active power filter for power quality improvement. Int Rev Electr Eng 2011;6(3):1118–29.

[53] Hencken K, Baur G, Trautmann D. Transverse momentum distribution of vector mesons produced in ultraperipheral relativistic heavy ion collisions. Phys Rev Lett 2006; 96:012303.

[54] Hossain F. Solar energy integration into advanced building design for meeting energy demand and environment problem. J Energy Res 2016;17:49−55.
[55] Hossain Faruque Fara Nowshin. Integration of Wind into Running Vehicles to Meet Its Total Energy Demand. Energy, Ecology, and Environment 2016;2(1):35−48 (Springer Nature).
[56] Green MA. Limits on the open-circuit voltage and efficiency of silicon solar cells imposed by intrinsic auger processes. IEEE Trans Electron Dev 1996;31:671−8.
[57] Güçlü MC, Li J, Umar AS, Ernst DJ, Strayer MR. Electromagnetic Lepton pair production in relativistic heavy-ion collisions. Ann Phys 1999;272:7−48.
[58] Hossain F. In situ geothermal energy technology: an approach for building cleaner and greener environment. J Ecol Eng 2016;17:49−55.
[59] Robyns B, Francois B, Degobert P, Hautier JP. Vector control of induction machines. London: Springer-Verlag; 2012.
[60] Soon JJ, Low KS. Optimizing photovoltaic model parameters for simulation. In: IEEE international Symposium on industrial electronics; 2012. p. 1813−8.
[61] Stanbery BJ. Copper indium selenides and related materials for photovoltaic devices. Crit Rev Solid State Mater Sci 2002;27:73−117.
[62] Kamal E, Koutb M, Sobaih AA, Abozalam B. An intelligent maximum power extraction algorithm for hybrid wind-diesel-storage system. Int J Electr Power Energy Syst 2010;32(3):170−7.
[63] Liu Y, Xie J, Ong CN, Vecitis CD, Zhou Z. Electrochemical wastewater treatment with carbon nanotube filters coupled with in situ generated H_2O_2. Environ Sci Water Res Technol 2015;(6).
[64] Lo PY, Xiong HN, Zhang WM. Breakdown of Bose-Einstein distribution in photonic, crystals. Sci Rep 2015;5.
[65] Sherif RA, Karam NH. 40% efficient metamorphic GaInP/GaInAs/Ge multijunction solar cells. Appl Phys Lett 2007;90:183516.
[66] Shockley W, Queisser HJ. Detailed balance limit of efficiency of p-n junction solar cells. J Appl Phys 1961;32:510−9.
[67] Sivasankar G, Kumar VS. Improving low voltage ride through of wind generators using STATCOM under symmetric and asymmetric fault conditions. IREMOS 2013;6: 1212−8.
[68] Klein SA. Calculation of flat-plate collector loss coefficients. Sol Energy 1975;17: 79−80.
[69] Millikan RA. A direct photoelectric determination of Planck's h. Phys Rev 1916;7: 355−88.
[70] Langer L, Poltavtsev SV, Yugova IA, Salewski M, Yakovlev DR, Karczewski G, Wojtowicz T, Akimov IA, Bayer M. Access to long-term optical memories using photon echoes retrieved from semiconductor spins. Nat Photonics 2014;8:851−7.
[71] Mohd Zin AA, Pesaran HA, Khairuddin AB, Jahanshaloo L, Shariatu O. An overview on doubly fed induction generators' controls and contibutions to wind based electricity generation. Renew Sustain Energy Rev 2013;27:692−708.
[72] Li Q, Xu DZ, Cai CY, Sun CP. Recoil effects of a motional scatterer on single-photon scattering in one dimension. Sci Rep 2013;3.
[73] Nema P, Nema RK, Rangnekar S. A current and future state of art development of hybrid energy system using wind and PV-solar: a review. Renew Sustain Energy Rev 2009;13:2096−103. https://doi.org/10.1016/j.rser.2008.10.006.
[74] Sharma KG, Bhargava A, Gajrani K. Stability analysis of DFIG based wind turbines connected to electric grid. IREMOS 2013;6:879−87.
[75] Shay JL, Wagner S. Efficient CuInSe2/CdS solar cells. App Phys Lett 2007;27:89−90.
[76] Soedibyo A, Pamuji FA, Ashari M. Grid quality hybrid power system control of microhydro, wind turbine and fuel cell using fuzzy logic. IREMOS 2013;6:1271−8.

[77] Stix M. The sun, an introduction. 2nd ed. New York: Springer; 2002.
83a Tang CW. Two-layer organic photovoltaic cell. Appl Phys Lett 1986;48:183–5.
[78] Zhang WM, Lo PY, Xiong HN, Tu MWY, Nori F. General non- markovian dynamics of open quantum systems. Phys Rev Lett 2012;109:170402.
[79] Yang L, Wang S, Zeng Q, Zhang Z, Pei T, Li Y, Peng LM. Efficient photovoltage multiplication in carbon nanotubes. Nat Photonics 2011;5:672–6.
[80] Tu MWY, Zhang WM. Non-Markovian decoherence theory for a double-dot charge qubit. Phys Rev B 2008;78:235311.
[81] Yan WB, Fan H. Single-photon quantum router with multiple output ports. Sci Rep 2014;4.

FURTHER READING

[1] Prabhakar RJ, Ragavan K. Power management Based Current control technique for photovoltaic-battery assisted wind–hydro hybrid system. Int J Emerg Electric Power Syst 2013;14:351–62.

CHAPTER FIVE

Infrastructure and Transportation

5.1 GREEN INFRASTRUCTURE

Resilience, an important term introduced in sustainable infrastructure system, which is similar to "autonomous adaptation," responds to change in conditions by the adaptive capacity through applying proper planning and practicing flexible solutions on environment-friendly technological applications. It makes the environment green and tidy because of its adaptability, flexibility, and sustainability, which will not result in maladaptation in the process. Accordingly, a better way, namely "green infrastructure," has been introduced in this chapter to support local transformation to climate change, which can help dealing with, recovering, and managing the stability affected because of the environmental crises. Necessarily, to build adaptive capacity by resilience, implementation of sophisticated technologies including water treatment, green roofs, and urban forestry is the better way to boost the sustainability.

5.1.1 Background and Technology

Green infrastructure is being considered to provide community benefits when managing wet weather impacts because of its attributes of cost-effectiveness and resilience. It realizes economic, environmental, and social benefits by reducing and treating storm water at its source ranging from individual houses to entire city, which is much efficient compared with single-purpose gray storm water infrastructure conventional piped drainage and water treatment system [1,2]. Particularly, the community application of green infrastructure helps local government to realize adaptation, environmental, and sustainability goals within their rights. Using green infrastructure technologies with LEED (leadership of energy and environmental design) implementation inserted in all sectors of built environment as a method to prevent and adapt to climate impacts and realize environmental sustainability depends much on the situations and incentives. Therefore this report will motivate green infrastructure projects by (1) showing the environmental sustainability; (2) showing evidence of upfront payment or life-cycle cost savings when compared with alternatives we introduced in both public and

Sustainable Design and Build
ISBN: 978-0-12-816722-9
https://doi.org/10.1016/B978-0-12-816722-9.00005-7

private projects; (3) providing direct financial motivations to property owners for green infrastructure installations; (4) drawing up laws, regulations, and local ordinances requiring realization of green infrastructure on private property; or (5) incorporating public projects with green infrastructure to demonstrate viability and value.

5.1.2 Methods and Approaches

Storm water is an important part in water cycle; however, when rain falls onto the roofs, streets, parking lots, and concrete or other impermeable materials, it cannot reach the soil. Therefore the storm water drainage system uses gutters, storm sewers, and other engineering collection methods to collect the water and discharge to nearest water body, which causes the storm water runoff to become one of the major causes of water pollution and leads to a disaster for the infrastructure in urban areas because of the trash, bacteria, heavy metals, and other pollutants it carries [3—5]. Besides, high water flow caused by heavy rains may also lead to erosion and flooding. To mitigate these problems, green infrastructure could be a good choice. When rain falls onto green infrastructures, for example, natural areas, the water will permeate into the ground and be absorbed by the soil and plants and make the problems no longer exist. In the following sections, green alley, storm sewers, and streets or building storm water—storage tunnels infrastructure, and several sophisticated technologies will be introduced to present and demonstrate the possibility of green infrastructure to change the mechanism on conventional infrastructure system.

5.1.2.1 Green Alleys

To achieve rapid storm water runoff into sewers and avoid flood and heat production, impermeable materials such as asphalt or concrete are used as the surface of urban alleys, which may cause the problems mentioned previously. Green alleys, by contrast, can not only achieve these objectives, but also reduce adverse environmental impacts.

5.1.2.2 Permeable Pavement

Permeable pavement, whose objectives are similar to green alleys, can allow storm water to permeate into the ground and other management systems and generate runoff characteristics in urban areas which is akin to the characteristics in a meadow or a forest. In addition, with the help of proper subsoiling, which refers to the maintenance of a porous layer of soil underneath, the permeable pavement is expected to penetrate 3 inches of storm water from a 1-h storm in its 30–35 years of life which means a 70%–90% drainage-volume reduction. Besides, to maximize its ability, typically it should be designed with the capacity standard to manage a 10-year rainfall event within a 24-h period, which intends to make it be adapted to accounting for projected increases in frequency and intensity of storms in the future.

5.1.2.3 Downspout Disconnection and RainWater Collection

Another method of controlling storm water is to disconnect downspouts from homes and commercial buildings and reconnect them to a collection or slow dispersion system, for example, a cistern for storage use, a rain garden for slow dispersion, or a rainwater harvesting design system for both collecting and storing storm water for later use. Among those aforementioned uses, rain garden is one of the most excellent and functional systems. Rain gardens can be installed at unpaved space because of its versatile features. With its ability to mimic natural hydrology by permeating, evaporating, transpiring, and evapotranspiration runoff, rain gardens can be used to collect and absorb storm water from rooftops, sidewalks, and streets and further be developed as bioretention or bioinfiltration cells. Except for the rain garden itself, other supplementary can also help with the management of rainwater [6,7]. Bioswales, for example, can be used for water treatment and retention during the transferring of runoff from one place to another by installing in long narrow spaces, which can be found between the sidewalk and the curb with vegetated, mulched, or xeriscaped channel in the rain garden. Another example is the planter box. A planter box is another form of rain garden, which is constructed with vertical walls and either open or closed bottoms. It not only has the same characteristics of a rain garden but also fits for space-limited urban areas acting as streetscape.

5.1.2.4 Urban Forestry

Trees, as contributors to improve urban life quality, have plenty of benefits including blocking and penetrating rainwater to avoid flooding and improving water quality, absorbing and transforming air pollutants, providing windbreaks to protect buildings from strong wind, reducing heat island effect by providing shadows and evaporation, cutting down carbon dioxide indirectly through lowering cooling demand for electricity, and served as mulch after dead [8–10]. Therefore planting and maintaining trees in urban area can be a good choice considering its benefits for resilience, adaptation, and climate mitigation. According to estimation, a typical medium-sized tree can block around 2380 gallons of rainfall per year. When it comes to mitigating urban heat island impacts, trees typically absorb 70%–90% of sunlight in summer and 20%–90% in winter (because of the seasonal variation between deciduous trees and evergreens) and further reduce the maximum surface temperature of the roofs and walls by 11–25°C [11,12]. The new shade trees, planted around houses, can lead to annual cooling energy savings of 1% per tree and annual heating energy use decreased by almost 2% per tree. Besides, direct energy savings from shading by trees could reduce carbon emissions around 1.5%–5% due to decreases in cooling energy use. As have mentioned previously, trees can absorb air pollutants, these includes particulate matter, sulfur dioxide, ground-level ozone, nitrogen oxides, and carbon monoxide. As estimated by a research, urban trees in the United States remove 784,000 tons of pollutant per year, which creates $3.8 billion economic value [13,14]. By increasing the urban tree coverage rate such as that in New York City by 10%, the ground-level ozone can be cut down by 3%, besides, a quarter ton of NOx and over one ton of particulate matter will be cut down per day with one million additional trees in city [15–17].

5.1.2.5 Green Parking

As many green infrastructure elements, for example, permeable pavements, rain gardens, bioswales mentioned previously can be easily coordinated into parking-lot design; it is a good choice to combine green infrastructure and parking lots. The benefits of green parking include reducing urban heat island and a more walkable built environment.

5.1.2.6 Green Roofs

Green roofs, another green infrastructure technology, refers to a roof covered with waterproof membrane with plants or trees suitable for the local climate and 3–15 inches of soil, sand, or gravel. For large storms, green roofs can detain 30% runoff water, whereas for storms less than one inch, the number rises to 90%. In addition, seasonal and physiological evapotranspiration rates of plants also affect effectiveness of runoff control, with summer growing season being better than winter. As the green roof is made up of plants and trees, it has similar benefits as trees including the transformation of water and air pollutants mentioned previously. With the help of a 1000-ft^2 green roof, around 40 pounds of particulate matter can be removed annually, which equals to the annual emission of 15 passenger cars. In addition, to the benefits of removing pollutants, green roofs can also provide climate change mitigation through managing temperature. Compared to traditional roofs, green roofs can cut down building electricity consumption by 2%–6% especially for summer cooling by maintaining temperature. Therefore the green roof would be a great way to control climate change and keep urban area environment friendly.

5.1.2.7 Better Approach to Green Management

Except for the implementation of green infrastructure technologies, green management practices, including urban design and planning, and smart growth approaches incorporated in green infrastructure system are also important [18,19]. Higher density housing, one of the green management practices, provides green open spaces, large-scale urban forestry projects in neighborhoods, green belts around cities, and coastal wetlands buffering against flooding and demonstrates the importance of green management. Besides, in flood zones, local building codes may require raising buildings or bridges above current and future flood levels or setting the first floors at floodable positions, which is also an application of green management. As green infrastructure is a method for creating more resilient metropolitan communities, improving environmental sustainability, smart growth, and climate adaptation in urban areas concurrently, smart growth practice may be the tool of urban design and planning to improve resource efficiencies, increase building density, realize mixed land uses, build more open space, achieve public transit–oriented development, and enhance quality of life. Climate adaptation, in this process, is better to be introduced to increase the capacity of local communities to better evaluate and regulate risks, impacts, and opportunities from irreversible climate change and extreme weather including wildfire, public health threats, floods, droughts, sea-level rise and so forth.

5.1.2.8 Conclusion

To conclude, green infrastructure, mentioned in this chapter, contributes to dealing with storm water runoff, managing water and air pollutants, increasing biodiversity, providing less heat stress, managing climate adaptation, realizing sustainability, improving human life quality, and so forth. It is used to provide an ecological framework for social, economic, and environmental health of the urban areas. After these sophisticated technologies mentioned previously have been realized, the green infrastructure shall be a good solution to achieve environmental and sustainability goals and further create a better resilient community.

5.2 INVISIBLE ROADS AND SUSTAINABLE TRANSPORTATION ENGINEERING

A new technology **invisible road** has been proposed in this chapter to confirm zero emission of greenhouse gas from transportation infrastructure system to green infrastructure and transportation system. Simple underground **maglev** system is to be constructed for all transportation system to run the vehicle smoothly just over two feet over the earth surface by propulsive and impulsive force at flying stage. Vehicle will not require any energy because it will run by superconducting electromagnetic force created by the maglev technology. A wind energy modeling has also been added to meet the vehicle's energy demand when it will run on nonmaglev area. Naturally all maglev infrastructure network will be covered by evergreen herb except pedestrian walkways to absorb CO_2, ambient heat, and moisture (vapor) from the surrounding environment to make it cool. Indeed, the proposed maglev transportation infrastructure technology will be an innovative one in modern engineering science, which will also reduce the climate change dramatically.

5.2.1 Background and Technology

Urban and suburban areas massively depend on transportation infrastructure networks which are primarily constructional with concrete and asphalt, and it does not have enough vegetation to absorb heat caused by these asphalt and concrete [20,21]. Recent research found that transportation infrastructure on earth takes approximately 0.9% of the total planetary surface area of 196.9 million mi^2, which is equivalent to 1.77 million mi^2 infrastructure on earth, causing nearly 6% of global warming by reflecting heat (albedo) back to the space [22,23]. On the other hand, conventional energy utilization for the transportation sectors is not only costly but also causing adverse environmental impact [24,25]. A variety of studies have been performed to understand long-term climate variations by conventional energy use by the transportation sectors that is causing nearly 28% global energy consumption, which is equivalent to a mega ton CO_2, and responsible for 28% of global warming, and thus infrastructure and transportation fuel causes a total of 34% global warming [26,27]. To mitigate transportation infrastructure crisis and its adverse environmental impact, I therefore proposed a new technology of maglev transportation infrastructure system for building better transportation infrastructure system.

A recent study described the dynamic characteristics, magnetic suspension systems, vehicle stability, and suspension control laws of maglev/guideway coupling systems about the maglev transportation system [28,29], but the fact commercial application of this research modeling considering life-cycle cost analysis, technology implementation, and infrastructure development did not show any possibility to apply it commercially [30,31]. Therefore the approach of this research is to apply the maglev transportation infrastructure commercially for confirming a greener and cleaner transportation infrastructure system where all vehicles shall run just over two feet above the earth surface at flying stage by the act of propulsive and impulsive superconducting force. Because the vehicle will run by electromagnetic force, it will not require any energy while running over the maglev. To mitigate energy consumption when vehicles need to run on maglev area, additional technology has also been proposed to implement wind energy into the vehicles while it is in motion as a backup energy source. Thus a detail mathematical modeling using MATLAB Simulink software has been implemented for this wind energy utilization for the vehicles by performing turbine and drive train modeling [32–34]. A concerted research effort has been performed recently on climate science and found hat currently 402-ppm CO_2 is present in atmosphere causing global warming, which is required to cut down to 300-ppm CO_2 to confirm global cooling at comfortable stage [35,36]. Once maglev transportation infrastructure system implemented throughout the world, it will reduce 34% of CO_2 per year. Thus, it will take only $\left\{ \int_{300}^{402} (1 - 0.34)dx \right\} = 67.32$ years to cool the atmosphere, resulting in no more climate change after 68 years. Simply it will be the most innovative technology in modern science to mitigate the cost and global warming dramatically.

5.2.2 Simulations and Methods

To present maglev transportation infrastructure modeling, I have formulated the following calculation by using MATLAB software in terms of (1) guideway model system by adopting Bernoulli-Euler beam equation of series of simply supported beams and (2) calculation of magnetic forces for uplift levitation and lateral guidance with allowable levitation and guidance distance, considering lateral vibration control LQR algorithm, tuning parameters, and maglev dynamics.

5.2.2.1 Guideway Model

To prepare the guideway modeling considering the free-body diagram (Fig. 5.1), I have considered multiple magnets with equal intervals (d) that is to be traveling at various levels of speed v, where $m =$ beam weight, $c =$ damping coefficient, $EI_y =$ flexural rigidity in the y-direction, $EI_z =$ flexural rigidity in the z direction, $l =$ car length, $m_w =$ lumped mass of magnetic wheel, $m_v =$ distributed mass of the rigid car body, and $\theta_{i=x,y,z} =$ midpoint rotation components of the rigid car body. Considering these, I have formulated the equations of motion for the jth guideway girder carrying a moving maglev vehicle suspended by multiple magnetic forces as follows:

$$m\ddot{u}_{y,j} + c_y\dot{u}_{y,j} + EI_y u''''_{y,j} = \sum_{k=1}^{K}\left[G_{y,k}\left(i_k, h_{y,k}\right)\varphi_j(x_k,t)\right] \tag{5.1}$$

$$m\ddot{u}_{z,j} + c_z\dot{u}_{z,j} + EI_z u''''_{u,j} = p_0 - \sum_{k=1}^{K}\left[G_{z,k}\left(i_k, h_{z,k}\right)\varphi_j(x_k,t)\right] \tag{5.2}$$

and

$$\varphi_j(x_k,t) = \delta(x - x_k)\left[H\left(t - t_k - \frac{(j-1)L}{v}\right) - H\left(t - t_k - \frac{jL}{v}\right)\right] \tag{5.3}$$

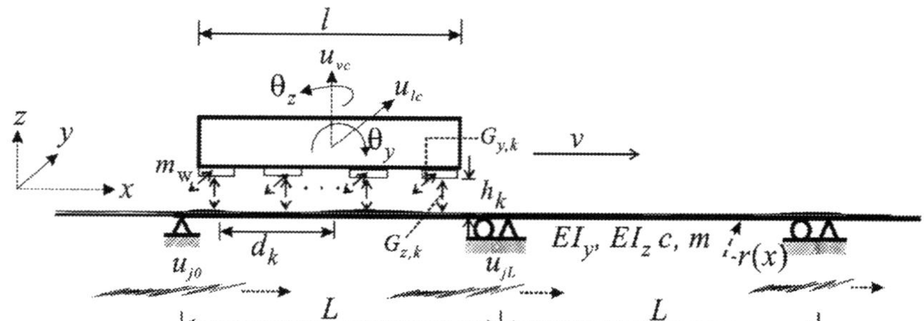

Figure 5.1 A free-body diagram shows the maglev guideway versus vehicle force, considering weight and motion where the superconducting guideway below the vehicle body. It is functioned by series of equal-distant concentrated masses to levitate the vehicle up to the superconducting guideway beam; the maglev bar gets stimulated by the lateral multisupport motion, which is induced by the superconducting force to allow traveling on longitudinal direction.

together with the following boundary conditions with lateral (y-direction) support movements:

$$\begin{aligned} u_{y,j}(0,t) &= u_{yj0}(t), u_{y,j}(L,t) = u_{yjL}(t), \\ EI_z u''_{z,j}(0,t) &= EI_z u''_{z,j}(L,t) = 0 \end{aligned} \tag{5.4}$$

$$\begin{aligned} u_{z,j}(0,t) &= u_{z,j}(L,t) = 0 \\ EI_y u''_{y,j}(0,t) &= EI_y u''_{y,j}(L,t) = 0 \end{aligned} \tag{5.5}$$

where $(\bullet)' = \partial(\bullet)/\partial x$, $(\bullet) = \partial(\bullet)/\partial t$, $u_{z,j}(x, t)$ = vertical deflection of the jth span, $u_{y,j}(x, t)$ = lateral deflection of the jth span, L = span length, K = number of magnets attached to the rigid levitation frame, $\delta(\bullet)$ = Dirac's delta function, $H(t)$ = unit step function, $k = 1, 2, 3, \ldots, K$th moving magnetic wheel on the beam, $t_k = (k-1)d/v$ = arrival time of the kth magnetic wheel into the beam, x_k = position of the k-th magnetic wheel on the guideway, and $(G_{y,k}, G_{z,k})$ = lateral guidance and uplift levitation forces of the kth lumped magnet in the vertical and lateral directions [37,38].

5.2.2.2 Magnetic Forces of Uplift Levitation and Lateral Guidance

Because the maglev vehicles will run over guideway by superconducting force with lateral ground motion (as shown in Fig. 5.1), guidance forces tuned by the maglev system need to be controlled for the lateral motion of the moving maglev vehicle. Therefore this study adopts the lateral guidance force $(G_{y,k})$ and the uplift levitation force $(G_{z,k})$ [39,40] to keep and guide the k-th magnet of the vehicle and this could be expressed as

$$G_{y,k} = K_0 \left(\frac{i_k(t)}{h_{z,k(t)}} \right)^2 K_{k,z} \tag{5.6}$$

$$G_{y,k} = K_0 \left(\frac{i_k(t)}{h_{z,k(t)}} \right)^2 (1 - K_{y,k}) \tag{5.7}$$

where $K_{y,k}$ and $K_{z,k}$ represent induced guidance factors and they are given by,

$$K_{y,k} = \frac{\chi_k \times h_{y,k}}{W(1+\chi_k)}, K_{z,k} = \frac{\chi_k \times h_{y,k}}{W(1+\chi_k)} \tag{5.8}$$

In Eqs. (5.6) and (5.7), $K_0 = \mu_0 N_0^2 A_0/4$ = coupling factor, $\chi_k = \pi h_{y,k} K_{z,k}/4h$, W = pole width, μ = vacuum permeability, N_0 = number of turns of the magnet windings, A_0 = pole face area, $i_n(t) = i_0 + i_n(t)$ = electric current, $i_n(t)$ = deviation of current, and (i_0, h_{y0}, h_{z0}) = desired current and air gaps around a specified nominal operating point of the maglev wheels at static equilibrium. And, the uplift levitation ($h_{y,k}$) and lateral guidance ($h_{z,k}$) gaps are respectively given by:

$$h_{y,k}(t) = h_{y0} + u_{l,k}(t) - u_{y,j}(x_k), \; u_{l,k}(t) = u_{lc}(t) + d_k\theta_z \tag{5.9}$$

$$h_{z,k}(t) = h_{z0} + u_{v,k}(t) - u_{z,j}(x_k) + r(x_k), \; u_{v,k}(t) = u_{vc}(t) + d_k\theta_y \tag{5.10}$$

where $u_{l,k}$, $u_{v,k}$ = displacements of the kth magnetic wheel in the y and z directions, u_{lc}, u_{vc} = midpoint displacements of the rigid car, θ_y,θ_z = midpoint rotations of the rigid car, $r(x)$ = irregularity of guideway, and d_k = location of the kth magnetic wheel to the midpoint of the rigid beam. As indicated in Eqs. (5.6)−(5.8), the motion-dependent nature and guidance factors ($K_{y,k}$, $K_{z,k}$) dominate the control forces of the maglev vehicle−guideway system. Then, the equations of motion of the 4 DOFs rigid maglev vehicle (Fig. 5.1) are written as

$$M_0\ddot{u}_{lc} = g(t) + \sum_{k=1}^{K} G_{y,k}, \quad I_T\ddot{\theta}_Z = g(t) \times l + \sum_{k=1}^{K}\left[G_{y,k}d_k\right] \tag{5.11}$$

$$M_0\ddot{u}_{vc} = p_0 + \sum_{k=1}^{K} G_{z,k}, \quad I_T\ddot{\theta}_y = -\sum_{k=1}^{K}\left[G_{z,k}d_k\right] \tag{5.12}$$

where $M_0 = m_v l + K m_w$ = lumped mass of the vehicle, $g(t)$ = control force to tune the lateral response of the maglev vehicle, I_T = total mass moment of inertia of the rigid car, and $p_0 = M_0 g$ = lumped weight of the maglev vehicle.

5.2.2.3 Wind Energy Modeling for the Vehicles

Though the vehicle will run by electromagnetic force, a wind turbine generator is to be used for powering vehicle as the additional source of energy to exit vehicle from road and park where maglev system is not available. Thus the model is developed by doubly fed induction generator (DFIG) for producing electricity for transportation vehicles [41−43]. The fundamental equation governing the mechanical power of the wind turbine is

$$P_w = \frac{1}{2}C_p(\lambda, \beta)\rho A V^3 \tag{5.13}$$

where ρ is the air density (kg/m^3), C_p is the power coefficient, A is the intercepting area of the rotor blades (m^2), V is the average wind speed (m/s), and λ is the tip speed ratio [10]. The theoretical maximum value of the power coefficient C_p is 0.593; C_p is also known as Betz's coefficient. Mathematically,

$$\lambda = \frac{R\omega}{V} \tag{5.14}$$

R is the radius of the turbine (m), ω is the angular speed (rad/s), and V is the average wind speed (m/s). The energy generated by wind can be obtained by the equation

$$Q_v = P \times (Time)(\text{kWh}) \tag{5.15}$$

It is well known that wind velocity cannot be obtained by a direct measurement from any particular motion [37,44]. In data taken from any reference, the motion needs to be determined for that particular motion; then, the velocity needs to be measured at a lower motion.

$$v(z)\ln\left(\frac{Z_r}{Z_0}\right) = v(Z_r)\ln\left(\frac{Z}{Z_0}\right) \tag{5.16}$$

where Z_r is the reference height (m), Z is the height at which the wind speed is to be determined, Z_0 is the measure of surface roughness (0.1−0.25 for crop land), $v(z)$ is wind speed at height z (m/s), and $v(z_r)$ is wind speed at the reference height z (m/s). The power output in terms of the wind speed shall be estimated using the following equation:

$$P_w(v) = \begin{cases} \dfrac{v^k - v_C^k}{v_R^k - v_C^k} \cdot P_R & v_C \le v \le v_R \\ P_R & v_R \le v \le v_F \\ 0 & v \le v_C \text{ and } v \ge v_F \end{cases} \tag{5.17}$$

where P_R is rated power, v_C is the cut-in wind speed, v_R is the rated wind speed, v_F is the rated cut-out speed, and k is the Weibull shape factor [18]. When the blade pitch angle is zero, the power coefficient is maximized for an optimal TSR [4]. The optimal rotor speed is to be calculated by

$$\omega_{\text{opt}} = \frac{\lambda_{\text{opt}}}{R} V_{wn} \tag{5.18}$$

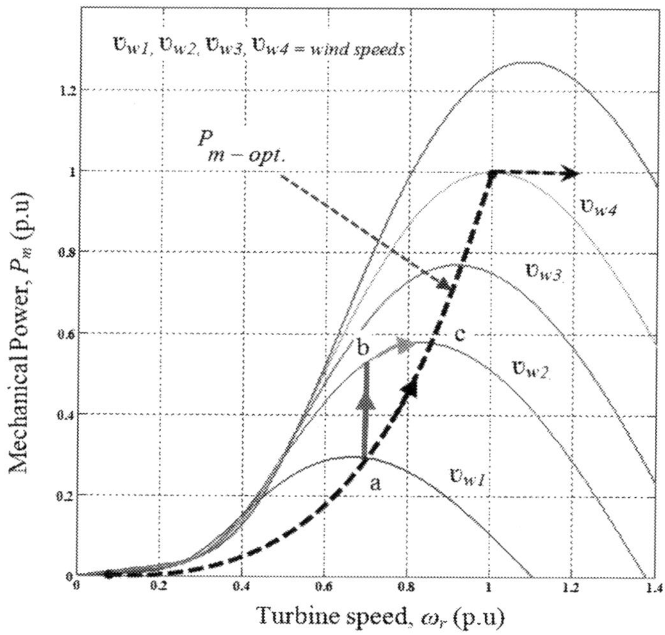

Figure 5.2 Relationship between mechanical power generation and turbine speeds at different wind speeds for implementation in a car.

which will give

$$V_{wn} = \frac{R\omega_{\text{opt}}}{\lambda_{\text{opt}}} \tag{5.19}$$

where ω_{opt} is the optimal rotor angular speed in rad/s, λ_{opt} is the optimal tip speed ratio, R is the radius of the turbine in meters, and V_{wn} is the wind speed in m/s.

The turbine speed and mechanical powers are depicted in the following graph (Fig. 5.2) with increasing and decreasing rates of wind speed while the vehicle is in motion [45,46]. When the wind is steady, the persistence forecasts yield good results [47,48]. When the wind speed is increased rapidly, sudden "ramps" in power output are generated, which is of tremendous benefit for capturing the energy.

5.2.2.4 Wind Energy Storage in Battery System

Standard Simulink/Sim Power Systems have been calculated by using MATLAB Simulink for the wind energy conversion that is to be stored in circuit-implemented inverter as a storage buffer, and all the electricity is to be supplied through the battery according to Peukert's Law to start the engine and to be used when the vehicle is not in motion [49,50].

5.2.2.5 Design of Traffic Control

Though underground maglev system has the capability to allow run up to 580 kph, the vehicles' high speed shall be calculated based on traffic flow, composition, volume, number and location of access points, and local environment importantly, allotting sufficient number of lanes considering Greenshields after road and highway capacity analysis (Fig. 5.3).

Because the maglev technology is invisible, to alert the drivers and pedestrian, the maglev roads, highways, and its exits are to be constructed by landscaping by covering the guideway by herb (green grass), and in between lanes, at least two feet is to be left blank (no landscaping) to differentiate the lanes.

5.2.3 Results and Discussion

Based the mathematical modeling described previously, I have performed load-resistant factor design (LRFD) calculation considering the following equation and selected W24 × 84 beam, which is for the continuous maglev underground runs (metal track guideway) that need to be structurally sound to carry enough current, load, and levitating force of the vehicles.

$$Fy \propto \frac{nl^2}{h} \tag{5.20}$$

$$Fx \propto \frac{-1}{ktv} \quad Fx \propto \frac{-nl^2}{h} \tag{5.21}$$

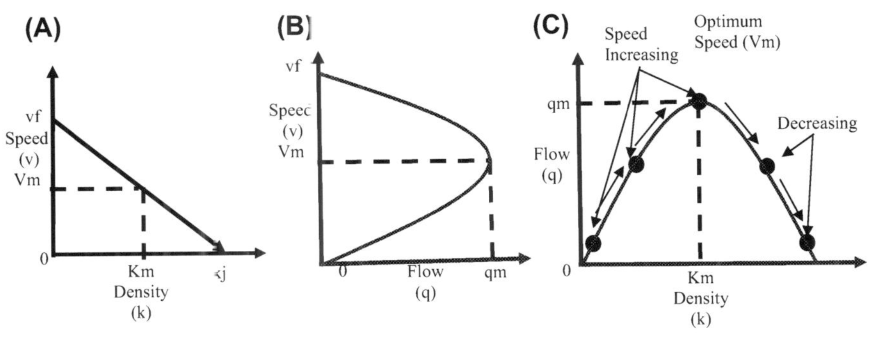

Figure 5.3 The Greenshield's fundamental diagrams (A) speed versus vehicle density, (B) flow versus vehicle density, and (C) speed versus flow analysis.

Where Fy = vehicle weight, n = total number of coils in maglev, I = current on each coil, h = height of levitation, t = thickness of conduction track, and k = conductivity of track.

To construct underground maglev guideway just two feet below the earth's surface, it will be needed to have a U-shaped cross section to fix the pole position [51,52]. Naturally heavy-duty waterproofing membrane is to be used to protect the maglev underground runs for avoiding floods and moisture. It is well researched that the propulsion coils run in elliptical loops along both walls of the guideway, generating magnetic force when electricity runs through them [53,54]. So, levitation and guidance coils will be formed, which will create their own magnetic force once the applied superconducting magnets pass on it where propulsion and levitation are the key factor to run the vehicle. In propulsion, as the direction of the current changes back and forth in the propulsion coils above the wall of the guideway, the north and south poles will reserve repeatedly, propelling the vehicle by alternating force of attracting and repulsion (Fig. 5.4). In levitation as the vehicle passes, an electric current is induced in the coil along the guideway, and the vehicle will be levitated by the force of attraction, which will pull up on the magnet in the vehicle, as well as by repulsion, which will push up on the magnet [38,55].

To create levitation and lateral balance in the vehicle, an electromagnetic induction is to be used. To confirm the most efficient and economical way to produce the powerful magnetic field by using the superconducting coils, I have assumed that permanent currents of about 700,000 A go through these superconducting coils [15], hence creating a strong magnetic field of almost 5 T, i.e. 100,000 times stronger than the earth magnetic field by implementing the following block diagram (Fig. 5.5).

Simply it can be explained that when an electric current flow through the propulsion coils, a magnetic field is produced. The forces of attraction and repulsion between the coils and the superconducting magnets on the vehicle propel the vehicle forward in a flying stage up to four feet height where two feet shall be considered covered underground and other two feet just over the earth's surface (Fig. 5.6). The vehicle's speed is to be adjusted by altering the timing of the polarity shift in the propulsion coils' magnetic field between north and south with the possibility of maximum speed of 580 kph [56,57]. As the vehicle passes just two feet above the guideway (one feet from the earth surface), an electric current is induced in the levitation and guidance coils, creating opposite magnetic poles in the upper and lower loops. The upper loops become the polar opposite

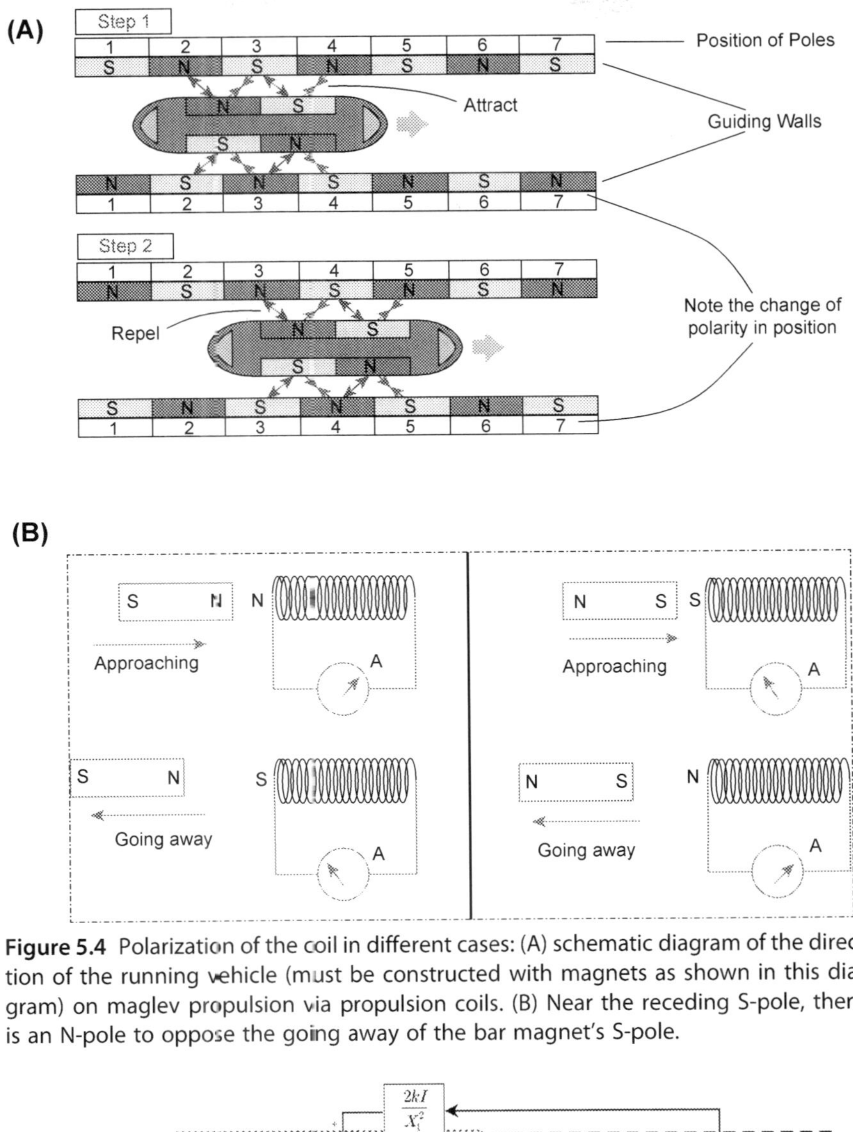

Figure 5.4 Polarization of the coil in different cases: (A) schematic diagram of the direction of the running vehicle (must be constructed with magnets as shown in this diagram) on maglev propulsion via propulsion coils. (B) Near the receding S-pole, there is an N-pole to oppose the going away of the bar magnet's S-pole.

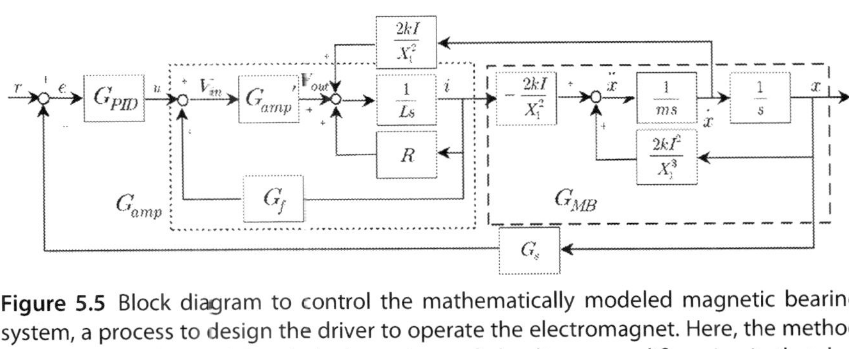

Figure 5.5 Block diagram to control the mathematically modeled magnetic bearing system, a process to design the driver to operate the electromagnet. Here, the method to determine the peripheral device values of the linear amplifier circuit that has the desired output by applying a generic algorithm and the method to identify the magnetic bearing system is shown.

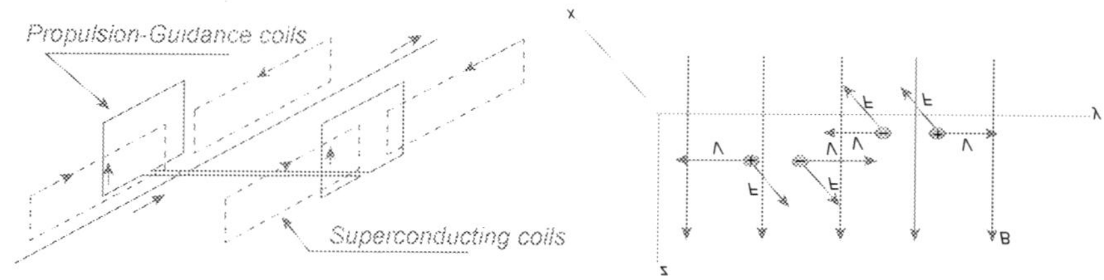

Figure 5.6 The maglev vehicle's force and directional diagram as shown by propulsion guidance coils and superconducting coils.

of the vehicle's magnets, producing attraction, which pulls the vehicle up. The lower loops have the same pole as the magnets. This generates repulsion, which pushes the vehicle in the same direction up. The two forces combine to levitate the vehicle while maintaining its lateral balance between the walls of the guideway.

Subsequently, a niobium—titanium alloy to be used to create superconducting magnets for maglev, but to reach superconductivity, they must be kept cold. To keep the alloy cool, liquid helium is used at a temperature of −269°C because alloy retains superconductivity at temperatures up to −263°C, though the maglev system can operate better at 6°C to produce sufficient magnetic force.

In addition to underground maglev, the WTGS for the backup energy source is to be implemented by the optimal operation of the whole system and based on a robustness test performed by adding a wind speed signal and power coefficient.

These conditions permit application of the wind profile that is considered to be a wind speed signal with a mean value of 8 m/s and a rated wind speed of 10 m/s; the whole system is tested under standard conditions with a stator voltage of approximately 50% for 0.5 s between 4 and 4.5 s, approximately 25% between 6 and 6.5 s, and 50% between 8 and 8.5 s (Fig. 5.7). Thus the machine is considered to be functioning in ideal conditions (no perturbations and no parameter variations). Moreover, to guarantee a unity power factor at the stator side, the reference for the reactive power is to be set to zero [5]. As a result of increasing wind speed, the generator shaft speed achieved maximum angular speed by tracking the maximum power point speed. Thus the wind turbine always works optimally because the pole placement technique is to be used to design the tracking control [7]. Consequently, decoupling among the components of the rotor current was also performed to confirm that the control system worked effectively. The bidirectional active and reactive power transfer between the rotor and power system is

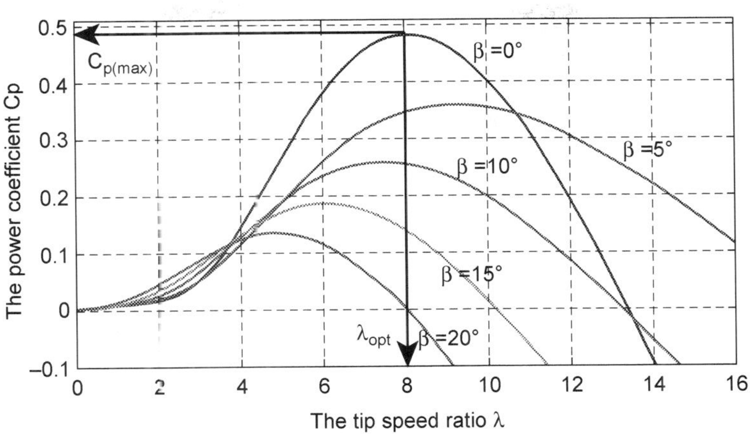

Figure 5.7 The maximum values of C_p are achieved for the curve associated with $\beta = 2$ degrees. From this curve, the maximum value of C_p ($C_{p,\max} = 0.5$) is obtained for $\lambda_{opt} = 0.91$. This value (λ_{opt}) represents the optimal speed ratio.

exchanged by the generator according to the super synchronous operation, achieving the nominal stator power, and the reactive power can be controlled by the load-side converter to obtain the unit's power factor to generate energy for powering vehicles [58,59].

5.2.3.1 Construction Cost Estimate Comparison

The order of magnitude cost estimate was obtained by using the HCSS (HeavyBid) software standard union rate of New York State locals with a project of 10% general condition, 10% overhead and profit, and 3% contingency over the hard cost of labor, materials, and equipment comparing between maglev infrastructure and tradition infrastructure system for a sample of 100 miles in length and 128 feet in width (12-feet-wide 4 lanes for each directions with two-sided 10-feet service space and a 6-feet median in the center of the road). To determine that the underground guideway (w24 × 84) can last long, I have calculated again the LRFD to provide the shoring for both sides for the entire 100 miles and 128-feet width (12-feet-wide 4 lanes each directions with two-sided 10-feet service space and a 6-feet median in the center of the road) construction cost, considering standard excavation up to 6-feet deep, with appropriate shoring and minimum embedment depth L4 of 5 feet and standard soil pressure $\Upsilon_s = 120\ \text{lbf/ft}^3$, angle of pressure $\Phi = 21°$, and the soil pressure coefficient $c = 800\ \text{lbf/ft}^2$. To prepare the conceptual estimate, we need to determine the length of soldier piles.

I have counted $6'$ on center (OC) soldier piles at both the sides by illustration and using the following LRFD method that soldier piles must be set at to support the necessary excavation and/or earth pressure against collapse.

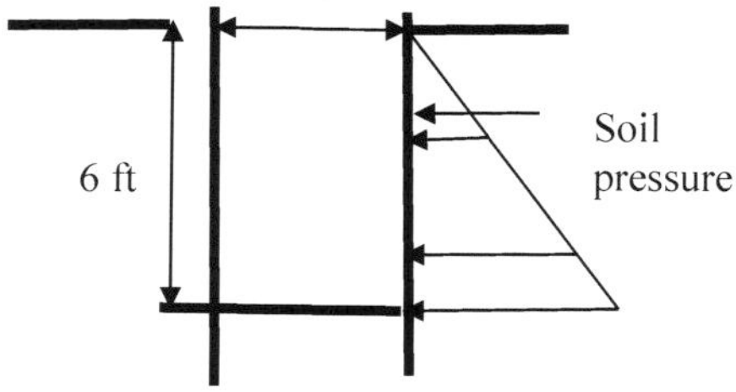

$$\text{Active Earth Pressure } K_a = \tan^2\left(45^\circ - \frac{\Phi}{2}\right) \tag{5.22}$$

$$\text{Passive Earth Pressure } \mathrm{K}_b = \tan^2\left(45^\circ + \frac{\Phi}{2}\right) \tag{5.23}$$

Use Eqs. (5.22) and (5.23) to find the lateral earth pressure the solid piles must support.

$$\begin{aligned} P_{EM} &= \Upsilon_s h k_{a,\text{piles}} \\ &= \left(120\ \frac{\text{lbf}}{\text{ft}^3}\right)(6.0)\tan^2\left(45^\circ - \frac{21^\circ}{2}\right) \\ &= 340.128\ \text{lbf/ft}^2 \end{aligned}$$

To determine the type of steel beams required for the soldier piles, we have taken the bending moments about the tributary area of the piles.

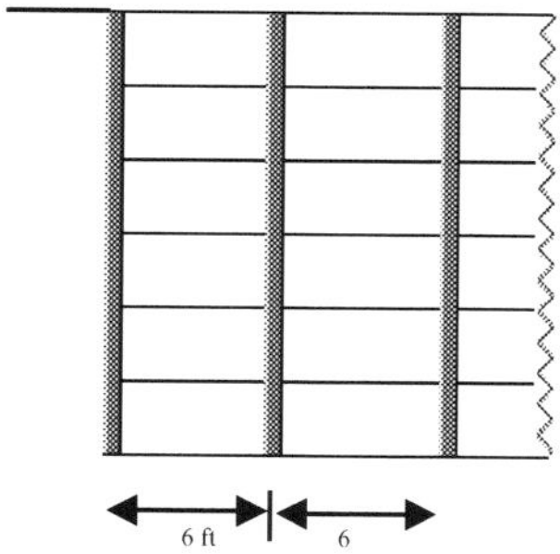

Soil pile spacing = 6ft
Side elevation

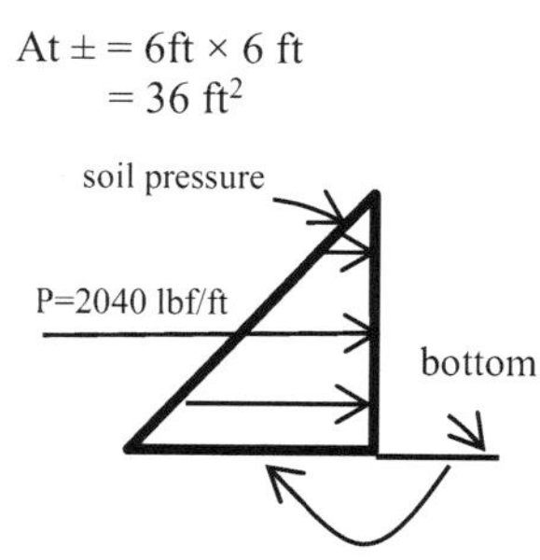

M = 2040.768 ft-lbf/ft

$$M = \left(\frac{\left(340.128\ \frac{\text{lb}}{\text{ft}^2}\right)(6\ \text{ft})}{2} \right)\left(\frac{6\ \text{ft}}{3}\right) = 2040.768\ \text{ft-lbf/ft}$$

The moment is a distributed moment applied to the base of the tributary area of each soldier pile. Therefore the moment is 2040.768 ft-lbf/ft. The total moment on the soldier pile (at the base) is

$$\begin{aligned} M_0 &= M(6\ \text{ft}) \\ &= \left(2040.768\ \frac{\text{ft} - \text{lbf}}{\text{ft}}\right)(6\ \text{ft}) \\ &= 12,244.61\ \text{ft} - \text{lbf} \end{aligned}$$

Now,

$$\begin{aligned} Z_{\text{req}} = \frac{M_0}{\phi b\ Fy} & \frac{\left(12\ \frac{\text{in}}{\text{ft}}\right)(12,244.61\ \text{ft-lbf})}{(0.9)\left(50,000\ \frac{\text{lbf}}{\text{in}^2}\right)} \\ &= 3.27\ \text{in}^3 \end{aligned}$$

From AISC tables, the soldier piles have been selected as to be W12 × 26, and the perpendicular support w8 × 12 members are of 6 ft long.

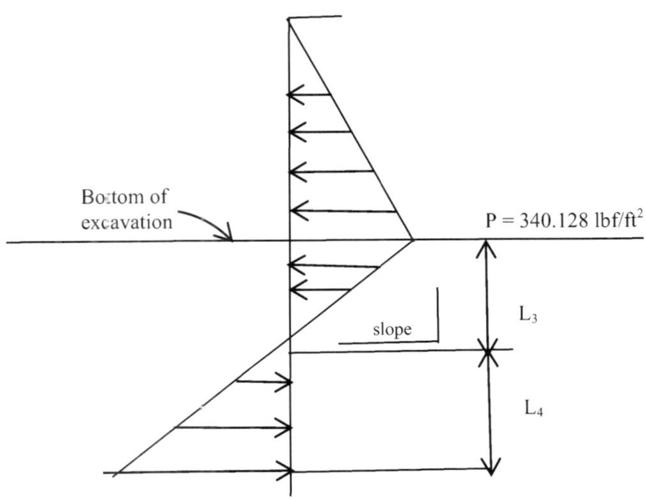

Then we have determined the depth required below subgrade by calculating the passive earth pressure coefficient using Eq. (5.23).

$$K_p = \tan^2\left(45° + \frac{\Phi}{2}\right)$$
$$= \tan^2\left(45° + \frac{21°}{2}\right)$$
$$= 2.12$$

Then we have calculated the active earth pressure coefficient using Eq. (5.22).

$$K_a = \tan^2\left(45° - \frac{\Phi}{2}\right)$$
$$= \tan^2\left(45° - \frac{21°}{2}\right)$$
$$= 0.4724$$

To determine the slopes of the excavation, depth is required because below the bottom of the excavation, pressure is considered to be both passive and have the same slope. The slope of the pressure profile above the reversal point is calculated from the standard equation for the slope, using L_3 as the rise and Υhk_a as the run (a value equal to the lateral earth pressure, expressed this way for the purposes of cancelation). Thus the slope of the pressure profile below the reversal point can be calculated similarly, using L_4 as the rise and the product of $\Upsilon L_4 k_p$ as the run. Because the slopes are the same, the two equations can be equated. Rearranging to solve for L_3,

$$\frac{L_3}{\Upsilon hk_a} = \frac{L_4}{\Upsilon L_4 k_p}$$

$$\mathrm{L}_3 = \frac{hk_a}{k_P} = \frac{(6\ \text{ft})(0.4724)}{2.12}$$
$$= 1.337\ \text{ft}$$

The necessary embedment depth is

$$1.337\ \text{ft} + 5\ \text{ft} = 6.337\ \text{ft}$$

The total required soldier pile length is

$$6.337\ \text{ft} + 6\ \text{ft} = 12.337\ \text{ft}\ (13\ \text{ft assumed})$$

So, I have determined that the solder pile (W12 × 26) should be 13-feet long, and the perpendicular support (w8 × 12) should be 6-feet long for a structurally sound maglev construction.

5.2.4 Construction Cost Estimate Comparison

The order of magnitude cost estimate was performed by using HCSS (Heavy Bid) software standard union rate of New York State locals with a project of 10% general condition, 10% overhead and profit, and 3% contingency over the hard cost of labor, materials, and equipment comparing between maglev infrastructure and tradition infrastructure system for a sample of 100 miles in length and 128 feet in width (12-feet-wide 4 lanes in each directions with two-sided 10-feet service space and a 6-feet median in the center of the road). To construct the long-lasting and sophisticated underground maglev, I have performed LRFD calculation and selected W24 × 84 beams so that the continuous maglev underground runs (structural beam) are structurally sound. Then I have calculated the required shoring concept for 100-mile-long and 128-feet-wide construction cost considering standard excavation up to 6-feet deep, with appropriate shoring with minimum embedment depth *L4* of 5 feet and standard soil pressure $\Upsilon_s = 120$ lbf/ft^3, angle of pressure $\Phi = 21^\circ$, and the soil pressure coefficient $c = 800$ lbf/ft^2 to determine the length of soldier piles. So, I have calculated by using LRFD methods again that the selected solder pile (W12 × 26) should be of length 13 feet, and the perpendicular support (w8 × 12) should be 6-feet long as the support maglev construction.

5.2.4.1 Cost of Maglev Infrastructure

The proposed maglev infrastructure therefore requires shoring, excavation, structural steel, and concrete operation, and thus I have calculated the estimate considering following components.

Shoring at 13′ deep with w24 × 26 steel soldier piles at 6′ OC for both sides at \$2/lf; top rail w8 × 12 for both sides at \$2/lf; 6′ length w8 × 12 perpendicular support 20 OC at \$2/lf; and protection board of 1,372,800 ft^2 for both sides at \$4/ft^2. The total cost would be \$23,724,800.

Excavation $(52{,}800'_{\text{length}} \times 128_{\text{width}} \times 6_{\text{deep}} \times 1.3_{\text{fluff factor}})/27$ is 19,524,266.67 yd^3 at \$56/yd^3 cost for digging, stock piling, and backfilling, and the total cost would be \$1,093,358,933.

Cost of materials includes 100-mile maglev system with structural steel (w24 × 84) support for 8 lanes is \$354,816,000; 2 × 2 structural concrete strip footing at $150/yd^3 is $93,866,666; reinforcement bars at 100 lb/yd^3 cost $62,577,778; and concrete form at $2/ft^2 costs $16,896,000. Thus the total cost of material is $ $528,156,445.

Cost of labor includes 200 iron workers for 2704 working days at $100/h; 100 concrete cement workers for 2704 working days at $90/h; 100 laborers for 2704 working days at $70/h; and 50 equipment operators for 2704 working days at $100/h. Thus the total labor cost is $886,912,000 considering standard 8 h a day.

Equipment cost includes 10 small renting at $1000/day; 10 small tool renting at $250/day; and 271 concrete pumps at $2000/each. Thus the total equipment cost is $34,342,000.

Other cost includes engineering service at $5/ft^2 and survey team at $4400/day for each working days; thus the total cost is $349,817,600. The net construction cost by adding 10% general condition, 10% overhead and profit, and 3% contingency into the excavation, material, labor, equipment, and other cost would be $$3,587,063,487.

5.2.4.2 Cost of Traditional Road Infrastructure

A typical highway consists of 8″ asphalt surface course, 4″ binder course, 4″ base course, and 12″ aggregate with standard wire mesh or framing, and thus we have calculated the estimate considering the following components:

Excavation $(5{,}2800_{\text{length}} \times 128_{\text{width}} \times 2.33_{\text{deep}} \times 1.3_{\text{fluff factor}})/27$ is 7,581,924 yd^3 at $56/yd^3 cost for digging, stock piling, and backfilling, and the total cost would be $424,587,744.

Cost of materials includes $50/yd^3; 4″ base course is 834,370 yd^3 at $50/yd^3; wire mesh or framing is (528,000 × 128) at $1/ft^2, and 12″ subbase aggregate is 2,503,111 yd^3 at $25/yd^3. Thus the total cost of material is $380,472,775.

Cost of labor includes 200 asphalt cement workers for 2704 working days at $100/h; 200 labor foremen for 2704 working days at $100/h; 200 laborers for 2704 working days at $70/h; 200 equipment operators for 2704 working days at $100/h; 100 truck drivers for 2704 working days at $100/h; and 200 small roller engineers for 2704 working days at $100/h. Thus the total cost is $2,249,728,000.

Equipment cost includes 200 rollers renting at $1000/week; 200 milling machines renting at $10,000/week; and 100 trucks renting at $500/week. Thus the total cost is $502,171,429.

Other cost includes detailing and shop drawing at $10/ft^2; engineering service at $5/ft^2; survey team at $4400/day for each working days; banking service of 301,037 yd^3 at \$1000/yd^3; and maiden concrete divider is 106,468 yd^3 at \$818/yd^3. Thus the total cost is \$1,326,694,600. The net construction cost by addirg 10% general condition, 10% overhead and profit, and 3% contingency into the excavation, material, labor, equipment, and other cost would be \$6,805,115,863.

5.2.4.3 Cost saving

In this article, I have calculated cost saving by using standard 100-mile highway of width 128 feet (12-feet-wide 4 lanes at each directions with two-sided 10-feet service space and a 6-feet median in the center of the road) as an experimental tool to compare construction cost between conventional and maglev infrastructure system. Total cost estimate for traditional infrastructure is \$6,805,115,863, and the maglev infrastructure system costs only \$\$3,587,063,487 for the same 100-mile highway, and the net cost saving is \$3,218,052,377 (Table 5.1). Consequently, it will reduce nearly 50% cost once maglev infrastructure system is used for the construction of invisible infrastructure which is also benign to the environment.

Table 5.1 This cost comparison is prepared by using HCSS cost data 2016 for material by using selective manufacturers and labor rate in accordance with international union of wage of each specified trade workers considering US location. The equipment rental cost is estimated considering current rental market in conjunction with the standard practice of construction of the production

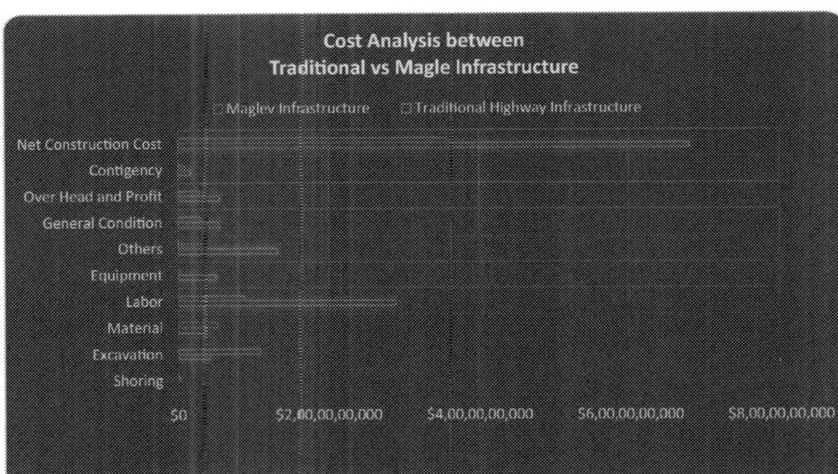

5.2.5 Conclusions

Traditional transportation infrastructure construction and maintenance throughout the world are not only expensive but is also consuming 5.6×10^{20} J/year (560 EJ/year) fossil fuel which indeed is dangerous of a cliché when discussing about climate [11,43]. To mitigate this issue, better infrastructure transportation planning needs to be achieved, in which environmental sustainability and climate adaptation has been confirmed to create more resilient and vibrant communities. Interestingly, **invisible infrastructure transportation** technology proposed in this chapter, for urban infrastructure transportation system, implicated by electromagnetic system and superconducting magnets, will thus be the emergent technology in modern science. It is because the technology is cheaper, and it will run by repulsive force and attractive force at the levitated (flying) stage while it runs on maglev system and will run by air (wind energy) while it is on nonlevitated area without consuming fossil fuel. Indeed, the maglev infrastructure transportation system would be the innovative technology ever to console infrastructure, transportation, energy, and global warming crisis.

5.3 ZERO-EMISSION VEHICLE

Massive development in transportation sectors has accelerated fossil fuel energy consumption, and greenhouse gas emissions from vehicles account for nearly 30% of global warming. We need clean energy for transportation sectors to meet their energy demand and avoid global warming. In this study, a wind energy system model was developed by integrating advanced technological and mathematical aspects to obtain a potential solution for the total energy demand of transportation sectors. Detailed analysis of the theoretical wind energy systems was modeled by a series of mathematical equations that were then proposed for use in transportation sectors to naturally meet their energy demand. To better explain this technology and its application in transportation sectors, and taking the practical scale and components into consideration, a sample experimental model of a car was also described as a hypothetical experiment. Interestingly, both the theoretical modeling and experimental analysis of the car confirm that a turbine can be a promising tool to use wind energy to generate electricity from self-renewing resources to power the car; importantly, wind energy is clean and globally abundant. The proposed wind energy system could be an innovative large-scale technology in energy science which can enable vehicles to produce energy from the wind when the vehicle is in motion, thus meeting 100% of the vehicle's energy demand.

5.3.1 Background and Technology

Conventional energy utilization for the transportation sectors is not only costly but also causes adverse environmental impact [60—62]. Interestingly, wind power technology can be implemented into all vehicles to meet its total energy demand while it is in motion, which indeed shall mitigate the energy and climate change crisis. Thus wind energy has the high potential to produce energy for transportation vehicles once it is used by a sophisticated technology. Modern wind turbines are small, simple, and highly sophisticated devices compared with turbines from the mid-twentieth century, that is, when turbines were extremely tall and large where the turbine engine capacity would be roughly 5—10 kW; rotors now often exceed 0.25 m in diameter and are positioned on towers exceeding 1 m in height. Thus, modern turbines are indeed much cheaper than previous ones, which could be the best tool to implement into the vehicle to produce energy naturally. In this article, an innovative theory was analyzed to implement this wind energy into vehicles while the vehicles are in motion. The wind turbine modeling and drive train modeling (by one-mass model) was conducted using the MATLAB Simulink software package. The kinetic energy conversion process was divided into two main interacting subsystems, and the detailed process of conversion procedure was modeled; results of the energy conversion chain analysis were also analyzed using the MATLAB software. Subsequently, the control structure, design, and generator model were also analyzed using a series of mathematical calculations to prepare a theoretical model for total process energy capture by wind turbine and utilization by running vehicles. Finally this model was hypothetically applied into a car as an experimental tool which revealed that wind energy—generated transportation vehicles could indeed be a cutting-edge technology to reduce or eliminate the cost of energy for all transportation vehicles.

5.3.2 Theoretical Modeling of Wind Energy

A wind turbine generator to power electronic equipment [63,64] is governed by the operation of variable-speed wind turbines. The reasons for using variable-speed operating wind turbines include possibilities for reducing stress and control of active and reactive power. MATLAB Simulink model wind generation modules that are unique to one of the following with a series of mathematical equations were used for modeling the turbines (Conceptual Diagram 5.1).

In this chapter, the model that was developed was used to study and simulate the DFIG for producing electricity for transportation vehicles.

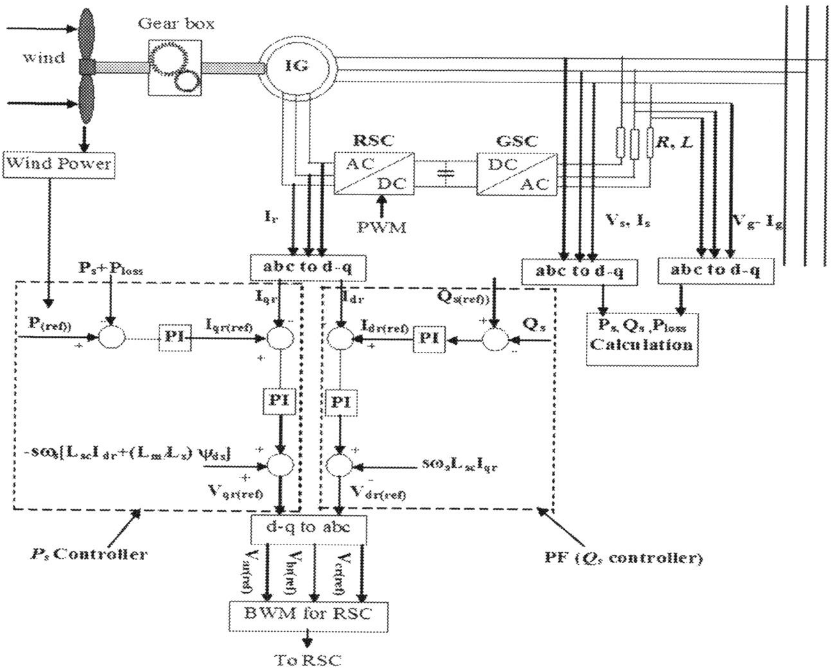

Conceptual Diagram 5.1 The aforementioned sketch illustrates the conceptual MATLAB Simulink model of the wind generator energy module.

A vector control strategy involving the DFIG order, power storage box active stator is presented in the following matrix.

$$\frac{d}{dt}\begin{bmatrix}\Delta\omega_t\\ \Delta\omega_r\\ T_g\end{bmatrix}=\begin{bmatrix}\frac{-D_t-D_{tg}}{2H_t} & \frac{D_{tg}}{2H_t} & \frac{-1}{2H_t}\\ \frac{D_{tg}}{2H_g} & \frac{-D_g-D_{tg}}{2H_g} & \frac{1}{2H_g}\\ K_{tg}\omega_e & -K_{tg}\omega_e & 0\end{bmatrix}\begin{bmatrix}\Delta\omega_t\\ \Delta\omega_r\\ T_g\end{bmatrix}+\begin{bmatrix}\frac{T_m}{2H_t}\\ \frac{-T_e}{2H_g}\\ 0\end{bmatrix} \tag{5.24}$$

The fundamental equation governing the mechanical power of the wind turbine is

$$P_w=\frac{1}{2}C_p(\lambda,\beta)\rho A V^3 \tag{5.25}$$

where ρ is the air density (kg/m^3), C_p is the power coefficient, A is the intercepting area of the rotor blades (m^2), V is the average wind speed (m/s), and λ is the tip speed ratio. The theoretical maximum value of the power coefficient C_p is 0.593; C_p is also known as Betz's coefficient [65,66]. Mathematically,

$$\lambda = \frac{R\omega}{V} \tag{5.26}$$

R is the radius of the turbine (m), ω is the angular speed (rad/s), and V is the average wind speed (m/s). The energy generated by wind can be obtained by

$$Q_v = P \times (Time)(kWh) \tag{5.27}$$

It is well known that wind velocity cannot be obtained by a direct measurement from any particular motion. In data taken from any reference, the motion needs to be determined for that particular motion; then, the velocity needs to be measured at a lower motion [67,68].

$$v(z)\ln\left(\frac{Z_r}{Z_0}\right) = v(Z_r)\ln\left(\frac{Z}{Z_0}\right) \tag{5.28}$$

where Z_r is the reference height (m), Z is the height at which the wind speed is to be determined, Z_0 is the measure of surface roughness (0.1–0.25 for crop land), $v(z)$ is wind speed at height z (m/s), and $v(z_r)$ is wind speed at the reference height z (m/s). The power output in terms of the wind speed can be estimated using the following equation:

$$P_w(v) = \begin{cases} \dfrac{v^k - v_C^k}{v_R^k - v_C^k} \cdot P_R & v_C \le v \le v_R \\ P_R & v_R \le v \le v_F \\ 0 & v \le v_C \text{ and } v \ge v_F \end{cases} \tag{5.29}$$

where P_R is rated power, v_C is the cut-in wind speed, v_R is the rated wind speed, v_F is the rated cut-out speed, and k is the Weibull shape factor. The angular speed of the generator must be changed to extract the maximum power; this process is known as maximum power point tracking (MPPT). When the blade pitch angle is zero, the power coefficient is maximized for an optimal TSR [69–71]. The optimal rotor speed is given by

$$\omega_{opt} = \frac{\lambda_{opt}}{R} V_{wn} \tag{5.30}$$

which will give

$$V_{wn} = \frac{R\omega_{\text{opt}}}{\lambda_{\text{opt}}} \tag{5.31}$$

where ω_{opt} is the optimal rotor angular speed in rad/s, λ_{opt} is the optimal tip speed ratio, R is the radius of the turbine in meters, and V_{wn} is the wind speed in m/s.

5.3.2.1 Drive Train Modeling

The basic concept is for a drive train to transfer high aerodynamic torque at the rotor to the low-speed shaft of the generator through a gearbox. Because the generator is coupled to the rotor to reduce complexity, a model of the generator is needed and is presented later in this section. Consequently, the drive train can be modeled using a one-mass model [72–74] based on the torsional multibody dynamic model as per the following matrix:

$$\begin{bmatrix} \dot{\omega}_t \\ \dot{\omega}_g \\ \dot{T}_{ls} \end{bmatrix} = \begin{bmatrix} -\frac{K_t}{J_t} & 0 & -\frac{1}{J_t} \\ 0 & -\frac{K_g}{J_g} & \frac{1}{n_g J_g} \\ \left(B_{ls} - \frac{K_{ls}K_r}{J_r}\right) & \frac{1}{n_g}\left(\frac{K_{ls}K_r}{J_g} - B_{ls}\right) & -K_{ls}\left(\frac{J_r + n_g^2 J_g}{n_g^2 J_g J_r}\right) \end{bmatrix} \times \begin{bmatrix} \omega_t \\ \omega_g \\ T_{ls} \end{bmatrix} + \begin{bmatrix} \frac{1}{J_r} \\ 0 \\ \frac{K_{ls}}{J_r} \end{bmatrix} T_m + \begin{bmatrix} 0 \\ -\frac{1}{J_g} \\ \frac{K_{ls}}{n_g J_g} \end{bmatrix} T_g \tag{5.32}$$

5.3.2.2 One-Mass Model

The turbine inertia can be calculated from the combined weight of the blades and the hub. Therefore, the turbine can be viewed as a large disk with a small thickness. If proper data are not available, the following simple equation can be used to estimate the mass moment of inertia of a disk with a small thickness the turbine can be considered (Conceptual diagram 5.2).

$$J_t\dot{\omega}_t = T_a - K_t\omega_t - T_g \tag{5.33}$$

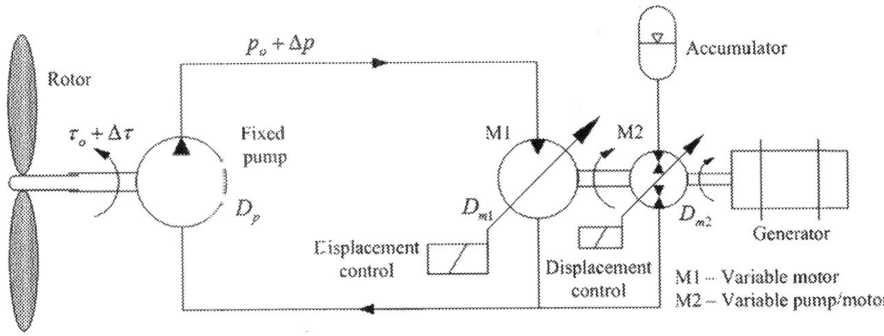

Conceptual Diagram 5.2 Schematic diagram of a one-mass model of a wind turbine system: a full-scale converter wind turbine with synchronous machine shows the detailed process of energy accumulation.

and

$$J_t = J_r + n_g^2 J_g \tag{5.34}$$

$$K_t = K_r + n_g^2 K_g \tag{5.35}$$

$$T_g = n_g T_{em} \tag{5.36}$$

where J_t is the moment of inertia in kg m^2, ω_t is the low shaft angular speed in rad/s^2, K_t is the turbine damping coefficient in Nm/rad s which represents the aerodynamic resistance, and K_g is the generator damping coefficient in Nm/rad s.

Therefore, the maximum power is calculated with the following equation with the one-mass model diagram and series of equations. This shows that the theoretical maximum power extracted from the wind is 0.5925 times its kinetic power.

$$\frac{dP_k}{dV_2} = \frac{1}{4}\rho A V_1^3\left(-3\frac{V_2^2}{V_1^3} - 2\frac{V_2}{V_1^2} + \frac{1}{V_1}\right) = 0 \tag{5.37}$$

$$\Rightarrow \frac{1}{4}\rho A\left(-3V_2^2 - 2V_2V_1 + V_1^2\right) = 0$$

$$\Rightarrow \frac{1}{4}\rho A\left(-3V_2^2 - 3V_2V_1 + V_2V_1 + V_1^2\right) = 0$$

$$\Rightarrow \frac{1}{4}\rho A\left[-3V_2(V_2 + V_1) + V_1(V_2 + V_1)\right] = 0$$

$$\Rightarrow \frac{1}{4}\rho A(-3V_2 + V_1)(V_2 + V_1) = 0$$

Here, ρ, A, and $(V_2 + V_1)$ cannot be zero; therefore,

$$(-3V_2 + V_1) = 0 \Rightarrow \frac{V_2}{V_1} = \frac{1}{3}$$

Now, putting this value of V_2/V_1 in Eq. (5.37), we get

$$P_k = \frac{1}{4}\rho A V_1^3\left(1 - \frac{1}{3^3} - \frac{1}{3^2} + \frac{1}{3}\right)$$

$$\Rightarrow P_k = 0.5925 \times \frac{1}{2}\rho A V_1^3$$

5.3.2.3 Wind Energy Conversion

Given the aforementioned conditions, the airflow mass has certain energy called kinetic energy [75–77]. The kinetic energy is shown in Fig. 5.8, which presents a wind energy conversion system (WECS) that uses a DFIG; these are the aerodynamic subsystem (wind speed, wind turbine, and gearbox) and the electrical subsystem (DFIG).

5.3.2.4 Aerodynamic Subsystem

Signals in the simulations may use logs of real speed at the real location of the wind turbine generation system (WTGS).

The choice for the wind speed model is described and proposed in [78,79]. The deterministic and stochastic parts are added together to obtain the total equivalent wind speed (V) and expression in the following form:

$$V(t) = V_0 + \sum_{i=1}^{n} A_i \, sin(\omega_i t + \varphi_i) \tag{5.38}$$

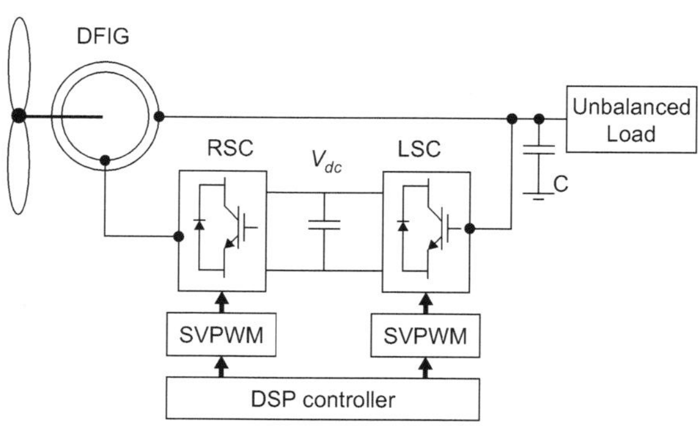

Figure 5.8 Typical wind energy conversion chain.

where V_0 is the mean component and A_i, ω_i, and φ_i are the magnitude, pulsation, and initial phase of each turbulent mode, respectively. In this article, the turbulence experienced by the rotating wind turbine blades is taken into account. The WTGS converts power from the kinetic energy of the wind and is expressed as C_p, which is called the power coefficient or Betz's factor.

The aerodynamic power is given by [80]

$$P_{aer} = \frac{1}{2} C_p(\lambda, \beta) \rho R \pi^2 V^3 \tag{5.39}$$

where ρ is the air density, R is the blade length, and V is the wind velocity. The percentage is represented by a coefficient $C_p(\lambda)$, which is a function of the wind speed, turbine speed, and pitch angle of the specific wind turbine blades [81,82].

Although this equation seems simple, C_p depends on the ratio between the turbine shaft speed Ω_t and the wind speed V. This ratio is called the tip speed ratio:

$$\lambda = \frac{\Omega_T \cdot R}{V} \tag{5.40}$$

Typical C_p versus TSR curves for different values of pitch angle β are shown in Fig. 5.9. In a wind turbine, there is an optimal value for the TSR for which C_p is maximal; thus, this TSR value maximizes the power for a given wind speed. The peak power for each wind speed occurs at the point at

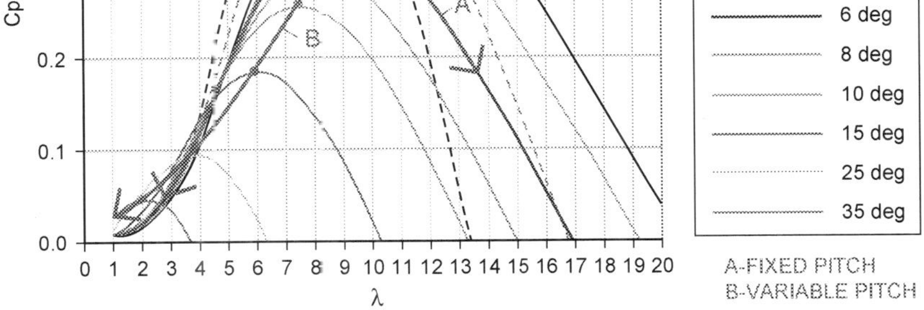

Figure 5.9 Power coefficient variation with TSR and pitch angle.

which C_p is maximized. To maximize the generated power, it is therefore desirable for the generator to have a power characteristic that follows the maximum C_{pmax} line.

The turbine torque is the ratio of the aerodynamic power to the turbine shaft speed [83]. The turbine is normally coupled to the generator shaft through a gearbox whose gear ratio G is chosen to set the generator shaft speed within a desired speed range. Neglecting the transmission losses, the torque and shaft speed of the wind turbine, referred to the generator side of the gearbox, are given by

$$\begin{cases} T_g = \dfrac{T_{aer}}{G} \\ \Omega_t = \dfrac{\Omega_g}{G} \end{cases} \tag{5.41}$$

where T_g is the driving torque of the generator and Ω_g is the generator shaft speed.

5.3.2.5 Electrical Subsystem

Variable-speed operation is obtained by injecting a controllable voltage into the rotor at the desired slip frequency [84,85]. The equations for the DFIG are identical to a squirrel-cage induction generator except that the rotor voltages are not zero [17,42,55] and can be expressed as follows [5]:

$$\begin{cases} v_{ds} = \dfrac{d\phi_{ds}}{dt} + \dfrac{R_s}{L_s}\phi_{ds} - \omega_s \cdot \phi_{qs} - M\dfrac{R_s}{L_s}i_{dr} \\ v_{ds} = \dfrac{d\phi_{qs}}{dt} + \dfrac{R_s}{L_s}\phi_{qs} + \omega_s \cdot \phi_{ds} - M\dfrac{R_s}{L_s}i_{qr} \\ \left(v_{dr} - \dfrac{M}{L_s}v_{ds}\right) = \sigma L_r\dfrac{di_{dr}}{dt} - M\dfrac{R_s}{L_s^2}\phi_{ds} + \dfrac{M}{L_s}\omega\phi_{qs} + \left(R_r + \dfrac{M^2}{L_s^2}R_s\right)i_{dr} - \sigma L_r\omega_r i_{qr} \\ \left(v_{qr} - \dfrac{M}{L_s}v_{qs}\right) = \sigma L_r\dfrac{di_{qr}}{dt} - M\dfrac{R_s}{L_s^2}\phi_{qs} - \dfrac{M}{L_s}\omega\phi_{ds} + \left(R_r + \dfrac{M^2}{L_s^2}R_s\right)i_{qr} + \sigma L_r\omega_r i_{dr} \end{cases} \tag{5.42}$$

where R_s and R_r are the stator and rotor resistances, respectively; L_s and L_t are the stator and rotor inductances, respectively; M and σ are the mutual inductance and leakage coefficient, respectively; $\Omega = p\Omega_g$ is the electrical speed; and p is the pair pole number.

The stator and rotor flux can be expressed as

$$\begin{cases} \phi_{ds} = L_s i_{ds} + M i_{dr} \\ \phi_{sq} = L_s i_{qs} + M i_{qr} \\ \phi_{rd} = L_r i_{dr} + M i_{ds} \\ \phi_{rd} = L_r i_{qr} + M i_{qs} \end{cases} \tag{5.43}$$

where i_{ds}, i_{qs}, i_{dr}, and i_{qr} are the direct, quadrate stator, and rotor currents, respectively. The active and reactive powers at the stator, in addition to those provided for the grid, are defined as

$$\begin{cases} P_s = v_{ds} \cdot i_{ds} + v_{qs} \cdot i_{qs} \\ Q_s = v_{qs} \cdot i_{ds} - v_{ds} \cdot i_{qs} \end{cases} \tag{5.44}$$

The electromagnetic torque is expressed as

$$T_{em} = p\left(i_{qs}\phi_{ds} - i_{ds}\phi_{qs}\right) \tag{5.45}$$

5.3.2.6 Control Structure

The wind turbine electric system time responses are much faster than those of the mechanical parts in the WECS. This makes it possible to dissociate the wind turbine and the DFIG control designs and thus describe a cascade control structure based on two subsystem controls:

1. The wind turbine subsystem control concerns the aerodynamic subsystem, which provides the reference inputs for the DFIG subsystem control.
2. The DFIG subsystem control concerns the electric generator via the power converter.

Subsequently, these two control levels will be considered separately, as shown in Fig. 5.10.

It is worth mentioning that there is no need for a reference voltage, torque limiter, or saturation block because of the inherent limits of the references generated at the transients. After obtaining the voltage, pulse width modulation (PWM) is used to generate the gating pulses with a fixed switching frequency for the load-side converter.

5.3.2.7 Wind Turbine Subsystem Control

Fig. 5.11 shows the four distinct regions in a typical WECS, where $V_{\max}$ is the wind speed at which the maximum allowable rotor speed is reached and $V_{\text{cut-off}}$ is the furling wind speed at which the turbine needs to be shut down for protection.

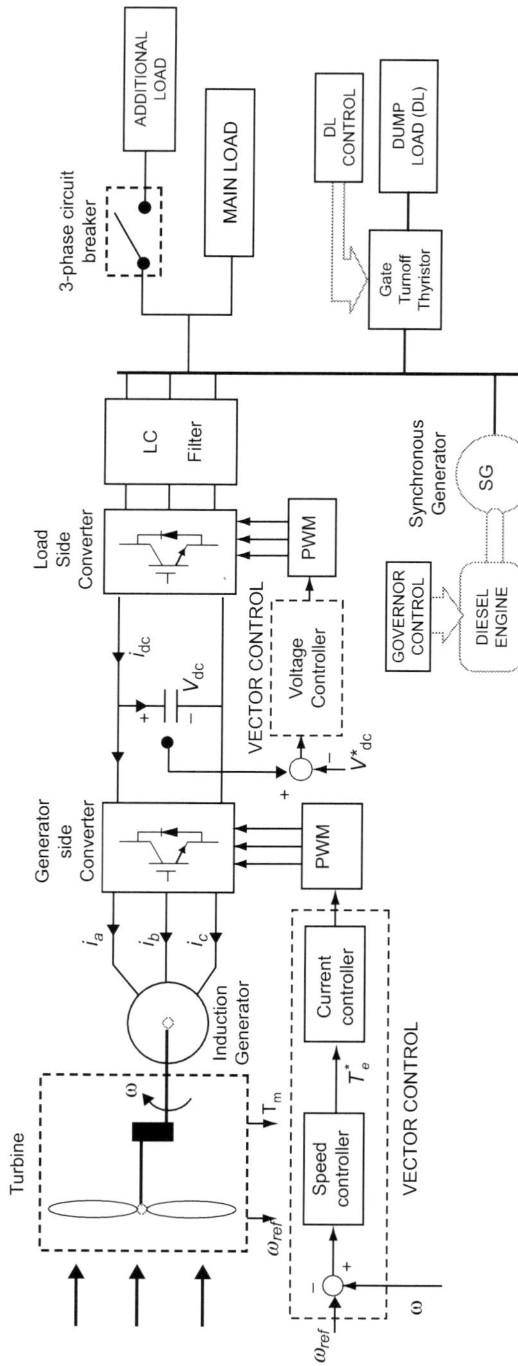

Figure 5.10 The block diagram of the wind turbine electric system.

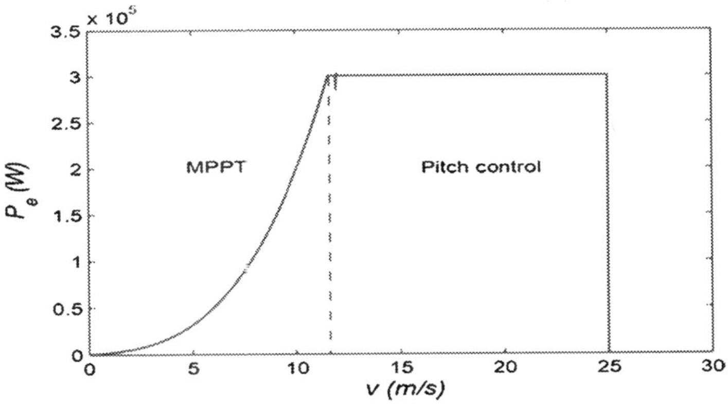

Figure 5.11 Area of operations and control WECS respectively MPPT and pitch control consideration starter initiated at zero.

In practice, there are two possible regions of turbine operation, namely high- and low-speed regions [86,87]. High-speed operation (IV) is frequently bounded by the speed limit of the machine. Conversely, regulation in the low-speed region (II) is usually not restricted by speed constraints.

However, the system has nonlinear nonminimum phase dynamics in this region. Generally, wind turbine control objectives are functions of the wind speed. For low wind speed, the objective is to optimize the capture of wind power through tracking the optimal rotor speed signals [88,89]. Once the wind speed increases above its nominal value, the control objective moves to the rated regulating power. Numerous methods for MPPT have been proposed in the literature [90–92].

The method proposed in this chapter is simple and is based on the tip speed of the wind turbine. Therefore, an anemometer is required for measuring the wind speed on the wind turbine. Assuming that the optimal value of the TSR λ can be obtained from Fig. 5.9, the optimal speed of the turbine can be determined using Eq. (5.46):

$$\Omega_{t,\mathrm{opt}} = \frac{\lambda_{\mathrm{opt}} \cdot V}{R} \tag{5.46}$$

For this MPPT method, the speed controller continuously adjusts the generator shaft speed to impose the reference electromagnetic torque of the DFIG with the aim of tracking, as shown in Fig. 5.12. The turbine shaft speed is then controlled to obtain a maximum power coefficient. The MPPT method significantly increases the efficiency of the wind turbine. For each wind speed, there is a certain rotational speed at which the power

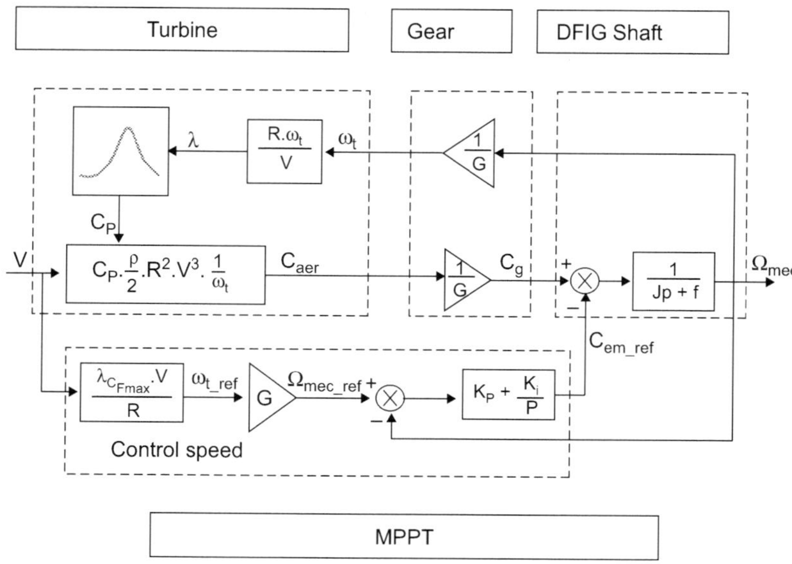

Figure 5.12 The block diagram of the MPPT with enslaved speed.

curve of a given wind turbine reaches a maximum value (C_p reaches its maximum value). Starting the description of the WECS with the aerodynamic subsystem, it should be mentioned that the present work focuses on region II. A block diagram of the MPPT control system for the wind turbine is shown in Fig. 5.12.

This control block diagram of variable-speed fixed-pitch WECS in region II generally aims at regulating the power harvested from wind by modifying the generator speed; in particular, the control objective captures the maximum power efficiency (MPE) of the power rotational speed curves for the 7.8-kW wind turbine that is considered in the present chapter under different wind speeds. Connecting all of the MPPs from each power curve, the optimal power curve is obtained, and the control system should follow the tracking characteristic curve (TCC) of the wind turbine. Each wind turbine has a TCC similar to the one in the following figure. When operating in region IV, which occurs above the rated wind speed, the turbine must limit the captured wind power such that safe electrical and mechanical loads are not exceeded (Fig. 5.13).

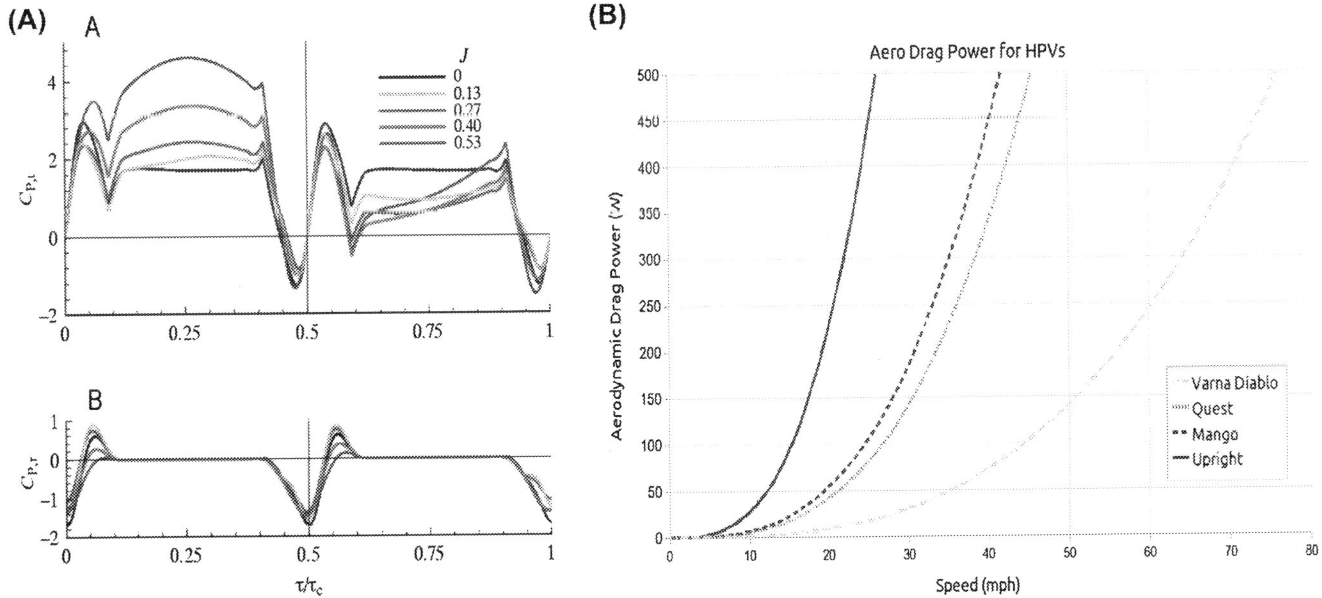

Figure 5.13 (A) Initiation of aerodynamics at various speeds. (B) Aerodynamic powers at various speed characteristics for different wind speed with indication of maximum power with a tracking curve.

5.3.2.8 DFIG Subsystem Control

The principle of this method consists of orientating the stator flux in such a manner that the stator flux vector points in the *d*-axis direction [92,93]. This approach is realized by setting the quadratic component of the stator flux to the null value:

$$\phi_s = \phi_{ds} \Rightarrow \phi_{qs} = 0 \tag{5.47}$$

In the Park reference frame, this approach is shown in Fig. 5.14. Using the aforementioned condition and supposing that the grid system is steady with a single voltage V_s, which leads to a constant flux in the stator ϕ_s, we can easily deduce the voltage as

$$\begin{cases} v_{ds} = 0 \\ v_{qs} = \omega_s \cdot \phi_s = V_s \end{cases} \tag{5.48}$$

The per-phase stator resistance is neglected (the realistic approximation for medium power machines used in WECS).

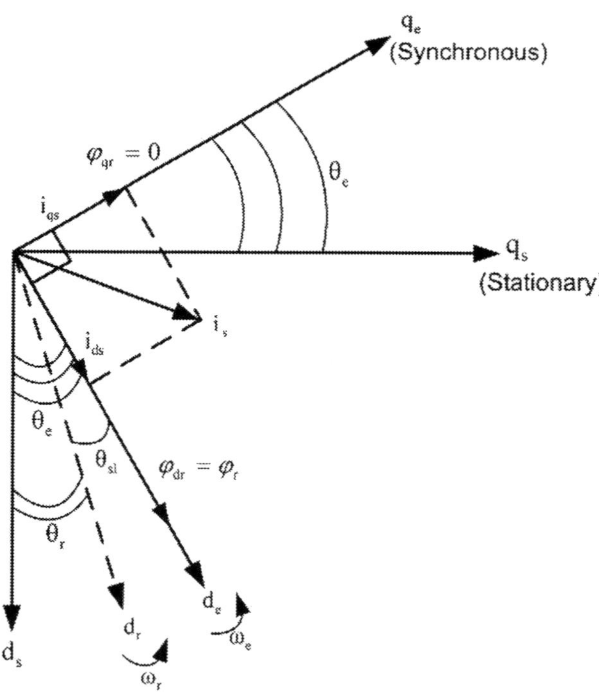

Figure 5.14 Stator flux orientation determined considering synchronous angles with stationary direction.

The stator voltage vector is consequently a quadratic advance compared with the stator flux vector. By using Eqs. (5.48) and (5.42), we obtain the rotor voltages:

$$\begin{cases} v_{dr} = \sigma L_r \dfrac{di_{dr}}{dt} + R_r i_{dr} - \sigma L_r \omega_r i_{qr} + \dfrac{M}{L_s}\dfrac{d\phi_{ds}}{dt} \\ v_{qr} = \sigma L_r \dfrac{di_{qr}}{dt} + R_r i_{qr} + \sigma L_r \omega_r i_{dr} + g\dfrac{M}{L_s} V_s \end{cases} \tag{5.49}$$

where V_s is the stator voltage magnitude, which is assumed to be constant, and g is the slip range. We can rewrite the rotor voltages as follows:

$$\begin{cases} v_{dr} = \sigma L_r \dfrac{di_{dr}}{dt} + R_r i_{dr} + fem_d \\ v_{qr} = \sigma L_r \dfrac{di_{qr}}{dt} + R_r i_{qr} + fem_q \end{cases} \tag{5.50}$$

With fem_d and fem_q, the crossed coupling terms between the d-axis and q-axis are as follows:

$$\begin{cases} fem_d = -\sigma L_r \omega_r i_{qr} \\ fem_q = \sigma L_r \omega_r i_{dr} + s\dfrac{M}{L_s} V_s \end{cases} \tag{5.51}$$

Consequently, regarding (5.47), the fluxes of (5.43) are simplified as follows:

$$\begin{cases} \phi_{ds} = L_s i_{ds} + M i_{dr} \\ 0 = L_s i_{qs} + M i_{qr} \end{cases} \tag{5.52}$$

From (5.52), we can deduce the currents to be

$$\begin{cases} i_{ds} = \dfrac{\phi_{ds} - M i_{dr}}{L_s} \\ i_{qs} = -\dfrac{M}{L_s} i_{qr} \end{cases} \tag{5.53}$$

Using Eqs. (5.44), (5.48), and (5.53), the stator active and reactive powers can then be linked to these rotor currents as follows:

$$\begin{cases} P_s = -V_s \cdot \dfrac{M}{L_s} i_{qr} \\ Q_s = -V_s \cdot \dfrac{M}{L_s}\left(i_{dr} - \dfrac{\phi_{ds}}{M} \right) \end{cases} \tag{5.54}$$

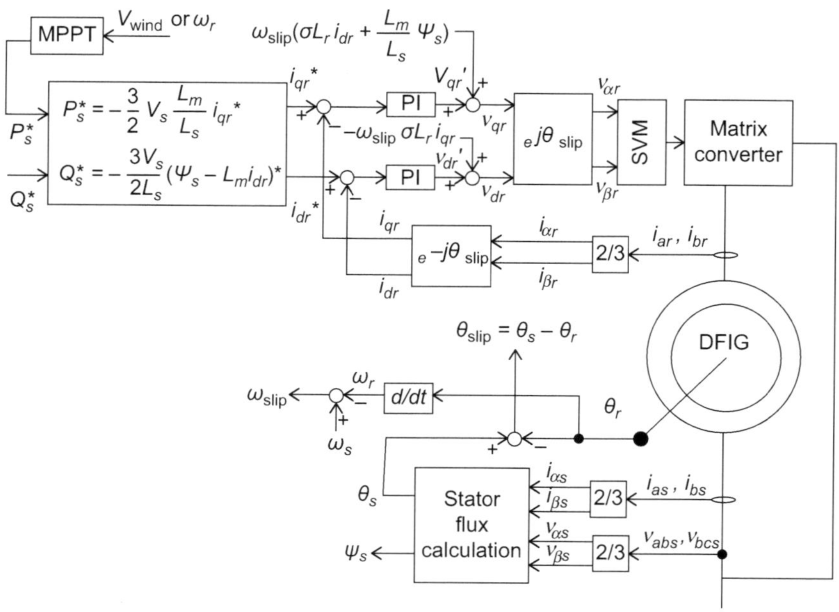

Figure 5.15 Simplified coupled model of the DFIG rotor current at the operation of matrix converter when stator fluxes are in MPPT level.

Because of the constant stator voltage, the stator active and reactive powers are controlled via i_{qr} and i_{dr}.

Therefore, field-oriented control of the DFIG can then be performed, with the rotor currents considered as the variables to be controlled. Consequently, by using the precedent block diagram in Fig. 5.15, the proposed control system is presented in Fig. 5.16.

Decoupled control is guaranteed without feed-forward compensation because the fuzzy logic controller (FLC) inherently eliminates the cross-coupling terms between the two axes (Eq. 5.51). In steady state for a lossless generator, we can use the following energy balance: $P_s + P_r = P_m$, where $P_s = T_{em}\,\omega_s$ and $P_m = T_q\,\omega$.

From the MPPT method, the electromagnetic torque is used to calculate the reference value for the stator's active power, which follows a predefined turbine power-speed characteristic to track the maximum power point [59,65].

The turbine shaft speed is then controlled to give the maximum power coefficient. It follows that $P_s = T_{em}\,\omega_s$, where T_{em}^* is the reference electromagnetic torque that is deduced from the MPPT control strategy. The MPPT method significantly increases the efficiency of the wind turbine.

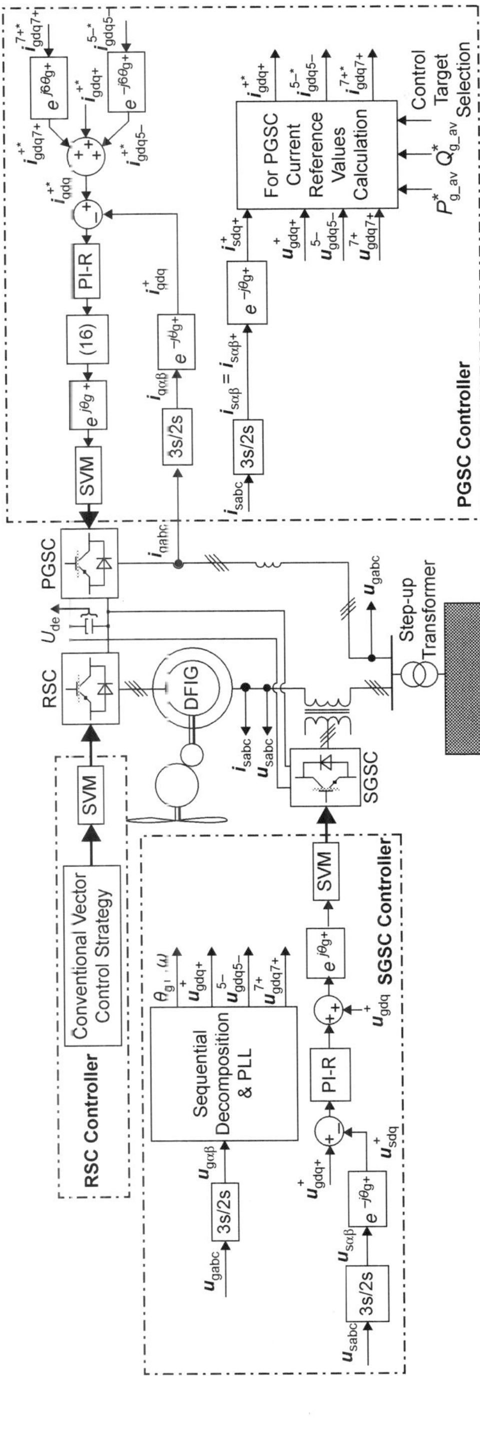

Figure 5.16 Schematic diagram of the proposed control system of RSC controller, SGSC controller, and PGSC controller.

Using Eq. (5.54), we can calculate the rotor current references, which allows setting the desired reference active and reactive powers, as follows:

$$\begin{cases} i_{qr}^* = -\dfrac{L_s}{MV_s} P_s^* \\ i_{dr}^* = -\dfrac{L_s}{MV_s}\left(Q_s^* - \dfrac{V_s^2}{\omega_s L_s}\right) \end{cases} \quad (5.55)$$

5.3.2.9 Controller Design

Because of the robustness of the FLC for many nonlinear procedures and characteristics, this chapter suggests the design of a FLC with a Mamdani fuzzy inference system [94,95]. The FLC includes four parts: (1) a fuzzification block that determines the input membership values; (2) a fuzzy inference system (FIS) that evaluates which control rules are appropriate at each time by using the fuzzy knowledge-based block [70,85]; (3) a deffuzification block that calculates the output of the rules leading to the defuzzification technique [1,2,10]. Fig. 5.17 shows the block diagram of the fuzzy controller.

For a successful design, proper selection of the gains in the FLC is crucial, and in many cases, this is performed through trial and error to achieve the best possible control performance.

In this section, FLC is used to control the wind turbine subsystem and the DFIG subsystem. For the proposed FLC of a wind turbine, we use the scheme shown in Fig. 5.12. There are two input signals to the FLC, the generator shaft speed error and the change of the error, and they are as follows:

$$\begin{cases} e_{\Omega_g}(n) = \Omega_g^*(n) - \Omega_g(n) \\ \Delta e_{\Omega_g}(n) = \Omega_g^*(n) - \Omega_g(n-1) \end{cases} \quad (5.56)$$

The input and output linguistic variables of the fuzzy controller are quantized based on the three fuzzy subsets. The fuzzy sets have been determined to be NG (negative great), EX (zero), and PG (positive great).

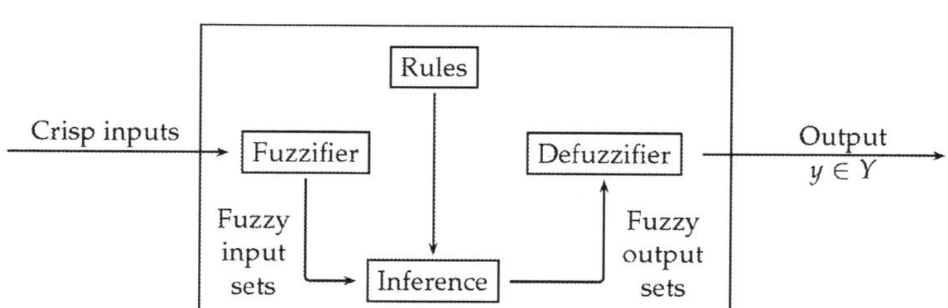

Figure 5.17 Block diagram of the fuzzy controller showing the crisp inputs and outputs through interference and defuzzifier.

The input/output variables used in this chapter are fuzzified and evaluated by triangular, trapezoidal, and symmetrical membership functions (MFs). Thus the input of the generator speed shaft fuzzy controller is T_{em}^*. For the proposed FLC of the DFIG, the diagram scheme shown in Fig. 5.16 was used; the inputs of the direct and quadrate axis rotor current fuzzy controllers are the d- and q-axis rotor current errors:

$$\begin{cases} e_{i_{dr}}(n) = i_{dr}^*(n) - i_{dr}(n) \\ e_{i_{qr}}(n) = i_{qr}^*(n) - i_{qr}(n) \end{cases} \tag{5.57}$$

Their changes in error are

$$\begin{cases} \Delta e_{_{dr}}(n) = i_{dr}^*(n) - i_{dr}(n-1) \\ \Delta e_{_{qr}}(n) = i_{qr}^*(n) - i_{qr}(n-1) \end{cases} \tag{5.58}$$

The input and output linguistic variables of the two fuzzy controllers have been quantized in the following five fuzzy subsets. The fuzzy sets have been defined as NL (negative large), NS (negative small), ZQ (zero), PS (positive small), Pm (positive medium), and PL (positive large).

Short-term fluctuations in the wind speed may result in a change in the output power and a shift in the operating point [21]. Thus to avoid these undesirable effects reducing the number of MFs is a solution to smooth the changes in the output power; increasing the number of MFs will produce a delay because of the increased number of computational steps required for some bands [15,24].

To trade-off between accuracy and complexity, rigorous simulation studies found that five MFs are sufficient to produce the desired results in the required bands.

5.3.2.10 Generator Modeling

Either induction and synchronous generators can be used for wind turbine systems [96,97]. Using a directly driven permanent magnet synchronous generator (PMSG) not only increases the reliability but also decreases the weight of the nacelle [98,99]. The PMSG model was designed based on a d-q synchronous reference frame. The PMSG voltage equation is

$$V_q = -R_s i_q - L_q \frac{di_q}{dt} - \omega L_d i_d + \omega \lambda_m \tag{5.59}$$

$$V_d = -R_s i_d - L_d \frac{di_q}{dt} + \omega L_q i_q \tag{5.60}$$

The electronic torque is

$$T_e = 1.5p[\lambda i_q + (L_d - L_q)i_d i_q] \tag{5.61}$$

where L_q is the q-axis inductance, L_d is the d-axis inductance, i_q is the q-axis current, i_d is the d-axis current, V_q is the q-axis voltage, V_d is the d-axis voltage, ω_r is the angular velocity of the rotor, λ is the amplitude of the flux induced, and p is the number of pairs of poles. In the case of a squirrel-cage induction generator (SCIG), the following equation in a stationary d-q frame of reference can be used for dynamic modeling:

$$\begin{bmatrix} V_{qs} \\ V_{ds} \\ V_{qr} \\ V_{dr} \end{bmatrix} = \begin{bmatrix} R_s + pL_s & 0 & pL_m & 0 \\ 0 & R_s + pL_s & 0 & pL_m \\ pL_m & -\omega_r L_m & R_r + pL_r & -\omega_r L_r \\ \omega_r L_m & pL_m & \omega_r L_r & R_r + pL_r \end{bmatrix} \begin{bmatrix} i_{qs} \\ i_{ds} \\ i_{qr} \\ i_{dr} \end{bmatrix} \tag{5.62}$$

From the stator side, the equations are as follows:

$$\begin{aligned} \lambda_{ds} &= L_s i_{ds} + L_m i_{dr} \\ \lambda_{qs} &= L_s i_{qs} + L_m i_{dr} \\ L_s &= L_{is} + L_m \\ L_r &= L_{lr} + L_m \\ V_{ds} &= R_s i_{ds} + \frac{d}{dt}\lambda_{ds} \\ V_{qs} &= R_s i_{qs} + \frac{d}{dt}\lambda_{qs} \end{aligned} \tag{5.63}$$

From the rotor side, the equations are as follows:

$$\begin{aligned} \lambda_{dr} &= L_r i_{dr} + L_m i_{ds} \\ \lambda_{qr} &= L_r i_{qr} + L_m i_{qs} \\ V_{dr} &= R_r i_{dr} + \frac{d}{dt}\lambda_{dr} + \omega_r \lambda_{qr} \\ V_{qr} &= R_r i_{qr} + \frac{d}{dt}\lambda_{qr} - \omega_r \lambda_{dr} \end{aligned} \tag{5.64}$$

For the air gap flux linkage, the equations are as follows:

$$\begin{aligned} \lambda_{dm} &= L_m(i_{ds} + i_{dr}) \\ \lambda_{qr} &= L_m(i_{qr} + i_{qs}) \end{aligned} \tag{5.65}$$

where R_s, R_r, L_m, L_{ls}, L_{lr}, ω_r, i_d, i_q, V_d, V_q, λ_d, and λ_q are the stator winding resistance, and fluxes of the d-q model [78]. The output power and torque of the turbine (T_t) in terms of rotational speed can be obtained by substituting Eq. (5.40) into Eq. (5.39):

$$P_w = \frac{1}{2}\rho A C_p(\lambda, \beta)\left(\frac{R\omega_{\text{opt}}}{\lambda_{\text{opt}}}\right)^3 \tag{5.66}$$

$$T_t = \frac{1}{2}\rho A C_p(\lambda, \beta)\left(\frac{R}{\lambda_{\text{opt}}}\right)^3 \omega_{\text{opt}} \tag{5.67}$$

The power coefficient (C_p) is a nonlinear function that is expressed by the fitting equation[51] with the form

$$C_p(\lambda, \beta) = c_1\left(c_2\frac{1}{\lambda_i} - c_3\beta - c_4\right)e^{-c_5\frac{1}{\lambda_i}} + c_6\lambda \tag{5.68}$$

with

$$\frac{1}{\lambda_i} = \frac{1}{\lambda + 0.08\beta} - \frac{0.035}{\beta^3 + 1} \tag{5.69}$$

The values of constants c_1–c_6 are discussed in the following sections.

The output of the wind energy generator module is processed by an energy conversion circuit diagram—implemented inverter from the standard Simulink/Sim Power Systems. The resulting MATLAB Simulink circuit model for the wind generator is a particular case of the more general model of an electrical generator that is presented in Fig. 5.13 and Fig. 5.18.

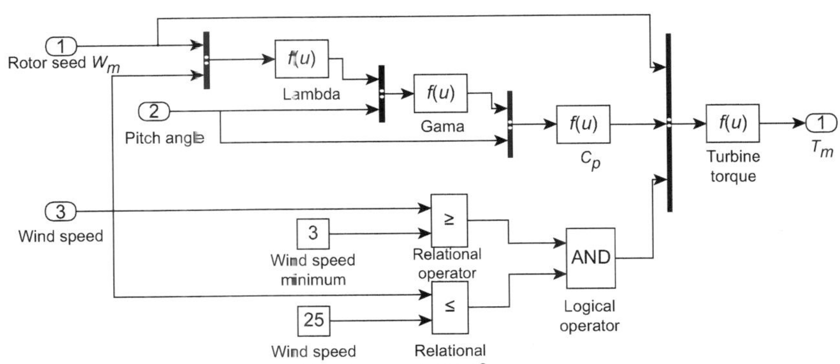

Figure 5.18 Equivalent circuit diagram of a small wind generator considering all applications of rotor winding resistance and separate generator excitation winding; current through the winding generates a main field, induced voltage, and terminal voltage.

5.3.3 Simulation and Discussion

In this chapter, a complete model was presented for a WTGS based on an induction generator. A cascade control algorithm was properly designed to ensure optimal operation of the whole system and was based on an FLC with stator flux orientation and the MPPT technique. Furthermore, the output power is smoothed despite wind fluctuations, which is commonly required in a grid connected to a wind farm. This technique has been successfully applied to generate a reference for tracking the active power by using the rotor-side converter control. In addition, this strategy can reduce the stress in the pitch control system by adjusting the value of the power coefficient according to wind speed variations to extract the maximum wind power and limit it to its rated value. In simulations, a robustness test was performed by adding a wind speed signal and voltage dips. The simulation results demonstrate the inherent ability of the FLC to address this type of noise while operating under fault conditions at the rated wind speed.

Fig. 5.9 shows that the maximum values of C_p are achieved for the curve associated with $\beta = 2$ degrees. From this curve, the maximum value of C_p ($C_{p,\max} = 0.5$) is obtained for $\lambda_{\text{opt}} = 0.91$. This value ($\lambda_{\text{opt}}$) represents the optimal speed ratio. These conditions permit application of MPPT control. For the DFIG control, we use indirect power control with an FLC. The wind profile was considered to be a wind speed signal with a mean value of 8 m/s and a rated wind speed of 10 m/s; the whole system is tested under standard conditions with a stator voltage of approximately 50% for 0.5 s between 4 and 4.5 s, approximately 25% between 6 and 6.5 s, and 50% between 8 and 8.5 s [26,38]. Thus, the machine is considered to be functioning in ideal conditions (no perturbations and no parameter variations). Moreover, to guarantee a unity power factor at the stator side, the reference for the reactive power is set to zero. The stator active and reactive powers are controlled according to the MPPT and FLC.

As a result of increasing wind speed, the generator shaft speed achieved maximum angular speed by tracking the maximum power point speed. Thus the wind turbine always works optimally. Consequently, decoupling among the components of the rotor current was also performed to confirm that the control system worked effectively.

The bidirectional active and reactive power transfer between the rotor and power system is exchanged by the generator according to the super synchronous operation, achieving the nominal stator power, and the reactive power can be controlled by the load-side converter to obtain the unit's power factor to generate energy for powering vehicles.

5.3.3.1 Theoretical Experiment on a Car

The theoretical assumption of this wind energy is thus modeled on a car as an experimental tool with the aim to ultimately use wind energy for all vehicles. The design of the wind turbines, materials used in all mechanical and electrical construction of the car, and all related issues have been considered, including all the machine's operational systems, wind strength, wind shears, and intensity and frequency of turbulence fluctuations (Fig. 5.19).

The turbine speed and mechanical powers are depicted in the following graph with increasing and decreasing rates of wind speed while the car is in motion. When the wind is steady, the persistence forecasts yield good results. However, when the wind speed is increased rapidly, sudden "ramps" in power output are generated, which are of tremendous benefit for capturing the energy (Fig. 5.20).

Specifically, the following matrix presents the mathematical calculations for wind energy—capturing strategies when the car is running. S_{r1-b} represents the direct wind energy capture, S_{r2-b} represents the rotation of the generator, S_{r3-b} represents the converting process of wind energy, and S_{r4-b} is the total electricity production to run the car. The matrix calculation shows that 100% of the wind energy is used for conversion into electrical energy.

$$S_{f1-b} = \begin{bmatrix} 0 & 0 & 1 \\ \cos(-(90-\varpi)) & \sin(-(90-\varpi)) & 0 \\ -\sin(-(90-\varpi)) & \cos(-(90-\varpi)) & 0 \end{bmatrix} = \begin{bmatrix} 0 & 0 & 1 \\ \sin\varpi & -\cos\varpi & 0 \\ \cos\varpi & \sin\varpi & 0 \end{bmatrix} \quad (5.70)$$

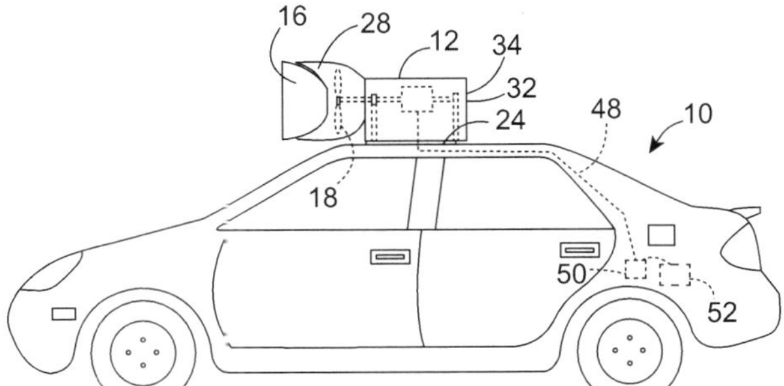

Figure 5.19 A conceptual model of a wind turbine for energy production to power a car while it is in motion.

$$S_{f2-b} = \begin{bmatrix} \cos 120^\circ & \sin 120^\circ & 0 \\ -\sin 120^\circ & \cos 120^\circ & 0 \\ 0 & 0 & 1 \end{bmatrix} \begin{bmatrix} 0 & 0 & 1 \\ \sin \varpi & -\cos \varpi & 0 \\ \cos \varpi & \sin \varpi & 0 \end{bmatrix}$$

$$= \begin{bmatrix} -\frac{1}{2} & \frac{\sqrt{3}}{2}\sin \varpi & -\frac{\sqrt{3}}{2}\cos \varpi \\ -\frac{\sqrt{3}}{2} & -\frac{1}{2}\sin \varpi & \frac{1}{2}\cos \varpi \\ 0 & \cos \varpi & \sin \varpi \end{bmatrix}$$

$$S_{f3-b} = \begin{bmatrix} \cos(-120^\circ) & \sin(-120^\circ) & 0 \\ -\sin(-120^\circ) & \cos(-120^\circ) & 0 \\ 0 & 0 & 1 \end{bmatrix} \begin{bmatrix} 0 & 0 & 1 \\ \sin \varpi & -\cos \varpi & 0 \\ \cos \varpi & \sin \varpi & 0 \end{bmatrix}$$

$$= \begin{bmatrix} -\frac{1}{2} & \frac{\sqrt{3}}{2}\sin \varpi & \frac{\sqrt{3}}{2}\cos \varpi \\ \frac{\sqrt{3}}{2} & \frac{1}{2}\sin \varpi & \frac{1}{2}\cos \varpi \\ 0 & \cos \varpi & \sin \varpi \end{bmatrix}$$

$S_{f4-b} = S_{f1-b}; S_{f5-b} = S_{f2-b}; S_{f6-b} = S_{f3-b}$

In a WECS, the accuracy with which the peak power output of electricity is obtained from a point of the MPPT control system by the controller depends on the track. This experiment thus fed the induction generator stand (DFIG) that is used for extracting the maximum power with the use of a WECS. With the MPPT control method, the MPPT WECSs are presented as controllers that are used for extracting the maximum possible energy from wind power generation, which, interestingly, is related to the speed of the car.

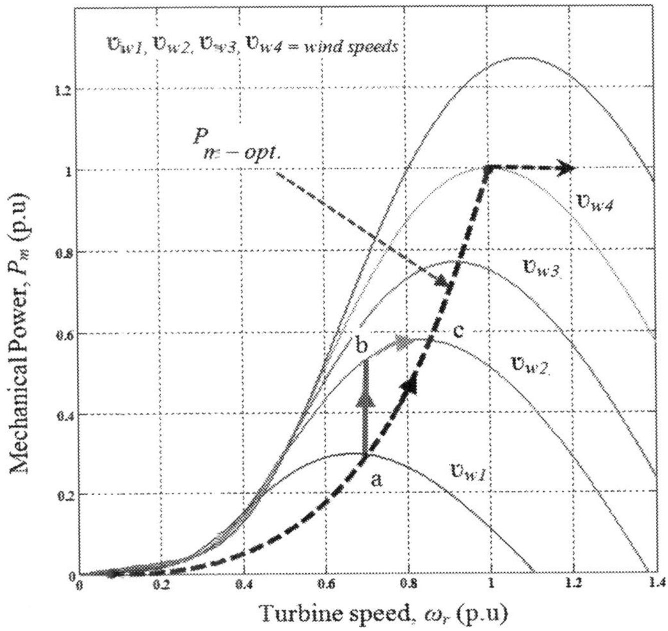

Figure 5.20 Relationship between mechanical power generation and turbine speeds at different wind speeds for implementation in a car.

The analysis was then clarified with the following figure that depicts the relationship between the wind speed (resulting from the motion of the car) in miles per hour (MPH) and kWh power production. Interestingly, it is revealed that at a maximum mean of 8 kWh and at an average speed of 10 MPH, energy is produced at a wind speed of 2 MPH immediately after the engine is started by the battery. For a standard car to become fully energized, 20 kWh is required (Fig. 5.21).

5.3.3.2 Battery Modeling

A battery is used as a backup power source to store the power when power production exceeds the demand. For a standard car to become fully energized, 20 kWH is required. Thus if the car runs at 10 MPH for 2 h, it will fully charge and be able to run for an average of 200 miles; consequently, if the car runs at 60 MPH, it will only take 20 min to fully charge and be able to run for the same number of miles.

In this model, a battery is used as a storage buffer, and all the electricity is supplied through the battery according to Peukert's Law to predict battery

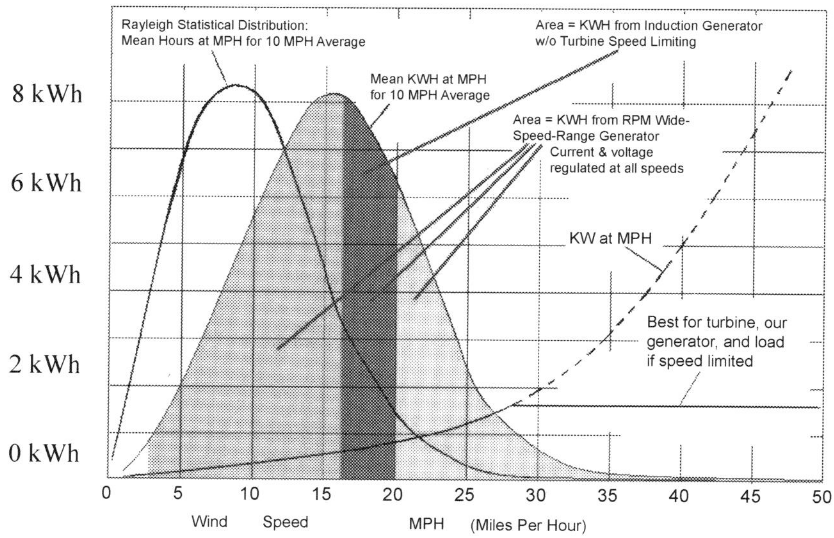

Figure 5.21 Conversion of wind energy in the car is described in this figure and shows the relationship between the wind speed and energy production considering mean hours (10 MPH average), kWh from the induction generator, and kWh calculated from the revolutions per minute of speed range generator current and voltage.

discharge considering the nonlinear properties of the battery [19]; this law is stated as follows:

$$t_{\text{discharge}} = H\left(\frac{C}{IH}\right)^{k} \tag{5.71}$$

where *t* is the battery discharge time, *C* is the battery capacity (Ampere hour value), *I* is the current that is drawn, *H* is the rated discharge time, and *k* is Peukert's coefficient. Peukert's coefficient is an empirical value that can be determined using the following formula:

$$k = \frac{\log T_2 - \log T_1}{\log I_1 - \log I_2} \tag{5.72}$$

where I_1 and I_2 are the two discharge current rates and T_1 and T_2 are the corresponding discharge durations. The battery capacity decreases with time; i.e. the time for charging and discharging will change.

Thus the value of *k* should be determined after a certain number of recharge cycles. The value of *k* for a lead acid battery is 1.3–1.4 (Fig. 5.22).

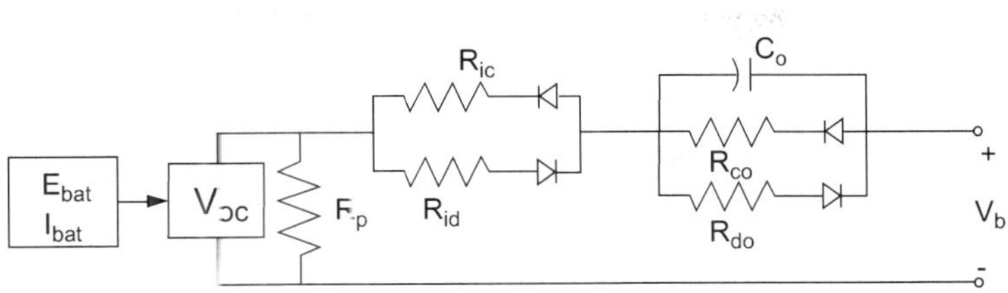

Figure 5.22 Model of a car with a battery as the backup power source to start the engine and for when the car is not in motion.

The charging time for a completely discharged battery is given by

$$t_{\text{charging}} = \frac{\text{Ampere hour of battery}}{\text{Charging Current}} \tag{5.73}$$

Conceptual estimate for design and construction of a renewable wind energy–powered car

List of components	Materials cost	Labor cost	Equipment cost	GC and OH cost	Total cost
Wind turbine	$3000	$900	$500	$880	$5280
Instrumentation	$500	$500	$400	$280	$1680
Electrical and mechanical control	$750	$600	$800	$430	$2580
Supply for 20 years cost is $0.00, but maintenance cost is $200/year					$4000
Total cost					$13,540

Note: This estimate was prepared using current (March 2016) material costs from top manufacturers and labor costs to install materials as per international union labor wages. The equipment purchase was calculated according to current market costs in conjunction with production rates of standard construction practices.

5.3.3.3 Savings in Terms of Energy Costs

Twenty years of operation of a conventionally powered standard car (15,000 miles/year) requires 750 gallons of gasoline per year at an average rate of 20 miles/gallon (750 × 33.70 = 25,275 kWh/year), which is equivalent to 505,500 kWh of energy for 20 years. The cost of this conventional energy is (750 gallons × 20 years × $2.25 or $0.067/kWh for total of 505,500 kWh) $33,750. This comparison between conventional energy usage and wind turbine energy production for a car clearly shows cost savings of $20,210 when a wind energy system is used as the energy source for a car.

The energy cost of $13,000 (furnish and install) for a turbine can be eliminated once the turbine system is commercially installed in cars, and the maintenance costs will be covered by the manufacturer's warranty.

Therefore, this estimate suggests that most of the $33,750 cost for running a car for 20 years can be saved once it is powered by wind energy. Subsequently, all other transportation energy costs can also be saved once vehicles are powered by wind energy.

5.3.4 Conclusions

Transportation sectors throughout the world are powered by fossil fuels; these sectors draw on finite resources that are already dwindling, resulting in increased costs. Fossil fuel retrieval is becoming ever more environmentally damaging. It is now generally accepted that meeting the steadily increasing demand for energy in transportation sectors solely on the basis of conventional generation technologies puts an unacceptably high stress on the environment. In combination, these factors have led to a major impetus for renewable energy—based power generation during the preceding two to three decades, and the trend is set to continue. Thus when considering efficiency, WECSs have been receiving the most attention among the various renewable energy systems. Extraction of the maximum possible power from available wind power is indeed an innovative source because the development of smaller wind turbines and advanced technology could enable wind energy to serve as a power source for transportation sectors.

Interestingly, this article revealed that natural wind speed is not required because the turbine will start to rotate once the vehicle is in motion. This emergent new technology, together with the favorable framework for renewable energy, can play a prime role in renewable energy utilization for massive use in transportation sectors. The transportation sector needs 5.6×10^{20} J/year (560 EJ/year); currently, this energy is obtained by burning fossil fuel, which accounts for nearly 30% of the total annual global energy demand.

Therefore if vehicles could be powered by wind, it would be a new era of science to use wind energy in transportation sectors. Wind energy—powered vehicles would not only eliminate the energy cost for the transportation sector but also play a vital role in drastically reducing global warming.

5.4 FLYING TRANSPORTATION TECHNOLOGY

To have pleasant road trips, avoid commute, and do less time on trips to arrive at the desired destination much faster, a model of flying transportation technology has been proposed, in general, trying to find a new, better, safe, and economic ways for transportation. In this research, a 3D numerical simulation of external flow for a flying car with different NACA 9618 rectangular wings has been designed. To provide theoretical basis and optimize its geometry to make it possess a minimum takeoff velocity and better performance in the air, first its aerodynamic characteristics have been calculated carefully and comprehensively. After that a 3D standard *k-omega* turbulence model has been proposed to capture the intrinsic flow physics during the takeoff phase. Then in the numerical study, a thoroughly implicit finite volume scheme of the compressible, Reynolds-Averaged, Navier—Stokes equations is being proposed to use. Necessarily the detailed parametric analytical model is being conjectured so that a flying car can be facilitated with high wings at different deployment during the takeoff phase for accomplishing its better performance. For the flying transportation vehicle run by wind, a wind energy modeling has also been introduced into the vehicle to use natural wind power by integrating small wind turbine into the vehicle to produce energy while vehicle is on a motion. Combining these technologies, indeed, would be the most innovative technology for the development of flying transportation vehicle which not only can save trillion dollars money yearly for the construction of roads, highways, bridges, and tunnels but also will reduce the climate change significantly.

5.4.1 Background and Technology

There is approximately 0.9% of the total planetary surface area been constructed as infrastructure, which equals to 1.77 million mi^2. These infrastructures lead to roughly 6% of climate change because of the heat reflection back to space and cost approximately 5575 trillion US dollars (100,000,000 mi^2 × 5280 sf × 5280 sf × \$200 per sf cost). Based on research, 1% infrastructure is added globally each year, and 2% existing infrastructure is repaired; therefore the cost of these two purposes is 55.75 trillion and 12 trillion US dollars, separately, which is not a small number. Besides, traditional transportation energy usage and traffic jam are not only expensive but also result in adverse environmental impact and a loss of time and energy [100,101]. Therefore to improve this situation, flying car can be a significant technology. Since early days of motoring, the flying car concept has already been studied by several researchers. However, the optimized aerodynamic design of the flying car is still a dispiriting but attracting task. These researches showed not only the attempts of the designers to build a flying car for around a century but also reveal that more efforts should be applied to make the flying car come true.

Recently, as the computational fluid dynamic (CFD) methods have been integrated into large-scale engineering disciplines, which may attribute to the positive trends in computational power and affordability, it may also be applied to the flying car design. However, the aerodynamics design of a flying car is more sophisticated than a race car or a traditional aircraft because of the complex prediction of transient unsteady flow features of the wings in deployment and takeoff phase. Therefore in this research, a simple propulsive and levitative force model has been introduced to provide a solution to the drag force, aerodynamics, CFDs limitations, and takeoff velocity and realize a reliable, safer, and faster flying transportation with the help of MATLAB software.

5.4.2 Thoughts and Simulation

To ensure the high performance of the flying car in the air, the major concern is to create an upward force with low takeoff velocity, minimum drag force, and better stability and control. Accordingly, the amount of upward force should be considered. It mainly depends on two things, the shape of the wings, including surface area, aspect ratio, and cross-section composition, and the vehicle's orientation. In this research, first a model will be designed to show the physical mode, and the Mach number contours to carry a successful 3D CFD *k-omega* turbulence mode of the flying car (Fig 5.23). Then, some 3D numerical studies on external flow features of a flying car with different wing positions and deployment history will be conducted to validate the CFD as a preliminary design tool for a reliable flying car.

5.4.2.1 Numerical Method of Solution

As mentioned, a 3D standard *k-omega* turbulence model, which is an empirical model based on model transport velocity for the turbulence kinetic energy and a specific dissipation rate, is introduced to deal with the numerical simulations as shown in the following equations [4,12].

$$\psi = \left[-\left(\frac{v_\infty r_0^2}{4}\right)\frac{r_0}{r} + \left(\frac{3v_\infty r_0}{4}\right)r - \left(\frac{v_\infty}{2}\right)r^2\right]\sin^2\theta \tag{5.74}$$

$$v_r = v_\infty\left[1 - \frac{3}{2}\left(\frac{r_0}{r}\right) + \frac{1}{2}\left(\frac{r_0}{r}\right)^3\right]\cos\theta \tag{5.75}$$

$$v_\theta = -v_\infty\left[1 - \frac{3}{4}\left(\frac{r_0}{r}\right) - \frac{1}{4}\left(\frac{r_0}{r}\right)^3\right]\sin\theta \tag{5.76}$$

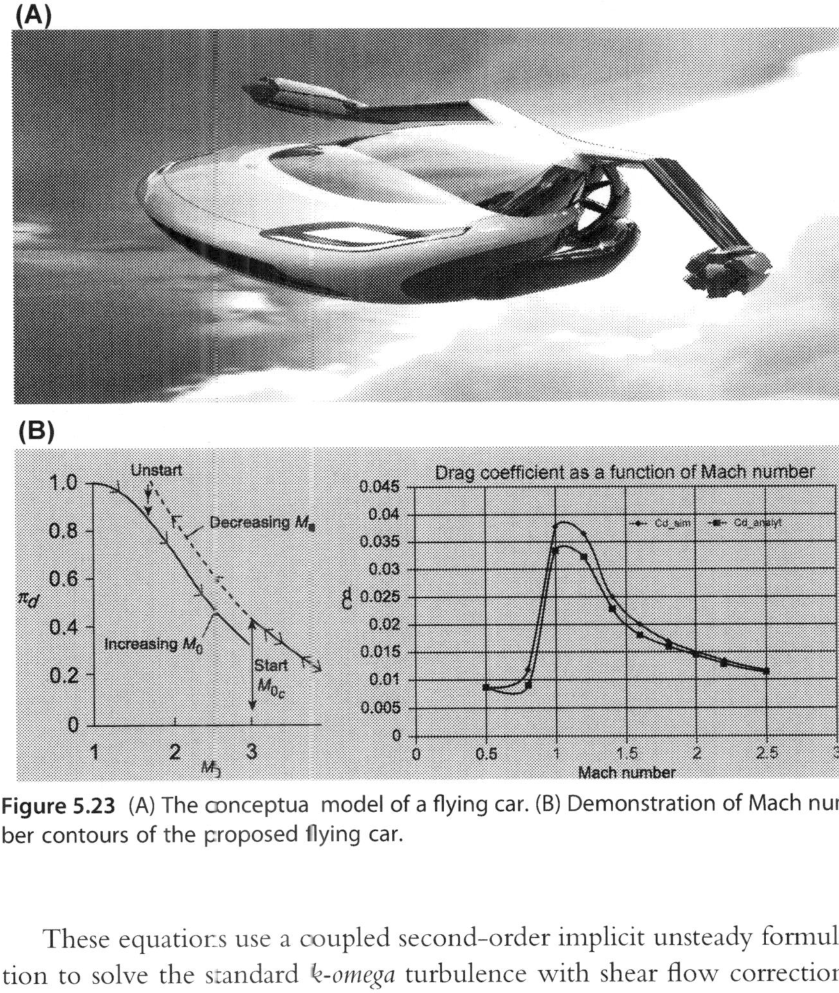

Figure 5.23 (A) The conceptua model of a flying car. (B) Demonstration of Mach number contours of the proposed flying car.

These equations use a coupled second-order implicit unsteady formulation to solve the standard *k-omega* turbulence with shear flow corrections. Therefore this numerical solution will be treated as a fully implicit finite volume scheme of the compressible, Reynolds-averaged, Navier–Stokes forces to neutralize initial wall temperature and total pressure of the flying car. Being considered about the mach number ratio with the velocity calculation, these codes may successfully validate the baseline solutions to design the wings with an NACA series of 9618 airfoils to realize better aerodynamic characteristics (Fig 5.24).

5.4.2.2 Wind Energy Modeling for the Flying Vehicles

Traditionally, a single-wind-turbine generator is used for powering vehicle by providing energy. Therefore to satisfy the higher energy needs of flying

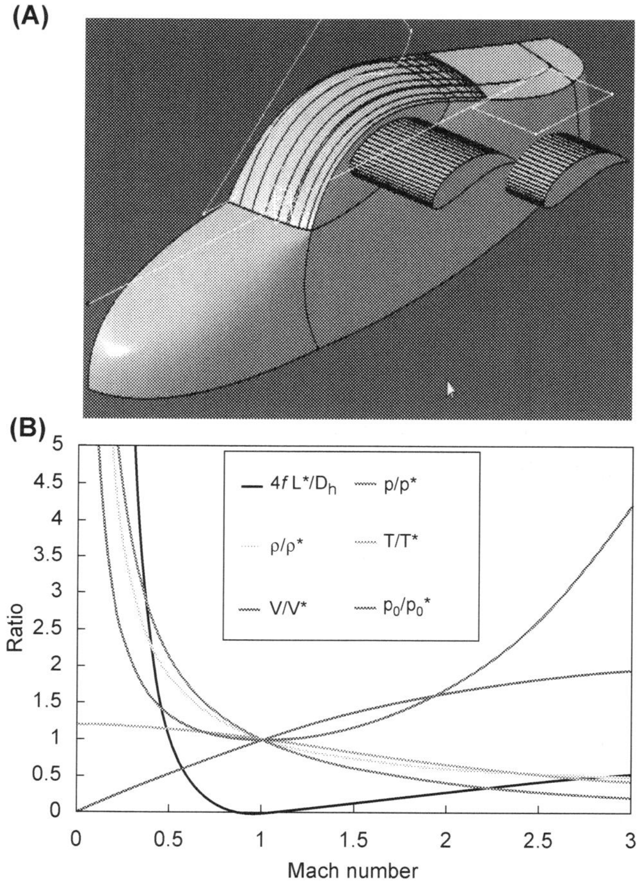

Figure 5.24 (A) The idealized 3D symmetrical model (half) of a flying car. (B) Aerodynamic characteristics considering mach number ratio, considering the velocity.

vehicles, DFIG may be used to produce electricity [3,7,11]. The elementary equation governing the mechanical power of the wind turbine is

$$P_w = \frac{1}{2} C_p(\lambda, \beta) \rho A V^3 \tag{5.77}$$

In this equation, V refers to the average wind speed (m/s), ρ refers to the air density (kg/m^3), C_p refers to the Betz's coefficient, which has a theoretical maximum value of 0.593, λ refers to the tip speed ratio, and A refers to the intercepting area of the rotor blades (m^2).

$$\lambda = \frac{R\omega}{V} \tag{5.78}$$

The aforementioned equation is used to calculate the tip speed ratio; in this equation, ω is the angular speed (rad/s), V is the average wind speed (m/s), and R is the radius of the turbine (m).

Then, the energy generated by wind can be obtained by

$$Q_w = P \times (Time)(\text{kWh}) \tag{5.79}$$

As the wind velocity cannot be obtained by a direct measurement from any particular motion [43,55,102], if the velocity is required, a lower motion is needed.

$$v(z)\ln\left(\frac{Z_r}{Z_o}\right) = v(Z_r)\ln\left(\frac{Z}{Z_0}\right) \tag{5.80}$$

where Z_r is the reference height (m), Z is the height at which the wind speed is to be determined, Z_0 is the measure of surface roughness (0.1−0.25 for crop land), $v(z)$ is wind speed at height z (m/s), and $v(z_r)$ is wind speed at the reference height z (m/s). The power output in terms of the wind speed shall be estimated using the following equation:

$$P_w(v) = \begin{cases} \dfrac{v^k - v_C^k}{v_R^k - v_C^k} \cdot P_R & v_C \le v \le v_R \\ P_R & v_R \le v \le v_F \\ 0 & v \le v_C \text{ and } v \ge v_F \end{cases} \tag{5.81}$$

where P_R is rated power, v_C is the cut-in wind speed, v_R is the rated wind speed, v_F is the rated cut-out speed, and k is the Weibull shape factor. The angular speed of the generator is to be changed to extract the maximum power, whose process is known as MPPT. When the blade pitch angle equals to zero, the power coefficient will be maximized to get an optimal TSR. The optimal rotor speed is to be calculated by the following equation:

$$\omega_{\text{opt}} = \frac{\lambda_{\text{opt}}}{R} V_{wn} \tag{5.82}$$

which will give

$$V_{wn} = \frac{R\omega_{\text{opt}}}{\lambda_{\text{opt}}} \tag{5.83}$$

where ω_{opt} is the optimal rotor angular speed in rad/s, λ_{opt} is the optimal tip speed ratio, R is the radius of the turbine in meters, and V_{wn} is the wind speed in m/s.

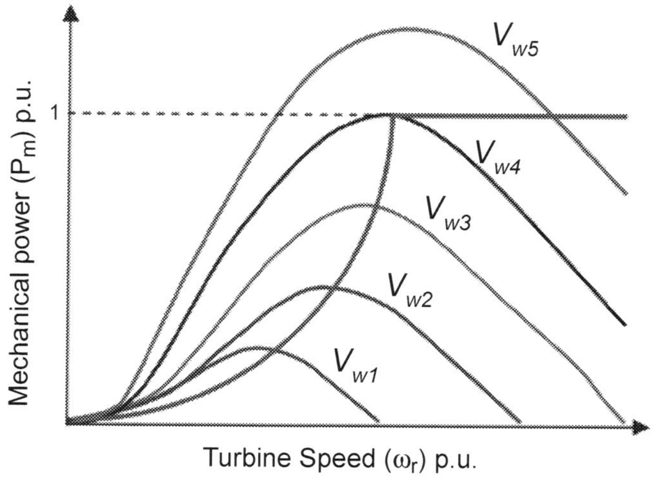

Figure 5.25 Relationship between mechanical power generation and turbine speeds at different wind speeds for implementation in a car.

In the following figure, the turbine speed and mechanical powers are described by increasing and decreasing rates of wind speed while the car is working [68,94]. The outcomes can be generated from the figure that when the wind is steady, the persistence forecasts lead to good results; meanwhile, tremendous benefit for capturing the energy can be achieved if the wind speed is increased rapidly (Fig. 5.25).

5.4.2.3 Wind Energy Conversion

Under the aforementioned conditions, the airflow mass has to be controlled at a certain level of kinetic energy, which will be presented by a WECS that is to be used by DFIG, including the aerodynamic subsystem (wind speed, wind turbine, and gearbox) and the electrical subsystem (DFIG) (Fig. 5.26).

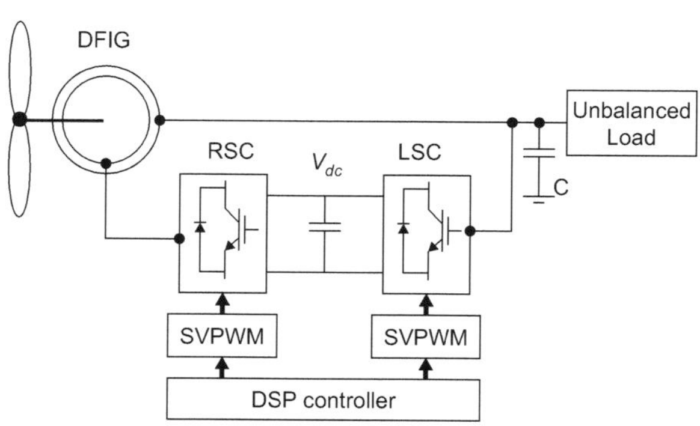

Figure 5.26 Wind energy conversion chain diagram considering DFIG, RSC, LSC, SVPWM, and DSP controller.

5.4.2.4 Generator Modeling

To provide energy for the wind turbine systems, either synchronous or induction generators models can be used. If being used, the directly driven PMSG can not only increase the reliability but also decrease the weight of the nacelle [34,54]. The following PMSG voltage equations show the principle of the PMSG model designed on the basis of a d-q synchronous reference frame.

$$V_q = -R_s i_q - L_q \frac{di_q}{dt} - \omega L_d i_d + \omega \lambda_m \tag{5.84}$$

$$V_d = -R_s i_d - L_d \frac{di_q}{dt} + \omega L_q i_q \tag{5.85}$$

where the electronic torque is

$$T_e = 1.5\rho[\lambda i_q + (L_d - L_q) i_d i_q] \tag{5.86}$$

where L_q is the q-axis inductance, L_d is the d-axis inductance, i_q is the q-axis current, i_d is the d-axis current, V_q is the q-axis voltage, V_d is the d-axis voltage, ω_r is the angular velocity of the rotor, λ is the amplitude of the flux induced, and p is the number of pairs of poles.

Another model involves the use of SCIG. For reference, the following equation is used for dynamical modeling under the circumstance of a stationary d-q frame of reference.

$$\begin{bmatrix} V_{qs} \\ V_{ds} \\ V_{qr} \\ V_{dr} \end{bmatrix} = \begin{bmatrix} R_s + pL_s & 0 & pL_m & 0 \\ 0 & R_s + pL_s & 0 & pL_m \\ pL_m & -\omega_r L_m & R_r + pL_r & -\omega_r L_r \\ \omega_r L_m & pL_m & \omega_r L_r & R_r + pL_r \end{bmatrix} \begin{bmatrix} i_{qs} \\ i_{ds} \\ i_{qr} \\ i_{dr} \end{bmatrix} \tag{5.87}$$

From the stator side, the equations shall be as follows:

$$\begin{aligned} \lambda_{ds} &= L_s i_{ds} + L_m i_{dr} \\ \lambda_{qs} &= L_s i_{qs} + L_m i_{dr} \\ L_s &= L_{ls} + L_m \\ L_r &= L_{lr} + L_m \\ V_{ds} &= R_s i_{ds} + \frac{d}{dt}\lambda_{ds} \\ V_{qs} &= R_s i_{qs} + \frac{d}{dt}\lambda_{qs} \end{aligned} \tag{5.88}$$

From the rotor side, the equations shall be as follows:

$$\begin{aligned}
\lambda_{dr} &= L_r i_{dr} + L_m i_{ds} \\
\lambda_{qr} &= L_r i_{qr} + L_m i_{qs} \\
V_{dr} &= R_r i_{dr} + \frac{d}{dt}\lambda_{dr} + \omega_r \lambda_{qr} \\
V_{qr} &= R_r i_{qr} + \frac{d}{dt}\lambda_{qr} - \omega_r \lambda_{dr}
\end{aligned} \tag{5.89}$$

For the air gap flux linkage, the equations are to be expressed as follows:

$$\begin{aligned}
\lambda_{dm} &= L_m(i_{ds} + i_{dr}) \\
\lambda_{qr} &= L_m(i_{qr} + i_{qs})
\end{aligned} \tag{5.90}$$

where R_s, R_r, L_m, L_{ls}, L_{lr}, ω_r, i_d, i_q, V_d, V_q, λ_d, and λ_q are the stator winding resistance and fluxes of the d-q model. These outputs of the wind energy generator modeling are to be prepared by an energy conversion circuit diagram—implemented inverter from the standard Simulink/Sim Power Systems by using the software MATLAB Simulink to calculate.

5.4.2.5 Battery Modeling

When the power produced exceeds the demand to start engine or when the car is not working, a battery modeling is needed to store the power and transfer them into backup power source [30,39]. In the model introduced in this research, a battery will be used as a storage buffer to supply all the electricity and further predict battery discharge based on the nonlinear properties of the battery, considering the following Peukert's Law equation:

$$t_{\text{discharge}} = H\left(\frac{C}{IH}\right)^k \tag{5.91}$$

where t is the battery discharge time, C is the battery capacity (Ampere hour value), I is the current that is drawn, H is the rated discharge time, and k is Peukert's coefficient. Peukert's coefficient is an empirical value that shall be determined using the following formula:

$$k = \frac{\log T_2 - \log T_1}{\log I_1 - \log I_2} \tag{5.92}$$

where I_1 and I_2 are the two discharge current rates and T_1 and T_2 are the corresponding discharge durations. However, after a certain number of recharge cycles, the time for charging and discharging will change because of

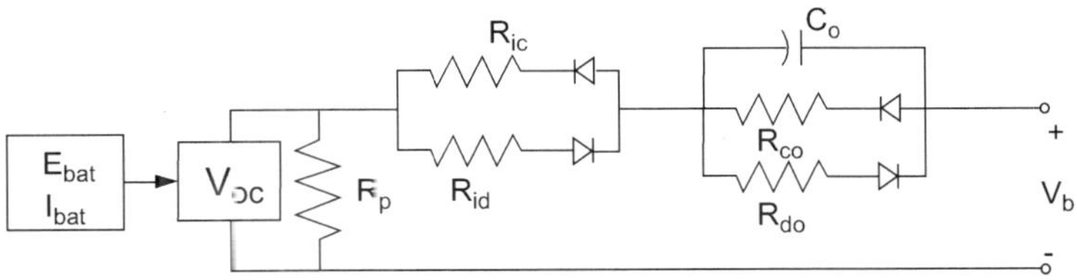

Figure 5.27 Model of a car with a battery as the backup power source to start the engine and for when the car is not in motion.

the decreasing battery capacity. Therefore the value of t should be redetermined by using 1.3–1.4. Meanwhile, the charging time for a completely discharged battery is to be calculated by the following equation:(Fig. 5.27)

$$t_{\text{charging}} = \frac{\text{Ampere hour of battery}}{\text{Charging Current}} \tag{5.93}$$

5.4.3 Optimization and Discussion

From the models proposed, it is indicated that the flying car can generate desirable lift at a satisfied takeoff velocity with different angles of attack and wing positions. In addition, a detailed parametric analytical study shows that even if at a low takeoff velocity because of the contours of static pressure at various speeds, the flying vehicle can still produce satisfied lift with the help of the deployable high wings (Fig 5.28). Mathematical analyses imply that at a given range of free-stream velocity of a flying car with different geometrical options will generate upward force, which indicates that these velocities are deployable and can generate lift higher than the weight of the flying car at the lower takeoff velocity under different flying conditions, orientations, and wing positions.

The analysis of external flow features and aerodynamic characteristics of a flying vehicle indicates that the speed of the flying car can be determined from the results of steady free-stream velocity, and the aerodynamic characteristics of the flying car which can give it the ability to take off, fly, and control properly at various degrees and mach numbers (Fig 5.28). By using these principles, this chapter simulates the condition of the flying vehicle when traveling at takeoff velocity to study the pattern of the car when rising off the ground by creating lift using additional wing force. Answers to this condition include streamlining the car body and using suitable wings to create desired lift heavier than the weight of the car by installing the rectangular NACA 9816 wings at different positions of the body with different angles of attack (Fig 5.29).

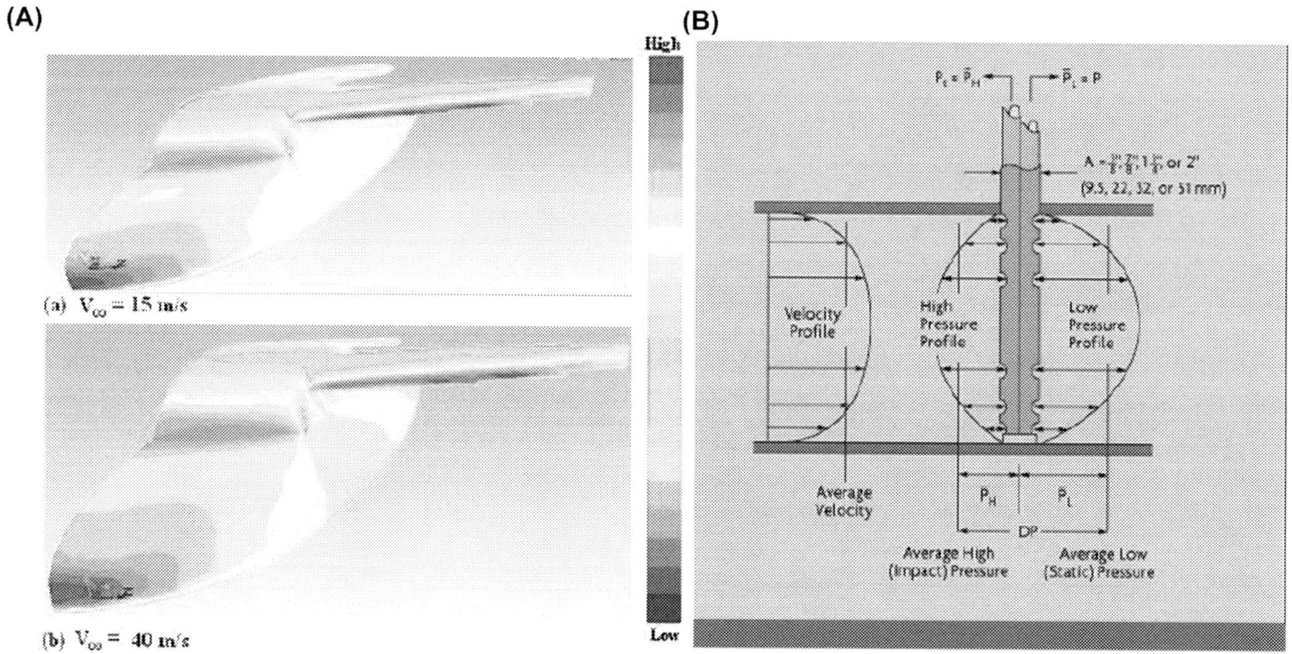

Figure 5.28 (A) Contours of static pressure at various speeds of flying car; it shows the aerodynamic characteristics of the flying car at two different speeds of lifting, that is, 15 and 40 m/s. (B) Contours of the static pressure at various velocity of a car with short high-wing model for getting better aerodynamic efficiency.

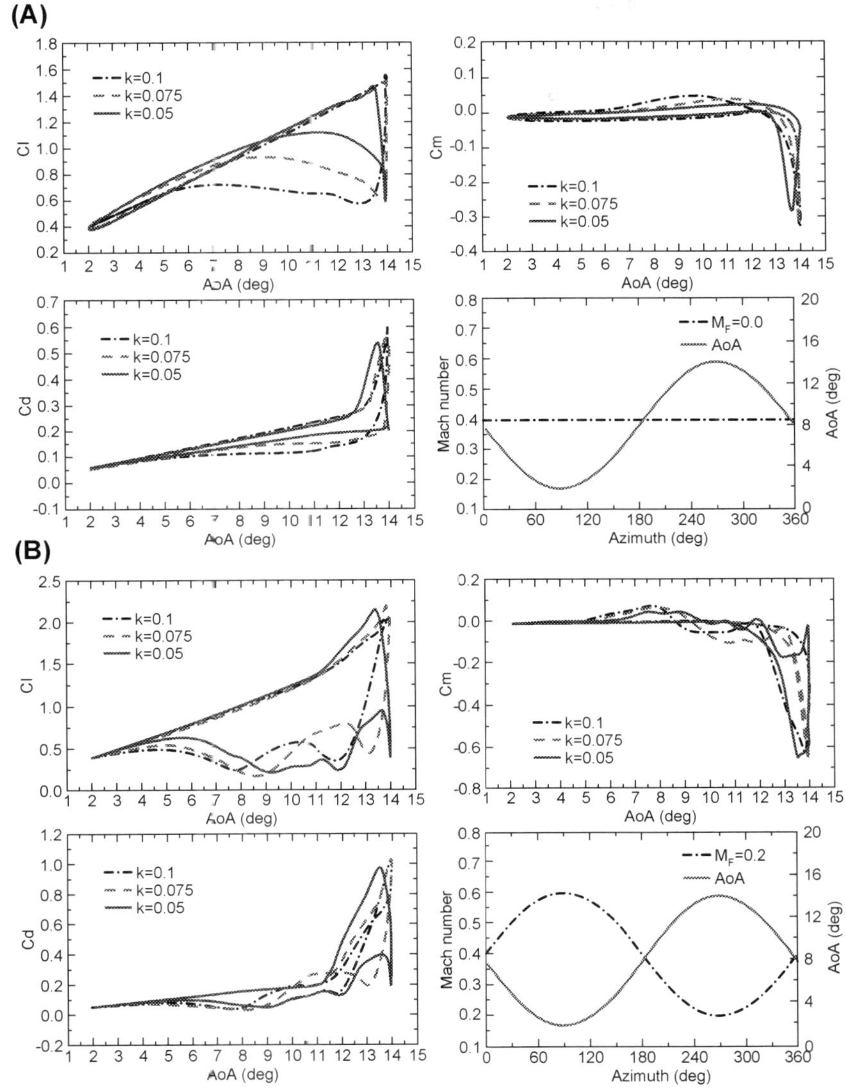

Figure 5.29 The speed of the flying car determined from (A) steady free-stream velocity and (B) variable free-stream velocity considering aerodynamics characteristics to control the takeoff, fly, and optimization of the velocity.

After satisfying the energy needed by the flying vehicle, a cleaner energy source of the WTGS should also be considered based on the analysis of an FLC with stator flux orientation and the MPPT technique. Therefore a robustness test was introduced in the simulations by adding a wind speed signal and voltage dips. The results of these simulations indicated that the inherent ability of the FLC can handle this issue while operating under fault conditions at rated wind speed. The MPPT control can also be applied into these conditions, whereas for the DFIG control, an FLC is used for indirect power control. In the simulations, the wind profile is considered to be a wind speed signal with a mean value of 8 m/s and a rated wind speed of 10 m/s; the whole system is tested under standard conditions with a stator voltage of approximately 50% for 0.5 s between 4 and 4.5 s, approximately 25% between 6 and 6.5 s, and 50% between 8 and 8.5 s [4,8,21], which indicates that under ideal conditions with no perturbations and no parameter variations, the machine is functioning. Besides, the stator active and reactive powers are also controlled by MPPT and FLC; therefore to ensure a unity power factor at the stator side, the reference for the reactive power needs to be set to zero. Through tracking the maximum power point speed, the generator shaft speed can realize maximum angular speed after the wind speed increases. After the tracking, to ensure the efficiency of the control system, a decoupling among the components of the rotor current needs to be conducted. In this process, the bidirectional active and reactive power transfer between the rotor and power system is exchanged by the generator through the super synchronous operation and the realization of the nominal stator power; meanwhile, the reactive power can be controlled by the load-side converter to obtain the unit's power factor to generate energy for powering vehicles [15,16,56].

5.4.4 Conclusion

Contemporarily, conventional transportation infrastructure system worldwide not only wastes large amount of money each year but also causes adverse environmental impacts. Therefore the flying vehicle technology introduced in this research may be an innovative solution to balance the relationship between traffic, cost, energy, and environment. Considering the higher energy demand and need for a friendlier environment, a wind turbine will be installed into the flying car to produce energy by itself when the vehicle is working. To optimize the use of the turbine, an analysis with a series of mathematical calculations of conversion process, control structure, and generator modeling is conducted in this research. The simulation results of NACA 9618 model by using MATLAB showed the high extendable nature of commercial use of the flying car by providing the reliability of taking off, flying, safety control of velocity and aerodynamics, and mach numbers. Nowadays, infrastructure construction costs trillion dollars, and transportation sector consumes 5.6×10^{20} J/year (560 EJ/year) fossil fuel. Besides, they are also responsible for nearly 34% of the total annual climate changes [103]. If developed properly, this technology can be the new era of science to remit severe transportation, infrastructure, and energy problems.

ACKNOWLEDGMENTS

This research was supported by Green Globe Technology under grant RD-02018-03. Any findings, conclusions, and recommendations expressed in this chapter are solely those of the author and do not necessarily reflect those of Green Globe Technology.

REFERENCES

[1] Eltamaly AM, Farh HM. Maximum power extraction from wind energy system based on fuzzy logic control. Elec Power Syst Res 2013;97:144–50.
[2] Belgacem K, Mezouar A, Massoum A. Fuzzy logic control of double-fed induction generator wind turbine. International Review on Modelling & Simulations 2013.
[3] Abdelmalek S, Barazane L, Larabi A, Belmili H. Contributions to diagnosis and fault tolerant control based on Proportional Integral Observer: application to a doubly-fed induction generator. In: 2015 4th international Conference on electrical engineering (ICEE), Boumerdes; 2015. p. 1–5. https://doi.org/10.1109/INTEE.2015.7416608.
[4] Abdullah Asuhaimi B, Mohd Z, Mahmoud Pesaran HA, Khairuddin AB, Jahanshaloo L, Shariatu O. An overview on doubly fed induction generators' controls and contibutions to wind based electricity generation. Renewable and Sustainable Energy Rreviews 2013;27:692–708.
[5] Bhandari B, Raj Poudel S, Lee K-T, Ahn S-H. Mathematical modeling of hybrid renewable energy system: a review on small hydro-solar-wind power generation. International Journal of Precision Engineering and Manufacturing-Green Technology 2014;1(2):157–73.
[6] Abdullah MA, Yatim AHM, Chee Wei T. A study of maximum power point tracking algorithms for wind energy system. In: Proc. Of IEEE first Conference on clean energy and technology(CET); 2011. p. 321–6.
[7] Junyent-Ferré A, Gomis-bellmunt O, Sumper A, Sala M, Mata M. Modeling and control of the doubly fed induction generator wind turbine. Simulat Model Pract Theor 2010;18:1365–81.
[8] Gaillard A, Poure P, Saadate S, Machmoum M. Variable speed D wind energy system for power generation and Harmonic current mitigation. Renew Energy 2009;34: 1545–53.
[9] Beltran B, Benbouzid M, Ahmed-Ali T, Mangel H. DFIG-based wind turbine robust control using high-order sliding modes and a high gain observer. International Review on Modelling and Simulations (IREMOS) 2011;4(3):1148–55.
[10] Robyns B, Francois B, Degobert P, Hautier JP. Vector control of induction machines. London: Springer-Verlag; 2012.
[11] Cardoso J, Casella I, Filho A, Costa F, Capovilla C. SCIG wind turbine wireless controlled using morphological filtering for power quality enhancement. Renew Energy 2016;92:303–11
[12] Hossain M. Invisible transportation infrastructure technology to mitigate energy and environment. Energy, Sustainability and Society 2017;7(1).
[13] Durbin D-A. Flying car gets closer to reality with test flight. boston.com. Associated Press; 2012-04-02.

[14] Gaillard A, Poure P, Saadate S. FPGA-based reconfigurable control for switch fault tolerant operation of WECS with DFIG without redundancy. Renew Energy 2013.
[15] Chang S. Evaluating disaster mitigations: Methodology for urban infrastructure systems. Nat Hazards Rev 2003:186–96. https://doi.org/10.1061/(ASCE)1527-6988(2003)4:4(186).
[16] Chien-Hsiung T, Lung-Ming F, Chang-Hsien T, Yen-Loung H, Jik-Chang L. Computational aero-acoustic analysis of a passenger car with a rear spoiler. Appl Math Model September 2009;33(9):3661–73.
[17] Mitchell DL, Gubser DU. "Magnetohydroynamic" Ship propulsion with superconducting magnets. J Supercond 1988;1(No. 4):349.
[18] Bhandari B, Lee KT, Lee GY, et al. Int. J. of Precis. Eng. and Manuf.-Green Tech 2015;2:99.
[19] Boroujeni HZ, Othman MF, Shirdel AH, Rahmani R, Movahedi P, Toosi ES. Improving waveform quality in direct power control of DFIG using fuzzy controller. Neural Comput Appl 2015.
[20] Casella I, Capovilla C, Sguarezi Filho A, Jacomini R, Azcue-Puma J, Ruppert E. An ANFIS power control for wind energy generation in smart grid Scenario using wireless coded OFDM-16-QAM. Journal of Control, Automation and Electrical Systems 2013;25(1):22–31.
[21] Chakib R, Essadki A, Cherkaoui M. Modeling and control of a wind system based on a DFIG by active Disturbance rejection control. International Review on Modelling and Simulations (IREMOS) 2014;7(4):626.
[22] Chierchie F, Paolini E. Real-time Digital PWM with zero Baseband Distortion and low switching frequency. IEEE Transactions on Circuits and Systems I: Regular Papers 2013;60(10):2752–62.
[23] El-Anwar O, Ye J, Orabi W. Efficient optimization of Post-disaster Reconstruction of transportation networks. J Comput Civ Eng 2016;30(3):04015047.
[24] Mamdani EH, Assilina S. An experiment in linguistic synthesises with a fuzzy logic controller. Int J Man Mach Stud 1975;7:1–13.
[25] Kamal E, Koutb M, Sobaih AA, Abozalam B. An intelligent maximum power extraction algorithm for hybrid wind-diesel-storage system. Int J Electr Power Energy Syst 2010;32(3):170–7.
[26] Godfrey Derek S, Kamali G, Prasanth S, Sanal Kumar VR. Conceptual design and analysis of Ferrari F430 flying car. With Proceedings of "International Journal of Research in Engineering and Technology (IJRET)" 2012;1(6):2277–4378.
[27] Gupta N, Singh SP, Dubey SP, Palwalia DK. Fuzzy logic controlled three-phase three-wired shunt active power filter for power quality improvement. International Review of Electrical Engineering (IREE) 2011;6(3):1118–29.
[28] Barambones O. Power output maximization for wave power generation plants using an adaptive sliding mode control. In: 2013 international Conference on renewable energy research and applications (ICRERA), Madrid; 2013. p. 178–83. https://doi.org/10.1109/ICRERA.2013.6749747.
[29] Barambones O. A robust control of double- feed induction generator for wind power generation. In: 2009 35th annual Conference of IEEE Industrial electronics; 11/2009.
[30] Thompson SJ, Congressional Research Service. High speed ground transportation (HGST): Prospects and public Policy. Apr. 6, 1989. p. 5.
[31] Pugh TAM, Robert MacKenzie A, Duncan Whyatt J, Nicholas Hewitt C. Effectiveness of green infrastructure for improvement of air quality in urban street Canyons. Environ Sci Technol 2012;46(14):7692–9. https://doi.org/10.1021/es300826w.
[32] Lli S, Haskew TA, Jackson J. Integrated power characteristic study of DFIG and its frequency converter in wind power generation. Renew Energy 2010;35:42–5.

[33] Sivasankar G, Suresh Kumar V. Improving low voltage Ride through of wind generators using STATCOM under symmetric and Asymmetric fault conditions. International Review on Modelling & Simulations 2013.
[34] Sivasankar G, Suresh Kumar V. Improving low voltage Ride through capability of wind generators using dynamic voltage Restorer. J Electr Eng 2014;65(4).
[35] Wang L, Truong D-N. Stability enhancement of DFIG-based Offshore wind farm fed to a multi-machine system using a STATCOM. IEEE Trans. on Power Sys 2013;28(3):2882–9.
[36] Wang L, Truong D-N. Stability enhancement of a power system with a PMSG-based and a DFIG-based off shore wind farm using a SVC with an adaptive-network-based fuzzy inference system. IEEE Trans. on Power Sys 2013;60(7):2799–807.
[37] Engelhardt S, Kretschmann J, Fortmann J, Shewarega F, Erlich I, Feltes C. Negative sequence control of DFG based wind turbines. 2011 IEEE Power and Energy Society General Meeting 2011.
[38] Faida H, Saadi J. Modelling, control strategy of DFIG in a wind energy system and feasibility study of a wind farm in Morocco. International Review on Modelling and Simulations (IREMOS) 2010;3(6):1350–62.
[39] Xu H, Ma X, Sun D. Reactive current Assignment and control for DFIG based wind turbines during grid voltage Sag and Swell conditions. Journal of Power Electronics 2015.
[40] Wang Z, Sun Y, Li G, Ooi BT. Magnitude and frequency control of grid-connected doubly fed induction generator based in synchronized model for Wind power generation IET Journal on Renewable Power Generation 2010;4:232–41.
[41] Chikouche T, Hadjeri S, Mezouar A, Terras T. A new state-space nonlinear control approach of a doubly fed induction motor using variable gain PI and fuzzy logic controllers. International Review on Modelling and Simulations (IREMOS) 2013;6(1):59–67.
[42] De J, Oliveir JGde. Electric machine Topologies in energy storage systems. Energy Storage 2010.
[43] de Sherbinin A, Levy MA, Zell E, Weber S, Jaiteh M. Using satellite data to develop environmental indicators. Environ Res Lett 2014.
[44] Eltamaly AM, Alolah AI, Abdel-Rahman MH. Improved simulation strategy for DFIG in wind energy applications. International Review on Modelling and Simulations (IREMOS) 2011;4(2):525–32.
[45] Hossain MF. Solar energy integration into advanced building design for meeting energy demand and environment problem. Int J Energy Res 2016.
[46] Hossain M. Solar energy integration into advanced building design for meeting energy demand and environment problem. Int J Energy Res 2016;40(9):1293–300.
[47] Cooper KR. Bluff-body aerodynamics as applied to vehicles. J Wind Eng Ind Aerod December 1993;49(1– 3):1–21.
[48] Kvasnica I, Kvasnica P. Multiphysics base of Modelling of the MEMS Micromirror and simulation by modal control. International Review of Automatic Control (IREACO) 2016;9(5):383.
[49] Rouabhi R, Abdessemed R, Chouder A, Djerioui A. Power quality enhancement of grid connected doubly-fed induction generator using sliding mode control. International Review of Electrical Engineering (IREE) 2015;10(2):266.
[50] Ryan DL. 'Flying car' at the Museum of science. Boston.com. (The Boston Globe); 2009-03-18.
[51] Loucif M, Boumédiène A. Modeling and direct power control for a DFIG under wind speed variation. In: 2015 3rd international Conference on control, engineering & Information technology (CEIT), Tlemcen; 2015. p. 1–6. https://doi.org/10.1109/CEIT.2015.7233042.

[52] Pichan M, Rastegar H, Monfared M. Two fuzzy-based direct power control strategies for doubly-fed induction generators in wind energy conversion systems. Energy 2013;51:154–62.

[53] Mone G. The driving Airplane gets real. Popular Sci 2008-10:42–8.

[54] Page L. Terrafugia flying car gets road-safety exemptions. The Register July 2011;4.

[55] Dürr M, Cruden A, Gair S, McDonald JR. Dynamic model of a lead acid battery for Use in a Domestic fuel Cell system. J Power Sources 2006;161(2):1400–11.

[56] Ghennam T, Berkouk EM, Francois B. A vector hysteresis current control applied on three-level inverter. Application to the active and reactive power control of doubly fed induction generator based wind turbine. International Review of Electrical Engineering (IREE) 2007;2(2):250–9.

[57] Gopal Sharma K, Bhargava A, Gajrani K. Stability analysis of DFIG based wind turbines connected to electric grid. International Review on Modelling and Simulations (IREMOS) 2013;6(3):879–87.

[58] Tsourakisa G, Nomikosb BM, Vournasa CD. Effect of wind parks with doubly fed asynchronous generators on small-signal stability. Elec Power Syst Res 2009;79: 190–200.

[59] Haines TB. First roadable airplane takes flight. Aircraft Owners and Pilots Association (AOPA); March 19 , 2009.

[60] He P, Wen F, Ledwich G, Xue Y. Small signal stability analysis of power systems with high penetration of wind power. Journal of Modern Power Systems and Clean Energy 2013.

[61] Lala H, Karmakar S. Continuous wavelet transform and artificial neural network based fault diagnosis in 52 bus hybrid distributed generation system. In: 2015 IEEE Students Conference on engineering and systems (SCES), Allahabad; 2015. p. 1–6. https://doi.org/10.1109/SCES.2015.7506463.

[62] Hossain M, Fara N. Integration of wind into running vehicles to meet its total energy demand. Energy, Ecology and Environment 2016;2(1):35–48.

[63] Soedibyo S, Pradipta A, Suyanto S, Ridwan M, Zulkarnain G, Ashari M. Active and reactive power control in 20 kV grid connected distributed generation system. International Review of Automatic Control (IREACO) 2017;10(3):211.

[64] Tamalouzt S, Rekioua T, Abdessemed R. Direct torque and reactive power control of Grid Connected Doubly Fed Induction Generator for the wind energy conversion. In: 2014 international Conference on electrical sciences and technologies in Maghreb (CISTEM); 2014.

[65] Jeong S, Nguyen T, Le Q, Lee D. High-performance control of three-phase four-wire DVR systems using Feedback Linearization. Journal of Power Electronics 2016;16(1):351–61.

[66] Watanakul N. An application of wind turbine generator on hybrid power conditioner to improve power quality. International Review of Electrical Engineering (IREE) 2012;7(5):5487–95.

[67] Sheahan TP. Industrial superconductivity, a report to the Office of Industrial Programs, U.S. Department of energy. October 1987.

[68] Tien H, Scherer C, Scherpen J, Muller V. Linear parameter Varying control of doubly fed induction machines. IEEE Trans Ind Electron 2016;63(1):216–24.

[69] Kerrouche K, Mezouar A, Boumedien L. A simple and efficient maximized power control of DFIG variable speed wind turbine. In: 3rd international Conference on systems and control, Algiers; 2013. p. 894–9. https://doi.org/10.1109/ICoSC.2013.6750963.

[70] Lu L. Investigation on wind power potential on Hong Kong islands-an analysis of wind power and wind turbine characteristics. Renew Energy 2002.
[71] Shahidehpour M, Eremia M. Wind power generation. Handbook of electrical power system dynamics modeling stability and control. 2013.
[72] Slootweg JG. Reduced order modeling of wind turbines, wind power in power systems. Wiley; 2005. p. 555−85.
[73] Kalantar M, Mousavi GS. Dynamic behavior of a stand-alone hybrid power generation system of wind turbine, microturbine, solar array and battery storage. Appl Energy 2010;87(10):3051−64.
[74] Kamal E, Oueidat M, Aitouche A, Ghorbani R. Robust Scheduler fuzzy controller of DFIG wind energy systems. IEEE Transactions on Sustainable Energy 2013;4(3): 706−15.
[75] Werner JP, Luterbacher J, Smerdon JE. A Pseudoproxy evaluation of Bayesian Hierarchical Modelling and Canonical Correlation analysis for climate field Reconstructions over Europe. J Clim 2012.
[76] Dawei WU, Zheng WU Landing ZANG, Jizhong HU. Safety and Airworthiness design of Ultra-Light and Very Light Amphibious aircrafts, the 2nd international Symposium on aircraft Airworthiness (ISAA 2011). Procedia Engineering 2011;17: 212−25.
[77] Zerikat M, Chekroun S, Mechernene A. Development and implementation of high-performance variable structure tracking for induction motor using fuzzy-logic controller. International Review of Electrical Engineering (IREE) 2010;5(1): 160−6.
[78] Kazemi MV, Moradi M, Kazemi RV. Minimization of powers ripple of direct power controlled DFIG by fuzzy controller and improved discrete space vector modulation. Electr Power Syst Res 2012;89:23−30.
[79] Khodakarami J, Ghobadi P. Urban pollution and solar radiation impacts. Renew Sustain Energy Rev 2016;57:965−76.
[80] Loucif M, Boumediene A. Modeling and direct power control for a DFIG under wind speed variation. In: 2015 3rd international Conference on control engineering & Information technology (CEIT); 2015.
[81] Kerrouche KD Mezouar A, Boumediene L, Belgacem K. Modeling and Optimum power control based DFIG wind energy conversion system. International Review of Electrical Engineering (I.R.E.E.) 2014;9(1).
[82] Kerrouche K, Mezouar A, Boumedien L. A simple and efficient maximized power control of DFIG variable speed wind turbine. In: 3rd international Conference on systems and control; 2013.
[83] Lin W, Hong C, Huang C, Ou T. Hybrid control of a wind induction generator based on Grey−Elman neural network. IEEE Trans Contr Syst Technol 2013; 21(6):2367−73.
[84] Mann ME, Bradley RS, Hughes MK. Global-scale temperature patterns and climate forcing over the past six centuries. Nature 1998;392:779−87.
[85] Boutotbat M, Mokrani L. Machmoum M. Control of a wind energy conversion system equipped by a DFIG for active power generation and power quality improvement. Renew Energy 2013;50:378−86.
[86] McBeath S. Competition car aerodynamics: a practical Handbook. Haynes Publishing; 2006. 9781844252305.
[87] Bharti OP, Saket RK, Nagar SK. Reliability analysis of DFIG based wind energy conversion system. 2017. p. 313−7. https://doi.org/10.1145/3057039.3057099.
[88] Phillips M. Flying car takes first Flight. The Middle Seat Terminal (The Wall St. Journal) March 18, 2009.

[89] Shaltout A, Gamal H. Power coordination of grid-connected wind turbine doubly fed induction generator augmented with battery storage. In: 2013 IEEE international Conference on smart energy grid engineering (SEGE); 2013.
[90] Sharma K, Bhargava A, Gajrani K. Stability analysis of DFIG based wind turbines connected to electric grid. International Review on Modelling and Simulations (IREMOS) 2013;6(3):879–87.
[91] Terrafugia's Transition street-legal airplane continues flight and drive testing". Terrafugia. Retrieved July 30, 2012.
[92] The "Yamato l," 38 meters long and 10 meters across, is the product of a 6-year, $350 million effort begun in 1985. The vessel will be built at Mitsubishi's Kobe plant. Supercond Week Jan. 29, 1990;4(5):5.
[93] Brekken TKA, Mohan N. Control of a doubly fed induction wind generator under unbalanced grid voltage conditions. IEEE Trans. On Energy Conversion March 2007;22(1):129–35.
[94] Vieira JPA, Nunes MVA, Bezerra UH, Barra Jr W. New fuzzy control strategies applied to the DFIG converter in wind generation systems. Trans. America Latina, IEEE 2007;5(3).
[95] Chang YH, Liu TH, Chen DF. Design and implementation of a robust controller for a Micro permanent magnet synchronous speed control systems. In: Second international Conference on innovative computing, Informatio and control (ICICIC 2007), Kumamoto; 2007. p. 99. https://doi.org/10.1109/ICICIC.2007.245.
[96] Zhang X, Duan D, Jing H, Xu D. A strategy of dead-time compensation to DFIG. International Review of Electrical Engineering-IREE 2012;7(3):4395–403.
[97] Widyan M. Operational performance of synchronous generator hybrid-Excited by PMDC and PV generators. International Review of Electrical Engineering (IREE) 2014;9(4):863.
[98] Hucho W-H. Chapter 1-Introduction to automobile aerodynamics. Aerodynamics of Road Vehicles 1987:1–46.
[99] Cai Y, Chen SS. Dynamic characteristics of magnetically-levitated vehicle systems. A Mech. Rev 1997;50(11):647–70.
[100] Zheng XJ, Wu JJ, Zhou YH. Effect of spring non-linearity on dynamic stability of a controlled maglev vehicle and its guideway system. J Sound Vib 2005;279:201–15.
[101] Zheng XJ, Wu JJ, Zhou YH. Numerical analyses on dynamic control of five-degree-of-freedom maglev vehicle moving on flexible guideways. J Sound Vib 1997;235: 43–61.
[102] Kamal E, Aitouche A, Ghorbani R, Bayart M. Robust fuzzy fault tolerant control of wind energy conversion systems subject to sensor faults. IEEE Trans. Sustain. Energy 2012;3(2):231–41.
[103] Kamal E, Oueidat M, Aitouch A, Ghorbani R. Robust Scheduler fuzzy controller of DFIG wind energy systems. IEEE Trans. Sustain. Energy 2013;4(3).

FURTHER READING

[1] Santiprapan P, Areerak K, Areerak K. The enhanced – DQF algorithm and optimal controller design for shunt active power filter. International Review of Electrical Engineering (IREE) 2015.
[2] Welsh J. Flying car maker offers 'show special' discount. Driver's Seat. Wall St. Journal April 5, 2012.

CHAPTER SIX

Water

6.1 NATURAL WATER RESOURCES

Water is an indispensable resource. Natural water resources are sources of water that are used for domestic, agricultural, industrial, recreational, and environmental activities. All living things require water to survive and thus it is told "water means live and live means water." The Earth's surface consists of 25% of land and 75% of water, and all creatures need water to survive. Among the surface water, 97% is saline while only 3% is freshwater. What's more, over two-thirds of the freshwater exist in the form of glaciers or ice caps. However, most of the rest of the freshwater which is unfrozen proves to be groundwater, and the final remainder is above ground or in the air. Fresh water is a renewable resource, yet the world's supply of groundwater is steadily decreasing, with depletion occurring most prominently in whole world which is threatening our survival near future (Fig. 6.1).

6.1.1 Chemical and Physical Properties of Water

H_2O, a polar inorganic compound, represents as liquid with attributes of tasteless, odorless, and nearly colorless with a little bit blue under normal condition. It is the simplest hydrogen chalcogenide which is described as a universal solvent because of its ability to dissolve many substances, e.g., many salts, sugar, simple alcohols, proteins, polysaccharides, DNA, oxygen, and carbon dioxide.

Water molecule is a polar molecule with an electrical dipole because the oxygen atom has a higher electronegativity than hydrogen atoms, which means the oxygen atom carries a slight negative charge, while hydrogen atoms are slightly positive. Besides, each water molecule contains one oxygen atom and two hydrogen atoms connected by covalent bonds (Fig. 6.2).

More importantly, it is the only common substance that can exist as solid, liquid, and gas in normal terrestrial conditions. Water can be referred to the liquid state of a substance that exists at standard ambient temperatures and pressures, its solid state (ice), or its gaseous state (steam or water vapor). It also exists in the forms of ice packs and icebergs, fog, glaciers, aquifers,

Sustainable Design and Build
ISBN: 978-0-12-816722-9
https://doi.org/10.1016/B978-0-12-816722-9.00006-9

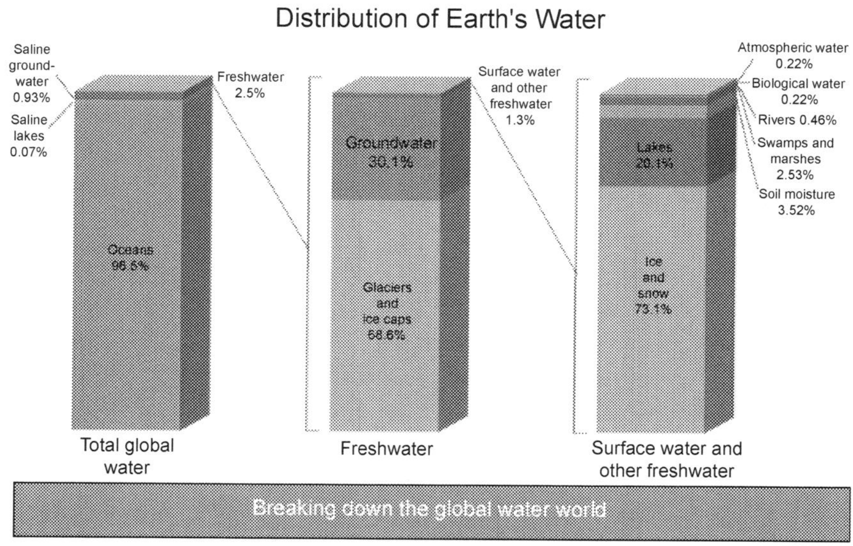

Figure 6.1 A graphical distribution of the locations of water on the Earth. Only 3% of the Earth's water is freshwater. Most of it is in ice caps and glaciers (69%) and groundwater (30%), while all lakes, rivers, and swamps combined only account for a small fraction (0.3%) of the Earth's total freshwater reserves.

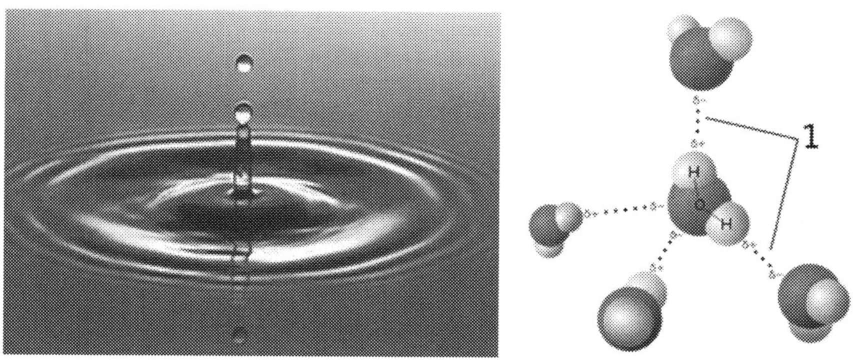

Figure 6.2 Model of hydrogen bonds [1] between molecules of water shows that oxygen atom has higher electronegativity than hydrogen atom.

atmospheric humidity, fog, and dew, which are the main components of the Earth's streams situated in difference locations.

Under different circumstances, water will have different attributes. When standard pressure is 1 atm, water is in the form of liquid between 0°C (32°F) and 100°C (212°F). However, the melting point will change to −5°C at 600 atm and −22°C at 2100 atm, which is used as the principle to explain the movement of glaciers, ice skating, and buried lakes of Antarctica. What's more, the melting point will rapidly increase again, leading to several exotic forms of ice that do not exist at lower pressures when pressure is higher than 2100 atm.

The boiling point will also change when pressure changes. For instance, the boiling point is 374°C at 220 atm but 68°C (154°F) at the top of the Mount Everest with the atmospheric pressure of 0.34 atm. The effect of this change is applied on pressure cooking, steam engine design, deep-sea hydrothermal vents, etc.

Water attributes change under extreme conditions. When the pressure is extremely low (lower than 0.006 atm), water cannot exist in liquid form and passes directly from solid to gas by sublimation, which is used for freeze drying of food. When the pressure is extremely high (higher than 221 atm), the liquid and gas states are no more distinguishable, a new state called supercritical steam replaces.

Another characteristic of water is that it becomes less dense as it freezes compared with most liquids. The maximum density of water in its liquid form (at 1 atm) is 1000 kg/m^3 (62.43 lb/cu ft), which occurs at 3.98°C (39.16°F) while the density of ice is 917 kg/m^3 (57.25 lb/cu ft). Thus, through calculation, water expands 9% in volume as it freezes, which reflects the fact that ice floats on liquid water.

6.1.2 Surface Water

Surface water refers to water locates in river, lake, or freshwater wetland. It is naturally refilled and lost by precipitation and discharge to the oceans, evapotranspiration, groundwater recharge, and evaporation. Although precipitation is the only natural input to any surface water system, many other factors also influence the total quantity of water available at any given time and proportions of water loss in the system, for example, wetlands and artificial reservoirs, the runoff characteristics of the land in the watershed, storage capacity in lakes, the timing of the precipitation, the permeability of the soil beneath these storage bodies, and local evaporation rates.

However, human activities can have large or even destructive impacts on these factors. By constructing reservoirs or draining wetlands, human can control the storage capacity of the surface water systems. Besides, by paving areas and channelizing the streamflow, human can increase runoff quantities and velocities.

The total quantity of water available at any given time is an important issue to considerate. Some water users have a desultory need for water, farms, for example, only need water when planting crops, which maybe only one season in the whole year. To satisfy the farm's need for water, a surface water system may require a large storage capacity to gather water during the whole year to release it in a short period of time. For other users who need continuous supply of water such as power plants, the storage capacity of the surface water system only needs to satisfy the minimum requirement of the average streamflow.

6.1.3 Under River Flow

During the flow process of a river, the total volume of water transporting downstream normally consists of the visible free water flow and the flow across rocks and sediments that underlie the river, whose floodplain is called the hyporheic zone. For rivers in large valleys, the latter flow may largely exceed the visible flow. The hyporheic zone often plays the roles of forming a dynamic interface between surface water and groundwater from aquifers and exchanging flow that may be fully charged or depleted between rivers and aquifers, which are particularly significant in karst areas where potholes and underground rivers are common.

6.1.4 Groundwater

Groundwater can be referred to the freshwater located in the subsurface pore space of soil and rocks and the water flows within aquifers below the water table. The critical difference between groundwater and surface water is that, compared to input, the volume of groundwater storage is much larger than surface water because of the former's slow rate of turnover. This difference let human to use groundwater unsustainably for a long time if there is no serious issue. However, the average rate of seepage above a groundwater source is the upper bound for average consumption of water from that source in the long term. The natural input and output of groundwater is the seepage from surface water and springs or seepage to the oceans, respectively. Therefore, if the surface water source is subject to substantial evaporation, the groundwater source below may become saline. This circumstance may happen artificially under irrigated farmland or naturally under endorheic bodies of water. Soil salinization may also happen in coastal areas because of the reverse of direction of seepage to ocean due to human use. What's more, the pollution that results from human activities may cause the loss of groundwater which can be stopped by building reservoirs or detention ponds.

6.1.5 Water Uses

6.1.5.1 Domestic

According to estimation, 8% of water use worldwide is for domestic purposes, which include cooking, laundry, toilet flushing, drinking water, bathing, gardening, and cleaning [2]. Excluding water for gardening, basic domestic water requirement is at around 50 L per person per day according to estimation [3,4]. Drinking water, more special than other uses, refers to water that is of sufficiently high quality so that it can be consumed or used without risk of harmless, which can also be called potable water. In most developed countries, the water provided for domestic, commercial, or industrial purposes is at drinking water standard, although only a small proportion is consumed for food preparation.

6.1.5.2 Agricultural

Compared with domestic use, agricultural use of water occupies a large amount of water use. According to an estimation, 70% water use worldwide is irrigation, with 15%–35% of irrigation withdrawals being unsustainable [3,5]. To satisfy one person's daily food need, 2000–3000 L of water will be consumed, which is a considerable amount compared to drinking 2–5 L [6,7]. Therefore, according to calculation, if enough food is provided for over 7 billion people now on the earth, the water required could fill a canal 10 m deep, 100 m wide, and 2100 km long.

6.1.5.3 Industrial

Another major water use is industrial, which occupies 22% worldwide water use [2,8]. Major application of industrial water use includes thermoelectric power plants, manufacturing plants, ore and oil refineries, and hydroelectric dams. Water in these processes can play the roles of cooling, solvent, and chemical reagents. Although water withdrawal may be high in some industries, these consumptions are still much lower than that of agriculture. Industrial water uses can also include renewable power generation. Hydroelectric power derives energy from the force of water flowing downhill, driving a turbine connected to a generator, which is a low cost, nonpolluting, renewable energy source. It can be used for load following and can provide continuous power. Another way to produce hydroelectricity depends on the sun. First, heat from the sun evaporates water, the vapor then condenses as rain in higher altitudes and flows downhill and produces electricity. Except for the two plants above, pumped-storage hydroelectric plants use grid electricity to pump water uphill when energy demand is low, then use the stored water to produce electricity when demand is high.

Hydroelectric power plants require the creation of artificial lake. Due to the larger surface area exposed to the elements, evaporation from this lake is higher compared with a natural river, as well as the water consumption. The process of driving water through the turbine and tunnels or pipes briefly removes the water from natural environment, creating water withdrawal, whose impact on wildlife varies largely depending on the design of the plant.

Another industrial water use is pressurized water, which is often used in water blasting, water jet cutters, and high-pressure water guns. These uses not only work well and safe but also make no harm to the environment. Cooling of machinery from overheating is also an industrial water use, which only occupies a small proportion of water consumption. In many large-scale industrial processes such as oil refining, fertilizer production, and other chemical plants, thermoelectric power production, and natural gas extraction from shale rock, water is also being used. Discharge of untreated water coming from these processes is pollution, which includes discharged solutes (chemical pollution) and increased water temperature (thermal pollution). These industrial water use mentioned above normally require pure water and apply a variety of purification techniques both in water supply and discharge. Most of the pure water is produced on-site from natural freshwater or municipal gray water. Industrial consumption of water is generally much lower than withdrawal because laws require industrial gray water to be treated and disposed to the environment. One exception is the thermoelectric power plant using cooling towers with high consumption of water equal to their withdrawal because most of the withdrawn water is evaporated as part of the cooling process.

6.1.5.4 Recreation

Sustainable management of water resources including protection of aquatic ecosystems, flood protection, adequate sanitation, provision of safe and reliable supplies for drinking water, irrigation, etc. poses huge challenges worldwide. Recreational water use is generally a very small but increasing percentage in total water use. Normally, recreational water use is tied to reservoirs. If a reservoir is kept fuller than it would otherwise be for recreation, then the water retained could be categorized as recreational usage. Common recreational usage includes whitewater boating, angling, water skiing, and swimming. Recreational water use is usually nonconsumptive. Golf courses are often accused of using excessive amount of water, especially in drier regions. To avert environmentalist's thoughts of wasting water, some governments categorize golf course usage as agricultural. However, it is not clear whether recreational irrigation has an obvious influence on water resources till now because of the inapplicability of reliable data. Besides, many golf courses use either primarily or exclusively treated effluent water, which will not affect potable water. In Arizona, an organized lobby concentrated on educating the public on how golf affects the environment has been established in the form of the Golf Industry Association. One disadvantage of recreational water use is that it may reduce the availability of water for other users at specific times and places. For example, water retained in a reservoir to allow boating in the late summer is not available to farmers during the spring planting season. Water released for whitewater rafting may not be available for hydroelectric generation during the time of peak electrical demand.

6.1.5.5 Environmental

Environmental water use is also a small but increasing percentage of total water use. The source of environmental water includes water stored in retention tank and released for environmental purposes or water retained in waterways through regulatory limits of abstraction. Then the usages of environmental water include fish ladders, water releasing from reservoirs timed to help fish spawn, restore more natural flow regimes, watering of natural or artificial wetlands, and create wildlife habitat. Similar to recreational usage, environmental usage is nonconsumptive but may influence other users. For example, water released from a reservoir to help fish spawn may not be available to farms upstream, and water retained in a river to maintain waterway health would not be available to water abstractors downstream.

6.1.6 Increasing Water Scarcity

People once thought that water was an infinite resource 50 years ago. It may because there were less than half of the current number of people on the planet which consumed less sources and did not need much water to produce these sources. However, nowadays, the condition of water resource is much more intense because of the explosive increase of people and food they need. Even more water will be needed in the future because the population worldwide is forecasted to rise to 9 billion by 2050. In 2007, International Water Management Institute in Sri Lanka conducted a research of water management in agriculture sector to find out whether the world had provided enough water for current population [9,10]. It assessed the current availability of water for agriculture on a global scale and mapped out locations suffering from water scarcity. The result of this assessment indicated that one-third of the world's population cannot have clean drinking water, and a fifth of the world's people, which is more than 1.2 billion, live in areas of physical water scarcity, which means a lack of water to meet all demands. Besides, 1.6 billion people live in areas of economic water scarcity, which means that the authorities cannot provide enough water to people because of lack of investment or human capacity in water. The research found it may be possible to meet the water requirement for people in the future, but the continuation of today's food production and environmental trends would lead to crises in many parts of the world. To avoid the happening of this situation, farmers have to strive to increase productivity while governments find methods to use water more efficiently (Fig. 6.3).

Irrigation is necessary to grow and increase productivity of crops. Various irrigation methods involve different relationships between water consumption, capital cost of equipment and structures, and crop yield. Some types of irrigation methods which are simple, cheap, and less efficient include furrow and overhead sprinkler, during whose process, much water evaporates, runs off, or drains below the root zone. Other irrigation methods which are more efficient and more expensive include drip or trickle irrigation, surge

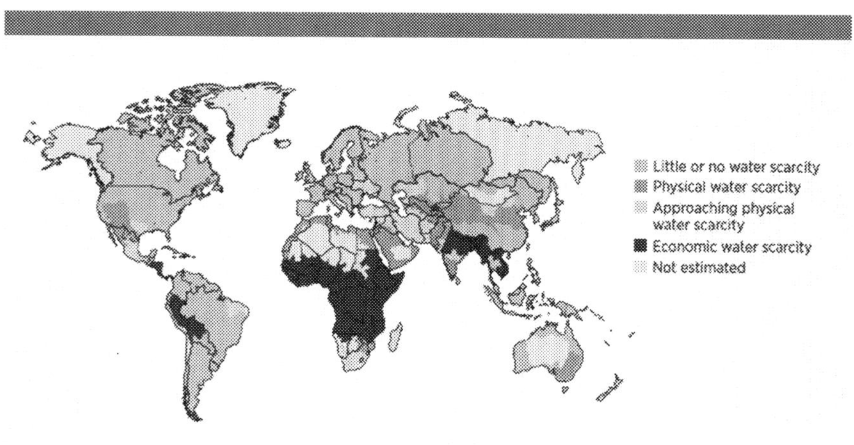

Figure 6.3 Shows that water scarcity already affects every continent where around 1.2 billion people, or almost one-fifth of the world's population, live in areas of physical scarcity, and 500 million people are approaching this situation.

irrigation, and some types of sprinkler systems where the sprinklers are operated near ground level. Issues often insufficiently considered when applying irrigation methods are salinization of groundwater and contaminant accumulation leading to water quality declines, which is the reason of the sharp rising of water scarcity.

6.1.6.1 Population Growth

As have mentioned above, the world population is increasing rapidly. In 2000, the world population was 6.2 billion. This number will increase to 9.7 billion by 2050 according to the estimation of the UN whose growth mainly comes from developing countries suffering water stress [11,12]. Therefore, the water demand will also increase unless water conservation or recycling increases [11,13]. The World Bank explained the result of the UN's research that the access to water for producing food will be one of the main challenges in the near future [14,15]. It will need to consider the impact of climate change and other environmental and social variables when trying to get the access to water.

6.1.6.2 Expansion of Business Activity

Business activities including industrialization, tourism, entertainment, etc. expand quickly nowadays, which increase demand of water supply and sanitation, and further lead to higher pressure on water resources and natural ecosystem.

6.1.6.3 Rapid Urbanization

Urbanization is also increasing rapidly. In low-density communities, small private wells and septic tanks are commonly used to provide water, which is not so useful in high-density urban areas. To deliver water to individuals and concentrate wastewater, urbanization requires large investment from individuals and business in water infrastructure. In 60% of European cities with more than 100,000 people, the replenishing rate of groundwater is less than consume rate [16,17]. Even if some water can be utilized, a large amount of cost will be needed to capture it.

6.1.7 Environmental Impact

In 2025, water shortage problems will be more obvious among poorer areas with limited resources and rapid population growth, such as the Middle East, parts of Asia, and Africa. Besides, at that time, to provide clean water and adequate sanitation, large urban and peri-urban areas will need to construct new infrastructures, which may lead to growing conflicts between government and agricultural water users, who currently consume most of the water. Compared with these poorer areas, more developed areas of North America, Europe, and Russia will not suffer the threat, not only because of their relative wealth but also because that their population better can align with available water resources.

6.1.8 Water Pollution

One of the major concerns today is water pollution and many countries have tried hard to find solutions to solve this problem. Among all pollutants threatening water supplies, the most widespread one is the discharge of raw sewage, sludge, garbage, or even toxic pollutants into natural water, which is most common in underdeveloped countries and some quasi-developed countries such as India, Iran, China, and Nepal. Treated sewage can form sludge, which may be placed in landfills, spread out on land, incinerated or dumped at sea. Except for sewage, nonpoint source pollution such as agricultural runoff, urban stormwater runoff, and chemical wastes are also a major source of pollution in some countries.

6.1.9 Climate Change

The misuse of water resources worldwide significantly impacts the climate change because of the close connections between the climate and hydrological cycle. Increased hydrologic variability and change in climate has and will continue to have a profound impact on the water sector through the hydrologic cycle, water availability, water demand, and water allocation at the global, regional, basin, and local levels. Rising temperatures will increase evaporation and further lead to increase in precipitation. Besides, it will also affect water quality by increasing eutrophication and other not well-understood reasons presently. Climate change could also result in much demand for garden sprinklers, swimming pools, farm irrigation, etc.

6.2 WATER AND WASTEWATER TREATMENT

Water treatment refers to processes that improve the water quality by removing contaminants and undesirable components or reducing their concentration to make it more suitable for specific end use including river flow maintenance, irrigation, water recreation, drinking, industrial water supply, or other uses. It is to convert wastewater into an effluent that can be further returned to the water cycle with minimal impact on the environment or directly reused, which can also be called as water reclamation. The treatment process takes place in a wastewater treatment plant (WWTP), often referred to as a water resource recovery facility (WRRF) or a sewage treatment plant. In this chapter, both water treatment and wastewater treatment will be discussed where most advancement of technological applications are applied.

6.2.1 Applications of Nanotechnology in Water and Wastewater Treatment

Providing clean and affordable water to meet human needs is a grand challenge of the 21st century. Worldwide, water supply struggles to keep up with the fast growing demand, which is exacerbated by population growth, global climate change, and water quality deterioration. The need for technological innovation to enable integrated water management cannot be overstated. Nanotechnology holds great potential in advancing water and wastewater treatment to improve treatment efficiency as well as to augment water supply through safe use of unconventional water sources. Here we review recent development in nanotechnology for water and wastewater treatment. The discussion covers candidate nanomaterials, properties, and mechanisms that enable the applications, advantages, and limitations as compared to existing processes, and barriers and research needs for commercialization. By tracing these technological advances to the physicochemical properties of nanomaterials, the present review outlines the opportunities and limitations to further capitalize on these unique properties for sustainable water management.

6.2.2 Introduction

Water is the most essential substance for all life on earth and a precious resource for human civilization. Reliable access to clean and affordable water is considered one of the most basic humanitarian goals and remains a major global challenge for the 21st century. Our current water supply faces enormous challenges, both old and new. Worldwide, some 780 million people still lack access to improved drinking water sources [18,19]. It is urgent to implement basic water treatment in the affected areas (mainly in developing countries) where water and wastewater infrastructure are often nonexistent. In both developing and industrialized countries, human activities play an ever-greater role in exacerbating water scarcity by contaminating natural water sources. The increasingly stringent water quality standards, compounded by emerging contaminants, have brought new scrutiny to the existing water treatment and distribution systems widely established in developed countries. The rapidly growing global population and the improvement of living standard continuously drive up the demand. Moreover, global climate change accentuates the already uneven distribution of freshwater, destabilizing the supply. Growing pressure on water supplies makes using unconventional water sources (e.g., stormwater, contaminated freshwater, brackish water, wastewater, and seawater) a new norm, especially in historically water-stressed regions. Furthermore, current water and wastewater treatment technologies and infrastructure are reaching their limit for providing adequate water quality to meet human and environmental needs. Recent advances in nanotechnology offer leapfrogging opportunities to develop next-generation water supply systems. Our current water treatment, distribution, and discharge practices, which heavily rely on conveyance and centralized systems, are no longer sustainable. The highly efficient, modular, and multifunctional processes enabled by nanotechnology are envisaged to provide high performance, affordable water and wastewater treatment solutions that rely less on large infrastructures [20,21]. Nanotechnology-enabled water and wastewater treatment promises to not only overcome major challenges faced by existing treatment technologies but also to provide new treatment capabilities that could allow economic utilization of unconventional water sources to expand the water supply. Here, we provide an overview of recent advances in nanotechnologies for water and wastewater treatment. The major applications of nanomaterials are critically reviewed based on their functions in unit operation processes. The barriers for their full-scale application and the research needs for overcoming these barriers are also discussed. The potential impacts of nanomaterials on human health and ecosystem as well as any potential interference with treatment processes are beyond the scope of this review and thus will not be addressed in detail here.

6.2.3 Current and Potential Applications for Water and Wastewater Treatment

Nanomaterials are typically defined as materials smaller than 100 nm in at least one dimension. At this scale, materials often possess novel size-dependent properties different from their large counterparts, many of which have been explored for applications in water and wastewater treatment. Some of these applications utilize the smoothly scalable size-dependent properties of nanomaterials, which relate to the high-specific surface area, such as fast dissolution, high reactivity, and strong sorption. Others take advantage of their discontinuous properties, such as superparamagnetism, localized surface plasmon resonance (LSPR), and quantum confinement effect. These applications are discussed below based on nanomaterial functions in unit operation processes (Table 6.1). Most applications discussed below are still in the stage of laboratory research. The pilot-tested or field-tested exceptions will be noted in the text.

6.2.4 Adsorption

Adsorption is commonly employed as a polishing step to remove organic and inorganic contaminants in water and wastewater treatment. Efficiency of conventional adsorbents is usually limited by the surface area or active sites, the lack of selectivity, and the adsorption kinetics. Nanoadsorbents offer significant improvement with their extremely high specific surface area and associated sorption sites, short intraparticle diffusion distance, and tunable pore size, and surface chemistry.

6.2.5 Carbon-Based Nanoadsorbents

6.2.5.1 Organic Removal

Carbon nanotubes (CNTs) have shown higher efficiency than activated carbon on adsorption of various organic chemicals [22,23]. Its high adsorption capacity mainly stems from the large specific surface area and the diverse contaminant—CNT interactions. The available surface area for adsorption on an individual CNT is its external surfaces [24,25]. In the aqueous phase, CNTs form loose bundles/aggregates due to the hydrophobicity of their graphitic surface, reducing the effective surface area. On the other hand, CNT aggregates contain interstitial spaces and grooves, which are high-adsorption energy sites for organic molecules [26,27]. Although activated carbon possesses comparable measured specific surface area as CNT bundles, it contains a significant number of micropores inaccessible to bulky organic molecules such as many antibiotics and pharmaceuticals [28,29]. Thus, CNTs have much higher adsorption capacity for some bulky organic molecules because of their larger pores in bundles and more accessible sorption sites.

A major drawback of activated carbon is its low adsorption affinity for low-molecular-weight polar organic compounds. CNTs strongly adsorb many of these polar organic compounds due to the diverse contaminant—CNT interactions including hydrophobic effect, $\pi-\pi$ interactions, hydrogen bonding, covalent bonding, and electrostatic interactions [25,30].

Table 6.1 Current and potential applications of nanotechnology in water and wastewater treatment

Applications	Representative nanomaterials	Desirable nanomaterial properties	Enabled technologies
Adsorption	Carbon nanotubes	High specific surface area, highly assessable adsorption sites, diverse contaminant-CNT interactions, tunable surface chemistry, easy reuse	Contaminant preconcentration/detection, adsorption of recalcitrant contaminants
	Nanoscale metal oxide	High specific surface area, short intraparticle diffusion distance, more adsorption sites, compressible without significant surface area reduction, easy reuse, some are superparamagnetic	Adsorptive media filters, slurry reactors
	Nanofibers with core shell structure	Tailored shell surface chemistry for selective adsorption, reactive core for degradation, short internal diffusion distance	Reactive nanoadsorbents
Membranes and membrane processes	Nanozeolites	Molecular sieve, hydrophilicity	High permeability thin film nanocomposite membranes
	Nano-Ag	Strong and wide-spectrum antimicrobial activity, low toxicity to humans	Antibiofouling membranes
	Carbon nanotubes	Antimicrobial activity (unaligned carbon nanotubes)	Antibiofouling membranes
		Small diameter, atomic smoothness of inner surface, tunable opening chemistry, high mechanical and chemical stability	Aligned carbon nanotube membranes
	Aquaporin	High permeability and selectivity	Aquaporin membranes
	Nano-TiO_2	Photocatalytic activity, hydrophilicity, high chemical stability	Reactive membranes, high-performance thin film nanocomposite membranes
	Nanomagnetite	Tunable surface chemistry, superparamagnetic	Forward osmosis

Photocatalysis	Nano-TiO_2	Photocatalytic activity in UV and possibly visible light range, low human toxicity, high stability, low cost	Photocatalytic reactors, solar disinfection systems
	Fullerene derivatives	Photocatalytic activity in solar spectrum, high selectivity	Photocatalytic reactors, solar disinfection systems
Disinfection and microbial control	Nano-Ag	Strong and wide-spectrum antimicrobial activity, low toxicity to humans, ease of use	Point-of-use (POU) water disinfection, antibiofouling surface
	Carbon nanotubes	Antimicrobial activity, fiber shape, conductivity	POU water disinfection, antibiofouling surface
	Nano-TiO_2	Photocatalytic reactive oxygen species generation, high chemical stability, low human toxicity and cost	POU to full scale disinfection and decontamination
Sensing and monitoring	Quantum dots	Broad absorption spectrum, narrow, bright, and stable emission which scales with the particle size and chemical component	Optical detection
	Noble metal nanoparticles	Enhanced localized surface plasmon resonances, high conductivity	Optical and electrochemical detection
	Dye-doped silica nanoparticles	High sensitivity and stability, rich silica chemistry for easy conjugation	Optical detection
	Carbon nanotubes	Large surface area, high mechanical strength and chemical stability, excellent electronic properties	Electrochemical detection, sample preconcentration
	Magnetic nanoparticles	Tunable surface chemistry, superparamagnetism	Sample preconcentration and purification

The π electron rich CNT surface allows $\pi-\pi$ interactions with organic molecules with C–C bonds or benzene rings, such as polycyclic aromatic hydrocarbons (PAHs) and polar aromatic compounds [31,32]. Organic compounds that have $-COOH$, $-OH$, $-NH_2$ functional groups could also form hydrogen bond with the graphitic CNT surface which donates electrons [33,34]. Electrostatic attraction facilitates the adsorption of positively charged organic chemicals such as some antibiotics at suitable pH [29,35].

6.2.5.2 Heavy Metal Removal

Oxidized CNTs have high adsorption capacity for metal ions with fast kinetics. The surface functional groups (e.g., carboxyl, hydroxyl, and phenol) of CNTs are the major adsorption sites for metal ions, mainly through electrostatic attraction and chemical bonding [36,37]. As a result, surface oxidation can significantly enhance the adsorption capacity of CNTs. Several studies show that CNTs are better adsorbents than activated carbon for heavy metals (e.g., Cu^{2+}fl, Pb^{2+}, Cd^{2+}, and Zn^{2+}) and the adsorption kinetics is fast on CNTs due to the highly accessible adsorption sites and the short intraparticle diffusion distance [33,38].

Overall, CNTs may not be a good alternative for activated carbon as wide-spectrum adsorbents. Rather, as their surface chemistry can be tuned to target specific contaminants, they may have unique applications in polishing steps to remove recalcitrant compounds or in preconcentration of trace organic contaminants for analytical purposes. These applications require small quantity of materials and hence are less sensitive to the material cost.

Produced by exfoliating graphite with strong acids and oxidizers, graphite oxide is a potentially low-cost adsorbent. It was recently reported that sand granules coated with graphite oxide were efficient in removing Hg^{2+} and a bulky dye molecule's, Rhodamine B, its performance was comparable with commercialy activated carbon [39,40].

6.2.5.3 Regeneration and Reuse

Regeneration is an important factor that determines the cost-effectiveness of adsorbents. Adsorption of metal ions on CNTs can be easily reversed by reducing the solution pH. The metal recovery rate is usually above 90% and often close to 100% at $pH < 2$ [33,41]. Moreover, the adsorption capacity remains relatively stable after regeneration. Lu et al. reported that Zn^{2+} adsorption capacity of SWNT and MWNT decreased less than 25% after

10 regeneration and reuse cycles, while that of activated carbon was reduced by more than 50% after one regeneration [33,42]. A statistical analysis based on the best-fit regression of Zn^{2+} adsorption capacity and the number of regeneration and reuse cycles suggested that CNT nanoadsorbents can be regenerated and reused up to several hundred times for Zn^{2+} removal while maintaining reasonable adsorption capacity [43,44].

6.2.5.4 Metal-Based Nanoadsorbents

Metal oxides such as iron oxide, titanium dioxide and alumina are effective low-cost adsorbents for heavy metals and radionuclides. The sorption is mainly controlled by complexation between dissolved metals and the oxygen in metal oxides [45,46]. It is a two-step process: fast adsorption of metal ions on the external surface, followed by the rate-limiting intraparticle diffusion along the micropore walls [47,48]. Their nanoscale counterparts have higher adsorption capacity and faster kinetics because of the higher specific surface area, shorter intraparticle diffusion distance, and larger number of surface reaction sites (i.e., corners, edges, vacancies). For instance, as the particle size of nanomagnetite decreased from 300 to 11 nm, its arsenic adsorption capacity increased more than 100 times [49,50]. Much of this observed increase in adsorption was attributed to the increase in specific surface area as the 300- and 20-nm magnetite particles have similar surface area normalized arsenic adsorption capacity (~6 μmol/m^2 or 3.6 atoms/nm^2). However, when particle size was reduced to below 20 nm, the specific surface area normalized adsorption capacity increased, with 11-nm magnetite nanoparticles absorbing three times more arsenic (~18 μmol/m^2 or 11 atoms/nm^2), suggesting a "nanoscale effect." This "nanoscale effect" was attributed to the change of magnetite surface structure, which creates new adsorption site vacancies [51,52].

In addition to high adsorption capacity, some iron oxide nanoparticles, e.g., nanomaghemite and nanomagnetite, can be superparamagnetic. Magnetism is highly volume dependent as it stems from the collective interaction of atomic magnetic dipoles. If the size of a ferro- or ferrimagnet decreases to the critical value (~40 nm), the magnet changes from multiple domains to single domain with higher magnetic susceptibility [34,53]. As the size further decreases, magnetic particles become superparamagnetic, losing permanent magnetic moments while responding to an external magnetic field, which allows easy separation and recovery by a low-gradient magnetic field. These magnetic nanoparticles can be either used directly as adsorbents or as the core material in a core—shell nanoparticle structure where the shell

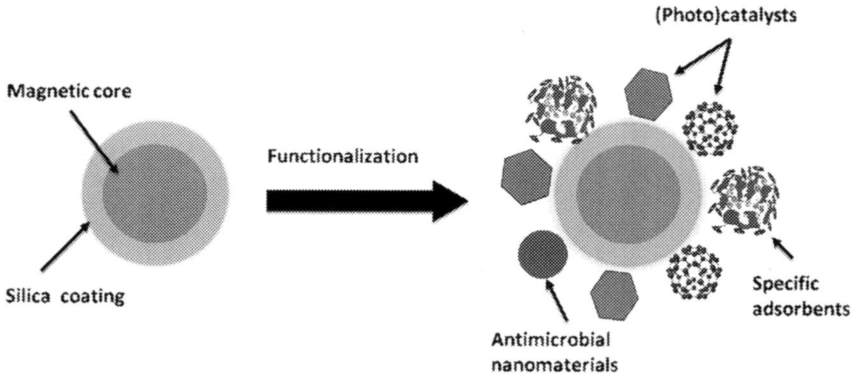

Figure 6.4 Multifunctional magnetic nanoparticles. Magnetic nanoparticles are used as the core material in a core–shell nanoparticle structure where the shell provides the desired function while the magnetic core realizes magnetic separation. Silica coating helps functionalization due to the rich silica chemistry.

provides the desired function while the magnetic core realizes magnetic separation (Fig. 6.4).

Metal oxide nanocrystals can be compressed into porous pellets without significantly compromising their surface area when moderate pressure is applied [33,54]. The pore volume and pore size can be controlled by adjusting the consolidation pressure. Thus, they can be applied in forms of both fine powders and porous pellets, which are the likely forms to be used in industry.

Metal-based nanomaterials have been explored to remove a variety of heavy metals, such as arsenic, lead, mercury, copper, cadmium, chromium, nickel, and have shown great potential to outcompete activated carbon [55,56]. Among them, the application for arsenic removal has attracted much attention. Although a good adsorbent for many organic and inorganic contaminants, activated carbon has limited capacity for arsenic, especially for As(V). Several metal oxide nanomaterials including nanosized magnetite and TiO_2 have shown arsenic adsorption performance superior to activated carbon [12,57,58]. Metal (hydr)oxide nanoparticles also can be impregnated onto the skeleton of activated carbon or other porous materials to achieve simultaneous removal of arsenic and organic cocontaminants, which favor point-of-use (POU) applications [59,60].

Regeneration and reuse. Metal oxide nanoadsorbents can be easily regenerated by changing solution pH [56,61]. In many cases, the adsorption capacity of metal oxide nanoadsorbents is well maintained after several regeneration and reuse cycles [62,63]. However, reduced adsorption capacity after regeneration has also been reported [64,65].

Above all, metal-based nanoadsorbents can be produced at relatively low cost. The high adsorption capacity, low cost, easy separation, and regeneration make metal-based nanoadsorbents technologically and economically advantageous.

6.2.6 Polymeric Nanoadsorbents

Dendrimers are tailored adsorbents that are capable of removing both organics and heavy metals. Their interior shells can be hydrophobic for sorption of organic compounds while the exterior branches can be tailored (e.g., hydroxyl- or amine-terminated) for adsorption of heavy metals. The sorption can be based on complexation, electrostatic interactions, hydrophobic effect, and hydrogen bonding [66,67]. A dendrimer-ultrafiltration (UF) system was designed to recover metal ions from aqueous solutions [15,68]. The system achieved almost complete removal of Cu^{2+} ions with initial concentration of 10 ppm and Cu^{2+} to PAMAM dendrimer-NH_2 ratio of 0.2. After adsorption, the metal ion—laden dendrimers were recovered by UF and regenerated by decreasing pH to 4.

6.2.7 Potential Application in Water Treatment

Nanoadsorbents can be readily integrated into existing treatment processes in slurry reactors or adsorbers. Applied in the powder form, nanoadsorbents in slurry reactors can be highly efficient because all surfaces of the adsorbents are utilized and the mixing greatly facilitates the mass transfer. However, an additional separation unit is required to recover the nanoparticles. Nanoadsorbents can also be used in fixed or fluidized adsorbers in the form of pellets/beads or porous granules loaded with nanoadsorbents. Fixed-bed reactors are usually associated with mass transfer limitations and head loss; but they do not need future separation process. Applications of nanoadsorbents for arsenic removal have been commercialized, and their performance and cost have been compared with other commercial adsorbents in pilot tests [69]. ArsenX^{np} is a commercial hybrid ion exchange medium comprising iron oxide nanoparticles and polymers. ADSORBSIA is a nanocrystalline titanium dioxide medium in the form of beads from 0.25 to 1.2 mm in diameter. Both nanoadsorbents were highly efficient in removing arsenic, and ArsenXnp required little backwash [69,70]. The estimated treatment cost for ArsenXnp is \$0.25 ~ \$0.35/1000 gal if the medium is regenerated, similar to \$0.37/1000 gal of Bayoxide E33, a high-performance granular iron oxide adsorbent [69,71]. ArsenX^{np} and ADSORBSIA have been employed in small- to medium-scale drinking water treatment systems and were proven to be cost-competitive.

6.2.8 Membranes and Membrane Processes

The basic goal of water treatment is to remove undesired constituents from water. Membranes provide a physical barrier for such constituents based on their size, allowing use of unconventional water sources. As the key component of water treatment and reuse, they provide high level of automation, require less land and chemical use, and the modular configuration allows flexible design [21]. A major challenge of the membrane technology is the inherent tradeoff between membrane selectivity and permeability. The high energy consumption is an important barrier to the wide application of pressure-driven membrane processes. Membrane fouling adds to the energy consumption and the complexity of the process design and operation. Furthermore, it reduces the lifetime of membranes and membrane modules.

The performance of membrane systems is largely decided by the membrane material. Incorporation of functional nanomaterials into membranes offers a great opportunity to improve the membrane permeability, fouling resistance, mechanical and thermal stability, as well as to render new functions for contaminant degradation and self-cleaning.

6.2.9 Nanofiber Membranes

Electrospinning is a simple, efficient, and inexpensive way to make ultrafine fibers using various materials (e.g., polymers, ceramics, or even metals) [72,73]. The resulting nanofibers have high specific surface area and porosity and form nanofiber mats with complex pore structures. The diameter, morphology, composition, secondary structure, and spatial alignment of electrospun nanofibers can be easily manipulated for specific applications [73]. Although nanofiber membranes have been commercially employed for air filtration applications, their potential in water treatment is still largely unexploited. Nanofiber membranes can remove micron-sized particles from aqueous phase at a high rejection rate without significant fouling [74]. Thus, they have been proposed to be used as pretreatment prior to UF or reverse osmosis (RO). Functional nanomaterials can be easily doped into the spinning solutions to fabricate nanoparticle-impregnated nanofibers or formed in situ [73]. The outstanding features and tunable properties make electrospun nanofibers an ideal platform for constructing multifunctional media/membrane filters by either directly using intrinsically multifunctional materials such as TiO_2 or by introducing functional materials on the nanofibers. For example, by incorporating ceramic nanomaterials or specific capture agents on the nanofiber scaffold, affinity nanofiber membranes can be designed to remove heavy metals and organic pollutants during filtration.

6.2.10 Nanocomposite Membranes

A significant number of studies on membrane nanotechnology have focused on creating synergism or multifunction by adding nanomaterials into polymeric or inorganic membranes. Nanomaterials used for such applications include hydrophilic metal oxide nanoparticles (e.g., Al_2O_3, TiO_2, and zeolite), antimicrobial nanoparticles (e.g., nano-Ag and CNTs), and (photo)catalytic nanomaterials (e.g., bimetallic nanoparticles, TiO_2).

The main goal of adding hydrophilic metal oxide nanoparticles is to reduce fouling by increasing the hydrophilicity of the membrane. The addition of metal oxide nanoparticles including alumina, silica, zeolite, and TiO_2 to polymeric UF membranes has been shown to increase membrane surface hydrophilicity, water permeability, or fouling resistance [75—78]. These inorganic nanoparticles also help enhance the mechanical and thermal stabilities of polymeric membranes, reducing the negative impact of compaction and heat on membrane permeability [5,78].

Antimicrobial nanomaterials such as nano-Ag and CNTs can reduce membrane biofouling. Nano-Ag has been doped or surface grafted on polymeric membranes to inhibit bacterial attachment and biofilm formation on the membrane surface as well as inactivate viruses [79—81]. However, its long-term efficacy against membrane biofouling has not been reported. Appropriate replenishment of nano-Ag needs to be addressed for practical application of this technology. CNTs inactivate bacteria upon direct contact [82]. High bacterial inactivation (>90%) has been achieved using polyvinyl-N-carbazole-SWNT nanocomposite at 3 wt% of SWNT [83]. As CNTs are insoluble in water and not consumed, there is no need for replenishment. However, as direct contact is required for inactivation, long-term filtration experiments are needed to determine the impact of fouling on the antimicrobial activity of CNTs. Addition of oxidized MWNT at low weight percentage (up to 1.5 wt%) also increases the hydrophilicity and permeability of polysulfone membranes [84].

Photocatalytic nanoparticle–incorporated membranes (a.k.a. reactive membranes) combine their physical separation function and the reactivity of a catalyst toward contaminant degradation. Much effort has been devoted to develop photocatalytic inorganic membranes consisting of nanophotocatalysts (normally nano-TiO_2 or modified nano-TiO_2) [85]. Metallic/bimetallic catalyst nanoparticles such as nano zerovalent iron (nZVI) and noble metals supported on nZVI have been incorporated into polymeric membranes for reductive degradation of contaminants, particularly chlorinated compounds [86,87]. nZVI serves as the electron donor and the noble metals catalyze the reaction.

6.2.11 Thin Film Nanocomposite Membranes

Development of thin film nanocomposite (TFN) membranes mainly focuses on incorporating nanomaterials into the active layer of thin film composite (TFC) membranes via doping in the casting solutions or surface modification. Nanomaterials that have been researched for such applications include nanozeolites, nano-Ag, nano-TiO_2, and CNTs. The impact of nanoparticles on membrane permeability and selectivity depends on the type, size, and amount of nanoparticles added.

Nanozeolites are the most frequently used dopants in TFN and have shown potential in enhancing membrane permeability. The addition of nanozeolites leads to more permeable, negatively charged, and thicker polyamide active layer [88]. One study reported water permeability increased up to 80% over the TFC membrane, with the salt rejection largely maintained ($93.9 \pm 0.3\%$) [89]. TFN membranes doped with 250 nm nanozeolites at 0.2 wt% achieved moderately higher permeability and better salt rejection (>99.4%) than commercial RO membranes [90]. It was hypothesized that the small, hydrophilic pores of nanozeolites create preferential paths for water. However, water permeability increased even with pore-filled zeolites, although less than the pore-open ones, which could be attributed to defects at the zeolite–polymer interface. Nanozeolites were also used as carriers for antimicrobial agents such as Agfl, which imparts antifouling property to the membrane [91]. The zeolite TFN technology has reached the early stage of commercialization. QuantumFlux, a seawater TFN RO membrane, is now commercially available (www.nanoH2O.com).

Incorporation of nano-TiO_2 (up to 5 wt%) into the TFC active layer slightly increased the membrane rejection while maintaining the permeability [92]. When the concentration of nano-TiO_2 exceeded 5 wt%, the water flux increased in the cost of reducing rejection, suggesting defect formation in the active layer. Upon UV irradiation, TiO_2 can degrade organic contaminants and inactivate microorganisms. This helps reduce organic and biological fouling as well as remove contaminants that are not retained by the membrane. However, the close adjacency between the photocatalyst and the membrane may also lead to detrimental effects on polymeric membrane materials, which needs to be addressed for long-term efficacy [93].

CNTs (unaligned) also found their application in TFN membranes due to their antimicrobial activities. Tiraferri et al. covalently bonded SWNTs to a TFC membrane surface [94]. This approach is advantageous as it uses relatively small amount of the nanomaterial and minimizes perturbation of the active layer. The resulting TFN membrane exhibited moderate antibacterial properties (60% inactivation of bacteria attached on the membrane surface in 1 h contact time), potentially reducing or delaying membrane biofouling.

6.2.12 Biologically Inspired Membranes

Many biological membranes are highly selective and permeable. Aquaporins are protein channels that regulate water flux across cell membranes. Their high selectivity and water permeability makes their use in polymeric membranes an attractive approach to improve membrane performance. Aquaporin-Z from *Escherichia coli* has been incorporated into amphiphilic triblock polymer vesicles, which exhibit water permeability at least an order of magnitude over the original vesicles with full rejection to glucose, glycerol, salt, and urea [95]. One potential design is to coat aquaporin-incorporated lipid bilayers on commercial nanofiltration membranes. On this front, limited success was achieved [96]. Aligned CNTs have been shown both experimentally and theoretically to provide water permeation much faster than what the Hagene–Poiseuille equation predicts, owing to the atomic smoothness of the nanosized channel, and the one dimensional single file ordering of water molecules while passing through the nanotubes [97,98]. It was predicted that a membrane containing only 0.03% surface area of aligned CNTs will have flux exceeding current commercial seawater RO membranes [99]. However, high rejection for salt and small molecules is challenging for aligned CNT membranes due to the lack of CNTs with uniformly subnanometer diameter. Functional group gating at the nanotube opening has been proposed to enhance the selectivity of aligned CNT membranes [100]. By grafting carboxyl functional groups on sub-2-nm CNT openings, 98% rejection of $Fe(CN_6)^{3-}$ was achieved at low ionic strength by Donnan exclusion [35]. However, KCl rejection was only 50% at 0.3 mM and decreased to almost zero at 10 mM. Grafting bulky functional groups at the tube opening could physically exclude salts. However, steric exclusion will significantly reduce membrane permeability [101]. Thus at the current stage, aligned CNT membranes are not capable of desalination. To achieve reliable salt rejection, the CNT diameter must be uniformly smaller than 0.8 nm [102]. A key barrier for both aquaporin and aligned CNT membranes is the scale-up of the nanomaterial production and membrane fabrication. Large-scale production and purification of aquaporins are very challenging. To date, chemical vapor deposition (CVD) is the most common way to make aligned nanotubes. A continuous high-yield CVD prototype has been designed for producing vertically aligned CNT, paving the way for large-scale production [64]. A postmanufacturing alignment method using magnetic field was also developed [103].

Nanocomposite and TFN membranes have good scalability as they can be fabricated using current industrial manufacturing processes. The high water permeability can reduce the applied pressure or required membrane area and consequently cut cost. This strategy may greatly improve the energy efficiency for treatment of waters with low osmosis pressure, but it may have limited advantage in seawater RO, whose energy consumption is already close to the thermodynamic limit [8]. A recent review ranked current membrane nanotechnologies based on their potential performance enhancement and state of commercial readiness [99].

6.2.13 Forward Osmosis

Forward osmosis (FO) utilizes the osmotic gradient to draw water from a low osmotic pressure solution to a high osmotic pressure one (i.e., the draw solution). The diluted draw solution is then treated by RO or thermal processes to generate pure water. FO has two major advantages over the pressure-driven RO: it does not require high pressure, and the membrane is less prone to fouling.

The key to FO is to have a draw solute with high osmolality and easily separable from water. Chemicals currently employed for draw solutions include NaCl and ammonia bicarbonate. Therefore, RO or thermal treatment, both energy intensive, is required to recover water from the draw solution. Magnetic nanoparticles were recently explored as a new type of draw solute for its easy separation and reuse. Hydrophilic coating was employed to aid dissolution and increase osmotic pressure. An FO permeate flux higher than 10 L/m^2 h was achieved using 0.065 M poly(ethylene glycol) diacid-coated magnetic nanoparticles when deionized water was used as the feed solution [104]. Magnetic nanoparticles were also applied to recover draw solutes. In a recent study, magnetic nanoparticles (Fe_3O_4@SiO_2) were used to recover $Al_2(SO_4)_3$ (the draw solute) through flocculation [42].

6.2.14 Photocatalysis

Photocatalytic oxidation is an advanced oxidation process (AOP) for removal of trace contaminants and microbial pathogens. It is a useful pretreatment for hazardous and nonbiodegradable contaminants to enhance their biodegradability. Photocatalysis can also be used as a polishing step to treat recalcitrant organic compounds. The major barrier for its wide application is the slow kinetics due to limited light fluence and photocatalytic activity. Current research focuses on increasing photocatalytic reaction kinetics and photoactivity range (Table 6.2).

6.2.15 Nanophotocatalyst Optimization

TiO_2 is the most widely used semiconductor photocatalyst in water/wastewater treatment owing to its low toxicity, chemical stability, low cost, and abundance as raw material. It generates an electron—hole (e^-/h^+) pair upon absorbing a UV photon, which later either migrates to the surface and form reactive oxygen species (ROS) or undergoes undesired recombination. The photoactivity of nano-TiO_2 can be improved by optimizing particle size and shape, reducing e^-/h^+ recombination by noble metal doping, maximizing reactive facets, and surface treatment to enhance contaminant adsorption.

The size of TiO_2 plays an important role in its solid phase transformation, sorption, and e^-/h^+ dynamics. Among the crystalline structures of TiO_2, rutile is the most stable for particles larger than 35 nm, while anatase, which is more efficient in producing ROS, is the most stable for particles smaller than 11 nm [105,106]. A major cause for the slow reaction kinetics of TiO_2 photocatalysis is the fast recombination of e^- and h^+. Decreasing TiO_2 particle size lowers volume recombination of e^-/h^+ and enhances interfacial charge carrier transfer [107]. However, when particle size is reduced to several nanometers, surface recombination dominates, decreasing photocatalytic activity. Therefore, the photocatalytic activity of TiO_2 has a maximum due to the interplay of the aforementioned mechanisms, which lies in the nanometer range. TiO_2 nanotubes were found to be more

Table 6.2 TiO_2 photocatalyst optimization

Optimization objectives	Optimization approaches	Optimization mechanisms	Water treatment applications
Enhance photocatalytic reaction kinetics	Size	More surface reactive sites, higher reactant adsorption, lower electron–hole recombination	High-performance UV–activated photocatalytic
	Nanotube morphology	Shorter carrier-diffusion paths in the tube walls, higher reactant mass transfer rate toward tube surface	Reactors
	Noble metal doping	Better electron–hole separation, lower electron–hole recombination	
	Reactive crystallographic facets	Higher reactant sorption, better electron–hole separation, lower electron–hole recombination	
Expand photoactivity range	Metal impurity doping Anion doping Dye sensitizer doping Narrow band gap semiconductors doping	Impurity energy levels Band gap narrowing Electron injection Electron injection	Low energy cost solar/visible light–activated photocatalytic reactors

efficient than TiO_2 nanoparticles in decomposition of organic compounds [108]. The higher photocatalytic activity was attributed to the shorter carrier diffusion paths in the tube walls and faster mass transfer of reactants toward the nanotube surface.

Noble metal doping can reduce the e^-/h^+ recombination because the photoexcited electrons tend to migrate to the noble metals with lower Fermi levels while the holes stay in TiO_2 [109]. The photocatalytic activity of TiO_2 can also be promoted by creating highly reactive crystallographic facets. Because high-energy {001} facets diminish quickly during crystal growth, anatase TiO_2 is usually dominated by the low-energy {101} facets. Using specific capping agent (usually fluoride), the percentage of {001} facets can be increased from less than 10% to up to 89%, substantially enhancing hydroxyl radical production and organic compound decomposition [110,111]. The enhanced activity stems from the strong adsorption of reactants on high-energy facets and the spatial separation of electrons and holes on specific crystal facets [111,112]. The optimal percentage of {001} facets for photocatalysis is still debated [112]. Improving contaminant adsorption by modifying photocatalyst surface is another way to enhance photocatalytic activity due to the short lifetime of ROS. However, little has been done in this area.

Another actively pursued research area is to extend the excitation spectrum of TiO_2 to include visible light. The general strategy is doping metal impurities, dye sensitizers, narrow band gap semiconductors, or anions into nano-TiO_2 to form hybrid nanoparticles or nanocomposites [105,109]. Metals and anions create impurity energy levels or narrow the band gap; upon visible light excitation, dye sensitizers and narrow band gap semiconductors inject electrons into TiO_2 to initiate the catalytic reactions. Among these methods, anions (especially nitrogen) doping, was considered most cost-effective and feasible for industrial applications, although their stability and long-term efficacy have not been tested. Decreased nitrogen concentration during photocatalysis has been reported [105,113].

Other than TiO_2, WO_3 and some fullerene derivatives also have the potential to be used in photocatalytic water treatment. WO_3 has a narrower band gap than TiO_2, allowing it to be activated by visible light (<450 nm) [114]. Pt doping further enhances WO_3 reactivity by facilitating multielectron reduction of O_2 and improving e^-/h^+ separation [63]. Amino fullerenes generate $^1O_2$ under visible light irradiation (<550 nm) and has been studied to degrade pharmaceutical compounds and inactivate viruses [115,116]. Fullerol and C_{60} encapsulated with poly(N-vinylpyrrolidone) can produce $^1O_2$ and superoxide under UVA light [117]. Amino fullerenes are more amenable to immobilization than fullerol and are more effective for disinfection purposes due to their positive charge. $^1O_2$ has lower oxidation potential than hydroxyl radicals produced by TiO_2, while it is a more selective ROS and consequently less susceptible to quenching by nontarget background organic matter. Fullerenes are currently much more expensive and not as readily available as TiO_2.

6.2.16 Potential Applications in Water Treatment

The overall efficiency of a photocatalytic water treatment process strongly depends on the configuration and operation parameters of the photoreactor. Two configurations are commonly used: slurry reactors and reactors using immobilized TiO_2. Various dispersion/recovery or catalyst immobilization techniques are being pursued to maximize its efficiency. Extensive investigation on operating parameters has been carried out with these lab or pilot scale systems. A recent critical review outlines the effects of water quality and a wide range of operating parameters including TiO_2 loading, pH, temperature, dissolved oxygen, contaminant type and concentration, light wavelength and intensity [118]. Readers are referred to this review for details regarding process optimization. A commercial product, Purifics Photo-Cat system, has treatment capacity as high as 2 million gal/day with a small footprint of 678 ft^2. Pilot tests showed that the Photo-Cat system is highly efficient for removing organics without producing waste streams, and it operates with relatively low specific power consumption of about 4 kWh/m^3 [119—121]. Nano-TiO_2—facilitated solar disinfection (SODIS) has been extensively tested and appears to be a feasible option to produce safe drinking water in remote areas of developing countries. The SODIS system can be small scale for one person or scaled up to medium-size solar compound parabolic collectors.

Photocatalysis has shown great potential as a low-cost, environmental friendly, and sustainable water treatment technology. However, there are several technical challenges for its large-scale application, including (1) catalyst optimization to improve quantum yield or to utilize visible light; (2) efficient photocatalytic reactor design and catalyst recovery/immobilization techniques; (3) better reaction selectivity.

Metal oxide nanomaterials such as TiO_2 and CeO_2 as well as carbon nanotubes have been studied as catalysts in heterogeneous catalytic ozonation processes that provide fast and comparatively complete degradation of organic pollutants. Both radical-mediated and nonradical-mediated reaction pathways have been proposed [122]. The adsorption of ozone and/or pollutants on the catalyst surface plays a critical role in both mechanisms. Nanomaterials have large specific surface area and an easily accessible surface, leading to high catalytic activity. Some nanomaterials were also reported to promote decomposition of ozone into hydroxyl radicals, facilitating degradation process through radical-mediated routes [123]. For future industrial scale applications, a better understanding of the mechanism of nanomaterial-enabled catalytic ozonation is in critical need.

6.2.17 Disinfection and Microbial Control

The dilemma between effective disinfection and formation of toxic disinfection by-products (DBPs) poses a great challenge for the water industry. It is now well recognized that conventional disinfectants, such as chlorine disinfectants and ozone can form toxic DBPs (e.g., halogenated disinfection by-products, carcinogenic nitrosamines, bromate, etc.). UV disinfection emerged as an alternative for oxidative disinfection as it produces minimal DBPs, while it requires high dosage for certain viruses (e.g., adenoviruses). These limitations urge the development of alternative methods that can enhance the robustness of disinfection while avoiding DBP formation.

Our previous review on antimicrobial nanomaterials highlighted the potential of nanotechnology in disinfection and microbial control [124]. Many nanomaterials, including nano-Ag, nano-ZnO, nano-TiO_2, nano-Ce_2O_4, CNTs, and fullerenes, exhibit antimicrobial properties without strong oxidation and hence have lower tendency to form DBPs (Table 6.3). The antimicrobial mechanisms of these nanomaterials, their merits, limitations, and applicability for water treatment, and the critical research needs are thoroughly discussed in that review paper [124]. Thus, only a brief update mainly regarding nano-Ag and carbon-based nanomaterials will be provided here.

Table 6.3 Nanomaterial antimicrobial mechanisms

Nanomaterials	Antimicrobial mechanisms
Nano-Ag	Release of silver ions, protein damage, suppression of DNA replication, membrane damage
Nano-TiO_2	Production of reactive oxygen species (ROS)
Nano-ZnO	Release of zinc ions, production of H_2O_2, membrane damage
Nano-MgO	Membrane damage
Nano-Ce_2O_4	Membrane damage
nC_{60}	ROS-independent oxidation
Fullerol and amino fullerene	Production of ROS
Carbon nanotubes	Membrane damage, oxidative stress
Graphene-based nanomaterials	Membrane damage, oxidative stress

6.2.18 Antimicrobial Mechanisms

Nano-Ag is currently the most widely used antimicrobial nanomaterial. Its strong antimicrobial activity, broad antimicrobial spectrum, low human toxicity, and ease of use make it a promising choice for water disinfection and microbial control. It is now well accepted that the antimicrobial activity of nanosilver largely stems from the release of silver ions [125,126]. Silver ions can bind to thiol groups in vital proteins, resulting in enzyme damage [127]. It also has been reported that silver ions can prevent DNA replication and induce structural changes in the cell envelope [39]. Thus, the release rate and bioavailability of silver ions are crucial for the toxicity of nano-Ag. Studies have suggested that physicochemical properties of nano-Ag play an important role in its antimicrobial activity. However, the influence of the size, shape, coating, and crystallographic facet appears to be mainly related to different release kinetics of silver ions. The presence of common ligands reduces the bioavailability of silver ions and mitigates its toxicity [125]. A recent study found that low concentration (sublethal) of silver ions or nano-Ag enhances *E. coli* growth, suggesting a hermetic response that could be counterproductive to its antimicrobial applications [126].

CNTs kill bacteria by causing physical perturbation of the cell membrane, oxidative stress, or disruption of a specific microbial process via disturbing/oxidizing a vital cellular structure/component upon direct contact with bacterial cells. Graphene and graphite materials exhibit antimicrobial properties through similar mechanisms [128,129]. The cytotoxicity of CNTs strongly depends on their physicochemical properties. Short, dispersed, and metallic CNTs with small diameters are more toxic [129–131].

6.2.19 Potential Applications in Water Treatment

Antimicrobial nanomaterials are envisaged to find their applications in three critical challenges in water/wastewater systems: disinfection, membrane biofouling control, and biofilm control on other relevant surfaces. Nano-Ag has good potential for application in POU treatment. It can improve water quality for high-end use, or provide another barrier against waterborne pathogens for vulnerable population. Commercial devices utilizing nano-Ag are already available, e.g., MARATHON and Aquapure systems. Nano-Ag has also been incorporated into ceramic microfilters as a barrier for pathogens, which can be employed in remote areas in developing countries [132].

The antimicrobial properties, fibrous shape, and high conductivity of CNTs enable novel CNT filters for both bacteria and virus removal: The thin layer of CNTs effectively remove bacteria by size exclusion and viruses by depth filtration; the retained bacteria are largely inactivated by CNTs within hours [133]. With a small intermittent voltage (2–3 V), MWNTs can directly oxidize attached bacteria and viruses and lead to inactivation in seconds [134,135]. The applied electric potential also enhances viral transport to the anodic CNTs [134]. Such CNT filters can be used as high-performance POU devices for water disinfection with minimal to no power requirement.

The application of nanomaterials can also be used in other water treatment related surfaces such as storage tanks and distribution pipes to control pathogen contamination, biofilm formation, and microbial influenced corrosion. Affordable coating techniques that can economize nanomaterial use and maximize its efficacy while allowing for regeneration are in critical need. An alternative approach is to employ nanoscale biofouling-resistant surface structures, a strategy used by marine organisms (dolphins and sharks) and plants (lotus leafs). A common disadvantage of many nanomaterial-enabled disinfection approaches is the lack of disinfection residue, which is crucial for controlling microbial growth during water storage and distribution. Nevertheless, nanotechnology-enabled disinfection can reduce DBP formation as chlorine or other chemical disinfectants are only needed as secondary disinfectants. Long-term efficacy is another major uncertainty for all the aforementioned technologies. Antimicrobial nanomaterials that rely on release of biocidal ions will be eventually depleted. Controlled release and the replenish strategies are thus needed. A potential "on-demand" release strategy is to encapsulate antimicrobial agents into a matrix gated by materials responsive to the presence of microorganisms or biofilms. This "on-demand" mechanism can be further coupled with recognition mechanisms for targeted release (Fig. 6.5). For nanomaterials relying on direct contact, fouling may largely suppress or even eliminate their antimicrobial activity.

6.2.20 Sensing and Monitoring

A major challenge for water/wastewater treatment is water quality monitoring due to the extremely low concentration of certain contaminants, the lack of fast pathogen detection, as well as the high complexity of the water/wastewater matrices. Innovative sensors with high sensitivity and selectivity, and fast response are in great need.

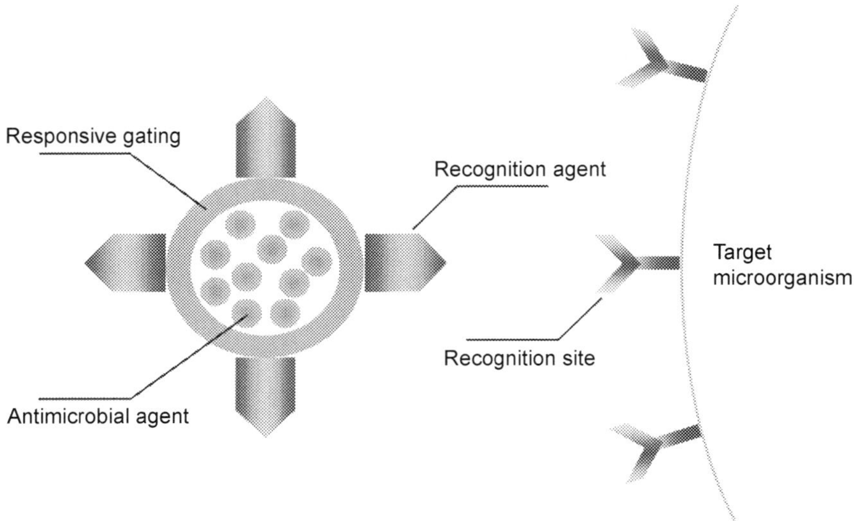

Figure 6.5 Schematic mechanism for "on-demand" microbial control. "On-demand" microbial control can be achieved by using recognition agents that target specific microorganism. The responsive gating material is designed to release the antimicrobial agent after the recognition event.

6.2.21 Pathogen Detection

Pathogen detection is of critical importance as it is directly related to public health. Conventional indicator systems such as coliform bacteria is slow and fail to monitor the presence of some important or emerging pathogens including viruses (hepatitis A and E, coxsackieviruses, echoviruses, adenoviruses, and Norwalk viruses), bacteria (Legionella and Helicobacter), and protozoan (Cryptosporidium and Giardia) [136]. Many of these pathogens are etiologic agents in outbreaks associated with drinking water. Additionally, pathogen detection is the key component of diagnosis-based water disinfection approach, in which disinfection is triggered by the detection of target microorganisms. Active research is going on developing nanomaterial-enabled pathogen sensors. These sensors usually consist of three major components: recognition agents, nanomaterials, and a signal transduction mechanism [137]. Recognition agents that specifically interact with antigens or other epitopes on the pathogen surface provide the selectivity. Sensitivity and fast response are achieved by the nanomaterial-related signal transduction upon the recognition event. A wide range of recognition agents have been utilized, including antibodies, aptamers, carbohydrates, and antimicrobial peptides [137]. Nanomaterials improve the sensitivity and speed of detection and achieve multiplex target detection owing to their unique physicochemical properties, especially electrochemical, optical, and magnetic properties. These sensors can be employed to detect whole cells as well as biomolecules [136,137].

The most commonly used nanomaterials in pathogen detection are magnetic nanoparticles, quantum dots (QDs), noble metals, dye-doped nanoparticles, and CNTs. Magnetic nanoparticles and CNTs have been extensively studied for sample concentration and purification. A commercial magnetic nanocomposite, Dynabead, is available for developing various pathogen detection kits.

QDs are fluorescent nanocrystals of semiconducting materials (e.g., CdSe) whose electronic characteristics depend on the size and shape of the individual crystals. QD particles with smaller sizes have wider band gaps and thus need more energy to excite and emit light with shorter wavelength. QDs have broad absorption spectra but narrow and stable fluorescent emission spectra. Thus QDs are especially suitable for multiplex detection using one excitation light source. The emission spectrum of QDs is brighter than an organic fluorophore and up to thousands of times more stable than conventional dyes [47,138].

Noble metal nanomaterials have been widely used in sensors mainly due to their enhanced LSPR, which depends on the size, shape, composition, and separation distance of nanoparticles, as well as the dielectric environment of the surrounding medium [139]. The high conductivity of noble metal nanoparticles also promotes the electron transfer between electrode surface and the analyst [140]. The presence of enhanced LSPR leads to high molar extinction coefficient and Rayleigh scattering, as well as enhanced local electromagnetic fields near the nanoparticle surface. Based on theoretical calculation, nano-Au spheres of w40 nm in diameter have an absorption cross-section with five orders of magnitude higher than conventional dyes, while 80-nm nano-Au spheres scatter light five orders of magnitude more than fluorescence dyes [141]. Noble metals were used mainly in colorimetric and surface-enhanced Raman spectroscopy (SERS) sensing. Colorimetric assays are fast and simple. The signal transduction relies on the color change of the nanoparticle suspension due to different interparticle distance or aggregation state [142]. It has been studied for detection of DNA, diagnosis of pathogen infection, and pathogen monitoring in water samples. However, the aggregation state of nanoparticles is sensitive to the solution chemistry and difficult to control. The coexisting water/wastewater constitutes will greatly affect results, reducing reproducibility. The SERS phenomenon is attributed to both electromagnetic effect and chemical mechanisms related to the charge transfer between the noble metals and the target molecules [143]. As a result, the efficiency of Raman scattering can be enhanced more than 1014 fold, which is even capable of detecting a single molecule [144].

Silica nanoparticles doped with either organic or inorganic luminescent dyes have been developed for ultrasensitive sensors. The large number of dye molecules confined in a single silica particle guarantees huge improvement in sensitivity. Moreover, the silica matrix protects the dye molecules from the external environment, largely suppressing photobleaching and photodegradation. The outstanding photostability makes dye-doped silica nanoparticles especially advantageous for applications that require high intensity or prolonged excitations [138]. The rich silica chemistry (e.g., silane chemistry) also helps future surface modification and conjugation.

The high conductivity along the length makes CNTs outstanding electrode materials. As a result, CNTs can greatly facilitate electrochemical detection by promoting electron transfer and electrode—analysts interactions [145]. They have been incorporated into electrodes via random or aligned coating, or used as a single CNT electrode [146]. Semiconducting CNTs can be used in nanoscale field effect transistors [147]. Besides their excellent electronic properties, the high adsorption capacity of CNTs increases detection sensitivity [66]. The major challenge for CNT-based sensors is the heterogeneity of CNTs. Separation of metallic and semiconducting SWNT has been extensively studied but still far from perfect. The production and purification processes of CNTs often introduce impurities, contaminants, and even degradation of the CNT structure. Therefore, better synthesis, purification, and separation are required to produce more homogeneous CNTs. Although most of these nanosensors possess excellent photostability and sensitivity, nonspecific binding is still a major challenge for their application in water and wastewater. Strategies to reduce nonspecific binding and prevent undesired nanoparticle aggregation are in critical need.

6.2.22 Trace Contaminant Detection

In trace organic or inorganic contaminant detection, nanomaterials can be used in both concentration and detection. CNTs have great potential for environmental analysis of trace metal or organic pollutants as they offer high adsorption capacity and recovery rate as well as fast kinetics as discussed above. The preconcentration factors for metal ions were found to be between 20 and 300 with fast adsorption kinetics [148]. CNTs have also been extensively studied for preconcentrating a variety of organic compounds, many of which were done in real water samples [149]. Adsorption of charged species to CNTs results in changes of conductance, providing the basis for the correlation between analyte concentration and current fluctuation [100]. Other nanomaterials such as nano-Au and QDs have also been used. Nano-Au was used to detect pesticides at ppb levels in a colorimetric assay; modified nano-Au was shown to detect Hg^{2+} and CH_3Hg^+ rapidly with high sensitivity and selectivity [43,150]. QD-modified TiO_2 nanotubes lowered the detection limits of PAHs to the level of pica-mole per liter based on fluorescence resonance energy transfer [151]. A nanosensor based on CoTe QDs immobilized on a glassy carbon electrode surface was reported to detect bisphenol A in water at concentrations as low as w10 nM within 5 s [152].

6.2.23 Multifunctional Devices

The advance in functional nanomaterials and their convergence with conventional technologies bring opportunities in designing a new family of nanotechnology-enabled multifunctional water treatment devices which are capable of performing multiple tasks in one device. Such multifunctional systems can enhance the overall performance and avoid excessive redundancy, miniaturizing the footprint. Therefore, the multifunctional concept is especially advantageous in decentralized and small-scale applications. Different functional nanomaterials can be integrated onto a common platform based on treatment requirement. Beside magnetic nanoparticles, membranes are a good and extensively studied platform to construct multifunctional devices. Notably, electrospun nanofibers have drawn much attention as an excellent nanomaterial carrier. Owing to the high performance, small footprint, and modular design of nanotechnology-enabled devices, it is envisaged that different functionalities can be assembled in layers of a cartridges or as modules arranged in series, allowing optimization/regeneration of each functionality separately [21]. The capacity and functionality of such nanotechnology-enabled systems can be easily manipulated by plugging in or pulling out modules.

6.2.24 Retention and Reuse of Nanomaterials

The retention and reuse of nanomaterials is a key aspect of nanotechnology-enabled device design due to both cost and public health concerns. It can be usually achieved by applying a separation device or immobilizing nanomaterials in the treatment system. A promising separation process is membrane filtration which allows continuous operation with small footprint and chemical use. Ceramic membranes are more advantageous than polymeric membranes in photocatalytic or catalytic ozonation applications as they are more resistant to UV and chemical oxidants [93]. The suspended particles in the receiving water are detrimental to reactor membrane hybrid systems as they can be retained by the membrane and significantly reduce the reaction efficiency. Thus, raw water pretreatment is usually required to reduce the turbidity. Nanomaterials also can be immobilized on various platforms such as resins and membranes to avoid further separation. However, current immobilization techniques usually result in significant loss of treatment efficiency. Research is needed to develop simple, low-cost methods to immobilize nanomaterials without significantly impacting its performance. For magnetic nanoparticles/nanocomposites, low-field magnetic separation is a possible energy-efficient option.

Little is known about the release of nanomaterials from nanotechnology-enabled devices. However, the potential release is expected to be largely dependent on the immobilization technique and the separation process employed. If no downstream separation is applied, nanomaterials coated on treatment system surfaces are more likely to be released in a relatively fast and complete manner, while nanomaterials embedded in a solid matrix will have minimum release until they are disposed of. For nanomaterials that release metal ions, their dissolution needs to be carefully controlled (e.g., by coating or optimizing size and shape). The detection of nanomaterial release is a major technical hurdle for risk assessment and remains challenging. Details regarding detection techniques are beyond the scope of this paper, and readers are referred to several recent reviews on this topic [153,154]. Few techniques can detect nanomaterials in complex aqueous matrices and they are usually sophisticated, expensive, and with many limitations. Fast, sensitive and selective nanomaterial analytical techniques are in great need.

6.2.25 Barriers and Research Needs

Although nanotechnology-enabled water/wastewater treatment processes have shown great promise in laboratory studies, their readiness for commercialization varies widely. Some are already on the market, while others require significant research before they can be considered for full-scale applications. Their future development and commercialization face a variety of challenges including technical hurdles, cost-effectiveness, and potential environmental, and human risk. There are two major research needs for full-scale applications of nanotechnology in water/wastewater treatment. First, the performance of various nanotechnologies in treating real natural and wastewaters needs to be tested. Future studies need to be done under more realistic conditions to assess the applicability and efficiency of different nanotechnologies as well as to validate nanomaterial-enabled sensing technologies. Secondly, the long-term efficacy of these nanotechnologies is largely unknown as most lab studies were conducted for relatively short period of time. Research addressing the long-term performance of water and wastewater treatment nanotechnologies is in great need. As a result, side-by-side comparison of nanotechnology-enabled systems and existing technologies is challenging.

Despite the superior performance, the adoption of innovative technologies strongly depends on the cost-effectiveness and the potential risk involved. The current cost of nanomaterials is prohibitively high with few exceptions such as nano-TiO_2, nanoscale ion oxide, and polymeric nanofibers. There are currently two approaches to address the cost issue. One proposed approach is to use low-purity nanomaterials without significantly compromising efficiency as much of the production cost is related to separation and purification [21]. Alternatively, the cost-effectiveness can be improved by retaining and reusing nanomaterials. Nanomaterials possess unique challenges for risk assessment and management as they are small particles instead of molecules or ions for which risk assessment framework and protocols are already in place. Better understanding and mitigating potential hazards associated with the use of nanomaterials in water and wastewater treatment will lead to broader public acceptance, which is crucial for new technology adoption. The compatibility between aforementioned nanotechnologies and current water and wastewater treatment processes and infrastructure also needs to be addressed. Most treatment plants and distribution systems in developed countries are expected to remain in place for decades to come. As a result, it is important to be able to implement nanotechnology with minimal changes to existing infrastructure in the near term. In the meantime, nanotechnology-enabled treatment processes can be employed in places where water treatment infrastructure does not exist or in POU devices.

6.2.26 Conclusions

Nanotechnology for water and wastewater treatment is gaining momentum globally. The unique properties of nanomaterials and their convergence with current treatment technologies present great opportunities to revolutionize water and wastewater treatment. Although many nanotechnologies highlighted in this review are still in the laboratory research stage, some have made their way to pilot testing or even commercialization. Among them, three categories show most promise in full-scale application in the near future based on their stages in research and development, commercial availability, cost of nanomaterials involved, and compatibility with the existing infrastructure: nanoadsorbents, nanotechnology enabled membranes, and nanophotocatalysts. All three categories have commercial products, although they have not been applied in large-scale water or wastewater treatment. Several other water treatment nanotechnologies have found their niche applications in POU systems.

The challenges faced by water/wastewater treatment nanotechnologies are important, but many of these challenges are perhaps only temporary, including technical hurdles, high cost, and potential environmental and human risk. To overcome these barriers, collaboration between research institutions, industry, government, and other stakeholders is essential. It is our belief that advancing nanotechnology by carefully steering its direction while avoiding unintended consequences can continuously provide robust solutions to our water/wastewater treatment challenges, both incremental and revolutionary.

ACKNOWLEDGMENTS

This topic is written by Xiaolei Qu, Pedro J.J. Alvarez, and Qilin Li at the Department of Civil and Environmental Engineering, Rice University, Houston, TX 77005, USA, and it was published as "Applications of nanotechnology in water and wastewater treatment" in Water Research. 47 (2013) 3931–3946.

6.3 RECENT DEVELOPMENTS IN PHOTOCATALYTIC WATER TREATMENT TECHNOLOGY

In recent years, semiconductor photocatalytic process has shown a great potential as a low-cost, environmental friendly, and sustainable treatment technology to align with the "zero" waste scheme in the water/wastewater industry. The ability of this advanced oxidation technology has been widely demonstrated to remove persistent organic compounds and microorganisms in water. At present, the main technical barriers that impede its commercialization remained on the postrecovery of the catalyst particles after water treatment. This paper reviews the recent R&D progresses of engineered-photocatalysts, photoreactor systems, and the process optimizations and modelings of the photooxidation processes for water treatment. A number of potential and commercial photocatalytic reactor configurations are discussed, in particular the photocatalytic membrane (PM) reactors. The effects of key photoreactor operation parameters and water quality on the photoprocess performances in terms of the mineralization and disinfection are assessed. For the first time, we describe how to utilize a multivariable optimization approach to determine the optimum operation parameters so as to enhance process performance and photooxidation efficiency. Both photomineralization and photodisinfection kinetics and their modelings associated with the photocatalytic water treatment process are detailed. A brief discussion on the life cycle assessment (LCA) for retrofitting the photocatalytic technology as an alternative waste treatment process is presented. This paper will deliver a scientific and technical overview and useful information to scientists and engineers who work in this field.

6.3.1 Introduction

Increasing demand and shortage of clean water sources due to the rapid development of industrialization, population growth, and long-term droughts have become an issue worldwide. With this growing demand, various practical strategies and solutions have been adopted to yield more viable water resources. The storage of rainwater for daily activities and increasing the catchment capacity for stormwater are just a few examples that could resolve the problems in short term. Water industries and governments in some arid areas with abundant of sunlight, less rainfall, and long-term drought have a challenge to seek viable water resources. It is estimated that around 4 billion people worldwide experience to have no or little access to clean and sanitized water supply, and millions of people died of severe waterborne diseases annually [155]. These statistical figures are expected to grow in the short future, as increasing water contamination due to overwhelming discharge of micropollutants and contaminants into the natural water cycle [156—158]. In view to suppress the worsening of clean water shortage, development of advanced with low-cost and high efficient water treatment technologies to treat the wastewater is desirable. One of a few attractive options is the possible reuse of on-site rural wastewater or the treated municipal wastewater from treatment plants for agricultural and industrial activities [159,160]. Because these wastewaters constitute one of the largest possible water resources, its reuse is anticipated to offset more clean water resource. Recycling wastewaters are usually associated with the presence of suspended solids, health-threat coliforms, and soluble refractory organic compounds that are both tedious and expensive to treat [161]. Currently available water treatment technologies such as adsorption or coagulation merely concentrate the pollutants present by transferring them to other phases, but still remain and not being completely "eliminated" or "destroyed" [162]. Other conventional water treatment methods such as sedimentation, filtration, chemical and membrane technologies involve high operating costs and could generate toxic secondary pollutants into the ecosystem [163]. These concentrated toxic contaminants are highly redundant and have been concerned worldwide due to the increasing environmental awareness and legislations. Chlorination has been the most commonly and widely used disinfection process. The DBPs generated from chlorination are mutagenic and carcinogenic to human health [164–166]. These have lead to the rapid R&D in the field of "AOPs" as the innovative water treatment technologies. The rationales of these AOPs are based on the in situ generation of highly reactive transitory species (i.e., H_2O_2, $OH^{\cdot}$, $O_2^{\cdot -}$, O_3) for mineralization of refractory organic compounds, water pathogens, and DBPs [13,167]. Among these AOPs, heterogeneous photocatalysis employing semiconductor catalysts (TiO_2, ZnO, Fe_2O_3, CdS, GaP, and ZnS) has demonstrated its efficiency in degrading a wide range of ambiguous refractory organics into readily biodegradable compounds, and eventually mineralized them to innocuous carbon dioxide and water. Among the semiconductor catalysts, titanium dioxide (TiO_2) has received the greatest interest in R&D of photocatalysis technology. The TiO_2 is the most active photocatalyst under the photon energy of 300 nm < l < 390 nm and remains stable after the repeated catalytic cycles, whereas Cds or GaP are degraded along to produce toxic products [155]. Other than these, the multifaceted functional properties of TiO_2 catalyst, such as their chemical and thermal stability or resistance to chemical breakdown and their strong mechanical properties have promoted its wide application in photocatalytic water treatment. A number of important features for the heterogeneous photocatalysis have extended their feasible applications in water treatment, such as (1) ambient operating temperature and pressure, (2) complete mineralization of parents and their intermediate compounds without secondary pollution, and (3) low operating costs. The fact that the highly ROS generated as a result of the photoinduced charge separation on TiO_2 surfaces for microbial inactivation and organic mineralization without creating any secondary pollution is well documented. So far, the application of such TiO_2 catalysts for water treatment is still experiencing a series of technical challenges. The postseparation of the semiconductor TiO_2 catalyst after water treatment remains as the major obstacle toward the practicality as an industrial process. The fine particle size of the TiO_2, together with their large surface area-to-volume ratio and surface energy, creates a strong tendency for catalyst agglomeration during the operation. Such particle agglomeration is highly detrimental in views of particles size preservation, surface area reduction, and its reusable life span. Other technical challenges include in the catalysts development with broader photoactivity range and its integration with feasible photocatalytic reactor system. In addition, the understanding of the theory behind the common reactor operational parameters and their interactions is also inadequate and presents a difficult task for process optimization. A number of commonly made mistakes in studying kinetic modeling on either the photomineralization or photodisinfection have also been seen over the years.

This review paper aims to give an overview of the understanding and development of photocatalytic water treatment technology, from fundamentals of catalyst and photoreactor development to process optimization and kinetics modeling, and eventually the water parameters that affects the process efficiency. A short outline of the feasible application of photocatalytic water technology via life cycle interpretation and the possible future challenges are also recommended.

6.3.2 Fundamentals and Mechanism of TiO_2 Photocatalysis

6.3.2.1 Heterogeneous TiO_2 Photocatalysis

The fundamentals of photophysics and photochemistry underlying the heterogeneous photocatalysis employing the semiconductor TiO_2 catalyst have been intensively reported in many literature [163,168]. The semiconductor TiO_2 has been widely utilized as a photocatalyst for inducing a series of reductive and oxidative reactions on its surface. This is solely contributed by the distinct lone electron characteristic in its outer orbital. When photon energy (hv) is greater than or equal to the band gap energy of TiO_2 is illuminated onto its surface, usually 3.2 eV (anatase) or 3.0 eV (rutile), the lone electron will be photoexcited to the empty conduction band in femtoseconds. Fig. 6.6 depicts the mechanism of the electron—hole pair formation when the TiO_2 particle is irradiated with adequate hv. The light wavelength for such photon energy usually corresponds to $l < 400$ nm. The photonic excitation leaves behind an empty unfilled valence band, and thus creating

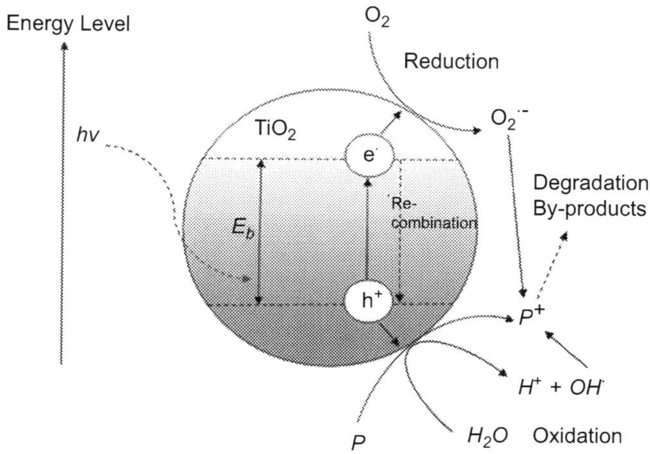

Figure 6.6 Photoinduced formation mechanism of electron–hole pair in a semiconductor TiO_2 particle with the presence of water pollutant (P).

the electron–hole pair (e^--h^+). The series of chain oxidative–reductive reactions (Eqs. 6.1–6.11) that occur at the photon-activated surface was widely postulated as follows:

$$\text{Photoexcitation: } TiO_2 + hv \rightarrow e^- + h^+ \tag{6.1}$$

$$\text{Charge-carrier trapping of } e^-: e^-_{CB} \rightarrow e^-_{TR} \tag{6.2}$$

$$\text{Charge-carrier trapping of } h^+: h^+_{VB} \rightarrow h^+_{TR} \tag{6.3}$$

$$\text{Electron-hole recombination: } e^-_{TR} + h^+_{VB}\left(h^+_{TR}\right) \rightarrow e^-_{CB} + \text{heat} \tag{6.4}$$

$$\text{Photoexcited } e^- \text{scavenging: } (O_2)_{ads} + e^- \rightarrow O_2^{\cdot-} \tag{6.5}$$

$$\text{Oxidation of hydroxyls: } OH^- + h^+ \rightarrow OH^{\cdot} \tag{6.6}$$

$$\text{Photodegradation by } OH^{\cdot}: R - H + OH^{\cdot} \rightarrow R'^{\cdot} + H_2O \tag{6.7}$$

Direct photoholes:

$$R + h^+ \rightarrow R^{+\cdot} \rightarrow \text{Intermediate(s)/Final Degradation products} \tag{6.8}$$

$$\text{Protonation of superoxides: } O_2^{\cdot-} + OH^{\cdot} \rightarrow HOO^{\cdot} \tag{6.9}$$

$$\text{Co-scavenging of } e^-: HOO^{\cdot} + e^- \rightarrow HO_2^- \tag{6.10}$$

$$\text{Formation of } H_2O_2: HOO^- + H^+ \rightarrow H_2O_2 \tag{6.11}$$

The e^-_{TR} and h^+_{TR} in Eq. (6.4) represent the surface-trapped valence band electron and conduction-band hole, respectively. It was reported that these trapped carriers are usually TiO_2 surface bounded and do not recombine immediately after photon excitation [169]. In the absence of electron scavengers (Eq. 6.4), the photoexcited electron recombines with the valence band hole in nanoseconds with simultaneous dissipation of heat energy. Thus, the presence of electron scavengers is vital for prolonging the recombination and successful functioning of photocatalysis. Eq. (6.5) depicts how the presence of oxygen prevents the recombination of electron—hole pair, while allowing the formation of superoxides radical ($O_2^{\cdot-}$). This $O_2^{\cdot-}$ radical can be further protonated to form the hydroperoxyl radical ($HO_2^{\cdot}$) and subsequently H_2O_2 as shown in Eqs. (6.9) and (6.10), respectively. The $HO_2^{\cdot}$ radical formed was also reported to have scavenging property, and thus, the coexistence of these radical species can doubly prolong the recombination time of the h^+_{TR} in the entire photocatalysis reaction. However, it should be noted that all these occurrences in photocatalysis were attributed to the presence of both dissolved oxygen (DO) and water molecules. Without the presence of water molecules, the highly reactive hydroxyl radicals ($OH^{\cdot}$) could not be formed and impede the photodegradation of liquid phase organics. This was evidenced from a few reports that the photocatalysis reaction did not proceed in the absence of water molecules. Some simple organic compounds (e.g., oxalate and formic acid) can be mineralized by direct electrochemical oxidation where the e^-_{TR} is scavenged by metals ions in the system without water presence [170]. Although the h^+_{TR} has been widely regarded for its ability to oxidize organic species directly, this possibility is remained inconclusive. The h^+_{TR} are powerful oxidants (+1.0 to +3.5 V against NHE), while e^-_{TR} are good reductants (+0.5 to −1.5 V against NHE), depending on the type of catalysts and oxidation conditions.

Many elementary mechanistic studies on different surrogate organic compounds (e.g., phenol, chlorophenol, oxalic acid) have been extensively investigated in the photodegradation over TiO_2 surface. Aromatic compounds can be hydroxylated by the reactive $OH^{\cdot}$ radical that leads to successive oxidation/addition and eventually ring opening. The resulting intermediates, mostly aldehydes and carboxylic acids, will be further carboxylated to produce innocuous carbon dioxide and water. Because the photocatalysis reaction occurs on the photon-activated surface of TiO_2, the understanding of the reaction steps that involve photodegradation of organics is essential in the formulation of kinetic expression. For heterogeneous photocatalysis, the liquid phase organic compounds are degraded to its

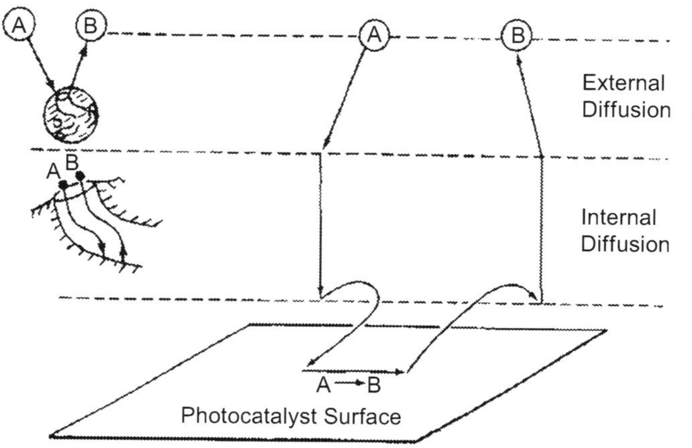

Figure 6.7 Steps in heterogeneous catalytic reaction [15].

corresponding intermediates and further mineralized to carbon dioxide and water, if the irradiation time is extended (Eq. 6.12).

$$\text{Organic Contaminants} \xrightarrow{TiO_2/hv} \text{Intermediate(s)} \rightarrow CO_2 + H_2O \qquad (6.12)$$

The overall photocatalysis reaction as portrayed by Eq. (6.12) can be divided into five independent steps, which are shown in Fig. 6.7 [15,171].

1. Mass transfer of the organic contaminant(s) (e.g., A) in the liquid phase to the TiO_2 surface.
2. Adsorption of the organic contaminant(s) onto the photon-activated TiO_2 surface (i.e., surface activation by photon energy occurs simultaneously in this step).
3. Photocatalysis reaction for the adsorbed phase on the TiO_2 surface (e.g., $A \rightarrow B$).
4. Desorption of the intermediate(s) (e.g., B) from the TiO_2 surface.
5. Mass transfer of the intermediate(s) (e.g., B) from the interface region to the bulk fluid.

In terms of rate determination, the overall rate of reaction is equal to the slowest step. When the mass transfer steps (1 and 5) are very fast compared with the reaction steps (2, 3, and 4), the organic concentrations in the immediate vicinity of the active sites are indistinguishable from those in the bulk liquid phase. In this scene, the mass transfer steps are not rate limiting and do not affect the overall rate of photocatalytic reaction. Vinodgopal and Kamat [172] reported the dependence of the photodegradation rate of the organic surrogate on surface coverage of the photocatalysts used.

This outlines the importance of molecule adsorption or surface contact with the catalyst during the photocatalytic degradation. If the mass transfer steps are rate limiting, a change in the aeration or liquid flow conditions past the TiO_2 photocatalyst may alter the overall photocatalytic reaction rate.

Similarly, the surface interaction of microorganisms with the catalyst used during the photodisinfection is essential for enhancing the inactivation rate. When the generated ROS contacts closely with the microorganisms, the cell wall will be the initial site of attack [173]. The lipopolysaccharide layer of the cell external wall is the initial site attacked by the photoinduced ROS. This is followed by the site attack on the peptidoglycan layer, peroxidation of the lipid membrane and the eventual oxidation on the proteins membrane. All these will cause a rapid leakage of potassium ions from the bacterial cells, resulting in direct reduction of cell viability. The decrease in cell viability is usually linked to the peroxidation of polyunsaturated phospholipid components of the cell membrane (i.e., loss of essential cell functions) and eventually leads to cell death. The formation of oxidative stress and its effects on the cell membrane can be observed using advanced atomic force microscopy or attenuated total reflection Fourier transform infrared spectroscopy. The rate of adsorption and the eventual photoinactivation is known to positively correlate to the bactericidal effect of TiO_2 catalyst. In this instance, the transfer of bacterial cell to the close vicinity of the surface-generated ROS site remains as the rate-limiting step in the photodisinfection reaction.

6.3.2.2 Homogeneous Photo-Fenton Reaction

The Fenton reaction is a process that does not involve any light irradiation as compared with the heterogeneous TiO_2 photocatalysis reaction, whereas the photo-Fenton does react up to a light wavelength of 600 nm. It was first recognized in the 1960s and remains one of the most applied AOPs for its ability to degrade high loading of organic compounds in highly saline conditions [7,174,175] Numerous studies on the photo-Fenton degradation of water pollutants such as chlorophenol, pesticides, and phenolic or aromatic compounds with organic loading of up to 25 g/L has been investigated [22,167,176–178]. A number of literature have provided a comprehensive review of the basic understanding and clarity of the principles underlying the Fenton reaction [175,179,180].

In the absence of a light source, hydrogen peroxide (H_2O_2) will decompose by Fe^{2+} ions that present in the aqueous phase, resulting in the formation of hydroxyl radicals. The photo-Fenton reaction is expedited when light source present, causing rapid H_2O_2 decomposition by ferrous or ferric ions and resulting in the formation of radicals. All these soluble iron hydroxy or iron complexes can absorb not only UV radiation but also visible light. However, the actual oxidizing species responsible for the photo-Fenton reaction is still under discussion [12,18]. These Fenton and photo-Fenton reaction could occur simultaneously with TiO_2 photocatalysis during UV–Vis irradiation period, post TiO_2 photocatalysis period or stand-alone photo-Fenton process. The Fenton reaction is seen to strongly correlate with the post TiO_2 photocatalysis reaction and thus is described in detail here. The mechanism for the Fenton reaction is shown in Eq. (6.13):

$$Fe^{2+}(aq) + H_2O_2 \rightarrow Fe^{3+}(aq) + OH^- + HO^{\cdot} \quad (6.13)$$

The Fe^{2+} can be reverted back to Fe^{3+} via different mechanisms:

$$Fe^{3+}(aq) + H_2O_2 \rightarrow Fe^{2+}(aq) + HO_2^{\cdot} + H^+ \quad (6.14)$$

$$Fe^{3+}(aq) + HO_2^{\cdot} \rightarrow Fe^{2+}(aq) + O_2 + H^+ \quad (6.15)$$

When a light source is present, the rate of photo-Fenton was reported to be positively enhanced compared with the dark condition. This is mainly due to the regeneration of Fe^{2+} (aq) from the photochemical effect of light and the concurrent generation of the $OH^{\cdot}$ radicals in the system. Such a reversion cycle of $Fe^{2+}(aq) \rightarrow Fe^{3+}(aq) \rightarrow Fe^{2+}(aq)$ continuously generates $HO^{\bullet}$, provided that the concentration of H_2O_2 in the system is substantial. The regeneration of the Fe^{2+} (aq) from Fe^{3+} (aq) is the rate-limiting step in the catalytic iron cycle, if small amount of iron is present. This photo-assisted reaction is termed as photo-Fenton reaction, where such reaction could be activated by irradiation wavelengths of up to 600 nm. It was known that this reaction is better functional under longer wavelengths as they are able to overcome the inner filter effects by photolyzing the ferric iron complexes.

Even if the photo-Fenton has higher photoactivity than the heterogeneous photocatalysis, its feasible operation is largely dependent on several water quality parameters. In the photo-Fenton reaction, the formation of the highly photoactive iron complexes is highly dependent on the water pH and ions content [181]. It was reported that the pH 2.8 was the frequent optimum pH for photo-Fenton reaction [8,18]. This is owing to the fact that at such low pH 2.8, the precipitation does not take place and further promotes the presence of dominant iron species of $[Fe(OH)]^{2+}$ in water. Such a low optimum pH 2.8, however, is not cost-effective for operation as it requires high chemical costs for pH rectification. The presence of different ions such as carbonate (CO_3^{2-}), phosphate (PO_4^{3-}), sulfate (SO_4^{2-}), and chlorine (Cl^-) also affects the iron equilibrium in water. These ions have the potential to raise the water pH and effectively lowered the photo-Fenton reaction rate. Both CO_3^{2-} and PO_4^{3-} have a double detrimental effect on the reaction, as they precipitate the iron and as well as scavenges the OH^- radicals. A higher pH of 4.0–5.0 was determined to be sufficient to sustain the photo-Fenton reaction with 2–6 mM of iron for the initiation of the treatment [177]. To date, the maximal iron loading reported was 450 mg/L [182,183].

Although H_2O_2 may be generated via the TiO_2 photocatalysis (Eq. 6.11), its relative amount in the system may be inadequate to drive the Fenton reaction. Many researchers have reported the addition of H_2O_2 in enhancing both the photo-Fenton and TiO_2 photocatalysis reactions. The H_2O_2 can inhibit the recombination of electron–hole pair, while further provides additional OH^- radicals through the following mechanisms:

$$H_2O_2 + e^- \rightarrow HO^{\cdot} + HO^- \tag{6.16}$$

$$O_2^{\cdot -} + H_2O_2 \rightarrow O_2 + HO^{\cdot} + HO^- \tag{6.17}$$

This combined TiO_2 photocatalysis and photo-dark-Fenton reaction is particularly useful for the disinfection process [184–186]. The addition of H_2O_2 to the photocatalysis and dark-Fenton system results in a residual disinfection to avoid microbial regrowth. Rincón and Pulgarin [187] performed trials with TiO_2 photocatalysis and photo-Fenton reaction for the disinfection of water contaminated with *E. coli*. They found that the bacterial inactivation rate was higher than the photocatalysis alone and the decrease in bacterial number continued in the dark conditions without significant regrowth within the following 60 h. However, it was found that such residual disinfection effect was highly dependent on the light intensity used during the irradiation period, as well as the relative concentrations of Fe^{3+} and H_2O_2. Further addition of H_2O_2 was found to decrease the overall reaction rate in several studies, owing to the formation of less penetrative $HO_2^{\cdot}$ radicals, as described by Eq. (6.18):

$$HO^{\cdot} + H_2O_2 \rightarrow HO_2^{\cdot} + H_2O \tag{6.18}$$

Other combined process of photo-Fenton and oxidative processes have also been proposed in the literature, such as ozone and ultrasound [16,188].

However, their significance as compared to the TiO_2/photo-Fenton will not be discussed in detail.

6.3.2.3 Advancements in Photocatalyst Immobilization and Supports

Since the discovery of photocatalytic effect on water splitting by Fujishima and Honda [105] using TiO_2 electrode, numerous research studies have evolved to synthesis TiO_2 catalyst of different scale, characterize its physical properties, and determine its photooxidation performances to the surface-oriented nature of photocatalysis reaction [30,189–192]. The TiO_2 catalyst in nanodimensions allows having a large surface area-to-volume ratio and can further promote the efficient charge separation and trapping at the physical surface [193,194]. The light opaqueness of this nanoscale TiO_2 catalysts was reported to have an enhanced oxidation capability compared with the bulk TiO_2 catalysts [195]. Although the nanoscale TiO_2 catalysts show considerable improvement in terms of their physical and chemical properties, their particle size and morphology remains the main problem in a large-scale water treatment process [196,197]. In this section, the current technical challenges that prevent the application of slurry TiO_2 photocatalytic system are discussed together with the possible engineering solutions to resolve the problem. We will have a brief discussion on the modified TiO_2 catalyst with dopants for enhanced photoactivity under solar irradiation.

6.3.2.4 Challenges in the Development of Photocatalytic Water Treatment Process

To date, the most widely applied photocatalyst in the research of water treatment is the Degussa P-25 TiO_2 catalyst. This catalyst is used as a standard reference for comparisons of photoactivity under different treatment conditions [198]. The fine particles of the Degussa P-25 TiO_2 have always been applied in a slurry form. This is usually associated with a high volumetric generation rate of ROS as proportional to the amount of surface active sites when the TiO_2 catalyst in suspension [199]. On the contrary, the fixation of catalysts into a large inert substrate reduces the amount of catalyst-active sites and also enlarges the mass transfer limitations. Immobilization of the catalysts results in increasing the operation difficulty as the photon penetration might not reach every single surface site for photonic activation [199]. Thus, the slurry type of TiO_2 catalyst application is usually preferred.

With the slurry TiO_2 system, an additional process step would need to be entailed for postseparation of the catalysts. This separation process is crucial to avoid the loss of catalyst particles and introduction of the new pollutant of contamination of TiO_2 in the treated water [200]. The catalyst recovery can be achieved through process hybridization with conventional sedimentation, cross-flow filtration, or various membrane filtrations [201–204]. Coupled with the pH control strategy close to the isoelectric point for induced coagulation, it was reported that the microfiltration (MF) hybridization can recover the remaining 3% of the catalyst particles for reuse [155]. Several important operating issues with slurry TiO_2 still remain even with a membrane integration process. These include the types of membrane, pore size, and blockage, regeneration or backwashing and fouling [205–207]. A number of studies have utilized micron-sized immobilizers for catalyst fixation that enhance surface contact with contaminants and prevent the membrane fouling or pores blocking with rapid backwashing [207,208]. These immobilizers include catalyst fixation onto activated carbon, mesoporous clays, fibers, or even onto the membrane itself [209–212]. The following subsections outline a few catalyst immobilization strategies that are suitable for the use of slurry reactor or membrane reactor or both.

6.3.2.5 Mesoporous Clays

Natural clays have been used intensively as the support for TiO_2, owing to their high adsorption capacity and cost-effectiveness. Fig. 6.8 shows the TiO_2 crystal being deposited on a clay material [209]. Different types of clays have been investigated, which includes bentonite, sepiolite, montmorillonite, zeolite, and kaolinite [209,213–216]. Although these clays are catalytically inactive, their superior adsorption capacity has been attractive for increasing the surface contact during photocatalysis reaction. It was proposed that the natural clays should not be used directly to immobilize TiO_2. This is owing to the presence of different surface- or lattice-bounded impurities that might diffuse and further affect the TiO_2 efficiency of the immobilized layer [209]. In addition, if these impurities are not removed, the polar molecules in the aqueous environment might initiate an internal reaction within the clay structure that result in clay swelling [209]. The swelling will be profound in certain type of clays, where Van der Waals forces held the entire clay in a turbostatic array. This is undesirable, particularly if the photocatalytic reactions take place in a reactor where the hydrodynamics may be strongly affected and consequently leading to the loss of photoactivity. Other factors that might need to be taken into

Figure 6.8 Nanocrystal of TiO_2 deposited on clay materials by SEM imaging. (A) 10 μm resolution; (B) 3 μm resolution.

a consideration if pillared clays are used as the immobilizer substrate include the density of the clays, particle size distribution range, and the complementary photoreactor system used. The use of mesoporous clays as the support for nanosized TiO_2 have been successfully demonstrated in a number of studies, including the slurry or membrane processes [217,218].

6.3.2.6 Nanofibers, Nanowires, or Nanorods

Glass, optical, carbon, titanate, and woven cloths fibers have also been studied as support materials in the photooxidation of various organic contaminants for water purification [199]. Most of these fibers have a protruded rod shape or longitudinal morphology. Using the nanofibers, nanowires, or nanorods, mass transfer limitation can be resolved by their thin longitudinal morphology. The use of less durable immobilizer fibers (e.g., of glass or woven cloths), however, may lead to a low durability as the deposited anatase crystals might wear-off, resulting in a loss of photoactivity over reaction time. Such immobilizer fibers also increase pressure drop in the reactor system. On the contrary, a good benefit of nanofibers with commercial success is that they can be fabricated into MF, UF, and PMs. The MF fiber membrane is of particular interest as it shows high-pollutant removal rate at low transmembrane pressure (<300 kPa). A commercial success of such fabricated MF and UF membranes has been demonstrated by Zhang and coworkers [203,219]. Further details on the fabricated nanofibers or nanowires MF membranes can be obtained in the literature.

6.3.2.7 Photocatalytic Membrane

Recently the use of the PMs has been targeted owing to the fact that photocatalytic reaction can take place on the membrane surface and the treated water could be continuously discharged without the loss of photocatalyst particles. The PMs can be prepared from different materials and synthesis methods. These include the TiO_2/Al_2O_3 composite membranes [61,220–223]. TiO_2 supported on polymer and metallic membranes or doted polymer membranes containing TiO_2 particles entrapped within the membrane structure during the membrane fabrication process [6,14,28,224–226]. Also, the possible TiO_2 organic and inorganic ceramic membranes have been investigated [210,224,227]. Fig. 6.9 shows different types of PMs for water treatment application [9,219]. In most studies, however, PMs may encounter various technical problems such as membrane structure deterioration, low photocatalytic activity, and loss of deposited TiO_2 layer over time. To prevent the problems associated with the TiO_2 membrane coating, an approach of using membranes without any deposited TiO_2 layer can be configured into a slurry membrane hybrid system, which will be outlined in Section 4.

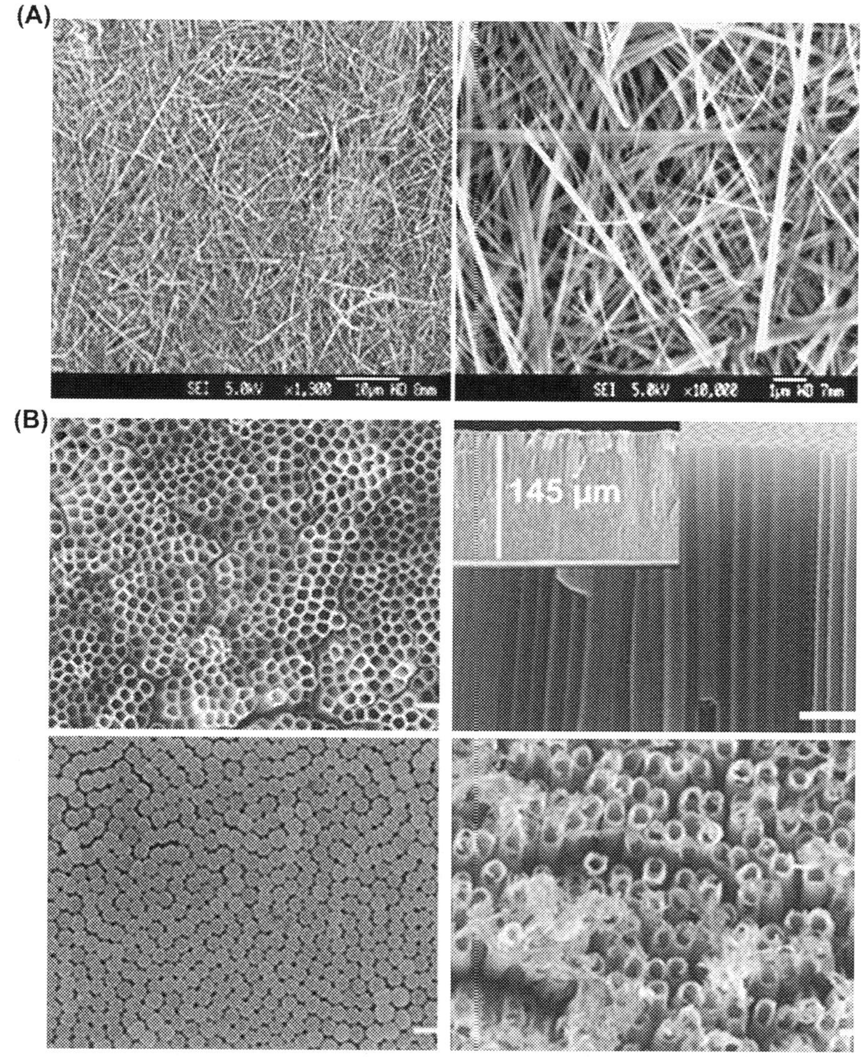

Figure 6.9 (A) FESEM images of the TiO_2 nanowire membrane [219]; (B) SEM images of TiO_2 nanotube layer formed in free-standing membrane [9].

6.3.2.8 Photocatalyst Modification and Doping

As TiO_2 photocatalytic reactions take place under ambient operating conditions, photoactivity is usually constrained by the narrow wavelength spectrum for photonic activation of catalysts. The higher-end of UV spectrum required for catalysts activation is usually accompanied by high operating costs. One attractive option is to utilize the vast abundance of outdoor solar irradiation for catalyst activation in a suitably designed photoreactor system. To broaden the photoresponse of TiO_2 catalyst for solar spectrum, various material engineering solutions have been devised, including composite photocatalysts with carbon nanotubes, dyed sensitizers, noble metals or metal ions incorporation, transition metals, and nonmetals doping [105,109,228–230].

The rationale in utilizing these material engineering strategies is to balance both the half-reaction rates of the photocatalytic reaction by adding electron acceptor, or modifying the catalyst structure and composition. Fig. 6.10 presents the use of different mechanisms to enhance the photoactivity of the catalysts. The presence of electron acceptors could scavenge the

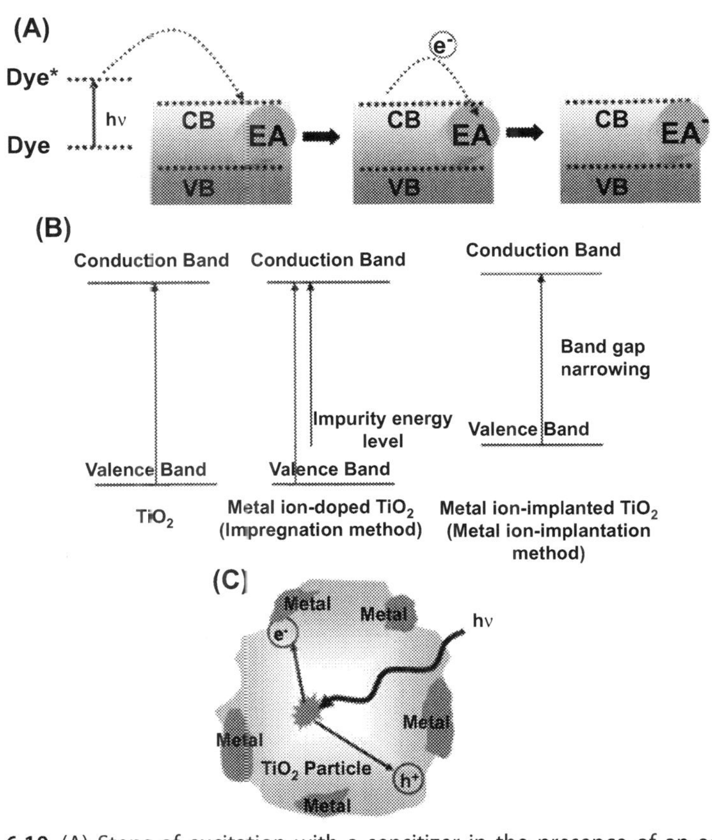

Figure 6.10 (A) Steps of excitation with a sensitizer in the presence of an adsorbed organic electron acceptor (EA); (B) Scheme of TiO_2 band structures, chemically ion-doped TiO_2, and physically ion-implanted TiO_2; (C) Electron capture by a metal in contact with a semiconductor surface [155].

excited electrons and altogether prevent the recombination of electron–hole pairs. Recent studies shown that modified TiO_2 catalysts have an enhanced photoactivity under solar irradiation [231–233]. CNTs coupling with TiO_2 have shown potential prolongation of electron–hole pairs by capturing the electron within their structure. As for dye-sensitized coupling, the excited dye molecules under solar illumination can provide additional electrons to the CB for enhancing the formation of electron–hole pairs (Fig. 6.10A) [229]. Dyes such as Methylene blue, Azure, Erythrosin, Rhodamin, and Crystal violet have been widely functionalized under solar irradiation [229].

Similarly, noble metals (e.g., Ag, Ni, Cu, Pt, Rh, and Pd) with Fermi level lower than TiO_2 catalyst have also been deposited on the TiO_2 surface for enhanced charge separation (Fig. 6.10B) [109]. These metals were reported to enhance electron transfer, but require good knowledge on the optimal deposited amount needed during the fabrication process. Although noble metals coupling could be efficient in prolonging the surface charge separation, their cost-effectiveness for an industrial application is usually replaced by more economical transition or nonmetal doping. The mechanism of such transition and nonmetals doping, however, is different from the noble metals coupling as the TiO_2 is incorporated into the TiO_2 crystal lattice [234–236]. Such incorporation introduces impurity in the band gap of TiO_2 and thus, reduces the photonic energy requirements (Fig. 6.10C). More recently, the use of nonmetal dopants (e.g., N, C, F, S, and etc.) can improve the photoactivity and feasibility of TiO_2 catalysts for industrial application [237,238]. Further research efforts may be needed to get a better understanding of the photoactivity kinetics, so as to improve the photooxidation efficiency for water treatment.

6.3.2.9 Photocatalytic Reactor Configuration

Photocatalytic reactors for water treatment can generally be classified into two main configurations, depending on the deployed state of the photocatalysts: (1) reactors with suspended photocatalyst particles and (2) reactors with photocatalyst immobilized onto continuous inert carrier [239]. Various types of reactors have been used in the photocatalytic water treatment, including the annular slurry photoreactor, cascade photoreactor, downflow contactor reactor, etc. [217,240,241]. The disparity between these two main configurations is that the first one requires an additional downstream separation unit for the recovery of photocatalyst particles while the latter permits a continuous operation.

Pareek et al. [242] addressed that the most important factors in configuring a photocatalytic reactor are the total irradiated surface area of catalyst per unit volume and light distribution within the reactor. Slurry-type photocatalytic reactor usually performs a high total surface area of photocatalyst per unit volume, while the fixed-bed configuration is often associated with mass transfer limitation over the immobilized layer of photocatalysts. The light photon distribution through either direct or diffuse paths within the reactors needs to be decided [18]. Direct photon utilization means that the photocatalysts are directly activated with light photon, rather with the assistance of various parabolic light deflectors to transfer the photons. To achieve uniformity in photon flux distribution within the reactor, a correct position of light source is essential to ensure maximal and symmetrical light transmission and distribution. The use of photoreactors with assisted parabolic light deflectors nowadays have become unfavorable, owing to the need of special configuration and high operating costs. This type of reactor needs to be specifically designed to ensure the maximal illuminated reactor volume with minimal pressure requirement for good catalyst mixing and dispersion. Until recently, the slurry photocatalytic reactor was still the preferred configuration owing to its high total surface area of photocatalyst per unit volume and ease of photocatalysts reactivation. The photocatalyst particles can be separated by settling tanks or external cross-flow filtration system to enable continuous operation of the slurry reactor. A technically promising solution for solving the downstream separation of photocatalyst particles after treatment is via the application of hybrid photocatalysis—membrane processes. Application of such a hybrid system prevents the use of a coagulation, flocculation, or sedimentation to separate the catalyst particles from the treated water stream. Other benefits include further energy saving and size of process installation and site area required.

The hybrid photocatalytic—membrane reactor system is generally known as the "photocatalytic membrane reactors" (PMRs). This is owing to the nature of the hybrid system where the membrane filtration unit could be configured into different positioning with the photocatalytic reactor. Fu et al. [243] designed a submerged membrane reactor (Fig. 6.11) with two different reaction zones; UV slurry TiO_2 zone with a movable baffle that separates the submerged membrane module. These PMRs can be generalized by (1) irradiation of the membrane module and (2) irradiation of a feed tank containing photocatalyst in suspension [206,244]. For the former configuration, the photocatalyst could be either deposited onto the membrane or suspended in the reaction water. The PMRs allow a continuous

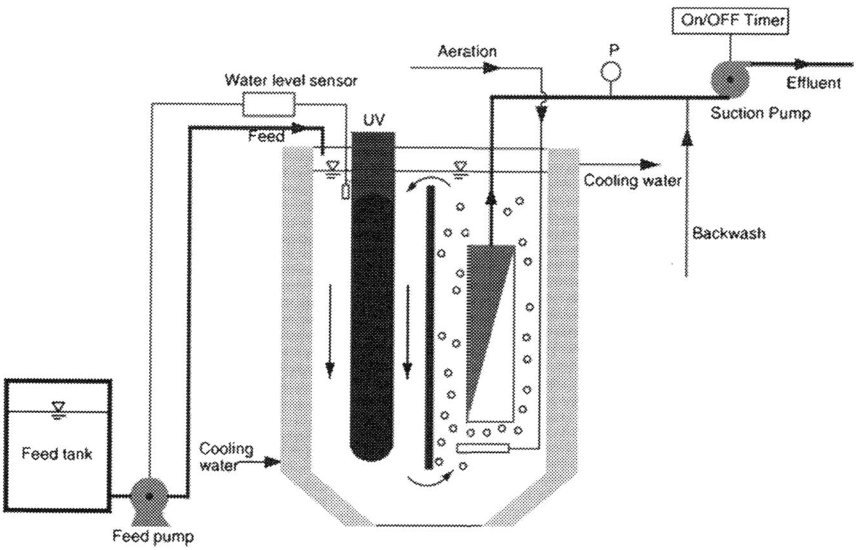

Figure 6.11 Schematic of submerged membrane photocatalytic reactor Fu et al. [243].

operation of the slurry-type reactor without any loss of photocatalyst particles as well as to control the water residence time independently. This enables the treated water to achieve the predefined level before being filtered through the hybrid membrane system. In the PMRs with immobilized PMs, the membrane module functionalized as the support for the photocatalyst particles and barrier against the different organic molecules in the reaction water. Similarly, the membrane also acts as a physical barrier against the photocatalyst particles and organic molecules or intermediate compounds to be degraded in the slurry PMRs.

In the PMRs with immobilized photocatalysts, the photocatalytic reaction takes place on the surface of the membrane or within its pores. The PMs used may be of MF [24,30,49,245–247], UF [206,218,238,248,249], and NF [4,206,226,244,250], depending on the targeted colloidal size and final water quality requirement. The MF membrane is useful when the colloidal size is in the range of 0.1–5 mm, while both UF and NF target a smaller particle size ranges. The photooxidation efficiency of the contaminants was reported to be higher when an immobilized PM was used, rather than in the case of PMRs with suspended catalyst particles [226]. It was, however, reported that immobilizing the photocatalyst particles might cause severe destruction to the membrane structure owing to their close contact with both UV light and hydroxyl radicals [93]. In view of this, the hybridization configuration of the membrane process using photocatalysts in suspension appears to be the more promising arrangement. This PMR configuration has been well described in the literature for water phase degradation of humic and fulvic acids, bisphenol A, phenol, 4-nitrophenol, 4-chlorophenol, gray water, para-chlorobenzoate, river water, and dyes [24,49,93,206,218,226,238,243,244,246–249,251].

With these PMRs, one of the main operational issues is the transmembrane pressure, which determines both the filtration rate and operating costs. It was known that the PMR treatment costs increase if the photocatalysts with small particle and colloidal size are used. With both the MF and UF membrane filtrations, the fine photocatalyst particles can cause membrane fouling and subsequently reduce membrane permeate flux. Fu et al. [243] utilized a spherical ball-shaped TiO_2 particle that promotes separation, recovery, and reused while prolonging the membrane life span as the particles do not cause pore blockage. Besides, the surface charge properties of the photocatalyst particles can also be manipulated to prevent membrane pore blockage. Xi and Geissen [207] integrated a thermoplastic membrane module of cross-flow MF and found that the low permeate flux occurred when the operating pH varied from the isoelectric point of the TiO_2 particles used. This is owing to the pH-induced coagulation–flocculation state of TiO_2 that declines the rate of permeate flux. This was resolved by maintaining the operating pH close to the isoelectric point of TiO_2 by adding certain electrolytes to the TiO_2 slurry. Even with such control strategies, the quality of permeate is low owing to the rapid penetration of small molecules through the membrane used.

Recently, different hybridization of PMRs with dialysis, pervaporation, and direct contact membrane distillation (MD) have been used (Fig. 6.12) [252–254]. Pervaporation is a physical process where usually a selective organophilic membrane was used to act as a selective barrier for the molecules to be degraded. Augugliaro et al. [255] observed that a synergistic effect occurs when pervaporation was used, where the intermediates from the degradation of 4-chlorophenol (i.e., hydroquinone, benzoquinone) could be selectively permeable without competing with 4-chlorophenol for photocatalytic reaction. Other types of organophilic membrane have also been investigated, such as polymeric UF membranes and polyethersulfonate NF membranes [75,256]. A stronger rejection impact on the membrane is usually associated with the use of the UF or NF membranes. The choice of membrane for an efficient hybridization depends on the organic molecular size, pH, or electrostatic interaction and Donnan exclusion phenomenon [257].

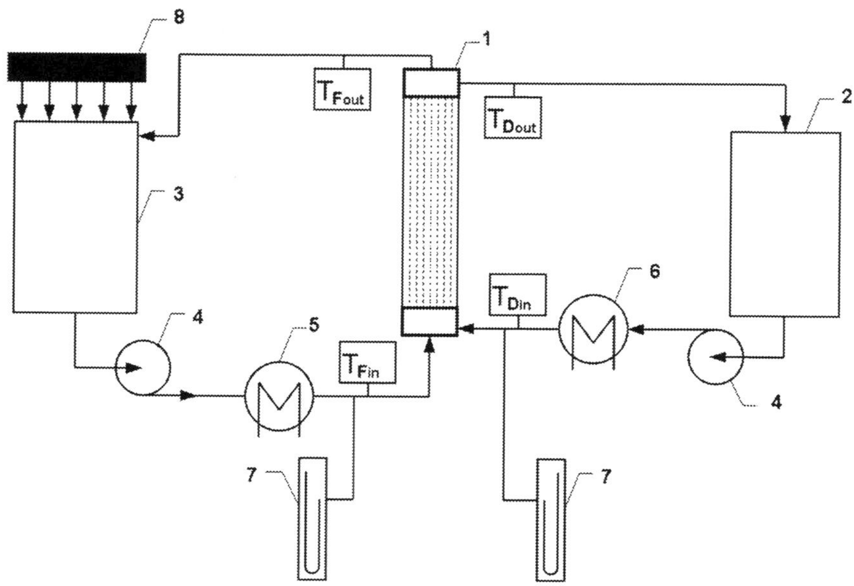

Figure 6.12 Schematic diagram of the apparatus for hybrid photocatalysis–membrane distillation process: (1) membrane module; (2) distillate tank; (3) feed tank (V [2.9 dm^3]); (4) pump; (5) and (6) heat exchangers; (7) manometers; (8) UV lamp; T_{Fin}, T_{Din}, T_{Fout}, T_{Dout}—inlet and outlet temperatures of feed and distillate, respectively [252].

The MD is a process where the feed volatile components in water are evaporated through a porous hydrophobic membrane to produce high-quality distillate products [252]. Its main advantage is that there is no membrane fouling when TiO_2 is present. During the process, the volatile stream is maintained inside the membrane pores. The difference in vapor pressure on both sides of the porous membrane remains as a driving force for the process. This force, however, largely depends on the temperatures and solution composition in the layer adjacent to the membrane [258,259]. It was reported that the feeding temperature in MD can range from 303 to 363K. Similar scaling-up operational constraints of low permeate flux and high energy demand have redundant its current full-scale development. However, some efforts to utilize alternative energy sources such as solar energy to enable MD application were found in the literature [10,260].

Among all the hybrid PMR systems, the pilot Photo-CatTM system (Fig. 6.13) (manufactured by Purifics Inc., Ontario, London) has shown the potential application. Benotti et al. [261] has evaluated its ability in the removal of 32 pharmaceuticals, endocrine-disrupting compounds, and

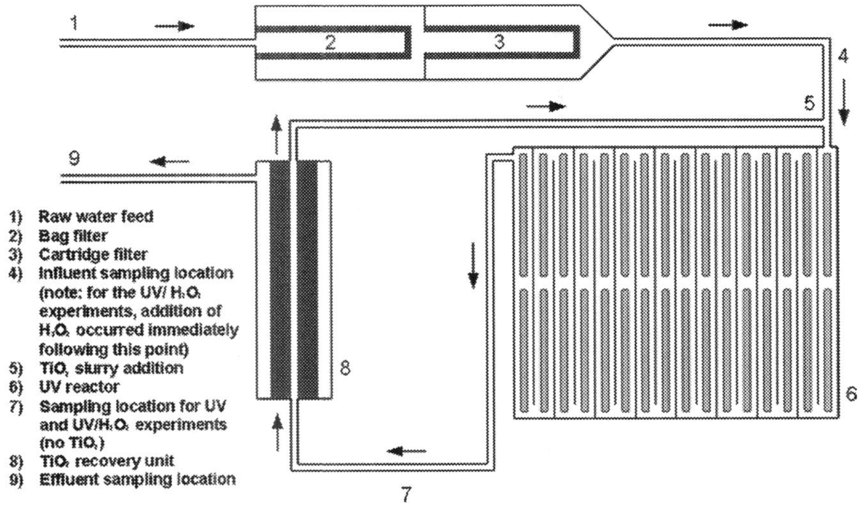

Figure 6.13 General schematic of photocatalytic reactor membrane pilot system [261].

estrogenic activity from water. They found that 29 targeted compounds and estrogenic activity of greater than 70% were removed, while only three compounds were less than 50% removed at the highest number of UV passes. In the Photo-CatTM system, the water stream passes through a pre-filter bag and a cartridge filter before being mixed with a nanoparticle TiO_2 slurry stream. The mixed stream then passes through the reactor within the 3 mm annulus of the 32 UV lamps aligned in series, which can be individually controlled for the varying water quality. The overall hydraulic residence time for the 32 UV passes between 1 and 32 s, depending on the number of UV lights being turned on. A cross-flow ceramic membrane TiO_2 recovery unit is hybridized downstream of the reactor to remove the TiO_2 from the flow stream, while allowing the treated water to exit. The retentate TiO_2 stream is recycled and remixed with the fresh TiO_2 slurry stream that enters the reactor stream.

6.3.2.10 Operational Parameters of the Photocatalytic Reactor

After the integration of the semiconductor catalyst with a photoreactor, the oxidation rates and efficiency of the photocatalytic system are highly dependent on a number of the operation parameters that govern the kinetics of photomineralization and photodisinfection. This section will discuss the significance of each operation parameters on the corresponding kinetics and some recent methods to optimize the photocatalytic system via response surface analysis. The following section outlines a range of photoreactor operating parameters that affects the photocatalytic performance of TiO_2 photocatalysts in water treatment.

6.3.2.11 TiO_2 Loading

Concentration of TiO_2 in the photocatalytic water treatment system affects the overall photocatalysis reaction rate in a true heterogeneous catalytic regime, where the amount of TiO_2 is directly proportional to the overall photocatalytic reaction rate [163]. A linear dependency holds until certain extent when the reaction rate starts to aggravate and becomes independent of TiO_2 concentration. This is attributed to the geometry and working conditions of the photoreactor where the surface reaction is initiated upon light photon absorption [11]. When the amount of TiO_2 increases above a saturation level (leading to a high-turbidity state), the light photon absorption coefficient usually decreases radially. However, such a light attenuation over the radial distance could not be well correlated with the Beer–Lambert Law owing to the strong absorption and scattering of light photons by the TiO_2 particles [31]. The excess TiO_2 particles can create a light screening effect that reduces the surface area of TiO_2 being exposed to light illumination and the photocatalytic efficiency. Therefore, any chosen photoreactor should be operated below the saturation level of TiO_2 photocatalyst used to avoid excess catalyst and ensure efficient photons absorption. In this sense, both catalyst loading and light scattering effect can be considered as a function of optical path length in the reactor.

A large number of studies have reported the effect of TiO_2 loadings on the process efficiency [93,163,171,217,241,262]. These results are mostly independent and a direct comparison cannot be made, as the working geometry, radiation fluxes, intensity, and wavelengths used were different. It was reported that the optimum catalyst loading for photomineralization and photodisinfection are varie, and mainly depends on the dimension of the photoreactor. In addition, the determination of photoreactor diameter is crucial in not only the effective photons absorption but also the water flow hydrodynamics [155]. Uniform flow region can ensure that a steady-state residence time is obtained, while turbulence flow removes catalyst deposition or reaction dead zone [263]. Reactor diameters smaller than 20–25 mm were not feasible for turbulent flow while diameter larger than 50–60 mm is impractical. This is because large diameters usually have lower saturated catalyst loading and efficiency. In this instance, the amount of catalyst should be considered. Usually the TiO_2 catalysts can be mixed uniformly with the targeted water prior to the introduction into the reactor system. During the dark homogenization period of the catalyst, a lower initial concentration of organic pollutants is observed owing to the strict adsorption of organics onto the catalysts surface [264]. Similarly, catalysts—bacteria interaction was reported in the photodisinfection treatment of microorganism [17].

6.3.2.12 pH

In heterogeneous photocatalytic water system, pH is one of the most important operating parameters that affect the charge on the catalyst particles, size of catalyst aggregates, and the positions of conductance and valence bands. Due to the nature of TiO_2 catalyst used, any variation in the operating pH is known to affect the isoelectric point or the surface charge of the photocatalyst used. Many reports have used the point of zero charge (PZC) of TiO_2 to study the pH impact on the photocatalytic oxidation performance [93,217,241,262,265]. The PZC is a condition where the surface charge of TiO_2 is zero or neutral that lies in the pH range of 4.5—7.0, depending on the catalysts used. At PZC of TiO_2, the interaction between the photocatalyst particles and water contaminants is minimal due to the absence of any electrostatic force. When operating pH < PZC(TiO_2), the surface charge for the catalyst becomes positively charged and gradually exerted an electrostatic attraction force toward the negatively charged compounds. Such polar attractions between TiO_2 and charged anionic organic compounds can intensify the adsorption onto the photon-activated TiO_2 surface for subsequent photocatalytic reactions [17,264,266]. This is particularly significant when the anionic organic compounds present in a low concentration level. At pH > PZC(TiO_2), the catalyst surface will be negatively charged and repulse the anionic compounds in water. Different pH will affect the surface charge density of the TiO_2 catalyst, according to the following water equilibrium equations (Eqs. 6.19 and 6.20);

$$\text{At pH} < \text{PZC: TiOH} + \text{H}^+ \leftrightarrow \text{TiOH}_+^2 \tag{6.19}$$

$$\text{At pH} > \text{PZC: TiOH} + \text{OH}^- \leftrightarrow \text{TiO}^- + \text{H}_2\text{O} \tag{6.20}$$

The surface charge density distribution for these TiO_2 catalyst clusters is highly dependent on the operating pH. It was reported that the distribution of TiOH is $\geq$80% at $3 < pH < 10$; $TiO^- \geq 20\%$ at pH > 10, and $TiOH_2^+ \geq 20\%$ at pH < 3. The equilibrium constants for these reactions at different pH are $pK_{TiOH_2^+} = 2.4$ and $pK_{TiOH} = 8.0$ [267]. During photocatalytic reaction, the initial operating pH usually drops slightly from the formation of multitude intermediate by-products that may pose different chemical functional groups and affect the water pH indifferently [268].

A similar electrostatic interaction enhancement for photodisinfection of microorganisms was observed during the photocatalytic process [17]. During the photodisinfection, the initial photo-induced damage to the microorganisms takes place on the lipopolysaccharide layer of the external cell wall and on the peptidoglycan layer. This is followed by lipid membrane peroxidation and the subsequent oxidation of the membrane protein and polysaccharides. An increased density of $TiOH_2^+$ (at low pH) can form electrostatic link with the bacteria of negatively charged surfaces, resulting in increasing rate of photodisinfection. Herrera Melián et al. [269] observed that the bacterial inactivation rate was enhanced at pH 5.0. It should be noted that the enhanced bactericidal activity of TiO_2 at a low pH is due solely to the TiO_2-mediated photokilling and not acidification of the cell. Heyde and Portalier [270] explained that the negligible *E. coli* reaction to acid conditions was from the presence of an acid tolerance response to the bacterium itself, which secreted acid-induced proteins for acid-shock protection. However, Rincón and Pulgarin [187] did not find any differences in *E. coli* inactivation rates when the initial pH varied between 4.0 and 9.0. To date, various types of microorganisms have been successfully inactivated using TiO_2 photocatalysis, which include *Lactobacillus acidophilus*, *E. coli*, *Saccharomyces cerevisiae*, *Chlorella vulgaris*, *Streptococcus faecalis* and *Streptococcus aureus*, *Enterobacter cloacae*, total coliforms, I, *Fusarium solami*, *Aspergillus niger*, and others [264,269,271–273].

Although the electrostatic link between the catalyst particles and microorganisms were reported to exist, a subsequent microbial cell adsorption and eventual penetration of the catalysts through the cell wall are highly dependent on the mean particle size. Since the relative sizes between the bacteria and catalysts are significantly different, the charged $TiOH_2^+$ clusters might not come into full contact with the bacteria. Sichel et al. [274] found that the TiO_2 catalyst is actually adsorbed onto fungal spores rather than the reverse setting. This is owing to the fact that the fungal spores are larger (a few order of magnitudes) than the catalyst particles. A small decrease in microbial loadings was observed during the initial dark homogenization period with catalyst particles. This is owing to the catalyst agglomeration and subsequent sedimentation of the agglomerates. The rate of bacterial adsorption in this instance was found to directly relate to the bactericidal activity of the catalyst used. It was also reported that the adsorption phenomena during the dark and irradiated phase will act indifferently and thus, a strong conclusion cannot be made.

Besides, the interaction between the catalyst particles itself also exists and is dependent on the operating pH. The particle size of the same catalysts can vary from 300 nm to 4 mm depending on the distance from the PZC of TiO_2 [155]. At pH = PZC, the neutral surface charge of the catalyst particles is unable to produce the interactive rejection for solid–liquid separation. Thus, this induces catalyst aggregation where the catalyst becomes larger, leading to catalyst sedimentation [275]. This physical property is usually manipulated in the hybridized PMR system, where the pH of the treated wastewater is neutralized to pH 7 for the subsequent recovery of catalyst particles. The larger TiO_2 clusters can settle faster than the smaller one. It was reported that with such neutralization strategy, almost 97% of the catalysts can be recovered in the settling tank. The remaining TiO_2 catalysts can be recovered via the downstream MF system. Similarly, the water pH will also affect the effective separation in the PMR system where inappropriate control of water pH promotes the electrostatic repulsion, the Donnan exclusion phenomena, and thus, the rejection tendency for the membrane used [276,277]. It must be stressed that appropriate pH control strategies must be implemented at every different location of a photocatalytic water treatment process for efficient photocatalytic reaction to proceed.

6.3.2.13 Temperature

Numerous studies have been conducted on the dependence of photocatalytic reaction on the reaction temperature [65,237,278–280]. Although heat energy is inadequate to activate the TiO_2 surface, the understanding on such dependency could be extrapolated when operating the process under natural sunlight illumination. Most of the previous investigations stated that an increase in photocatalytic reaction temperature (>80°C) promotes the recombination of charge carriers and disfavor the adsorption of organic compounds onto the TiO_2 surface [163]. At a reaction temperature greater than 80°C, the photocatalytic reaction is interpreted with Langmuir–Hinshelwood (L–H) mechanism where the adsorption of the reactants is disfavored resulting in KC becoming "1." This will reduce the L–H expression (Eq. 6.23) into the apparent rate equation $r = k_{apparent}C$. All these drastically reduce the photocatalytic activity of TiO_2 when the reaction temperature rises. The desorption of degraded products from the TiO_2 surface is the rate-limiting step when temperatures rise. On the contrary, a low temperature below 80°C actually favors adsorption, which is a spontaneous exothermic phenomenon, resulting in getting KC of L-H model 1, enhancing the adsorption of final reaction products. A further reaction in temperature down to 0°C will cause an increase in the apparent activation energy. As a consequence, the optimum reaction temperature for photomineralization is reported to be in the range of 20–80°C [155].

For photodisinfection using TiO_2 photocatalysis, the increase in the reaction temperature increased the inactivation rate of microorganisms [280]. This is consistent with the Van't Hoff–Arrhenius equation (Eq. 6.21), where the rate constant k is linearly proportional to the exponential (−1/T);

$$\ln\left(\frac{k_1}{k_2}\right) = \frac{Ea}{R}\left(\frac{1}{T_2} - \frac{1}{T_1}\right) \tag{6.21}$$

in which k_1 and k_2 are the constants for temperatures T_1 and T_2, Ea is the energy of activation, and R is the universal gas constant. The viability of a microbe to the catalyst activity depends on its incubation temperature, type, and resistance to temperature change. The order of resistance of microorganisms to conventional disinfection treatment is nonforming spores bacteria < viruses < forming spores bacteria < helminths < protozoa (oocysts). To date, there is no comprehensive study conducted to compare the effect of TiO_2 photodisinfection on each microorganism types under different operating temperature. Thus, the photodisinfection using TiO_2 catalyst is usually conducted below an ambient temperature of 80°C to prevent high water heating costs (high heat capacity) [281].

6.3.2.14 Dissolved Oxygen

DO plays an important role in TiO_2 photocatalysis reaction to assure sufficient electron scavengers present to trap the excited conduction band electron from recombination [217]. The oxygen does not affect the adsorption on the TiO_2 catalyst surface as the reduction reaction takes place at a different location from where oxidation occurs [155]. Other roles for DO may involve in the formation of other ROS and the stabilization of radical intermediates, mineralization, and direct photocatalytic reactions. The total amount of DO in a reactor depends on a few technical considerations. For a photoreactor, the total delivered DO not only acts as an electron sink but also provides sufficient buoyant force for complete suspension of TiO_2 particles. Photoreactor sparging with pure oxygen in TiO_2 slurry reactor is usually a cost-ineffective solution, as the amount of DO being held-up is a function of the photoreactor geometry. The difference between the two sparging media of air and oxygen is usually not very drastic as the mass transfer of oxygen to the close vicinity of the surface is the rate-dependent step [147]. Generally, the Henry's law can be assumed to give a good approximation of the amount of oxygen dissolved under the experimental conditions, provided the oxygen sparging rate and the photoreactor gas hold-up are known [262]. In this equilibrium law, it is also necessary to account for the decrease in oxygen solubility with increasing reaction temperature. As discussed, it is preferential to operate the photoreactor under ambient conditions to prevent the elevate cost of air or oxygen sparging for enhanced electron sink.

The presence of DO is also suggested to induce the cleavage mechanism for aromatic rings in organic pollutants that are present in the water matrices. Wang and Hong [282] proposed that during aromatic ring cleavage in the degradation of a dioxyl compound, the molecular oxygen will follow the cleaving attack of a second hydroxyl radical. The analogous partial pressure of oxygen applied during the reaction for a closed reactor system is also important. A significantly higher partial pressure of oxygen will result in a higher initial organics photomineralization rate than its structural transformation [237,282]. It was reported that at low oxygen pressure of 0.5 kPa, 75% of the original 2-CB compounds were transformed, but only 1% of them was mineralized to CO_2 after 5 h of UV irradiation [282]. Although elevated partial pressure of oxygen is imperative, it is difficult to quantify their specific influence on the surface activity of TiO_2, owing to the polyphasic nature of the photocatalytic liquid phase reaction [171].

The detrimental effect of DO on the photocatalysis reaction was reported by Shirayama et al. [283]. They observed an elevated photodegradation rate of chlorinated hydrocarbons in the absence of DO. This could be explained by the strong absorption characteristics for UV photons at intensities of 185 and 254 nm. The DO molecules act as an inner filter in this case and cause a sharp attenuation in UV light intensity mainly at UV-C germicidal region. There is no report found to determine the DO impact on a photocatalysis process using UV-A or UV-B light sources. To date, the effect of DO on the efficiency of photodisinfection rate has been paid little attention. The formation of various ROS under the series of redox reactions on the catalysts surface was assumed to be similar in both photomineralization and photodisinfection reaction. If sufficient nutrients are available, the constant sparging of DO will generally promote microbial growth and offset the photodisinfection rate. This indirectly prolongs the irradiation time necessary to achieve the desired inactivation level. It is thus recommended that the effect of DO on the microbial inactivation should be investigated thoroughly in a particular photoreactor system before improvising the DO sparging strategy.

6.3.2.15 Contaminants and Their Loading

Previous investigations have reported the dependency of the TiO_2 photocatalytic reaction rate on the concentration of water contaminants [217,241,262,265]. Under similar operating conditions, a variation in the initial concentration of the water contaminants will result in different irradiation time necessary to achieve complete mineralization or disinfection. Owing to the photonic nature of the photocatalysis reaction, excessively high concentration of organic substrates is known to simultaneously saturate the TiO_2 surface and reduce the photonic efficiency leading to photocatalyst deactivation [284].

Not all organic substrates will have such profound effect on the irradiation time, and this also depends on the corresponding chemical nature of the targeted compounds for TiO_2 photocatalysis reaction. For instances, 4-chlorophenol will undergo a degradation pathway with constant evolution of intermediate(s) product (i.e., hydroquinone and benzoquinone), while oxalic acid will undergo direct transformation to carbon dioxide and water [285]. In the case of 4-chlorophenol, such evolution of intermediate(s) will further prolong the irradiation time necessary for total mineralization owing to the direct competition over unselective TiO_2 surfaces. In the development of mathematical model that represents the kinetics of mineralization while relating to the TiO_2 loading required, commonly used water quality parameters such as chemical oxygen demand (COD), total organic carbon (TOC), or dissolved organic carbon (DOC) could be more appropriate to account for such competitiveness of intermediate(s) with its predecessor compounds. Also organic substrates with electron-withdrawing nature such as benzoic acid and nitrobenzene were found to strongly adhere and more susceptible to direct oxidation than those with electron donating groups [286]. Most of the TiO_2 studies conducted to date, utilize a range of model organic substrates with different substituent groups, but these rarely convey any useful information and merely test on the photoefficiency of a new photocatalyst or an integrated reactor column. Some field kinetics of the photomineralization of real wastewater has also been reported [287—290]. It was observed that owing to the persistency of the dissolved organic in the real wastewater, a slow photomineralization kinetics is attained with prolong irradiation times to achieve complete mineralization. Slow kinetics turnover of the photocatalytic water treatment as a stand-alone process means a higher initial cost on the reactor volume and site area is required. Recently, this heterogeneous photocatalytic technology has been coupled with biological treatment to increase its industrial feasibility [291]. Such coupling allows the retention time in biological treatment stages to be reduced, where the non-biodegradable compounds of the wastewater can be turned into biodegradable compounds with the aid of photocatalytic treatment.

Similarly, the photodisinfection efficiency of various microorganisms has been assessed for the possible application of photocatalytic technology to replace the chemical disinfectant methods. In general, the mechanism involved in the microbial disinfection includes the destruction of the microbial protein structures and inhibition of their enzymatic activities [173]. Compared with the persistency during organic photomineralization, a general classification in the bacterial resistance to the disinfectant used has also been proposed. Among all, the most resistant infectious type of microorganisms are prions, followed by coccidia (Cryptosporidium), bacterial endospores (Bacillus), mycobacteria (*Mycobacterium tuberculosis*), viruses (poliovirus), fungi (Aspergillus), Gram-negative (Pseudomonas), and eventually Gram-positive bacteria (Enterococcus) [155]. Their differences in resistance are explained by their cell wall permeability, size, and complexity of the specific microorganisms. Each microorganism might also be of infectious nature, which causes epidemic diseases when they multiply in water. Most bacteria can be killed easily with TiO_2 photocatalysis, but a complete inactivation might have to be ensured as they are highly infectious. Similarly, this infectious nature can also be found in viruses (adenoviruses, enteroviruses, hepatitis A and E viruses, noroviruses and sapoviruses, rotaviruses) and the most in protozoa (Acanthamoeba spp., *Cryptosporidium parvum*, *Cyclospora cayetanensis*, *Entamoeba histolytica*, *Giardia intestinalis*, *Naegleria fowleri*, *Toxoplasma gondii*) [292]. All these protozoa are highly infectious in low concentration, and the photocatalytic treatment should be targeted on these microorganisms as the surrogate indicators. This is to ensure that adequate photocatalytic treatment to prevent the outbreak of the epidemic diseases in the treated water, if photocatalytic treatment is chosen.

6.3.2.16 Light Wavelength

The photochemical effects of light sources with different wavelength emitting ranges will have a profound consequence on the photocatalytic reaction rate, depending on the types of photocatalysts used—crystalline phase, anatase-to-rutile composition, and any state of photocatalyst modifications. Using commercial Degussa P-25 TiO_2, which has a crystalline ratio of anatase 70/80: 20/30, a light wavelength at $\lambda < 380$ nm is sufficient for photonic activation [171,285]. The crystalline phase of rutile TiO_2 has a smaller band gap energy of EB ~ 3.02 eV, compared with the anatase TiO_2 of 3.2 eV [30,163,190]. This dictates that rutile TiO_2 can be activated with light wavelength of up to 400 nm, depending on the band gap threshold for the type of rutile TiO_2 used.

For UV irradiation, its corresponding electromagnetic spectrum can be classified as UV-A, UV-B, and UV-C, according to its emitting wavelength. The UV-A range has its light wavelength spanning from 315 to 400 nm (3.10−3.94 eV), while UV-B has a wavelength range of 280−315 nm (3.94−4.43 eV), and the germicidal UV-C ranges from 100 to 280 nm (4.43−12.4 eV) [5,293]. In most of the previous studies, the UV-A light provides light photons sufficient for photonic activation of the catalyst [93,120,286,294]. As with outdoor solar irradiation, the UV-C is usually absorbed by the atmosphere and not reaching the earth surface. Only the lamp-driven photoreactor system can utilize UV-C irradiation artificially for photonic activation of catalyst and reduction of viable microorganisms. The mechanism of UV-C cell destruction involves the direct induction on pyrimidine and purine and pyrimidine adducts on the cell DNA. However, not all microorganisms are susceptible to the UV-C radiation and some highly resistant microorganisms can survive through disinfection process. These include *Legionella pneumophila* and *C. parvum* oocysts [155].

The natural UV radiation that reaches the surface of the earth consists of both UV-A and UV-B spectrums. The photolysis mechanism for both UV irradiations on cell inactivation is dissimilar to the discussed UV-C mechanism. Both UV-A and UV-B irradiations can be absorbed by cellular components called intracellular chromophores. L-trytophan is the best-known intracellular chromophore and is thought to contain unsaturated bonds such as flavins, steroids, and quinines [295]. Among these UV irradiations, the UV-A irradiation is toxic only in the presence of oxygen. The ROS or oxidative stress generated from the chromophore light absorption can damage cells and cell components, leading to lipid peroxidation, pyrimidine dimer formation, and eventually DNA lesions. The contact between the ROS and DNA results in single-strand breaks and nucleic acid modifications. Such damages on the DNA are usually lethal or mutagenic irreversible. With the presence of TiO_2 catalyst as the light sensitizers, a high degree of cell damages is seen, as the amount of ROS generated increases accordingly. A few microorganisms that are resistant to UV-A photolysis have been inactivated successfully by TiO_2 photocatalysis, namely; *E. cloacae*, *E. coli*, *Pseudomonas Aeruginosa*, and *Salmonella typhimurium* [202].

A longer wavelength of solar irradiation (l > 400 nm) has also been used in solar disinfection (SODIS) study [272,274,296–298]. However, the photokilling mechanism is as yet unclear as it involves a variety of microbial and a larger mixed spectrum of UV-A and solar irradiation. A similar cell destruction mechanism to the one proposed for UV-A irradiation is thought to take place in this mixed light spectrum. In the SODIS, the pathogens in the drinking water contained in PET bottles were found to be inactivated within 6 h of sunlight exposure. However, significant research and developments on the disinfection using photocatalytic-mediated process need to be conducted to broaden the photoactivity of current TiO_2 catalysts used.

6.3.2.17 Light Intensity

The photonic nature of the photocatalysis reaction has outlined the dependency of the overall photocatalytic rate on the light source used. Light intensity is one of the few parameters that affect the degree of photocatalytic reaction on organic substrates. Fujishima et al. [168] indicated the initiation of TiO_2 photocatalysis reaction rates is not highly dependent on light intensity, where a few photons of energy (i.e., as low as 1 mW/cm) can sufficiently induce the surface reaction. To achieve a high photocatalytic reaction rate, particularly in water treatment, a relatively high light intensity is required to adequately provide each TiO_2 surface active sites with sufficient photon energy required. However, when using the nominal TiO_2 particles without modifications, the surface reaction is restricted to photons with wavelengths shorter than the absorption edge of approximately 400 nm. The organic conversion in the presence of UV wavelength (l < 400 nm) in many studies obeyed the linear proportionality correlation to the incident radiant flux. This was evidenced by Glatzmaier et al. [299] and Glatzmaier [300], where they observed that the destruction of dioxin and polychlorinated biphenyls was significantly enhanced in the presence of high-intensity photons. A similar finding was reported in Magrini and Webb [301] where the organic decomposition rate was reported to increase with the radiation intensity.

Later, it was discovered that the dependency of the reaction rate on radiant intensity behaves indifferently under different lighting conditions [26,302,303]. The linear dependency of the photocatalytic reaction rate on radiant flux (∅) changed to a square-root dependency ($\varnothing^{0.5}$) above certain threshold value. Such a shift in dependency form was postulated owing to the amount of photogenerated holes available during the electron–hole pair formation. In the TiO_2 catalyst used, the photo-induced generation of valence band holes are much less than the conduction-band electrons available. In this instance, the photo-generated holes are the rate-limiting step and the details of derivation of the square-root dependency can be obtained from Malato et al. [155]. At high intensities, the dependency of the photocatalytic reaction rate on radiant flux reduced to zero ($\varnothing^0$). This was explained by the saturated surface coverage of the catalyst, resulting in a mass transfer limitation in the adsorption and desorption and thus, preventing the effect of light intensity to set in. An increase in the fluid turbulency in this case might help to alleviate the mass transfer problem on the surface of the catalyst. The desorbed final products might also affect the dependency of reaction rate on radiant flux, as they might scavenge the electron acceptors and further promote the electron–hole pair recombination.

Rincón and Pulgarin [266] reported that the residual disinfecting ability of the photocatalyst largely depends on the duration of light intensity without any temporal interruptions. They investigated the effect of light intensities at 400 and 1000 W/m^2 on bacterial lethality and regrowth and found that the higher intensity without any temporal interruptions can cause irreversible damage to the *E. coli*. In the intermittent light irradiations with constant interruptions, the bacteria were seen to regrow during the subsequent 24 or 48 h. Some studies suggested that this regrowth is due to the dark-repair mechanism where the partially damaged cells recover in the presence of nutrients [304]. Others have suggested that the damaged but not totally inactivated cells could recover its viability through photorepairing under radiation of 300–500 nm or on the resynthesis and postreplication of cells [274,280]. It must be noted in this case that for photodisinfection using different light intensity, a final conclusive point cannot be made directly. The disinfection results of 400 W/m^2 at 2.5 h irradiation might not be the same as the result arose from 1000 W/m^2 for 1 h. Thus, to predict the minimum irradiation required at constant irradiance, preliminary studies into both the photoreactor performance and microbial consortia (different resistants) presents are important.

6.3.2.18 Response Surface Analysis

From the earlier discussions on the effect of the operation parameters on the photocatalytic reaction rate, it can be seen that these parameters would affect the system indifferently. In overall, it can be interpreted that a multivariable (MV) optimization approach is actually required to optimize a photoreactor system as parameter interaction might exist. Parameter interactions refer to the relationship between operating parameters such as TiO_2 loading on pH or pH on radiant flux. For the optimization of a photoreactor system, the conventional one-parameter-at-a-time approach is mostly used to unveil the effects of one parameter after another. Although this conventional optimization approach is widely acceptable, the reported outcomes could be of insignificant and have less predictive power if the condition for one operating parameter changes.

This has lead to the application of effective design of experiments (DOEs), statistical analysis, and response surface analysis for photocatalytic studies [305—309]. Using this approach, different permutations of experimental design are involved and the operational parameters and spans are defined. As compared with the conventional one-parameter-at-a-time approach, the MV optimization approach has predetermined experimental points that are dispersed uniformly throughout the study domain, i.e., only a small region is covered in the domain of conventional study. This allows the optimization process to be more time-effective and enhances the identification of parameter interactions, where they can be interpreted using commercial statistical software such as Design Expert software.

Chong et al. [306] proposed the use of Taguchi—DOE approach, together with analysis of variance, statistical regression, and response surface analysis to study the combined effects of four key operation parameters that affect the photocatalytic reaction rate in an annular photoreactor. They utilized nine experimental permutations to analyze the 81 possible parameter combinations. It was reported that the interaction between the TiO_2 loading and aeration rate had a positive synergistic effect on the overall reaction rate. A response surface model was developed to correlate the reaction rate dependency on the four different parameters according to the statistical regression as shown in Eq. (6.22);

$$R_0 = b_0 + \sum_{i=1}^{k} b_i X_i + \sum_{i=1}^{k} b_{ij} X_i^2 + \sum_{i_i<j}^{k} \sum_{j}^{k} b_{ij} X_i X_j \tag{6.22}$$

where R_0 is the predicted response output of the photomineralization rate, and i, j are linear, quadratic coefficients, respectively. The parameters of b and k are regression coefficient and the number of parameters studied in the experiment, respectively, and X_i, X_j ($I = 1, 4; j = 1, 4, i \neq j$) represent the number of independent variables in the study. This model (Eq. 6.22) is empirical and independent for each photoreactor system. Subsequent verification works are required to determine the accuracy and applicability of such model for the prediction of photoreaction rate under the variation of its parameters. Other DOE approach to photocatalytic reactor optimization has also been applied over the years, such as central composite design, Bayesian, and Plackett–Burman designs [310–312].

6.3.2.19 Kinetics and Modeling

Kinetics and mechanistic studies on the photomineralization or photodisinfection rate of the water contaminants are useful for process scale-up. The appropriate utilization of kinetic models for the interpretation of experimental data enables the design and optimization of photoreactor system with sufficient capacity and minimal nonilluminated reactor volume. In this section, the different kinetics and rate model for both photomineralization and photodisinfection will be discussed along with the common misconceptions in kinetic modeling.

6.3.2.20 Photomineralization Kinetics

In most of the photocatalytic studies, kinetic or mechanistic studies over the irradiated TiO_2 surfaces usually only involve a single constituent model organic compound. The kinetics of different organic compounds ranging from dye molecules, pesticides, herbicides, and phenolic compounds to simple alkanes, haloalkanes, aliphatic alcohols, and carboxylic acids has been investigated [75,155,163,171]. The nonselective nature of the $OH^{\cdot}$ radicals means that the disappearance rate of the studied compound with irradiation time should not be referenced as a standard for reactor design purpose. This is because that numerous intermediates are formed en-route to complete mineralization, and neglecting this aspect is a common mistake in portraying the photomineralization kinetics. In this instance, the organic concentrations can be expressed collectively in COD or TOC to yield an in-depth understanding on the photomineralization kinetics.

As the L–H model is surface area dependent, the reaction rate is expected to increase with irradiation times since less organic substrate will remain after increased irradiation times with higher surface availability. A zero rate of degradation is associated with the total decomposition achieved. Numerous assumptions for the L–H saturation kinetics type exist and for the applicability in the rate of photomineralization, any of the four possible situations is valid: (1) reactions take place between two adsorbed components of radicals and organics; (2) the reactions are between the radicals in water and adsorbed organics; (3) reactions take place between the radical on the surface and organics in water; (4) reaction occurs with both radical and organics in water.

Some researchers found that simpler zero- or first-order kinetics is sufficient to model the photomineralization of organic compounds. This was, however, only applicable for limited conditions where the solute concentration is inadequately low. In most kinetic studies, a plateau-type of kinetic profile is usually seen where the oxidation rate increases with irradiation time until the rate becomes zero [313,314]. The appearance of such kinetics regime usually fits the L–H scheme. According to the L–H model (Eq. 6.23), the photocatalytic reaction rate (r) is proportional to the fraction of surface coverage by the organic substrate (θ_x), k_r is the reaction rate constant, C is the concentration of organic species, and K is the Langmuir adsorption constant:

$$r = -\frac{dC}{dt} = k_r = \frac{k_r KC}{1 + KC} \tag{6.23}$$

The applicability of Eq. (6.23) depends on several assumptions, which include (1) the reaction system is in dynamic equilibrium; (2) the reaction is surface mediated [119]; the competition for the TiO_2 active surface sites by the intermediates and other ROS is not limiting [262]. If these assumptions are valid, the reactor scheme only consists of adsorption surface sites, organic molecules and its intermediates, electron–hole pairs, and the ROS. The rate constant (k_r) for most of the photocatalytic reaction in water is usually reported to be in the order of 10^6–10^9/(Ms) [155]. This k_r-value is the proportionality constant for the intrinsic reactivity of photoactivated surface with C. Others have also related that k_r is proportional to the power law of effective radiant flux (i.e., $\varnothing_e^n$) during the photomineralization reaction [313,315]. The K-parameter is the dynamic Langmuir adsorption constant (M^{-1}) that represents the catalysts adsorption capacity. Eq. (6.23) can be solved explicitly for t using discrete change in C from initial concentration to a reference point:

$$\ln\left(\frac{C}{C_0}\right) + K(C - C_0) = k_r K_t \tag{6.24}$$

The K-value can be obtained using a linearized form of Eq. (6.23), where 1/r is plotted against 1/C:

$$\frac{1}{r_0} = \frac{1}{k_r} + \frac{1}{k_r K C_0} \tag{6.25}$$

It was reported that the real K-value obtained from the linearized plot of 1/r against 1/C is significantly smaller [155]. This was explained by the differences in adsorption–desorption phenomena during dark and illuminated period. When the organics concentration is low (in mM), an "apparent" first-order rate constant (Eq. 6.26) could be expressed where k' (min^{-1}) = k_rK:

$$r = -\frac{dC}{dt} = k_rKC = k'C \tag{6.26}$$

Rearranging and integration of Eq. (6.26) yields the typical pseudo-first-order model as in Eqs. (6.27 and 6.28):

$$C = C_0e^{-k't} \tag{6.27}$$

$$\ln\left(\frac{C_0}{C}\right) = -k_rKt = -k't \tag{6.28}$$

The apparent rate constant, however, is only served as a comparison and description for the photocatalytic reaction rate in the reactor system. Fig. 6.14 shows a typical saturation kinetic plot for the degradation of organic dye molecules in an annular photoreactor system where the reaction rate increases to a point where the rate plateaus off. To interpret the maximal photomineralization rate, the tailing regime of the L–H saturation profile should be neglected. Only the slope of the tangent to the inflexion point should be used to obtain the maximal photomineralization rate. In this

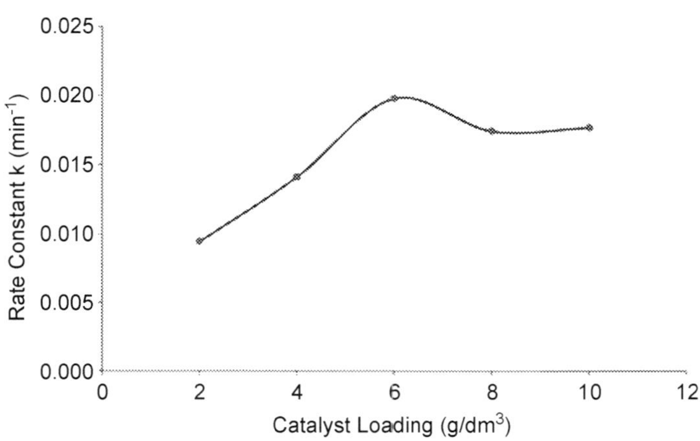

Figure 6.14 Typical saturation kinetics plot for the degradation of organic dye molecules in an annular photoreactor system [217].

instance, the unit for the slope has the same chemical reaction order as the zero-order rate constant.

A lump-sum L–H saturation kinetics profile has also been used to simplify the approximation for a specific photocatalytic reactor system [316]. In such an empirical lump-sum L–H approach, the degree of organics mineralization is actually expressed in terms of TOC (Eq. 6.29):

$$r_{TOC,O} = \frac{\beta_1[TOC]}{\beta_2 + \beta_3[TOC]} \tag{6.29}$$

This Eq. (6.29) allows the prediction of TOC degradation as a function of irradiation time. Similar reciprocal plots of 1/r against 1/[TOC] can be used to determine the empirical parameters, β_1, β_2, and β_3 as in Eq. (6.25). The irradiation time taken to achieve the fractional degradation of TOC can also be estimated when Eq. (6.29) is expressed as in Eq. (6.24). Such an empirical lump-sum LeH model has greatly summarized the needs for precise kinetics measurement and made a great approximation for any particular photoreactor system, provided sufficient data are collected for the determination of rate parameters.

6.3.2.21 Photodisinfection Kinetics

Since the first application of semiconductor catalysts for disinfection by Matsunaga et al. [273], few studies have been found in the literature, which comprehensively study the kinetic modeling of photodisinfection of microorganisms in a water treatment process. Empirical kinetic models have been mostly applied for interpretation of photodisinfection data because of process complexity and variability. The general expression for the empirical photodisinfection models for demand-free condition is expressed in Eq. (6.30). In this instance, demand-free conditions assume that the catalyst concentration is constant with irradiation time.

$$\frac{dN}{dT} = -kmN^{x}C^{n}T^{m-1} \tag{6.30}$$

where dN/dt = rate of inactivation; N = number of bacterial survivors at irradiation time t; k = experimental reaction rate; C = concentration of photocatalyst used; m, n, and x are the empirical constants. However, the most commonly employed disinfection model in photodisinfection studies to date is the simple mechanistic Chick–Watson (C–W) model (Eq. 6.31) [20,293].

$$\log\frac{N}{N_0} = -k'T \qquad (6.31)$$

In this C—W model, the photodisinfection rate is expressed as a linear function of the enumerated bacteria and catalyst loading. The combined kinetic parameter of CT between the catalyst concentration and irradiation time required to achieve complete inactivation is widely used as a reference for process design. Other than this, this CT value concept is usually used to compare the efficacy of different disinfectant used in water treatment [317]. This C—W model, however, may not always be applicable as many studies may have experienced a curvilinear or nonlinear photodisinfection profile. Hom [318] reproduced a useful empirical modification on C—W model after having observed that the disinfection plots of natural algal—bacterial systems were curvilinear, rather than typical log-linear type (Eq. 6.32). In the Hom model, the bacterial inactivation level is predicted in a nonlinear function of C and T, depending on the empirical parameters of n and m, respectively. Since this Hom model is a two-parameter model, it should be stressed that the model is only applicable to a photodisinfection profile with a maximum of two different nonlinear regions. For m-value greater than unity, the inactivation curve displays an initial "shoulder" while "tailing" is seen when the m-value is less than unity.

$$\log\frac{N}{N_0} = -k'C^nT^m \qquad (6.32)$$

Marugán et al. [186] have outlined the common nonlinearity found in the photodisinfection kinetics profile using Degussa P-25 TiO_2 catalyst fitted with common disinfection models (Fig. 6.15). In their study, they found that three different inactivation regions in the photodisinfection profile, namely (1) a lag or initial smooth decay, known as the "shoulder," followed by (2) a typical log—linear inactivation region and ends with (3) a long deceleration process at the end of the disinfection, which is known as the "tailing." The presence of "shoulder" was justified by the cumulative damage nature of photodisinfection treatment on the cytoplasmic membrane rather than an instantly lethal [317]. However, the "tailing" during the photodisinfection is not well understood. Benabbou et al. [319] proposed that the "tailing" region was related to the competition for photocatalysis between organic products released from constant cell lyses and the remaining intact cells. Others have proposed that the "tailing" deviations from the log—linear reduction were due to the presence of variations in the bacterial population resistant to the disinfectant used [320]. Nevertheless,

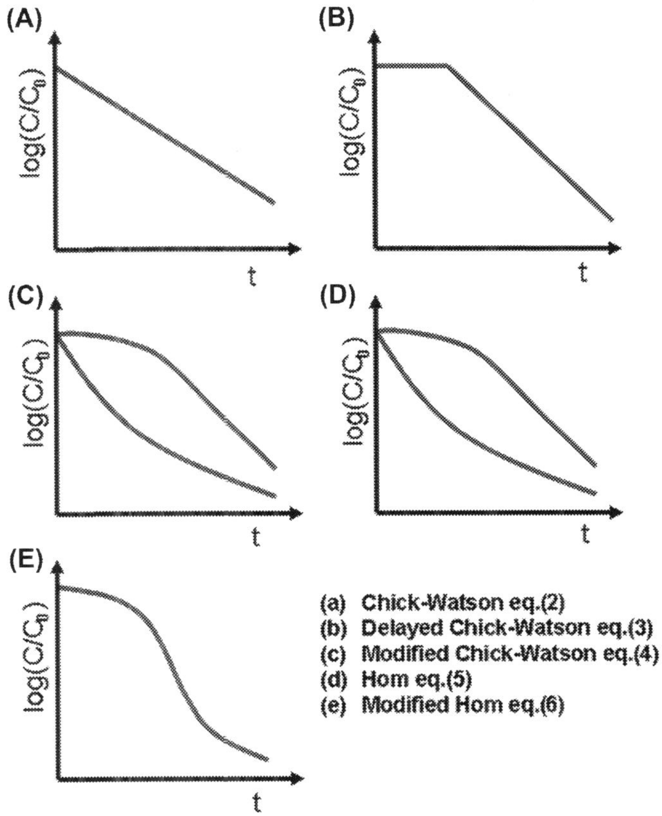

Figure 6.15 Common nonlinearity in the photodisinfection kinetics profile using TiO_2 catalyst, along with the appropriate disinfection models used [186].

the use of CT concept or Hom model can lead to an overdesign for a photodisinfection system [317]. A further modification to the Hom model was made to account for the simultaneous presence of shoulder, log–linear reduction and tailing (Eq. 6.33) [186,321].

$$\log \frac{N}{N_0} = -k_1[1 - \exp(-k_2 t)]^{k_3} \tag{6.33}$$

Eq. (6.33) is known as the modified Hom model, and it expands the applicability of the Hom model for the fitting of the initial shoulder, log–linear reduction, and prolonged tailing behaviors. Another detailed empirical model that can detail the different bacterial inactivation regions is a power law expression (Eq. 6.34), when m·1 for the generalized differential equation (Eq. 6.30) [322].

$$\frac{dN}{dt} = -kN^{x}C^{n} \tag{6.34}$$

Integration of Eq. (6.34) yields the Rational model (Eq. 6.35). If $x = 1$, this Rational model can be reverted back to the C—W model. In this instance, the Rational model assumes that $x \neq 1$;

$$\log\frac{N}{N_0} = -\frac{\log\left[1 + N_0^{x-1}(x-1)kC^{n}T\right]}{(x-1)} \tag{6.35}$$

The Rational model (Eq. 6.35) can describe both "shoulder" and "tailing" characteristics for x less than or greater than unity, respectively. Similarly, the Hom model (Eq. 6.32) can also be integrated according to the Rational model with the introduction of both x and $m \neq 1$ to yield the HomePower model (Eq. 6.36):

$$\log\frac{N}{N_0} = -\frac{\log\left[1 + N_0^{x-1}(x-1)kC^{n}T^{m}\right]}{(x-1)} \tag{6.36}$$

Anotai [2] reported that this Hom—Power model may provide a better fit than both the Hom and Rational models. However, the existence of four empirical parameters in the model may result in an overparameterization with null physical meanings for each parameter within the model. To reduce the number of null parameters, the Selleck model (Eq. 6.37) was proposed. This model assumes that the catalyst concentration remains constant during the irradiation period:

$$\frac{dS}{dt} = \frac{kCS}{1 + KCT} \tag{6.37}$$

where S is the survival ratio $= N/N_0$ at irradiation time t; k and K are the rate constants. In this model, the "shoulder" was assumed to be a result of cumulative effects of the chemical disinfectant on the microbial target during the contact time. A gradual decrease in the permeability of the outer cell membrane of *E. coli* as an action of the catalysts can be idealized by a series of first-order reaction steps on a single *E. coli* cell that leads to lethality. This model can be modified to account for the disinfection of coliform bacteria in water/wastewater effluent (Eq. 6.38) [322]:

$$\log\frac{N}{N_0} = -n\log\left[1 + \frac{CT}{k}\right] \tag{6.38}$$

Although empirical models provide a simple correlation of the photodisinfection data, the mathematical nature of their relevant terms yields null or misinterpret physical meaning. For instance, the m-value >1 in the Hom model actually suggests that photocatalytic reactivity increases with irradiation time in the ASF. If the vitalistic assumption of bacterial population resistance distribution is valid, this actually indicates that the most resistant bacteria are killed first prior to the least resistant one [320]. This shows that the rationale for the applicability of Hom model is contradictory.

It was proposed that the mechanistic models could convey a better physical meaning in their kinetic terms. It was hypothesized that the photodisinfection mechanism can be viewed as a pure physicochemical phenomenon and precedes in a similar way to a chemical reaction [320]. The classical C—W model is a typical example of the mechanistic photodisinfection kinetics model. In the model, the apparent exponential decay curves for bacterial survival ratios with irradiation times follow a similar decay mechanism of a chemical reaction and are thus applicable under both phenomena. This can be visualized when the C—W model is integrated mathematically, with N assuming to be the number of moles of reactant to yield the log—linear curve of a first-order reaction.

To formulate a mechanistic model for photodisinfection kinetics, the occurrence of both "shoulder" and "tailing" region should be rationalized idealistically. The presence of "shoulder" can be justified mechanistically by the single-hit multiple targets or a serial phenomenon event [322]. Under mechanistic assumption, the damage to the microbial cell is viewed as cumulative rather than instantly lethal. This dictates that a large number of critical molecules need to be denatured prior to cell inactivation [322]. Severin et al. [323] proposed that the cumulative inactivation of a single bacterium can be collectively represented by a series of integer steps. These disinfection steps were thought to pass on a bacterium from one level to another in a first-order reaction with respect to the catalyst used until a finite number of lethal (l) events was reached. The microorganisms which accumulate less than the postulated number of lethal steps are considered to survive the photodisinfection process. As for the "tailing" region, many have regarded that it was owing to the presence of a microbial subpopulation resistant to thermal sterilization. Najm [324] also suggested that the "tailing" is due to the intrinsic distribution of bacterial resistance to the sterilization method, making the bacteria more resistant, adapted, and inaccessible to heat treatment. Marugán et al. [186] reviewed that such "tailing" was not a common phenomenon in water disinfection treatment. So far, no justifiable explanation has been proposed to account for the occurrence of "tailing" in the water disinfection treatment with well-mixed conditions and equally resistant cloned-bacterial population. It was suggested that the "tailing" was a result of gradual deterioration in the rate of disinfection, and total inactivation was achievable only after a sufficient retention time. The random collisions between the catalyst and bacteria can be expressed by the Poisson probability, as the number of collisions supersede prior to the number of microbial death [322]. If the destruction rate of the microorganisms is assumed to be the same for the first and lth target at the kth bacteria site, thus the rate of destruction for the microorganisms is given by:

$$\frac{dN_k}{dt} = kCN_{k-1} - kCN_k \tag{6.39}$$

Solving for K = 0 to K = l − 1 gives the log fraction of microbial survival, not exceeding l − 1 at the end of the contact time:

$$\ln\frac{N}{N_0} = -kCT + \ln\left[\sum_{k=0}^{l-1}\frac{(kCT)^K}{K!}\right]$$

Lambert and Johnston [320] stated in the formulation of all mechanistic models that take into account the "shoulder" and "tailing," an intermediate population stage is usually suggested with different rates of disinfection for each microbial state. Typically, two different microbial states for the number of damaged bacterial and undamaged bacteria (i.e., $C_{dam} + C_{undam}$) were introduced to account for such nonlinearity in bacterial survivor curves. A similar L−H type of mechanistic model (Eqs. 6.41 and 6.42) can also be applied to represent the photodisinfection kinetics to yield more meaningful kinetic terms [12,70]. Johnston et al. [325] showed that the inoculum size of the bacteria has a large impact on resistance distribution and thus, the errors associated with this term should be given attention during the mechanistic modeling.

$$\frac{dC_{undam}}{dt} = \frac{-k_1 K_{undam} C_{undam}^{\eta_{undam}}}{1 + K_{undam} C_{undam}^{\eta_{undam}} + K_{undam} C_{undam}^{\eta_{undam}}} \tag{6.41}$$

$$\frac{dC_{dam}}{dt} = \frac{k_1 K_{undam} C_{undam}^{\eta_{undam}} - k_2 K_{dam} C_{dam}^{\eta_{dam}}}{1 + K_{undam} C_{undam}^{\eta_{undam}} + K_{dam} C_{dam}^{\eta_{dam}}} \tag{6.42}$$

where k_i is the true log−linear deactivation rate constant for the reaction between generated ROS with bacteria. The pseudoadsorption constant, K_i, represents the surface interaction between the catalyst and bacteria. This constant is similar to the adsorption equilibrium constant in the conventional

L−H model. Owing to the significant size differences between *E. coli* cells and the catalyst agglomerates, these constants do not represent strict adsorption phenomena but a more general surface interaction during photodisinfection. These adsorption constants allow the "shoulder" representation in the photodisinfection kinetics data. The inhibition coefficient, η_i, is a power coefficient that accounts for the "tailing" in the bacteria inactivation curve. Marugán et al. [186] proposed that η_i is needed to account for the inhibition produced by the increasing concentrations in the medium of cell lysis and oxidation products competing for the ROS. This is particularly important toward the end of photodisinfection, as high concentrations of these compounds and small numbers of viable bacteria are present in the suspension. In this instance, the η_i in the proposed L−H model essentially means that the reaction order with respect to the microbial population is higher than one.

Both Eqs. (6.41 and 6.42) have six independent parameters that describe photodisinfection rates. This constitutes a high risk of overfitting the experimental data, where the statistical significance of parameters and the plausibility of the model are low. By taking into account the intrinsic kinetics of ROS attack, the catalyst−bacteria interaction and the inhibition by-products are similar for both undamaged and damaged bacteria, the following are assumed:

$$k_1 = k_2 = k \tag{6.43}$$

$$k_{undam} = k_{dam} = K \tag{6.44}$$

$$\eta_{undam} = \eta_{dam} = \eta \tag{6.45}$$

With these, Eqs. (6.41) and (6.42) are reduced to Eqs. (6.46) and (6.47). The three independent parameters allow a simultaneous fitting of three different inactivation regimes of "shoulder," "log−linear," and "tailing." Fitting of Eqs. (6.46) and (6.47) to the experimental measurements of (Cundam + Cdam)/C_0 can be achieved using a nonlinear regression algorithm coupled with a fifth-order Runge−Kutta numerical approach. Marugán et al. [186] showed a good fitting of such mechanistic L−H model to the photodisinfection of *E. coli* under different loadings of Degussa P25 TiO_2 catalyst.

$$\frac{dC_{undam}}{dt} = \frac{-kKC_{undam}^{\eta}}{1 + KC_{undam}^{\eta} + KC_{dam}^{\eta}} \tag{6.46}$$

$$\frac{dC_{dam}}{dt} = k\frac{KC^{\eta}_{undam} - KC^{\eta}_{dam}}{1 + KC^{\eta}_{undam} + KC^{\eta}_{dam}} \tag{6.47}$$

6.3.2.22 Water Quality

In a full-scale plant for water treatment, a number of technical issues have to be sorted out. Water quality of the influents in the treatment plant can vary with time. In this section, the implication of TiO_2 photocatalytic processes for treatment of water sources of different qualities will be discussed. This enables a full understanding on the effects of various key water quality parameters on the suitability of applying such advanced TiO_2 photocatalytic processes.

6.3.2.23 Turbidity

Turbidity often refers to the insoluble particulates that are present in the targeted water [161]. The presence of such insoluble particulate matters are highly detrimental to the TiO_2 photocatalysis-based process, as they can affect the optical properties and further impede the penetration of UV light by strong scattering and absorption of the rays [326,327]. This will cause a variation in the predicted use of TiO_2 loading, UV penetration path, and light intensity. Also, excessive levels of turbidity can reduce both the photomineralization and photodisinfection efficiency of the pollutants that present in water owing to the shielding effects that attenuate the light penetration causing those pollutants to flee from the treatment [326–328]. Suspended solids will also shield target pollutants from oxidation reactions. In this instance, the suspended solids refer to the nonfilterable residue that is retained on the filter medium after filtration. All these factors ultimately decrease the overall photocatalytic efficiency for water treatment. To ensure rapid photocatalytic reaction rate, the turbidity of the targeted water should be kept below 5 nephalometric turbidity unit (NTU) for optimal UV light utilization and photocatalytic reaction [92,329]. Rincón and Pulgarin [280] observed that the water turbidity higher than 30 NTU will negatively affect the rate of photocatalytic disinfection. The limit of 5 NTU is arbitrary and depends on the receiving water bodies and the treatment levels required. Since TiO_2 photocatalytic processes are retrofitted to advanced water treatment stage, prior reduction in turbidity could be achieved via conventional treatment processes such as screening, filtration, sedimentation, coagulation, and flocculation. The standard for 1 NTU is a 1.0 mg/L of specified size of silica suspension and can be prepared for laboratory investigation. The turbidity standard can be calibrated with a photoelectric detector or nephalometry for the intensity of scattered light.

6.3.2.24 Inorganic Ions

The presence of inorganic ions in the targeted water for TiO_2 photocatalytic treatment is expected. To functionalize a TiO_2 water treatment process, the basic understanding of these inorganic ions on the photocatalytic performance is essential. Crittenden et al. [330] reported that when a photocatalyst is utilized in either slurry or fixed-bed configuration to treat real waters with different inorganic ions, photocatalyst deactivation was usually observed. This resulted from a strong inhibition from the inorganic ions on the surface of the TiO_2 semiconductor used. Thus, the presence of these inorganic ions together with their permissible levels on the photocatalytic performance of TiO_2 in water treatment has to be determined to ensure minimal disturbances on the efficient operation of the TiO_2-based treatment process. With such data, cost-effective fouling prevention with inorganic ions and photocatalyst regeneration strategies can be customized.

A number of studies have been conducted on the effects of different inorganic anions or cations, on both TiO_2 photomineralization and photodisinfection reactions [36,266,331–338]. It must be emphasized, however, that most of these studies have concentrated on how different inorganic ions affect the rates of photocatalytic reactions with a model surrogate organic compound. The surrogate model organic compound can be biased toward its photocatalytic performance owing to its underlying chemical properties and the main constituent groups that form the compound. For instance, the model compounds with either electron withdrawing or donating groups will contribute a different degree of interaction to degradation pathways [171]. Due to the zwitterionic nature of the TiO_2 particles used, it is also possible that the operating pH might have a profound effect on the selective inhibition of inorganic ions on the surface of the TiO_2 particles [332]. However, few discussions have been centered on considering how the chemical nature of the model organic used and the operating pH in the photocatalytic reactor on the inhibition of different inorganic ions in photocatalytic water treatment.

To date, the effects of both inorganic cations (i.e., Na^+, K^+, Ca^{2+}, Cu^{2+}, Mn^{2+}, Mg^{2+}, Ni^{2+}, Fe^{2+}, Zn^{2+}, Al^{3+}) and inorganic anions (i.e., Cl^-, NO^{3-}, HCO^{3-}, ClO^{4-}, SO_4^{2-}, HPO_4^{2-}, PO_4^{3-}) on the photocatalytic water treatment have been investigated [36,266,331–337,339]. A general consensus from these studies concludes that Cu^{2+}, Fe^{2+}, Al^{3+}, Cl^-, PO_4^{3-}

at certain levels may decrease photomineralization reaction rates while Ca^{2+}, Mg^{2+}, Zn^{2+} may have negligible effects. This is because the Ca^{2+}, Mg^{2+}, and Zn^{2+} are at their maximum oxidation states resulting in their inability to inhibit the photocatalysis reaction. The presence of Fe^{2+} can catalyze both the Fenton and photo-Fenton reactions. However, Choi et al. [340] observed that Fe^{2+} fouled the photocatalysts surface by introducing a rusty orange color change via the formation of $Fe(OH)_3$, while the PO_4^{3-} in the nominal pH range remains strongly adsorbed onto the TiO_2 surface and further inhibits its photoactivity [1,341]. Some research groups observed that NO_3^-, SO_4^{2-}, ClO_4^- and HCO_3^- inhibit the surface activity of the photocatalysts, while others suggest no such impact. Both NO_3^- and SO_4^{2-} have detrimental effect on the photodisinfection rate [36]. The Cu^{2+} can enhance the photocatalytic activity at its concentration up to 0.1 mM, while further increases in its concentration reduce the reaction rate [342]. Nitrogen-containing molecules are mineralized into NH_4^+ and mostly NO_3^-. Ammonium ions are relatively stable and their proportion depends mainly on the initial oxidation degree of nitrogen on the irradiation time. Pollutants containing sulfur atoms are mineralized into sulfate ions. Cl^- has no inhibition on the photocatalytic degradation of trichloroethylene at a concentration up to 3.0 mM. On contrary, the addition of oxyanion oxidants such as ClO_2^-, ClO_3^-, IO_4^-, $S_2O_8^-$, and BrO_3^- increased photoreactivity by scavenging conduction-band electrons and reducing the charge—carrier recombination [343]. The presence of salts is also known to diminish the colloidal stability as screening effects become more profound. This was followed in the double layer compression and surface charge neutralization, which increases the mass transfer limitations and reduces surface contacts between the pollutants and catalysts. Other inorganic ions also affect photodegradation rates, where the presence of SO_4^{2-} in a TiO_2-coated glass spiral reactor could double the disappearance rate of the pesticide monocrotophos [344]. The Mn^{2+} improved the photoactivity by simultaneously increasing the electron—hole pairs and preventing their recombination [345].

Several mechanisms for fouling effects of inorganic ions on TiO_2 photoactivity have been proposed [346,347]. These include UV screening, competitive adsorption to surface active sites, competition for photons, surface deposition of precipitates and elemental metals, radical and hole scavenging, and direct reaction with the photocatalyst. The NO_3^- ion was reported to UV screening the photocatalyst than inhibiting the TiO_2 surface [346]. The competition for surface active sites involves the constant displacement of hydroxide ions from TiO_2 surface and thus, further reduces the generation of radicals. Quantum yield is reduced as a result of the direct competition of inorganic ions for light photons (lowered number of photons entering the reactor). A similar decrease in photonic efficiency was observed when precipitates were formed and deposited onto the TiO_2 surface, blocking the accessibility of both photons and organic compounds. The ultimate inorganic anions that were determined to scavenge both the hole and radicals include Cl^-, HCO_3^-, SO_4^{2-}, and PO_4^{3-} [348]. A mechanism for Cl^- and HCO_3^- in inhibiting photocatalysis via hydroxyl radical and holes scavenging was proposed by Matthews and McEnvoy [349]; Lindner et al. [350], respectively:

$$Cl^- + OH^{\cdot} \rightarrow Cl^{\cdot} + OH^- \quad (6.48)$$

$$Cl^- + h^+ \rightarrow Cl^{\cdot} \quad (6.49)$$

The Cl^- accounted for its inhibitory effect on TiO_2 photocatalysis through a preferential adsorption displacement mechanism over the surface bound OH^- ions. This reduces the number of OH^- ions available on the TiO_2 surface, and the substituted Cl^- further increases the recombination of electron–hole pairs. Among other chlorinated molecules, Cl^- ions are readily released in the solution. This effect could be of benefit in a process where photocatalysis is associated with a biological depuration system, which is generally not efficient for chlorinated compounds [171]. Other ions such as PO_4^{3-} are known to avert adsorption of amino acids over the TiO_2 catalysts, while CO_3^{2-} and other ionic species react with OH^- radicals to compete with microorganisms and further reduce the efficiency. Thus, the presence of inorganic ions in water subjected to TiO_2 photocatalytic treatment is an important factor in determining its successful implementation.

To resolve TiO_2 photocatalyst fouling issue, preventative or regenerative strategies can be adopted, depending on the nature of the photocatalysts deactivation in the water matrix. Fouling preventative strategies by means of water pretreatment, complexation, and photocatalyst surface modifications can be addressed, while rinsing the TiO_2 surface with different chemical solutions constitutes the regenerative strategies. Water pretreatment with ion exchange resins is one of a few methods that can be employed to ensure minimal disturbances by inorganic ions. Burns et al. [346] discussed that the operational cost of water pretreatment with ion exchange resins can be minimized if the fouling ions were identified and selectively removed. Other preventative strategies such as complexation of fouling agents after "escorting" the ions through the reactor can be utilized, provided that the strong fouling ions are hard to remove from the feed water stream. Modifications of the TiO_2 surface to increase the hydrophobicity and adsorption capacity is an upstream preventative way to enhance the rate of photodegradation in the presence of fouling ions. However, this method is unstable and impractical compared with the others, where the modifications are reported to be displaced away with time. As for the regenerative strategies, different types of chemical rinsing were reported to potentially resolubilize surface deposits, precipitates, and reduced metals. The fouled ions usually do not form strong surface complexes and can be easily displaced by an ion exchange rinse. Abdullah et al. [1] reported that TiO_2 fouling with SO_4^{2-} and PO_4^{3-} can be displaced by NaOH, KOH, and $NaHCO_3$, while Cl^- can be easily regenerated with water. More complex mixtures of inorganic ions need to be investigated to better mimic real water matrices in photocatalytic water treatment or after investigation of water matrices with one known inorganic ion composition at a time.

6.3.2.25 Heavy and Noble Metals

Heavy metals that might be present in trace amount in the wastewater stream are highly toxic in some of their valence states [351]. Due to the pliable nature of biological treatment, these intoxicant metals can remain and permeate through the treatment process. To treat such metals, TiO_2 photocatalytic process has been reported to simultaneously convert these metals into nontoxic ionic states and further reduced them into their corresponding elemental form on the TiO_2 surface for metal recovery. Prairie et al. [352] reported that metals of Ag(I), Cr(IV), Hg(II), and Pt(II) were easily treated with TiO_2 of 0.1wt%, whereas Cd(II), Cu(II), and Ni(II) could not be removed. The extent of such metals conversion and recovery process are highly dependent on the standard reduction potential of the metals for the reduction reactions. It was reported that for an efficient removal of the metals, a positive potential of greater than 0.4 V or the flat band potential of TiO_2 was required [171,351]. Since the rates of both oxidation (organics) and reduction (metals) on the TiO_2 surface are intrinsically interrelated, the presence of sufficient organics in the water matrices was found to facilitate the metal recovery. The redox process for the metal reduction on the TiO_2 surface is given below:

$$Mn^{+} + H_2O \xrightarrow{TiO_2(hv)} M^0 + nH^{+} + \frac{n}{4}O_2 \qquad (6.50)$$

Herrmann et al. [353] reported that small crystallites of silver (3 and 8 nm) were initially deposited and began to agglomerate (few hundreds of nm) when the conversion increased. Since the photosensitive surface was not masked, a relatively large amount of silver was recovered, leaving behind concentration of lower than the detection limits of atomic absorption spectroscopy ($\leq$0.01 ppm) in the solution. The effect of various factors on the photoreduction of silver on TiO_2 surface were further investigated [354]. Angelidis et al. [355] also reported that even at low concentration of metal ions, the photodeposition of metals on the TiO_2 surface from solution was still effectively performed. However, they noted that the rate of photodeposition was enhanced when Pt loaded was used instead of unloaded TiO_2 particles. This metal photodeposition property on TiO_2 surface is particularly useful when the water legislation limit on the metal contents becomes more stringent.

6.3.2.26 Life Cycle Assessment of Photocatalytic Water Treatment Processes

In the current development of photocatalytic water/wastewater treatment processes, their possible application for the industry is still being investigated at pilot plant scale. A few pilot plants have been established to obtain feasibility data, such as the treatment efficiency, site area requirements for targeted volume, electrical energy consumption, process emissions, and chemical costs. The heterogeneous photocatalysis and photo-Fenton plants located at the INETI (Instituto Nacional de Engenharia, Technologia Industrial e Inovacao, Portugal) and PSA (Plataforma Solar de Almeria, Spain) are two renowned pilot plants that have delivered most of these data for technical analysis. Both plants consist of compound parabolic collectors (4.16 m^2 aperture area) exposed to sunlight, a reservoir tank, a recirculation pump, and connecting tubing and are operated in batch mode [356]. Further technical details of these plants can be found in the literature [357,358].

To assess these photocatalytic processes as emerging technologies for large-scale water/wastewater treatment, an LCA should be evaluated based on the currently available data. LCA is one of the most widely accepted tools that not only consider the environmental impact of the emergent photocatalytic water treatment but also its technical feasibility and costs. Andreozzi et al. [3] also pointed out that the potential application of ROS-based oxidation processes entails high costs for energy and reactants consumption. Thus, to consider the feasibility of photocatalytic water treatment on the whole, a comprehensive LCA based on viable technical data should be carried out. Muñoz et al. [359] carried out a simplified LCA based on small-scale laboratory data of heterogeneous photocatalysis and other AOPs. However, they found that small-scale laboratory data interpretation in their LCA study can lead to inconclusive results.

Using LCA, the environmental burdens from a product, process, or activity via materials and energy balances can be defined and reduced as well as the waste discharges, its impacts on the environment, and the environmental improvement opportunities over the whole life cycle [360]. This holistic LCA approach in decision-making over other environmental assessment approaches includes all the burdens and impacts, and focuses on the emissions and wastes generated [361].

Muñoz et al. [356] carried out an LCA based on the two pilot plants of INETI and PSA, which treated 1 m^3 of methyl phenyl glycerine to destroy nonbiodegradable and toxic compounds to a level that met the quality of aquatic ecosystems with both homogeneous photo-Fenton and heterogeneous photocatalysis. Other technically rational assumptions were also made to facilitate the estimation of energy and materials consumed and produced. Nine impact categories of the possible large-scale photocatalytic water treatment process were included in the analysis, namely, global warming potential, ozone depletion potential, human toxicity potential, freshwater aquatic toxicity potential, photochemical oxidant formation potential, acidification potential, eutrophication potential, nonrenewable energy consumption, and land use.

Fig. 6.16 shows the LCA results for possible large-scale water application using photocatalytic technology. The LCA results showed that the retrofitting of heterogeneous photocatalysis process to the existing biological wastewater treatment can lower eutrophication potential but require higher site area requirement and electricity consumption. These technical constraints are a direct result from the requirement for a large land area and the raw materials to build the parabolic collector infrastructure and high power needed to pump the wastewater through the system. However, the results from the impact categories cannot be compared directly with each other as they were expressed in different measurement units. From the engineering point of view, these constraints mainly arise from the low photoactivity of the catalyst used under solar irradiation. Further materials engineering solutions and studies should be carried out to resolve such technical issues to permit the scale-up of the technology to a commercially viable process.

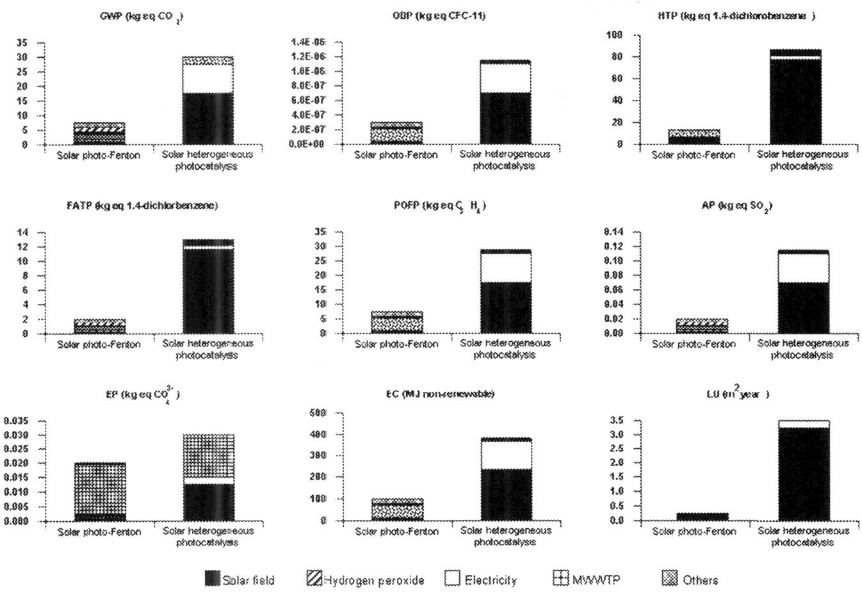

Figure 6.16 Life cycle impact assessment results for the alternatives under study [356].

6.3.2.27 Future Challenges and Prospects

Semiconductor photocatalytic technology using either UV light or solar has become more prominent owing to its advantages of the use of vast additive chemicals or disinfectants and its mineralization aspects. These are particularly important, as recalcitrant organics are mineralized rather than being transformed to another phase. Coupled with the ambient operation of the process, all these make photocatalytic water treatment technology a viable alternative for commercialization in the near future. Different water contaminants, ranging from hazardous contaminants of pesticides, herbicides, and detergents to pathogens, such as viruses, coliforms and spores, are effectively removed by this photocatalytic process.

The applicability of the heterogeneous photocatalytic technology for water treatment is constrained by several key technical issues that need to be further investigated. The first consideration would be whether the photocatalytic process is a pretreatment step or a stand-alone system. The nonselective reactivity on the nonbiodegradable water-soluble pollutants means the photocatalytic process can be used effectively as a pretreatment step to enhance biodegradation of recalcitrant organic pollutants prior to biological water treatment. In such a way, the residence time and reaction volume for the biological treatment could be significantly reduced. If the photocatalytic process is used as a stand-alone treatment system, the residence time required might be prolonged for total bacterial inactivation or mineralization. As discussed, this is hindered by the slow kinetics, low photoefficiency, and a need for continuous (without interruption) illumination to achieve the required total organic carbon removal or microbial inactivation. For the stand-alone system, the site area requirement might be proportional from any increased reaction volume required.

To promote the feasibility of photocatalytic water treatment technology in the near future, several key technical constraints ranging from catalyst development to reactor design and process optimization have to be addressed. These include (1) catalyst improvement for a high photoefficiency that can utilize wider solar spectra; (2) catalyst immobilization strategy to provide a cost-effective solid—liquid separation; (3) improvement in the photocatalytic operation for wider pH range and to minimize the addition of oxidant additives; (4) new integrated or coupling system for enhanced photomineralization or photodisinfection kinetics; and (5) effective design of photocatalytic reactor system or parabolic solar collector for higher utilization of solar energy to reduce the electricity costs. Currently, the utilization of solar energy is limited by the photoefficiency of the TiO_2 catalyst band gap to only 5% of the solar spectrum. The need for continuous illumination for efficient inactivation of pathogens has diverted solar utilization to artificial UV lamp—driven process. In addition, the low efficacy design of current solar collecting technology (0.04% capture of original solar photons) has encouraged the developmental progress of photocatalytic technology in water treatment industry. Further pilot plant investigations with different reactor configurations are needed to ensure that the photocatalytic water technology is well established and presents vast technoeconomic data for any LCA study. Finally, a large-scale photocatalytic treatment process with high efficacy, solar-driven, and low site area requirements can be realized in the short future with rapid evaluation of different possible pilot plant configurations.

ACKNOWLEDGMENTS

This topic is written by Meng Nan Chong[a,b], Bo Jin[a,b,c], Christopher W.K. Chow[c], Chris Saint[c] at [a]School of Chemical Engineering, The University of Adelaide, 5005 Adelaide, Australia; [b]School of Earth and Environmental Sciences, The University of Adelaide, Adelaide, SA 5005, Australia; [c]Australian Water Quality Centre, SA Water Corporation, 5000

Adelaide, SA, Australia. This paper was published as "Recent Developments in Photocatalytic Water Treatment Technology: A Review" in journal "Water Research". 44 (2010); 2997–3027.

6.4 RENEWABLE WATER ENGINEERING

The transpiration mechanism has been proposed for rerouting as it is the main cause of groundwater loss which is also causing significant global warming by releasing water vapor into the air. Since electrostatic force has the tendency to tug down the water, thus, a static electricity force creating plastic tank has been proposed to install at the bottom of plants to capture the transpiration water vapor and treated it *in site* by applying UV technology to meet the daily water demand throughout the world.

6.4.1 Background

Plants give O_2 and take CO_2 by the process of photosynthesis to keep the global environment in balance. Plants are simply the hero for the environment; unfortunately, hero plants are also the villain for the environment who plays the significant role in causing global warming. It is a fact that the body of plants needs water for the reaction of biochemical metabolism for its growth [255]. This water is taken up by the cohesion–tension mechanism of the soil (groundwater) through the roots, transported by osmosis through the xylem to the leaves of the plants [361]. Interestingly, only a mere 0.5%–3% of water is used by plants for their metabolism and the rest of water releases into the air through stomatal cells by transpiration process [4,52]. This process of transpiration is not only causing the largest loss of groundwater that is also causing global warming, since this water vapor is a notable cause for global warming. Recent studies on transpiration and groundwater relationship have been discussed terrestrial water fluxes where their water models revealed that streamflow getting lower due to the plants transpiration [4,255]. These are very interesting findings, but no mechanism has been studied yet to trap this transpiration water for meeting global water demand. In this research, therefore, a technology has been proposed to eliminate this water loss by diverting this transpiration mechanism by collecting this water vapor instead of allowing it to enter the air and transform it into water potable and clean energy. Simply static electricity creator plastic tank near the plants has been proposed to install to trap all the water vapor as the water vapor is attracted by the force of static electricity. Just because water vapor has positive and negative charges and the electrons that ended up on static electrical force has a positive charge, while water molecules have a negative charge on one side, the positive charge of static electric force and negative charges of water vapor pull each other closer together, the positive side tug the direction and force the water to come down to get collected in a tank and be treated *in site* to meet the daily water demand. Calculation revealed only of four standard oak trees can meet the total water for a small family throughout the year. Since the groundwater strata are getting to lower fast to finite level, and global water and global warming getting dangerous seriously to putting earth on vulnerable condition, thus these two vital needs must be resolved immediately. Interestingly, this new finding has the total solution to solve the global water, and environmental crisis for the survival of this planet which will indeed open to new door in science.

6.4.2 Methods and Simulation

6.4.2.1 Static Electric Force Generation

To capture the water vapor from air which is released by stomatal cells of the plants during the day time, a model been proposed to create *Hossain static electric force (*HsrF = h*)* by implementing the friction of insulator into the plastic tank to pulling down the water vapor into the plastic tank [1,362]. To create HsrF into the plastic tank, I have implemented abelian local symmetries calculation by using MATLAB software considering gauge field symmetry and the Goldstone scalar with respect to longitudinal mode of the vector [9,83]. Thus, for each spontaneously broken particle T^{α} of the local symmetry will be corresponding gauge field of $A^{\alpha}_{\mu}(x)$ where HsrF will started to work at a local U(1) phase symmetry [3,119]. Therefore, the model will be comprised as a complex scalar field Φ (x) of static electric charge q coupled to the EM field $A^{\mu}(x)$, which is expressed by h:

$$h = -\frac{1}{4} F_{\mu\nu}F^{\mu\nu} + D_{\mu}\Phi^{*} D^{\mu}\Phi - V(\Phi^{*}\Phi) \tag{6.51}$$

where

$$\begin{aligned} D_{\mu}\Phi(x) &= \partial_{\mu}\Phi(x) + iqA_{\mu}(x)\Phi(x) \\ D_{\mu}\Phi^{*}(x) &= \partial_{\mu}\Phi^{*}(x) - iqA_{\mu}(x)\Phi^{*}(x) \end{aligned} \tag{6.52}$$

and

$$V(\Phi^{*}\Phi) = \frac{\lambda}{2} (\Phi^{*}\Phi)^{2} + m^{2}(\Phi^{*}\Phi) \tag{6.53}$$

Suppose $\lambda > 0$ but $m^2 < 0$, so that $\Phi = 0$ is a local maximum of the scalar potential, while the minima forms a degenerate circle $\Phi = \frac{v}{\sqrt{2}} \times e^{i\theta}$,

$$v = \sqrt{\frac{-2m^2}{\lambda}}, \quad \text{any real } \theta \tag{6.54}$$

Consequently, the scalar field Φ develops a nonzero vacuum expectation value $\Phi \neq 0$, which spontaneously creates the U(1) symmetry of the static electric field. The breakdown would lead to a massless Goldstone scalar stemming from the phase of the complex field $\Phi(x)$. But for the local U(1) symmetry, the phase of $\Phi(x)$, not just the phase of the expectation value Φ but the x-dependent phase of the dynamical $\Phi(x)$ field. To analyze this static electricity force mechanism, I have used polar coordinates in the scalar field space, thus

$$\Phi\,(x) = \frac{1}{\sqrt{2}}\,\Phi_r(x) \times e^{i\Theta(x)}, \text{ real } \Phi_r(x) > 0, \text{ real } \Phi(x) \tag{6.55}$$

This field redefinition is singular when $\Phi(x) = 0$, so I never used it for theories with $\Phi \neq 0$, but it is alright for spontaneously broken theories where I can expect $\Phi x \neq 0$ almost everywhere. In terms of the real fields $\phi_r(x)$ and $\Theta\,(x)$, the scalar potential depends only on the radial field ϕ_r,

$$V(\phi) = \frac{\lambda}{8}\left(\phi_r^2 - v^2\right)^2 + \text{const}, \tag{6.56}$$

or in terms of the radial field shifted by its VEV, $\Phi_r(x) = v + \sigma(x)$,

$$\phi_r^2 - v^2 = (v + \sigma)^2 - v^2 = 2v\sigma + \sigma^2 \tag{6.57}$$

$$V = \frac{\lambda}{8}\left(2v\sigma - \sigma^2\right)^2 = \frac{\lambda v^2}{2} \times \sigma^2 + \frac{\lambda v}{2} \times \sigma^3 + \frac{\lambda}{8} \times \sigma^4 \tag{6.58}$$

At the same time, the covariant derivative $D_\mu\phi$ becomes

$$\begin{aligned} D_\mu\phi &= \frac{1}{\sqrt{2}}\left(\partial_\mu\left(\phi_r e^{i\Theta}\right) + iqA_\mu \times \phi_r e^{i\Theta}\right) \\ &= \frac{e^{i\Theta}}{\sqrt{2}}\left(\partial_\mu\phi_r + \phi_r \times i\partial_\mu\Theta + \phi_r \times iqA_\mu\right) \end{aligned} \tag{6.59}$$

$$\begin{aligned}\left|D_\mu\phi\right|^2 &= \frac{1}{2}\left|\partial_\mu\phi_r + \phi_r \times i\partial_\mu\Theta + \phi_r \times iqA_\mu\right|^2\\ &= \frac{1}{2}\left(\partial_\mu\phi_r\right) + \frac{\phi_r^2}{2} \times \left(\partial_\mu\Theta qA_\mu\right)^2\\ &= \frac{1}{2}\left(\partial_\mu\sigma\right)^2 + \frac{(v+\sigma)^2}{2} \times \left(\partial_\mu\Theta + qA_\mu\right)^2\end{aligned} \tag{6.60}$$

Altogether,

$$h = \frac{1}{2}\left(\partial_\mu \sigma\right)^2 - \nu(\sigma) - \frac{1}{4}F_{\mu\nu}F^{\mu\nu} + \frac{(v+\sigma)^2}{2} * \left(\partial_\mu\Theta + qA_\mu\right)^2 \tag{6.61}$$

To confirm the creating of this static electric force (h_{sef}) into the static electric field properties of this *HsrF*, it has been expanded in powers of the fields (and their derivatives) and focus on the quadratic part describing the free particles,

$$h_{sef} = \frac{1}{2}\left(\partial_\mu\sigma\right)^2 - \frac{\lambda v^2}{2} \times \sigma^2 - \frac{1}{4}F_{\mu\nu}F^{\mu\nu} + \frac{v^2}{2} \times \left(qA_\mu + \partial_\mu\Theta\right)^2 \tag{6.62}$$

Here this HsrF (h_{free}) function obviously will suggest a real scalar particle of positive $mass^2 = \lambda v^2$ involving the $A_\mu(x)$ and the $\Theta(x)$ fields to initiate to create tremendous static electricity force within the electric field of the plastic tank (Fig. 6.17).

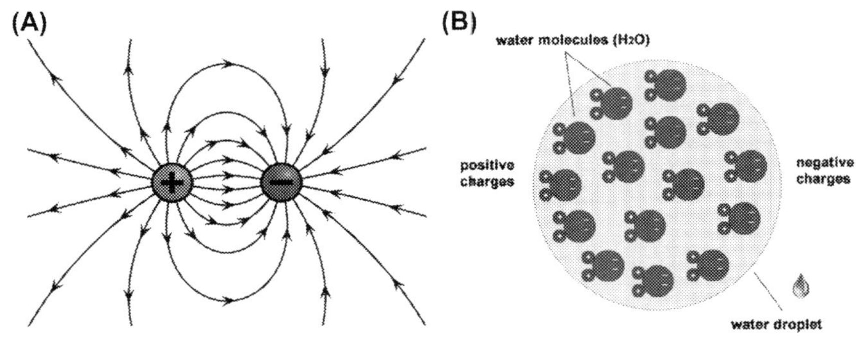

Figure 6.17 (A) The creation of static electricity force, and (B) its mechanism of conversion of static energy into an electromotive force of positive and negative charges that mobilize the "static" electricity to tug down the water molecules.

6.4.2.2 In Site Water Treatment

Since the collected water in the plastic tank is just nothing but the liquid form of vapor, thus it will not require any sedimentation, coagulation, and chlorination to clean the water. Only mixing physics (UV application) and filtration will be required to treat the water to meet the US National Primary Drinking Water Standard code [cc]. It is the simplest way to treat water by using *SODIS* system (*SOlar DISinfection*), where a transparent container is filled with water and exposed to full sunlight for several hours. As soon as the water temperature reaches 50°C with a UV radiation of 320 nm, the inactivation process will be accelerated to lead to complete microbiological disinfection immediately and the treated water shall be used to meet the total domestic water demand (Fig. 6.18).

6.4.3 Results and Discussion

To mathematically determine the electric static force proliferation around the plastic tank to confirm to tug down the water, I have initially solved the dynamic photon proliferation by integrating *HSEF* electric field creation thus, the local U(1) gauge invariant did allow to add a mass term for the gauge particle under $\varnothing' \rightarrow e^{i\alpha(x)}\varnothing$. In detail, it can be explained by a covariant derivative with a special transformation rule for the scalar field expressed by Refs [51,234]:

$$\begin{aligned} &\partial_\mu \rightarrow D_\mu = \partial_\mu = ieA_\mu \ [\text{covariant derivatives}] \\ &A'_\mu = A_\mu + \frac{1}{e}\ \partial_\mu\alpha \ [A_\mu\text{derivatives}] \end{aligned} \tag{6.63}$$

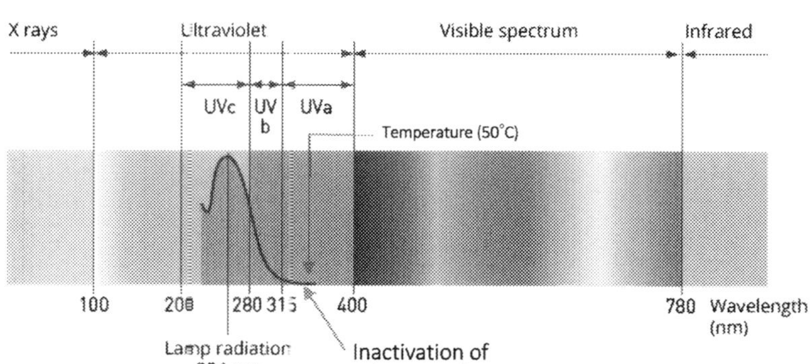

Figure 6.18 The photophysics radiation application for the purification of water which shows that once UV radiation of 320 nm applied into the water, it stars to disinfect all microorganisms immediately once temperature reaches at 50°C.

where the local U(1) gauge invariant *HSEF* for a complex scalar field is given by:

$$h = (D^{\mu})^{+}\left(D_{\mu}\varnothing\right) - \frac{1}{4}F_{\mu\nu}F^{\mu\nu} - V(\varnothing) \tag{6.64}$$

The term $\frac{1}{4}F_{\mu\nu}F^{\mu\nu}$ is the kinetic term for the gauge field (heating photon) and $V(\varnothing)$ is the extra term in the HSEF, then $V(\varnothing^{*}\varnothing) = \mu^{2}(\varnothing^{*}\varnothing) + \lambda(\varnothing^{*}\varnothing)^{2}$.

Therefore, the HSEF (h) under perturbations into the quantum field have initiated with the massive scalar particles ϕ_1 and ϕ_2 along with a mass μ. In this situation, $\mu^2 < 0$ had an infinite number of quantum, each has been satisfied by $\phi_1^2 + \phi_2^2 = -\mu^2/\lambda = v^2$ and the h through the covariant derivatives using again the shifted fields η and ξ defined the quantum field as $\phi_0 = \frac{1}{\sqrt{2}}[(\upsilon + \eta) + i\xi]$.

$$\begin{aligned}\text{Kinetic term: } h_{(\eta,\xi)} &= (D^{\mu}\phi)^{+}(D^{\mu}\phi)\\ &= (\partial^{\mu} + ieA^{\mu})\phi^{+}\left(\partial_{\mu} - ieA_{\mu}\right)\phi\end{aligned} \tag{6.65}$$

Thus, this expanding term in the h associated to the scalar field is suggesting that HSEF electric field is prepared to initiate the proliferation of static electricity force into its quantum field to tug down the water [2,6].

To confirm this tug downing water by static electricity force, hereby, with readily I have implemented the calculation of $\overline{\varphi}[s_0]$ for the confirmation of the expected value of s_0 for capturing water vapor [69,355]. Thus, the corrective functional asymptotic formulas are being used as follows:

$$\begin{aligned}\overline{\varphi}[s_0] &= 2s_0(\ln 4s_0 - 2) + \ln 4s_0(\ln 4s_0 - 2)\\ &\quad - \frac{\pi^2 - 9}{3} + s_0^{-1}\left(\ln 4s_0 - \frac{9}{8}\right) + \ldots (s_0 \gg 1);\end{aligned} \tag{6.66}$$

$$\overline{\varphi}[s_0] = \left(\frac{2}{3}\right)(s_0 - 1)^{\frac{2}{3}} + \left(\frac{5}{3}\right)(s_0 - 1)^{\frac{5}{3}} - \left(\frac{1507}{420}\right)(s_0 - 1)^{\frac{7}{3}} + \ldots (s_0 \gg 1). \tag{6.67}$$

The function $\frac{\overline{\varphi}[s_0]}{(s_0 - 1)}$ is thus described as $1 < s_0 < 10$; for larger s_0, and it contains natural logarithmic which is s_0 to confirm the tug down of 100% water vapor by the *HSEF* into the plastic tank.

Then the application of mixing physics (UV application) and filtration for treating this water into the plastic tank as per US National Primary

Drinking Water Standard code confirms that this water is potable which was analyzed by *SODIS* system (*SOlar DISinfection*).

In average, 100 gal water is required per day per person in a standard daily life [7,253]. Thus, it will require total (100 gal/day/person × 4 persons × 365 days) 146,000 gal water per year a small family of four persons. Since a standard oak tree can transpire 40,000 gal (151,000 L) per year, thus, tug down of 100% water vapor by HSEF described above will require only four standard oak trees to satisfy the total water demand for a small family.

6.4.4 Conclusions

Water and environmental vulnerability are the top two problems on earth where trees play a significant role in creating these problems by the process of transpiration. To mitigate these problems, transpiration mechanism has been proposed to transform and convert it into clean water to meet the global water demand and reduce the global warming by the utilization of electrostatic force to capture this transpiration water vapor and treat in site by UV application would indeed be a novel, integrated, and innovative field in science to console the global water and global warming crisis.

ACKNOWLEDGMENTS

This research was supported by Green Globe Technology under grant RD-02018-06 for building a better environment. Any findings, predictions, and conclusions described in this article are solely performed by the authors, and we confirm that there is no conflict of interest for publishing in a suitable journal.

REFERENCES

[1] Abdullah M, Low GKC, Matthews RW. Effects of common inorganic anions on rates of photocatalytic oxidation of organic carbon over illuminated titanium dioxide. J Phys Chem 1990;94:6820–5.

[2] Anotai J. Effect of calcium ion and chemistry and disinfection efficiency of free chlorine at pH 10 [Ph.D. dissertation]. Philadelphia: Drexel University; 1996.

[3] Andreozzi R, Caprio V, Insola A, Marotta R. Advanced oxidation processes (AOP) for water purification and recovery. Catal Today 1999;53:51–9.

[4] Augugliaro V, García-López E, Loddo V, Malato-Rodíguez S, Maldonado I, Marcí G, Molinari R, Palmisano L. Degradation of linomycin in aqueous medium: coupling of solar photocatalysis and membrane separation. Sol Energy 2005;79:402–8.

[5] Ebert K, Fritsch D, Koll J, Tjahjawiguna C. Influence of inorganic fillers on the compaction behaviour of porous polymer based membranes. J Membr Sci 2004;233(1–2):71–8.

[6] Artale MA, Augugliaro V, Drioli E, Golemme G, Grande C, Loddo V, Molinari R, Palmisano L, Schiavello M. Preparation and characterisation of membranes with entrapped TiO_2 and preliminary photocatalytic tests. Ann Chim 2001;91:127–36.

[7] Bacardit J, Stötzner J, Chamarro E. Effect of salinity on the photo-Fenton process. Ind Eng Chem Res 2007;46:7615–9.
[8] Elimelech M, Phillip WA. The future of seawater desalination: energy, technology, and the environment. Science 2011;333(6043):712–7.
[9] Albu SP, Ghicov A, Macak JM, Hahn R, Schmuki P. Self-organized, free-standing TiO_2 nanotube membrane for flow through photocatalytic applications. Nano Lett 2007;7:1286–9.
[10] Bouguecha S, Hamrouni B, Dhahbi M. Small scale desalination pilots powered by renewable energy sources: case studies. Desalination 2005;183:151–65.
[11] Bamba D, Atheba P, Robert D, Trokourey A, Dongui B. Photocatalytic degradation of the diuron pesticide. Environ Chem Lett 2008;6:163–7.
[12] Daus B, Wennrich R, Weiss H. Sorption materials for arsenic removal from water: a comparative study. Water Res 2004;38(12):2948–54.
[13] Esplugas S, Giménez J, Conteras S, Pascual E, Rodriguez M. Comparison of different advanced oxidation processes for phenol degradation. Water Res 2002;36:1034–42.
[14] Bellobono IR, Morazzoni F, Bianchi R, Mangone ES, Stanescu R, Costache C, Tozzi PM. Solar energy driven photocatalytic membrane modules for water reuse in agricultural and food industries. Pre-industrial experience using s-triazines as model molecules. Int J Photoenergy 2005;7:87–94.
[15] Fogler HS. Elements of chemical reaction engineering: chapter 10: catalysis and catalytic reactors. Prentice-Hall PTR Inc.; 1999. p. 581–685.
[16] Beltran-Heredia J, Torregrosa J, Dominguez JR, Peres JA. Comparison of the degradation of p-hydroxybenzoic acid in aqueous solution by several oxidation processes. Chemosphere 2001;42:351–9.
[17] Gogniat G, Thyssen M, Denis M, Pulgarin C, Dukan S. The bactericidal effect of TiO_2 photocatalysis involves adsorption onto catalyst and the loss of membrane integrity. FEMS Microbiol Lett 2006;258:18–24.
[18] Cassano AE, Alfano OM. Reaction engineering of suspended solid heterogeneous photocatalytic reactors. Catal Today 2000;58:167–97.
[19] WHO. Progress on drinking water and sanitation. 2012. 2012 Update.
[20] Cho M, Yoon J. Measurement of OH radical CT for inactivating Cryptosporidium parvum using photo/ferrioxalate and photo/TiO_2 systems. J Appl Microbiol 2008;104:759–66.
[21] Qu XL, Brame J, Li Q, Alvarez JJP. Nanotechnology for a safe and sustainable water supply: enabling integrated water treatment and reuse. Acc Chem Res 2013;46(3):834–43.
[22] Gernjak W, Maldonado MI, Malato S, Cáceres J, Krutzler T, Glaser A, Bauer R. Pilot-plant treatment of olive mill wastewater (OMW) by solar TiO_2 photocatalysis and solar photo-Fenton. Sol Energy 2004;77:567–72.
[23] Pan B, Xing BS. Adsorption mechanisms of organic chemicals on carbon nanotubes. Environ Sci Technol 2008;42(24):9005–13.
[24] Ryu J, Choi W, Choo KH. A pilot-scale photocatalyste membrane hybrid reactor: performance and characterization. Water Sci Technol 2005;51:491–7.
[25] Yang K, Xing BS. Adsorption of organic compounds by carbon nanomaterials in aqueous phase: polanyi theory and its application. Chem Rev 2010;110(10):5989–6008.
[26] Curcó D, Giménez J, Addarak A, Cervera-March S, Esplugas S. Effects of radiation absorption and catalyst concentration on the photocatalytic degradation of pollutants. Catal Today 2002;76:177–88.
[27] Pan B, Lin DH, Mashayekhi H, Xing BS. Adsorption and hysteresis of bisphenol A and 17 alpha-ethinyl estradiol on carbon nanomaterials. Environ Sci Technol 2008;42(15):5480–5.

[28] Bellobono IR, Morazzoni F, Tozzi PM. Photocatalytic membrane modules for drinking water purification in domestic and community appliances. Int J Photoenergy 2005;7:109–13.
[29] Ji LL, Chen W, Duan L, Zhu DQ. Mechanisms for strong adsorption of tetracycline to carbon nanotubes: a comparative study using activated carbon and graphite as adsorbents. Environ Sci Technol 2009;43(7):2322–7.
[30] Joo J, Kwon SG Yu T, Cho M, Lee J, Yoon J, Hyeon T. Large-scale synthesis of TiO_2 nanorods via nonhydrolytic solegel ester elimination reaction and their application to photocatalytic inactivation of *E. coli*. J Phys Chem B 2005;109:15297–302.
[31] Chen CC, Lu CS, Chung YC, Jan JL. UV light induced photodegradation of malachite green on TiO_2 nanoparticles. J Hazard Mater 2007;141:520–8.
[32] Lin DH, Xing BS. Adsorption of phenolic compounds by carbon nanotubes: role of aromaticity and substitution of hydroxyl groups. Environ Sci Technol 2008;42(19): 7254–9.
[33] Lu CS, Chiu H, Liu CT. Removal of zinc(II) from aqueous solution by purified carbon nanotubes: kinetics and equilibrium studies. Ind Eng Chem Res 2006;45(8): 2850–5.
[34] Yang K, Wu WH, Jing QF, Zhu LZ. Aqueous adsorption of aniline, phenol, and their substitutes by multi-walled carbon manotubes. Environ Sci Technol 2008; 42(21):7931–6.
[35] Fornasiero F, Park HG, Holt JK, Stadermann M, Grigoropoulos CP, Noy A, Bakajin O. Ion exclusion by sub-2-nm carbon nanotube pores. Proc Natl Acad Sci USA 2008;105(45):17250–5.
[36] Habibi MH, Hassanzadeh A, Mahdavi S. The effect of operational parameters on the photocatalytic degradation of three textile azo dyes in aqueous TiO_2 suspensions. J Photochem Photobiol A Chem 2005;172:89–96.
[37] Rao GP, Lu C, Su F. Sorption of divalent metal ions from aqueous solution by carbon nanotubes: a review. Sep Purif Technol 2007;58(1):224–31.
[38] Li YH, Ding J, Luan ZK, Di ZC, Zhu YF, Xu CL, Wu DH, Wei BQ. Competitive adsorption of Pb^{2+}, Cu^{2+} and Cd^{2+} ions from aqueous solutions by multiwalled carbon nanotubes. Carbon 2003;41(14):2787–92.
[39] Feng QL, Wu J, Chen GQ, Cui FZ, Kim TN, Kim JO. A mechanistic study of the antibacterial effect of silver ions on *Escherichia coli* and *Staphylococcus aureus*. J Biomed Mater Res 2000 52(4):662–8.
[40] Gao W, Majumder M, Alemany LB, Narayanan TN, Ibarra MA, Pradhan BK, Ajayan PM. Engineered graphite oxide materials for application in water purification. ACS Appl Mater Interfaces 2011;3(6):1821–6.
[41] Li YH, Di ZC, Ding J, Wu DH, Luan ZK, Zhu YQ. Adsorption thermodynamic, kinetic and desorption studies of Pb^{2+} on carbon nanotubes. Water Res 2005; 39(4):605–9.
[42] Liu ZY, Bai HW, Lee J, Sun DD. A low-energy forward osmosis process to produce drinking water. Energy Environ Sci 2011;4(7):2582–5.
[43] Lin YH, Tseng WL. Ultrasensitive sensing of Hg(2fl) and CH(3)Hg(fl) based on the fluorescence quenching of lysozyme type VI-stabilized gold nanoclusters. Anal Chem 2010;82(22):9194–200.
[44] Lu C, Chiu H, Bai H. Comparisons of adsorbent cost for the removal of zinc (II) from aqueous solution by carbon nanotubes and activated carbon. J Nanosci Nanotechnol 2007;7(4–5):1647–52.
[45] Douglas JS, Habibian H, Hung C-L, Gorshkov AV, Kimble HJ, Chang DE. Quantum many-body models with cold atoms coupled to photonic crystals. Nat Photonics 2015.

[46] Koeppenkastrop D, Decarlo EH. Uptake of rare-earth elements from solution by metal-oxides. Environ Sci Technol 1993;27(9):1796–802.
[47] Sukhanova A, Devy M, Venteo L, Kaplan H, Artemyev M, Oleinikov V, Klinov D, Pluot M, Cohen JHM, Nabiev I. Biocompatible fluorescent nanocrystals for immunolabeling of membrane proteins and cells. Anal Biochem 2004;324(1):60–7.
[48] Trivedi P, Axe L. Modeling Cd and Zn sorption to hydrous metal oxides. Environ Sci Technol 2000;34(11):2215–23.
[49] Rivero MJ, Parsons SA, Jeffrey P, Pidou M, Jefferson B. Membrane chemical reactor (MCR) combining photocatalysis and microfiltration for grey water treatment. Water Sci Technol 2006;53:173–80.
[50] Yean S, Cong L, Yavuz CT, Mayo JT, Yu WW, Kan AT, Colvin VL, Tomson MB. Effect of magnetite particle size on adsorption and desorption of arsenite and arsenate. J Mater Res 2005;20(12):3255–64.
[51] Auffan M, Rose J, Bottero JY, Lowry GV, Jolivet JP, Wiesner MR. Towards a definition of inorganic nanoparticles from an environmental, health and safety perspective. Nat Nanotechnol 2009;4(10):634–41.
[52] Auffan M, Rose J, Proux O, Borschneck D, Masion A, Chaurand P, Hazemann JL, Chaneac C, Jolivet JP, Wiesner MR, Van Geen A, Bottero JY. Enhanced adsorption of arsenic onto maghemites nanoparticles: as(III) as a probe of the surface structure and heterogeneity. Langmuir 2008;24(7):3215–22.
[53] Yavuz CT, Mayo JT, Yu WW, Prakash A, Falkner JC, Yean S, Cong LL, Shipley HJ, Kan A, Tomson M, Natelson D, Colvin VL. Low-field magnetic separation of monodisperse Fe_3O_4 nanocrystals. Science 2006;314(5801):964–7.
[54] Lucas E, Decker S, Khaleel A, Seitz A, Fultz S, Ponce A, Li WF, Carnes C, Klabunde KJ. Nanocrystalline metal oxides as unique chemical reagents/sorbents. Chem A Eur J 2001;7(12):2505–10.
[55] Seven O, Dindar B, Aydenir S, Metin D, Ozinel MA, Icli S. Solar photocatalytic disinfection of a group of bacteria and fungi aqueous suspensions with TiO_2, ZnO and Sahara desert dust. J Photochem Photobiol A Chem 2004;165:103–7.
[56] Sharma YC, Srivastava V, Singh VK, Kaul SN, Weng CH. Nano-adsorbents for the removal of metallic pollutants from water and wastewater. Environ Technol 2009; 30(6):583–609.
[57] Deliyanni EA, Bakoyannakis DN, Zouboulis AI, Matis KA. Sorption of As(V) ions by akaganeite-type nanocrystals. Chemosphere 2003;50(1):155–63.
[58] Mayo JT, Yavuz C, Yean S, Cong L, Shipley H, Yu W, Falkner J, Kan A, Tomson M, Colvin VL. The effect of nanocrystalline magnetite size on arsenic removal. Sci Technol Adv Mater 2007;8(1–2):71–5.
[59] Hossain MF. Photonic thermal energy control to naturally cool and heat the building. Adv Therm Eng 2018;131:576–86 [Elsevier].
[60] Hristovski KD, Westerhoff PK, Moller T, Sylvester P. Effect of synthesis conditions on nano-iron (hydr)oxide impregnated granulated activated carbon. Chem Eng J 2009;146(2):237–43.
[61] Zhang H, Quan X, Chen S, Zhao H, Zhao Y. Fabrication of photocatalytic membrane and evaluation its efficiency in removal of organic pollutants from water. Sep Purif Technol 2006;50:147–55.
[62] Hu J, Chen GH, Lo IMC. Selective removal of heavy metals from industrial wastewater using maghemite nanoparticle: performance and mechanisms. J Environ Eng 2006;132(7):709–15.
[63] Kim J, Lee CW, Choi W. Platinized WO(3) as an environmental photocatalyst that generates OH radicals under visible light. Environ Sci Technol 2010;44(17): 6849–54.

[64] de Villoria RG, Hart AJ, Wardle BL. Continuous highyield production of vertically aligned carbon nanotubes on 2D and 3D substrates. ACS Nano 2011;5(6):4850–7.
[65] Evgenidou E, Fytianos K, Poulios I. Semiconductor sensitized photodegradation of dichlorvos in water using TiO_2 and ZnO as catalysts. Appl Catal B Environ 2005; 59:81–9.
[66] Collins PG, Bradley K, Ishigami M, Zettl A. Extreme oxygen sensitivity of electronic properties of carbon nanotubes. Science 2000;287(5459):1801–4.
[67] Crooks RM, Zhao MQ, Sun L, Chechik V, Yeung LK. Dendrimer-encapsulated metal nanoparticles: synthesis, characterization, and applications to catalysis. Acc Chem Res 2001 34(3):181–90.
[68] Diallo MS, Christie S, Swaminathan P, Johnson JH, Goddard WA. Dendrimer enhanced ultrafiltration. 1. Recovery of Cu(II) from aqueous solutions using PAMAM dendrimers with ethylenediamine core and terminal NH_2 groups. Environ Sci Technol 2005;39(5):1366–77.
[69] Aragon M, Kottenstette R, Dwyer B, Aragon A, Everett R, Holub W, Siegel M, Wright J. Arsenic pilot plant operation and results. Anthony (New Mexico): Sandia National Laboratories; 2007.
[70] Sylvester P, Westerhoff P, Mooller T, Badruzzaman M, Boyd O. A hybrid sorbent utilizing nanoparticles of hydrous iron oxide for arsenic removal from drinking water. Environ Eng Sci 2007;24(1):104–12.
[71] Westerhoff P, De Haan M, Martindale A, Badruzzaman M. Arsenic adsorptive media technology selection strategies. Water Qual Res J Canada 2006;41(2):171–84.
[72] Cloete TE, Kwaadsteniet Md, Botes M, Lopez-Romero JM. Nanotechnology in water treatment applications. Caister Academic Press; 2010.
[73] Li D, Xia YN. Electrospinning of nanofibers: reinventing the wheel? Adv Mater 2004;16(14):1151–70.
[74] Ramakrishna S, Fujihara K Teo WE, Yong T, Ma ZW, Ramaseshan R. Electrospun nanofibers: solving global issues. Mater Today 2006;9(3):40–50.
[75] Bae TH, Tak TM. Effect of TiO_2 nanoparticles on fouling mitigation of ultrafiltration membranes for activated sludge filtration. J Membr Sci 2005;249(1–2):1–8.
[76] Bottino A, Capannelli G, D'Asti V, Piaggio P. Preparation and properties of novel organic-inorganic porous membranes. Sep Purif Technol 2001;22–23(1–3): 269–75.
[77] Maximous N, Nakhla G, Wong K, Wan W. Optimization of Al(2)O(3)/PES membranes for wastewater filtration. Sep Purif Technol 2010;73(2):294–301.
[78] Pendergast MTM, Nygaard JM, Ghosh AK, Hoek EMV. Using nanocomposite materials technology to understand and control reverse osmosis membrane compaction. Desalination 2010;261(3):255–63.
[79] De Gusseme B, Hennebel T, Christiaens E, Saveyn H, Verbeken K, Fitts JP, Boon N, Verstraete W. Virus disinfection in water by biogenic silver immobilized in polyvinylidene fluoride membranes. Water Res 2011;45(4):1856–64.
[80] Mauter MS, Wang Y, Okemgbo KC, Osuji CO, Giannelis EP, Elimelech M. Antifouling ultrafiltration membranes via post-fabrication grafting of biocidal nanomaterials. ACS Appl Mater Interfaces 2011;3(8):2861–8.
[81] Zodrow K, Brunet L, Mahendra S, Li D, Zhang A, Li QL, Alvarez PJJ. Polysulfone ultrafiltration membranes impregnated with silver nanoparticles show improved biofouling resistance and virus removal. Water Res 2009;43(3):715–23.
[82] Brady-Estevez AS, Kang S Elimelech M. A single-walled carbon-nanotube filter for removal of viral and bacterial pathogens. Small 2008;4(4):481–4.

[83] Ahmed F, Santos CM, Vergara R, Tria MCR, Advincula R, Rodrigues DF. Antimicrobial applications of electroactive PVK-SWNT nanocomposites. Environ Sci Technol 2012;46(3):1804−10.
[84] Choi JH, Jegal J, Kim WN. Fabrication and characterization of multi-walled carbon nanotubes/polymer blend membranes. J Membr Sci 2006;284(1−2):406−15.
[85] Choi H, Stathatos E, Dionysiou DD. Sol-gel preparation of mesoporous photocatalytic TiO_2 films and TiO_2/Al_2O_3 composite membranes for environmental applications. Appl Catal B Environ 2006;63(1−2):60−7.
[86] Wu L, Shamsuzzoha M, Ritchie SMC. Preparation of cellulose acetate supported zero-valent iron nanoparticles for the dechlorination of trichloroethylene in water. J Nanoparticle Res 2005;7(4−5):469−76.
[87] Wu LF, Ritchie SMC. Enhanced dechlorination of trichloroethylene by membrane-supported Pd-coated iron nanoparticles. Environ Prog 2008;27(2):218−24.
[88] Lind ML, Ghosh AK, Jawor A, Huang XF, Hou W, Yang Y, Hoek EMV. Influence of zeolite crystal size on zeolitepolyamide thin film nanocomposite membranes. Langmuir 2009;25(17):10139−45.
[89] Jeong BH, Hoek EMV, Yan YS, Subramani A, Huang XF, Hurwitz G, Ghosh AK, Jawor A. Interfacial polymerization of thin film nanocomposites: a new concept for reverse osmosis membranes. J Membr Sci 2007;294(1−2):1−7.
[90] Lind ML, Suk DE, Nguyen TV, Hoek EMV. Tailoring the structure of thin film nanocomposite membranes to achieve seawater RD membrane performance. Environ Sci Technol 2010;44(21):8230−5.
[91] Lind ML, Jeong BH, Subramani A, Huang XF, Hoek EMV. Effect of mobile cation on zeolite-polyamide thin film nanocomposite membranes. J Mater Res 2009;24(5): 1624−31.
[92] Fernándezs-Ináñez P, Sichel C, Poloo López MI, de Cara- Garcia M, Tello JC. Photocatalytic disinfection of natural well water contaminated by Fusarium solani using TiO_2 slurry in solar CPC photo-reactors. Catal Today 2009;144:62−8.
[93] Chin SS, Chiang K, Fane AG. The stability of polymeric membranes in TiO_2 photocatalysis process. J Membr Sci 2006;275(1−2):202−11.
[94] Tiraferri A, Vecitis CD, Elimelech M. Covalent binding of single-walled carbon nanotubes to polyamide membranes for antimicrobial surface properties. ACS Appl Mater Interfaces 2011;3(8):2869−77.
[95] Kumar M, Grzelakowski M, Zilles J, Clark M, Meier W. Highly permeable polymeric membranes based on the incorporation of the functional water channel protein Aquaporin Z. Proc Natl Acad Sci USA 2007;104(52):20719−24.
[96] Kaufman Y, Berman A, Freger V. Supported lipid bilayer membranes for water purification by reverse osmosis. Langmuir 2010;26(10):7388−95.
[97] Holt JK, Park HG, Wang YM, Stadermann M, Artyukhin AB, Grigoropoulos CP, Noy A, Bakajin O. Fast mass transport through sub-2-nanometer carbon nanotubes. Science 2006;312(5776):1034−7.
[98] Hummer G, Rasaiah JC, Noworyta JP. Water conduction through the hydrophobic channel of a carbon nanotube. Nature 2001;414(6860):188−90.
[99] Pendergast MM, Hoek EMV. A review of water treatment membrane nanotechnologies. Energy Environ Sci 2011;4(6):1946−71.
[100] Mauter MS, Elimelech M. Environmental applications of carbon-based nanomaterials. Environ Sci Technol 2008;42(16):5843−59.

[101] Nednoor P, Chopra N, Gavalas V, Bachas LG, Hinds BJ. Reversible biochemical switching of ionic transport through aligned carbon nanotube membranes. Chem Mater 2005;17(14):3595–9.
[102] Hinds B. Dramatic transport properties of carbon nanotube membranes for a robust protein channel mimetic platform. Curr Opin Solid State Mater Sci 2012;16(1):1–9.
[103] Mauter MS, Elimelech M, Osuji CO. Nanocomposites of vertically aligned single-walled carbon nanotubes by magnetic alignment and polymerization of a lyotropic precursor. ACS Nano 2010;4(11):6651–8.
[104] Ge QC, Su JC, Chung TS, Amy G. Hydrophilic superparamagnetic nanoparticles: synthesis, characterization, and performance in forward osmosis processes. Ind Eng Chem Res 2011;50(1):382–8.
[105] Fujishima A, Zhang X, Tryk DA. TiO_2 photocatalysis and related surface phenomena. Surf Sci Rep 2008;63(12):515–82.
[106] Zhang HZ, Banfield JF. Understanding polymorphic phase transformation behavior during growth of nanocrystalline aggregates: insights from TiO_2. J Phys Chem B 2000;104(15):3481–7.
[107] Zhang ZB, Wang CC, Zakaria R, Ying JY. Role of particle size in nanocrystalline TiO_2-based photocatalysts. J Phys Chem B 1998;102(52):10871–8.
[108] Macak JM, Zlamal M, Krysa J, Schmuki P. Self organized TiO_2 nanotube layers as highly efficient photocatalysts. Small 2007;3(2):300–4.
[109] Ni M, Leung MKH, Leung DYC, Sumathy K. A review and recent developments in photocatalytic water-splitting using TiO_2 for hydrogen production. Renew Sustain Energy Rev 2007;11(3):401–25.
[110] Han XG, Kuang Q, Jin MS, Xie ZX, Zheng LS. Synthesis of titania nanosheets with a high percentage of exposed (001) facets and related photocatalytic properties. J Am Chem Soc 2009;131(9):3152.
[111] Murakami N, Kurihara Y, Tsubota T, Ohno T. Shape controlled anatase titanium(IV) oxide particles prepared by hydrothermal treatment of peroxo titanic acid in the presence of polyvinyl alcohol. J Phys Chem C 2009;113(8):3062–9.
[112] Liu SW, Yu JG, Jaroniec M. Anatase TiO(2) with dominant high-energy {001} facets: synthesis, properties, and applications. Chem Mater 2011;23(18):4085–93.
[113] Kitano M, Funatsu K, Matsuoka M, Ueshima M, Anpo M. Preparation of nitrogen-substituted TiO(2) thin film photocatalysts by the radio frequency magnetron sputtering deposition method and their photocatalytic reactivity under visible light irradiation. J Phys Chem B 2006;110(50):25266–72.
[114] Kominami H, Yabutani K, Yamamoto T, Kara Y, Ohtani B. Synthesis of highly active tungsten(VI) oxide photocatalysts for oxygen evolution by hydrothermal treatment of aqueous tungstic acid solutions. J Mater Chem 2001;11(12):3222–7.
[115] Lee J, Mackeyev Y, Cho M, Wilson LJ, Kim JH, Alvarez PJJ. C(60) aminofullerene immobilized on silica as a visible light-activated photocatalyst. Environ Sci Technol 2010;44(24):9488–95.
[116] Lof R, Van Veenendaal M, Jonkman H, Sawatzky G. Band gap, excitons and Coulomb interactions of solid C 60. J Electron Spectrosc Relat Phenom 1995;72:83–7.
[117] Brunet L, Lyon DY, Hotze EM, Alvarez PJJ, Wiesner MR. Comparative photoactivity and antibacterial properties of C-60 fullerenes and titanium dioxide nanoparticles. Environ Sci Technol 2009;43(12):4355–60.
[118] Chong MN, Jin B, Chow CWK, Saint C. Recent developments in photocatalytic water treatment technology: a review. Water Res 2010;44(10):2997–3027.

[119] Al-Bastaki NM. Performance of advanced methods for treatment of wastewater: UV/ TiO_2, RO and UF. Chem Eng Process 2004;43(7):935–40.
[120] Benotti MJ, Stanford BD, Wert EC, Snyder SA. Evaluation of a photocatalytic reactor membrane pilot system of pharmaceuticals and endocrine disrupting compounds from water. Water Res 2009;43:1513–22.
[121] Westerhoff P, Moon H, Minakata D, Crittenden J. Oxidation of organics in retentates from reverse osmosis wastewater reuse facilities. Water Res 2009;43(16): 3992–8.
[122] Nawrocki J, Kasprzyk-Hordern B. The efficiency and mechanisms of catalytic ozonation. Appl Catal B Environ 2010;99(1–2):27–42.
[123] Orge CA, Orfao JJM, Pereira MFR, de Farias AMD, Neto RCR, Fraga MA. Ozonation of model organic compounds catalysed by nanostructured cerium oxides. Appl Catal B Environ 2011;103(1–2):190–9.
[124] Li QL, Mahendra S, Lyon DY, Brunet L, Liga MV, Li D, Alvarez PJJ. Antimicrobial nanomaterials for water disinfection and microbial control: potential applications and implications. Water Res 2008;42(18):4591–602.
[125] Xiu ZM, Ma J, Alvarez PJJ. Differential effect of common ligands and molecular oxygen on antimicrobial activity of silver nanoparticles versus silver ions. Environ Sci Technol 2011;45(20):9003–8.
[126] Xiu ZM, Zhang QB, Puppala HL, Colvin VL, Alvarez JJP. Negligible particle-specific antibacterial activity of silver nanoparticles. Nano Lett 2012;12(8): 4271–5.
[127] Liau SY, Read DC, Pugh WJ, Furr JR, Russell AD. Interaction of silver nitrate with readily identifiable groups: relationship to the antibacterial action of silver ions. Lett Appl Microbiol 1997;25(4):279–83.
[128] Liu SB, Zeng TH, Hofmann M, Burcombe E, Wei J, Jiang RR, Kong J, Chen Y. Antibacterial activity of graphite, graphite oxide, graphene oxide, and reduced graphene oxide: membrane and oxidative stress. ACS Nano 2011;5(9):6971–80.
[129] Vecitis CD, Zodrow KR, Kang S, Elimelech M. Electronic-structure-dependent bacterial cytotoxicity of single-walled carbon nanotubes. ACS Nano 2010;4(9): 5471–9.
[130] Kang S, Herzberg M, Rodrigues DF, Elimelech M. Antibacterial effects of carbon nanotubes: size does matter. Langmuir 2008;24(13):6409–13.
[131] Kang S, Mauter MS, Elimelech M. Physicochemical determinants of multiwalled carbon nanotube bacterial cytotoxicity. Environ Sci Technol 2008;42(19):7528–34.
[132] Peter-Varbanets M, Zurbrugg C, Swartz C, Pronk W. Decentralized systems for potable water and the potential of membrane technology. Water Res 2009;43(2): 245–65.
[133] Brady-Estevez AS, Schnoor MH, Kang S, Elimelech M. SWNT-MWNT hybrid filter attains high viral removal and bacterial inactivation. Langmuir 2010;26(24): 19153–8.
[134] Rahaman MS, Vecitis CD, Elimelech M. Electrochemical carbon-nanotube filter performance toward virus removal and inactivation in the presence of natural organic matter. Environ Sci Technol 2012;46(3):1556–64.
[135] Vecitis CD, Schnoor MH, Rahaman MS, Schiffman JD, Elimelech M. Electrochemical multiwalled carbon nanotube filter for viral and bacterial removal and inactivation. Environ Sci Technol 2011;45(8):3672–9.
[136] Theron J, Cloete TE, de Kwaadsteniet M. Current molecular and emerging nanobiotechnology approaches for the detection of microbial pathogens. Crit Rev Microbiol 2010;36(4):318–39.
[137] Vikesland PJ, Wigginton KR. Nanomaterial enabled biosensors for pathogen monitoring e a review. Environ Sci Technol 2010;44(10):3656–69.

[138] Yan JL, Estevez MC, Smith JE, Wang KM, He XX, Wang L, Tan WH. Dye-doped nanoparticles for bioanalysis. Nano Today 2007;2(3):44–50.
[139] Petryayeva E, Krull UJ. Localized surface plasmon resonance: nanostructures, bioassays and biosensing e a review. Anal Chim Acta 2011;706(1):8–24.
[140] Lei JP, Ju HX. Signal amplification using functional nanomaterials for biosensing. Chem Soc Rev 2012;41(6):2122–34.
[141] Jain PK, Lee KS, El-Sayed IH, El-Sayed MA. Calculated absorption and scattering properties of gold nanoparticles of different size, shape, and composition: applications in biological imaging and biomedicine. J Phys Chem B 2006;110(14):7238–48.
[142] Kelly KL, Coronado E, Zhao LL, Schatz GC. The optical properties of metal nanoparticles: the influence of size, shape, and dielectric environment. J Phys Chem B 2003;107(3):668–77.
[143] Moskovits M. Surface-enhanced Raman spectroscopy: a brief retrospective. J Raman Spectrosc 2005;36(6–7):485–96.
[144] Nie SM, Emery SR. Probing single molecules and single nanoparticles by surface-enhanced Raman scattering. Science 1997;275(5303):1102–6.
[145] McCreery RL. Advanced carbon electrode materials for molecular electrochemistry. Chem Rev 2008;108(7):2646–87.
[146] Yang WR, Ratinac KR, Ringer SP, Thordarson P, Gooding JJ, Braet F. Carbon nanomaterials in biosensors: should you use nanotubes or graphene? Angew Chem Int Ed 2010;49(12):2114–38.
[147] Heller I, Janssens AM, Mannik J, Minot ED, Lemay SG, Dekker C. Identifying the mechanism of biosensing with carbon nanotube transistors. Nano Lett 2008;8(2):591–5.
[148] Duran A, Tuzen M, Soylak M. Preconcentration of some trace elements via using multiwalled carbon nanotubes as solid phase extraction adsorbent. J Hazard Mater 2009;169(1–3):466–71.
[149] Cai YQ, Jiang GB, Liu JF, Zhou QX. Multiwalled carbon nanotubes as a solid-phase extraction adsorbent for the determination of bisphenol a, 4-n-nonylphenol, and 4-tertoctylphenol. Anal Chem 2003;75(10):2517–21.
[150] Lisha KP, Anshup, Pradeep T. Enhanced visual detection of pesticides using gold nanoparticles. J Environ Sci Health Part B Pestic Food Contam Agric Wastes 2009;44(7):697–705.
[151] Yang LX, Chen BB, Luo SL, Li JX, Liu RH, Cai QY. Sensitive detection of polycyclic aromatic hydrocarbons using CdTe quantum dot-modified TiO(2) nanotube array through fluorescence resonance energy transfer. Environ Sci Technol 2010;44(20):7884–9.
[152] Yin HS, Zhou YL, Ai SY, Chen QP, Zhu XB, Liu XG, Zhu LS. Sensitivity and selectivity determination of BPA in real water samples using PAMAM dendrimer and CoTe quantum dots modified glassy carbon electrode. J Hazard Mater 2010;174(1–3):236–43.
[153] da Silva BF, Perez S, Gardinalli P, Singhal RK, Mozeto AA, Barcelo D. Analytical chemistry of metallic nanoparticles in natural environments. TrAc Trends Anal Chem 2011;30(3):528–40.
[154] Tiede K, Boxall ABA, Tear SP, Lewis J, David H, Hassellov M. Detection and characterization of engineered nanoparticles in food and the environment. Food Addit Contam 2008;25(7):795–821.
[155] Malato S, Fernández-Ibáñez P, Maldonado MI, Blanco J, Gernjak W. Decontamination and disinfection of water by solar photocatalysis: recent overview and trends. Catal Today 2009;147:1–59.

[156] Richardson SD. Environmental mass spectrometry: emerging contaminants and current issues. Anal Chem 2008;80:4373–402.
[157] Suárez S, Carballa M, Omil F, Lema JM. How are pharmaceutical and personal care products (PPCPs) removed from urban wastewaters? Rev Environ Sci Biotechnol 2008;7:125–38.
[158] Wintgens T, Salehi F, Hochstrat R, Melin T. Emerging contaminants and treatment options in water recycling for indirect potable use. Water Sci Technol 2008;57: 99–107.
[159] Bradley BR, Daigger GT, Rubin R, Tchobanoglous G. Evaluation of onsite wastewater treatment technologies using sustainable development criteria. Clean Technol Environ Policy 2002;4:87–99.
[160] Lapña L, Cerezo M, Garía-Augustin P. Possible reuse of treated municipal wastewater for Citrus spp. plant irrigation. Bull Environ Contam Toxicol 1995;55:697–703.
[161] Viessman Jr W, Hammer MJ. Water supply and pollution control. 6th ed. California (USA): Addison Wesley Longman Inc.; 1998.
[162] Padmanabhan PVA, Sreekumar KP, Thiyagarajan TK, Satpute RU, Bhanumurthy K, Sengupta P, Dey GK, Warrier KGK. Nano-crystalline titanium dioxide formed by reactive plasma synthesis. Vacuum 2006;80:11–2.
[163] Gaya UI, Abdullah AH. Heterogeneous photocatalytic degradation of organic contaminants over titanium dioxide: a review of fundamentals, progress and problems. J Photochem Photobiol C Photochem Rev 2008;9:1–12.
[164] Coleman HM, Marquis CP, Scott JA, Chin SS, Amal R. Bactericidal effects of titanium dioxide-based photocatalysts. Chem Eng J 2005;113:55–63.
[165] Lu J, Zhang T, Ma J, Chen Z. Evaluation of disinfection by-products formation during chlorination and chloramination of dissolved natural organic matter fractions isolated from a filtered river water. J Hazard Mater 2009;162:140–5.
[166] Yang H, Cheng H. Controlling nitrite level in drinking water by chlorination and chloramination. Sep Purif Technol 2007;56:392–6.
[167] Pera-Titus M, García-Molina V, Baños MA, Giménez J, Esplugas S. Degradation of chlorophenols by means of advanced oxidation processes: a general review. Appl Catal B Environ 2004;47:219–56.
[168] Fujishima A, Rao TN, Tryk DA. Titanium dioxide photocatalysis. J Photochem Photobiol C Photochem Rev 2000;1:1–21.
[169] Furube A, Asahi T, Masuhara H, Yamashita H, Anpo M. Direct observation of a picosecond charge separation process in photoexcited platinum-loaded TiO_2 particles by femtosecond diffuse reflectance spectroscopy. Chem Phys Lett 2001;336:424–30.
[170] Byrne JA, Eggins BR. Photoelectrochemistry of oxalate on particulate TiO_2 electrodes. J Electroanal Chem 1998;457:61–72.
[171] Herrmann JM. Heterogeneous photocatalysis: fundamentals and applications to the removal of various types of aqueous pollutants. Catal Today 1999;53:115–29.
[172] Vinodgopal K, Kamat PV. Photochemistry on surfaces: photodegradation of 1, 3-diphenylisobenzofuran over metal oxide particles. J Phys Chem 1992;96:5053–9.
[173] Maness PC, Smolinski S, Blake DM, Huang Z, Wolfrum EJ, Jacoby WA. Bactericidal activity of photocatalytic TiO_2 reaction: toward an understanding of its killing mechanism. Appl Environ Microbiol 1999;65:4094–8.
[174] Machulek Jr A, Moraes JEF, Vautier-Giongo C, Silverio CA, Friedrich LC, Nascimento CAO, Gonzalez MC, Quina FH. Abatement of the inhibitory effect of chloride anions on the photo-Fenton process. Environ Sci Technol 2007;41: 8459–63.
[175] Neyens E, Baeyens J. A review of classic Fenton's peroxidation as an advanced oxidation technique. Water Res 2003;98:33–50.

[176] Fallmann H, Krutzler T, Bauer R, Malato S, Blanco J. Applicability of the photo-Fenton method for treating water containing pesticides. Catal Today 1999;54:309–19.

[177] Gernjak W, Krutzler T, Malato S. Photo-Fenton treatment of olive mill wastewater applying a combined Fenton/flocculation pretreatment. J Sol Energy Eng 2007;129:53–9.

[178] Huston PL, Pignatello JJ. Degradation of selected pesticide active ingredients and commercial formulations in water by the photo-assisted Fenton reaction. Water Res 1999;33:1238–46.

[179] Gogate PR, Pandit AB. A review of imperative technologies for wastewater treatment II: hybrid methods. Adv Environ Res 2004;8:553–97.

[180] Pignatello JJ, Oliveros E, MacKay A. Advanced oxidation processes for organic contaminant destruction based on the Fenton reaction and related chemistry. Crit Rev Environ Sci Technol 2006;36:1–84.

[181] De Laat J, Le GT, Legube B. A comparative study of the effects of chloride, sulphate and nitrate ions on the rates of decomposition of H_2O_2 and organic compounds by $Fe(II)/H_2O_2$ and $Fe(III)/H_2O_2$. Chemosphere 2004;55:715–23.

[182] Oliveros E, Legrini O, Hohl M, Mü ller T, Braun AM. Industrial wastewater treatment: large scale development of a light-enhanced Fenton reaction. Chem Eng Proc 1997;36:397–405.

[183] Torrades F, Pérez M, Mansilla HD, Peral J. Experimental design of Fenton and photo-Fenton reactions for the treatment of cellulose bleaching effluents. Chemosphere 2003;53:1211–20.

[184] Dominguez C, García J, Pedraz MA, Torres A, Galán MA. Photocatalytic oxidation of organic pollutants in water. Catal Today 1998;40:85–101.

[185] Marugán J, Lopez-Muñoz MJ, Gernjak W, Malato S. Fe/TiO_2/pH interactions in solar degradation of imidacloprid with TiO_2/SiO_2 photocatalysts at pilot-plant scale. Ind Eng Chem Res 2006;45:8900–8.

[186] Marugán J, van Grieken R, Sordo C, Cruz C. Kinetics of the photocatalytic disinfection of *Escherichic coli* suspensions. Appl Catal B Environ 2008;82:27–36.

[187] Rincón AG, Pulgarin C. Comparative evaluation of Fe^{3+} and TiO_2 photoassisted processes in solar photocatalytic disinfection of water. Appl Catal B Environ 2006;63:222–31.

[188] Torres RA, Nieto JI, Combet E, Pétrier C, Pulgarin C. Influence of TiO2 concentration on the synergistic effect between photocatalysis and high-frequency ultrasound for organic pollutant mineralization in water. Appl Catal B Environ 2008;80:168–75.

[189] Fujishima A, Honda K. Electrochemical photolysis of water at a semiconductor electrode. Nature 1972;238:37–8.

[190] Hosono E, Fujihara S, Kakiuchi K, Imai H. Growth of submicrometer-scale reactangular parallelepiped rutile TiO_2 films in aqueous $TiCl_3$ solutions under hydrothermal conditions. J Am Chem Soc 2004;126:7790–1.

[191] Kondo Y, Yoshikawa H, Awaga K, Murayama M, Mori T, Sunada K, Bandow S, Iijima S. Preparation, photocatalytic activities, and dye-sensitized solar-cell performance of submicron-scale TiO_2 hollow spheres. Langmuir 2008;24:547–50.

[192] Wang R, Hashimoto K, Fujishima A, Chikuni M, Kojima E, Kitamura A, Shimohigoshi M, Watanabe T. Photogeneration of highly amphiphilic TiO_2 surfaces. Adv Mater 1999;10:135–8.

[193] Nagaveni K, Sivalingam G, Hegde MS, Madras G. Solar photocatalytic degradation of dyes. High activity of combustion synthesized nano TiO_2. Appl Catal B Environ 2004;48:83–93.

[194] Nagaveni K, Sivalingam G, Hegde MS, Madras G. Photocatalytic degradation of organic compounds over combustion synthesized nano-TiO_2. Environ Sci Technol 2004;38:1600–4.
[195] Siddiquey IA, Furusawa T, Sato M, Honda K, Suzuki N. Control of the photocatalytic activity of TiO_2 nanoparticles by silica coating with polydiethoxysiloxane. Dyes Pigment 2008;76:754–9.
[196] Byrne JA, Eggins BR, Brown NMD, McKinley B, Rouse M. Immobilisation of TiO2 powder for the treatment of polluted water. Appl Catal B Environ 1998;17:25–36.
[197] Yu JC, Yu J, Zhao J. Enhanced photocatalytic activity of mesoporous and ordinary TiO_2 thin films by sulphuric acid treatment. Appl Catal B Environ 2002;36:31–43.
[198] Serpone N, Sauvé G, Koch R, Tahiri H, Pichat P, Piccinini P, Pelizetti E, Hidaka H. Standardization protocol of process efficiencies and activation parameters in heterogeneous photocatalysis: relative photonic efficiencies zr. J Photochem Photobiol A Chem 1996;94:191–203.
[199] Pozzo RL, Baltanás MA, Cassano AE. Supported titanium dioxide as photocatalyst in water decontamination: state of the art. Catal Today 1997;39:219–31.
[200] Yang GCC, Li CJ. Electrofiltration of silica nanoparticle containing wastewater using tubular ceramic membranes. Sep Purif Technol 2007;58:159–65.
[201] Doll TE, Frimmel FH. Cross-flow microfiltration with periodical back-washing for photocatalytic degradation of pharmaceutical and diagnostic residues-evaluation of the long-term stability of the photocatalytic activity of TiO_2. Water Res 2005;39:847–54.
[202] Fernández-Ibáñez P, Blanco J, Malato S, de las Nieves FJ. Application of the colloidal stability of TiO_2 particles for recovery and reuse in solar photocatalysis. Water Res 2003;37:3180–8.
[203] Zhang X, Du AJ, Lee P, Sun DD, Leckie JO. TiO_2 nanowire membrane for concurrent filtration and photocatalytic oxidation of humic acid in water. J Membr Sci 2008;313:44–51.
[204] Zhao Y, Zhong J, Li H, Xu N, Shi J. Fouling and regeneration of ceramic microfiltration membranes in processing acid wastewater containing fine TiO_2 particles. J Membr Sci 2002;208:331–41.
[205] Lee SA, Choo KH, Lee CH, Lee HI, Hyeon T, Choi W, Kwon HH. Use of ultrafiltration membranes for the separation of TiO_2 photocatalysts in drinking water treatment. Ind Eng Chem Res 2001;40:1712–9.
[206] Molinari R, Palmisano L, Drioli E, Schiavello M. Studies on various reactor configurations for coupling photocatalysis and membrane processes in water purification. J Membr Sci 2002;206:399–415.
[207] Xi W, Geissen S. Separation of titanium dioxide from photocatalytically treated water by cross-flow microfiltration. Water Res 2001;35:1256–62.
[208] Zhang X, Pan JH, Du AJ, Fu W, Sun DD, Leckie JO. Combination of one-dimensional TiO_2 nanowire photocatalytic oxidation with microfiltration for water treatment. Water Res 2009;43:1179–86.
[209] Chong MN, Vimonses V, Lei S, Jin B, Chow C, Saint C. Synthesis and characterisation of novel titania impregnated kaolinite nano-photocatalyst. Microporus Mesoporus Mater 2009;117:233–42.
[210] Kwak SY, Kim SH. Hybrid organic/inorganic reverse osmosis (RO) membrane for bactericidal anti-fouling. 1. Preparation and characterization of TiO_2 nanoparticle selfassembled aromatic polyamide thin-film-composite (TFC) membrane. Environ Sci Technol 2001;35:2388–94.

[211] Lee DK, Kim SC, Cho IC, Kim SJ, Kim SW. Photocatalytic oxidation of microcystin-LR in a fluidized bed reactor having TiO_2-coated activated carbon. Sep Purif Technol 2004;34:59–66.
[212] Zhu H, Gao X, Lan Y, Song D, Xi Y, Zhao J. Hydrogen titanate nanofibers covered with anatase nanocrystals: a delicate structure achieved by the wet chemistry reaction of the titanate nanofibers. J Am Chem Soc 2004;126:8380–1.
[213] Fukahori S, Ichiura H, Kitaoka T, Tanaka H. Capturing of bisphenol A photodecomposition intermediates by composite TiO_2 ezeolite sheets. Appl Catal B Environ 2003;46:453–62.
[214] Kun R, Mogyorósi K, Dekány I. Synthesis and structural and photocatalytic properties of TiO_2/montmorillonite nanocomposites. Appl Clay Sci 2006;32:99–110.
[215] Sun Z, Chen Y, Ke Q, Yang Y, Yuan J. Photocatalytic degradation of a cationic azo dye by TiO_2/bentonite nanocomposite. J Photochem Photobiol A Chem 2002;149: 169–74.
[216] Xie ZM, Chen Z, Dai YZ. Preparation of TiO_2/sepiolite photocatalyst and its application to printing and dyeing wastewater treatment. Environ Sci Technol 2009;32: 123–7.
[217] Chong MN, Lei S, Jin B, Saint C, Chow CWK. Optimisation of an annular photoreactor process for degradation of Congo red using a newly synthesized titania impregnated kaolinite nano-photocatalyst. Sep Purif Technol 2009;67:355–63.
[218] Sun D, Meng TT, Loong TH, Hwa TJ. Removal of natural organic matter from water using a nano-structured photocatalyst coupled with filtration membrane. Water Sci Technol 2004;49:103–10.
[219] Zhang X, Du AJ, Lee P, Sun DD, Leckie JO. Grafted multifunctional titanium dioxide nanotube membrane: separation and photodegradation of aquatic pollutant. Appl Catal B Environ 2008;84:262–7.
[220] Bosc F, Ayral A, Guizard C. Mesoporous anatase coatings for coupling membrane separation and photocatalyzed reactions. J Membr Sci 2005;265:13–9.
[221] Choi H, Stathatos E, Dionysiou DD. Solegel preparation of mesoporous photocatalytic TiO_2 films and TiO_2/Al_2O_3 composite membranes for environmental applications. Appl Catal B Environ 2005;63:60–7.
[222] Choi H, Stathatos E, Dionysiou DD. Photocatalytic TiO_2 films and membranes for the development of efficient wastewater treatment and reuse systems. Desalination 2007;202:199–206.
[223] Zhang H, Quan X, Chen S, Zhao H, Zhao Y. The removal of sodium dodecylbenzene sulfonate surfactant from water using silica/titania nanorods/nanotubes composite membrane with photocatalytic capability. Appl Surf Sci 2006;252:8598–604.
[224] Kim SH, Kwak SY, Sohn BH, Park TH. Design of TiO_2 nanoparticle self-assembled aromatic polyamide thin-filmcomposite (TFC) membrane as an approach to solve biofouling problem. J Membr Sci 2003;211:157–65.
[225] Kleine J, Peinemann KV, Schuster C, Warnecke HJ. Multifunctional system for treatment of wastewaters from adhesive-producing industries: separation of solids and oxidation of dissolved pollutants using doted microfiltration membranes. Chem Eng Sci 2002;57:1661–4.
[226] Molinari R, Pirillo F, Falco M, Loddo V, Palmisano L. Photocatalytic degradation of dyes by using a membrane reactor. Chem Eng Proc 2004;43:1103–14.
[227] Yang Y, Wang P. Preparation and characterizations of a new PS/TiO_2 hybrid membrane by solegel process. Polymer 2006;47:2683–8.
[228] Litter MI. Heterogeneous photocatalysis: transition metal ions in photocatalytic systems. Appl Catal B Environ 1999;23:89–114.

[229] Vinodgopal K, Wynkoop DE, Kamat PV. Environmental photochemistry on semiconductor surfaces: photosensitized degradation of a textile azo dye, Acid Orange 7, on TiO_2 particles using visible light. Environ Sci Technol 1996;30:1660–6.
[230] Yu Y, Yu JC, Yu JG, Kwok YC, Che YK, Zhao JC, Ding L, Ge WK, Wong PK. Enhancement of photocatalytic activity of mesoporous TiO_2 by using carbon nanotubes. Appl Catal A Gen 2005;289:186–96.
[231] Ishibai Y, Sato J, Nishikawa T, Miyagishi S. Synthesis of visible-light active TiO_2 photocatalyst with Pt-modification: role of TiO_2 substrate for high photocatalytic activity. Appl Catal B Environ 2008;79:117–21.
[232] Li H, Li J, Huo Y. Highly active TiO_2N photocatalysts prepared by treating TiO_2 precursors in NH_3/ethanol fluid under supercritical conditions. J Phys Chem B 2006;110:1559–65.
[233] Shaban YA, Khan SUM. Visible light active carbon modified n-TiO_2 for efficient hydrogen production by photoelectrochemical splitting of water. Int J Hydrog Energy 2008;33:1118–26.
[234] Asahi R, Morikawa T, Ohwaki T, Aoki K, Taga Y. Visible-light photocatalysis in nitrogen-doped titanium dioxides. Science 2001;293:269–71.
[235] Ihara T, Miyoshi M, Iriyama Y, Matsumoto O, Sugihara S. Visible-light-active titanium dioxide photocatalyst realized by an oxygen-deficient structure and by nitrogen doping. Appl Catal B Environ 2003;42:403–9.
[236] Irie H, Watanabe Y, Hashimoto K. Nitrogen concentration dependence on photocatalytic activity of $TiO_{2-x}N_x$ powders. J Phys Chem B 2003;107:5483–6.
[237] Chen D, Ray AK. Photodegradation kinetics of 4- nitrophenol in TiO_2 suspension. Water Res 1998;32:3223–34.
[238] Tsarenko SA, Kochkodan VM, Samsoni-Todorov AO, Goncharuk VV. Removal of humic substances from aqueous solutions with a photocatalytic membrane reactor. Colloid J 2006;68:341–4.
[239] Pozzo RL, Giombi JL, Baltanas MA, Cassano AE. The performance in a fluidized bed reactor of photocatalysts immobilized onto inert supports. Catal Today 2000;62: 175–87.
[240] Chan AHC, Chan CK, Barford JP, Porter JF. Solar photocatalytic thin film cascade reactor for treatment of benzoic acid containing wastewater. Water Res 2003;37: 1125–35.
[241] Ochuma IJ, Fishwick RP, Wood J, Winterbottom JM. Optimisation of degradation conditions of 1,8-diazabicyclo [5. 4.0] undec-7-ene in water and reaction kinetics analysis using a cocurrent downflow contactor photocatalytic reactor. Appl Catal B Environ 2007;73:259–68.
[242] Pareek V, Chong S, Tadé M, Adesina AA. Light intensity distribution in heterogeneous photocatalytic reactors. Asia Pac J Chem Eng 2008;3:171–201.
[243] Fu J, Ji M, Wang Z, Jin L, An D. A new submerged membrane photocatalysis reactor (SMPR) for fulvic acid removal using a nano-structured photocatalyst. J Hazard Mater 2006;131:238–42.
[244] Molinari R, Grande C, Drioli E, Palmisano L, Schiavello M. Photocatalytic membrane reactors for degradation of organic pollutants in water. Catal Today 2001;67: 273–9.
[245] Chin SS, Lim TM, Chiang K, Fane AG. Hybrid low pressure submerged membrane photoreactor for the removal of bisphenol A. Desalination 2007;202:253–61.
[246] Huang X, Meng Y, Liang P, Qian Y. Operational conditions of a membrane filtration reactor coupled with photocatalytic oxidation. Sep Purif Technol 2007;55:165–72.
[247] Meng Y, Huang X, Yang Q, Qian Y, Kubota N, Fukunaga S. Treatment of polluted river water with a photocatalytic slurry reactor using low-pressure mercury lamps coupled with a membrane. Desalination 2005;181:121–33.

[248] Sopajaree K, Qasim SA, Basak S, Rajeshwar K. An integrated flow reactor-membrane filtration system for heterogeneous photocatalysis. Part I. Experiments and modelling of a batch-recirculated photoreactor. J Appl Electrochem 1999;29:533–9.
[249] Sopajaree K, Qasim SA, Basak S, Rajeshwar K. An integrated flow-reactor membrane filtration system for heterogeneous photocatalysis. Part II. Experiments on the ultrafiltration unit and combined operation. J Appl Electrochem 1999;29:1111–8.
[250] Molinari R, Pirilla F, Loddo V, Palmisano L. Heterogeneous photocatalytic degradation of pharmaceuticals in water by using polycrystalline TiO_2 and a nanofiltration membrane reactor. Catal Today 2006;118:205–13.
[251] Jung JT, Kim JO, Choi WY. Performance of photocatalytic microfiltration with hollow fiber membrane. Mater Sci Forum 2007;544:95–8.
[252] Mozia S, Morawski AW, Toyoda M, Tsumura T. Effect of process parameters on photodegradation of Acid Yellow 36 in a hybrid photocatalysis membrane distillation system. Chem Eng J 2009;150:152–9.
[253] Azrague K, Aimar P, Benoit-Marqué F, Maurette MT. A new combination of a membrane and photocatalytic reactor for the depollution of turbid water. Appl Catal B Environ 2006;72:197–205.
[254] Camera-Roda G, Santarelli F. Intensification of water detoxification by integrating photocatalysis and pervaporation. J Sol Energy Eng 2007;129:68–73.
[255] Augugliaro V, Litter M, Palmisano L, Soria J. The combination of heterogeneous photocatalysis with chemical and physical operations: a tool for improving the photoprocess performance. J Photochem Photobiol C Photochem Rev 2006;7:127–44.
[256] Lee HS, Im SJ, Kim JH, Kim HJ, Kim JP, Min BR. Polyamide thin-film nanofiltration membranes containing TiO_2 nanoparticles. Desalination 2008;219(1–3):48–56.
[257] Meares P. Synthetic membranes: science, engineering and applications. 1st ed. Peidel, Dordrecht (Netherlands): Springer Publisher; 1986.
[258] Gryta M, Tomaszewska M, Grzechulska J, Morawski AW. Membrane distillation of NaCl solution containing natural organic matter. J Membr Sci 2001;181:279–87.
[259] Tomaszewska M, Gryta M, Morawski AW. The influence of salt in solution on hydrochloric acid recovery by membrane distillation. Sep Purif Technol 1998;14:183–8.
[260] Lawson KW, Lloyd DR. Membrane distillation. J Membr Sci 1997;124:1–25.
[261] Benotti MJ, Stanford BD Wert EC, Snyder SA. Evaluation of a photocatalytic reactor membrane pilot system for the removal of pharmaceuticals and endocrine disrupting compounds from water. Water Res 2009;43(6):1513–22.
[262] Chong MN, Jin B, Zhu HY, Chow CWK, Saint C. Application of H-titanate nanofibers for degradation of Congo red in an annular slurry photoreactor. Chem Eng J 2009;150:49–54.
[263] Malato S, Blanco J, Campos A, Cáceres J, Guillard C, Herrmann JM, Fernández-Alba AR. Effect of operating parameters on the testing of new industrial titania catalysts at solar pilot plant scale. Appl Catal B Environ 2003;42:349–57.
[264] Xu Y, Langford CH. Variation of Langmuir adsorption constant determined for TiO_2-photocatalyzed degradation if acetophenone under different light intensity. J Photochem Photobiol A Chem 2000;133:67–71.
[265] Toor AP, Verma A, Jotshi CK, Bajpai PK, Singh V. Photocatalytic degradation of Direct Yellow 12 dye using UV/TiO_2 in a shallow pond slurry reactor. Dyes Pigment 2006;68:53–60.
[266] Rincón AG, Pulgarin C. Effect of pH, inorganic ions, organic matter and H_2O_2 on *E. coli* K12 photocatalytic inactivation by TiO_2-implications in solar water disinfection. Appl Catal B Environ 2004;51:283–302.

[267] Körmann C, Bahnemann DW, Hoffman MR. Photolysis of chloroform and other organic molecules in aqueous titanium dioxide suspensions. Environ Sci Technol 1991;25:494–500.
[268] Stylidi M, Kondarides DI, Verykios XE. Pathways of solar light-induced photocatalytic degradation of azo dyes in aqueous TiO_2 suspensions. Appl Catal B Environ 2003;40:271–86.
[269] Herrera Melián JA, Doña Rodíguez JM, Viera Suárez A, Valés do Campo C, Arana J, Pérez Peña J. The photocatalytic disinfection of urban waste waters. Chemosphere 2000;41:323–7.
[270] Heyde M, Portalier R. Acid shock proteins of *Escherichia coli*. FEMS Microbiol Lett 1990;69:19–26.
[271] Ibáñez JA, Litter MI, Pizarro RA. Photocatalytic bactericidal effect of TiO_2 on *Enterobacter cloacae*: comparative study with other Gram (-) bacteria. J Photochem Photobiol A Chem 2003;157:81–5.
[272] Lonnen J, Kilvington S, Kehoe SC, Al-Touati F, McGuigan KG. Solar and photocatalytic disinfection of protozoan, fungal and bacterial microbes in drinking water. Water Res 2005;39:877–83.
[273] Matsunaga T, Tomoda R, Nakajima T, Wake H. Photoelectrochemical sterilization of microbial cells by semiconductor powders. FEMS Microbiol Lett 1985;29:211–4.
[274] Sichel C, Tello J, de Cara M, Fernández-Ibáñez P. Effect of UV solar intensity and dose on the photocatalytic disinfection of bacteria and fungi. Catal Today 2007; 129:152–60.
[275] Blanco J, Malato S, de las Nieves J, Fernández P. Method of sedimentation of colloidal semiconductor particles. European patent application EP-1-101-737-A1, European Patent Office Bulletin 21. 2001.
[276] Molinari R, Caruso A, Argurio P, Poerio T. Degradation of the drugs Gemfibrozil and Tamoxifen in pressurized and de-pressurized membrane photoreactors using suspended polycrystalline TiO_2 as catalyst. J Membr Sci 2008;319:54–63.
[277] Seffaj N, Persin M, Alami Younssi S, Albizane A, Bouhria M, Loukili H, Larbot A. Removal of salts and dyes by low $ZnAl_2O_4$-TiO_2 ultrafiltration membrane deposited on support made from raw clay. Sep Purif Technol 2005;47:36–42.
[278] Fu X, Clark LA, Zeltner WA, Anderson MA. Effects of reaction temperature and water vapour content on the heterogeneous photocatalytic oxidation of ethylene. J Photochem Photobiol A Chem 1996;97:181–6.
[279] Muradov NZ, Raissi AT, Muzzey D, Painter CR, Kemme MR. Selective photocatalytic degradation of airborne VOCs. Sol Energy 1996;56:445–53.
[280] Rincón AG, Pulgarin C. Photocatalytical inactivation of *E. coli*: effect of (continuous-intermittent) light intensity and of (suspended-fixed) TiO_2 concentration. Appl Catal B Environ 2003;44:263–84.
[281] Herrmann JG. Research to protect water infrastructure: EPA's water security research program. Proc SPIE 2005;5781:48.
[282] Wang Y, Hong CS. TiO_2-mediated photomineralization of 2-chlorobiphenyl: the role of O_2. Water Res 2000;34:2791–7.
[283] Shirayama H, Tohezo Y, Taguchi S. Photodegradation of chlorinated hydrocarbons in the presence and absence of dissolved oxygen in water. Water Res 2001;35: 1941–50.
[284] Saquib M, Muneer M. TiO_2-mediated photocatalytic degradation of a triphenylmethane dye (gentian violet), in aqueous suspensions. Dyes Pigment 2003;56:37–49.
[285] Bahnemann D. Photocatalytic water treatment: solar energy applications. Sol Energy 2004;77:445–59.

[286] Bhatkhnade DS, Kamble SP, Sawant SB, Pangarkar VG. Photocatalytic and photochemical degradation of nitrobenzene using artificial ultraviolet light. Chem Eng J 2004;102:283–90.

[287] Malato-Rodíguez S, Richter C, Gálvez JB, Vincent M. Photocatalytic degradation of industrial residual waters. Sol Energy 1996;56:401–10.

[288] Maruán J, Aguado J, Gernjak W, Malato S. Solar photocatalytic degradation of dichloroacetic acid with silica supported titania at pilot-plant scale. Catal Today 2007;129:59–68.

[289] Radjenovíc J, Sirtori C, Petrovíc M, Barceló D, Malato S. Solar photocatalytic degradation of persistent pharmaceuticals at pilot-scale: kinetics and characterization of major intermediate products. Appl Catal B Environ 2009;89:255–64.

[290] Vilar VJP, Maldonado MI, Oller I, Malato S, Boaventura RAR. Solar treatment of cork boiling and bleaching wastewaters in a pilot plant. Water Res 2009;43(16): 4050–62.

[291] Parra S, Malato S, Pulgarin C. New integrated photocatalytic biological flow system using supported TiO_2 and fixed bacteria for the mineralization of isoproturon. Appl Catal B Environ 2002;36:131–44.

[292] WHO. Guidelines for drinking-water quality first addendum to third edition 1 recommendations. WHO Library Cataloguingin-Publication Data; 2006.

[293] Cho M, Chung H, Choi W, Yoon J. Linear correlation between inactivation of *E. coli* and OH radical concentration in TiO_2 photocatalytic disinfection. Water Res 2004; 38:1069–77.

[294] Chen W, Duan L, Zhu DQ. Adsorption of polar and nonpolar organic chemicals to carbon nanotubes. Environ Sci Technol 2007;41(24):8295–300.

[295] Tyrrell RM, Keyse SM. New trends in photobiology the interaction of UVA radiation with cultured cells. J Photochem Photobiol B Biol 1990;4:349–61.

[296] Berney M, Weilenmann HU, Siminetti A, Egli T. Efficacy of solar disinfection of *Escherichia coli*, *Shigella flexneri*, *Salmonella Typhimurium* and Vibrio cholera. J Appl Microbiol 2006;101:828–36.

[297] Kehoe SC, Barer MR, Devlin LO, McGuigan KG. Batch process solar disinfection is an efficient means of disinfecting drinking water contaminated with *Shigella dysenteriae* type I. Lett Appl Microbiol 2004;38:410–4.

[298] McGuigan KG, Méndez-Hermida F, Castro-Hermida JA, Ares-Mazás E, Kehoe SC, Boyle M, Sichel C, Fernández- Ibáñez P, Meyer BP, Ramalingham S, Meyer EA. Batch solar disinfection inactivates oocysts of *Cryptosporidium parvum* and cysts of Giardia muris in drinking water. J Appl Microbiol 2006;101:453–63.

[299] Glatzmaier GC, Nix RG, Mehos MS. Solar destruction of hazardous chemicals. J Environ Sci Health A 1990;25:571–81.

[300] Glatzmaier GC. Innovative solar technologies for treatment of concentrated organic wastes. Sol Energy Mater 1991;24:672.

[301] Magrini KA, Webb JD. Decomposition of aqueous organic compounds as a function of solar irradiation intensity. In: Beard, Ebadian MA, editors. 12th ASME Int. Sol. Energy conference. New York: ASME; 1990. p. 159–62.

[302] Karunakaran C, Senthilvelan S. Photooxidation of aniline on alumina with sunlight and artificial UV light. Catal Commun 2005;6:159–65.

[303] Qamar M, Muneer M, Bahnemann D. Heterogeneous photocatalysed degradation of two selected pesticide derivatives, triclopyr and daminozid in aqueous suspensions of titanium dioxide. J Environ Manage 2006;80:99–106.

[304] Shang C, Cheung LM, Ho CM, Zeng M. Repression of photoreactivation and dark repair of coliform bacteria by TiO_2-modified UV-C disinfection. Appl Catal B Environ 2009;89:536–42.

[305] Calza P, Sakkas VA, Medana C, Baiocchi C, Dimou A, Pelizetti E, Albanis T. Photocatalytic degradation study of diclofenac over aqueous TiO_2 suspensions. Appl Catal B Environ 2006;67:197–205.
[306] Chong MN, Jin B, Chow CWK, Saint CP. A new approach to optimise an annular slurry photoreactor system for the degradation of Congo red: statistical analysis and modelling. Chem Eng J 2009;152:158–66.
[307] Korbahti BK, Rauf MA. Application of response surface analysis to the photolytic degradation of Basic Red 2 dye. Chem Eng J 2008;138:166–71.
[308] Liu HL, Chiou YR. Optimal decolorization efficiency of Reactive Red 239 by UV/TiO_2 photocatalytic process coupled with response surface methodology. Chem Eng J 2005;112:173–9.
[309] Lizama C, Freer J, Baeza J, Mansilla HD. Optimized photodegradation of reactive blue 19 on TiO_2 and ZnO suspensions. Catal Today 2002;76:235–46.
[310] Cho IH, Zoh KD. Photocatalytic degradation of azo dye (reactive Red 120) in TiO_2/UV system: optimization and modelling using a response surface methodology (RSM) based on the central composite design. Dyes Pigment 2007;75:533–43.
[311] Jaworski R, Pawlowski L, Roudet F, Kozerski S, Petit F. Characterization of mechanical properties of suspension plasma sprayed TiO_2 coatings using scratch test. Surf Coat Technol 2008;202:2644–53.
[312] De Soto W, Klein SA, Beckman WA. Improvement and validation of a model for photovoltaic array performance. Sol Energy Jan. 2006;80(1):78–88.
[313] Cunningham J, Sedlak P. Kinetic studies of depollution process in TiO_2 slurries: interdependences of adsorption and UV-intensity. Catal Today 1996;29:209–315.
[314] Minero C. Kinetic analysis of photoinduced reactions at the water semiconductor interface. Catal Today 1999;54:205–16.
[315] Monllor-Satoca D, ómez R, González-Hidalgo M, Salvador P. The "Directeindirect" model: an alternative kinetic approach in heterogeneous photocatalysis based on the degree of interaction of dissolved pollutant species with the semiconductor surface. Catal Today 2007;129:247–55.
[316] Minero C, Pelizetti E, Malato S, Blanco J. Large solar plant photocatalytic water decontamination: effect of operational parameters. Sol Energy 1996;56:421–8.
[317] Finch GR, Black EK, Labatiuk CW, Gyürék L, Belosevic M. Comparison of Giardia lamblia and Giardia muris cyst inactivation by ozone. Appl Environ Microbiol 1993; 59:3674–80.
[318] Hom LW. Kinetics of chlorine disinfection in an ecosystem. J Sanit Eng Div 1972;98: 183–94.
[319] Benabbou AK, Derriche Z, Felix C, Lejeune P, Guillard C. Photocatalytic inactivation of *Escherischia coli*: effect of concentration of TiO2 and microorganism, nature, and intensity of UV irradiation. Appl Catal B Environ 2007;76:257–63.
[320] Lambert RJW, Johnston MD. The effect of interfering substances on the disinfection process: a mathematical model. J Appl Microbiol 2000;91:548–55.
[321] Cho M, Chung H, Yoon J. Disinfection of water containing natural organic matter by using ozoneinitiated radical reactions. Appl Environ Microbiol 2003;69:2284–91.
[322] Gürék LL, Finch GR. Modelling water treatment chemical disinfection kinetics. J Environ Eng 1998;124:783–92.
[323] Severin BF, Suidan MT, Engelbrecht RS. Series-event kinetic model for chemical disinfection. J Environ Eng ASCE 1984;110:430–9.
[324] Najm I. An alternative interpretation of disinfection kinetics. J Am Water Works Assoc 2006;98:93–101.
[325] Johnston MD, Simons EA, Lambert RJW. One explanation for the variability of the bacterial suspension test. J Appl Microbiol 2000;88:237–42.

[326] Chin ML, Mohamed AR, Bhatia S. Performance of photocatalytic reactors using immobilized TiO_2 film for the degradation of phenol and methylene blue dye present in water stream. Chemosphere 2004;57:547–54.
[327] Tang C, Chen V. The photocatalytic degradation of reactive black 5 using TiO_2/UV in an annular photoreactor. Water Res 2004;38:2775–81.
[328] Rincón AG, Pulgarin C. Use of coaxial photocatalytic reactor (CAPHORE) in the TiO_2 photo-assisted treatment of mixed *E. coli* and *Bacillus* sp. and bacterial community present in wastewater. Catal Today 2005;101:331–44.
[329] Gelover S, Gómez LA, Reyes K, Teresa Leal M. A practical demonstration of water disinfection using TiO_2 films and sunlight. Water Res 2006;40:3274–80.
[330] Crittenden JC, Zhang Y, Hand DW, Perram DL, Marchand EG. Solar detoxification of fuel-contaminated groundwater using fixed-bed photocatalysts. Water Environ Res 1996;68:270–8.
[331] Chen HY, Zahraa O, Bouchy M. Inhibition of the adsorption and photocatalytic degradation of an organic contaminant in an aqueous suspension of TiO_2 by inorganic ions. J Photochem Photobiol A Chem 1997;108:37–44.
[332] Guillard C, Lachheb H, Houas A, Ksibi M, Elaloui E, Hermann JM. Influence of chemical structure of dyes, of pH and of inorganic salts on their photocatalytic degradation by TiO_2 comparison of the efficiency of powder and supported TiO_2. J Photochem Photobiol A Chem 2003;158:27–36.
[333] Leng W, Liu H, Cheng S, Zhang J, Cao C. Kinetics of photocatalytic degradation of aniline in water over TiO_2 supported on porous nickel. J Photochem Photobiol A Chem 2000;131:125–32.
[334] Özkan A, Özkan MH, Gürkan R, Akçay M, Sökmen M. Photocatalytic degradation of a textile azo dye, Sirius Gelb GC on TiO_2 or Ag-TiO_2 particles in the absence and presence of UV irradiation: the effects of some inorganic anions on the photocatalysis. J Photochem Photobiol A Chem 2004;163:29–35.
[335] Riga A, Soutsas K, Ntampegliotis K, Karayannis V, Papapolymerou G. Effect of system parameters and of inorganic salts on the decolorization and degradation of Procion H-exl dyes. Comparison of H_2O_2/UV, Fenton, UV/Fenton, TiO_2/UV and TiO_2/UV/H_2O_2 processes. Desalination 2007;211:72–86.
[336] Schmelling DC, Gray KA, Vamat PV. The influence of solution matrix on the photocatalytic degradation of TNT in TiO_2 slurries. Water Res 1997;31:1439–47.
[337] Wang K, Zhang J, Lou L, Yang S, Chen Y. UV or visible light induced photodegradation of AO7 on TiO_2 particles: the influence of inorganic anions. J Photochem Photobiol A Chem 2004;165:201–7.
[338] Wu CH, Huang KS, Chern JM. Decomposition of acid dye by TiO_2 thin films prepared by the solegel method. Ind Eng Chem Res 2006;45:2040–5.
[339] Wong CC, Chu W. The direct photolysis and photocatalytic degradation of alachlor at different TiO_2 and UV sources. Chemosphere 2003;50:981–7.
[340] Choi W, Termin A, Hoffman MR. The role of metal ion dopants in quantum-sized TiO_2: correlation between photoreactivity and charge Carrier recombination dynamics. J Phys Chem 1994;98:13669–79.
[341] Kerzhentsev M, Guillard C, Herrmann JM, Pichat P. Photocatalytic pollutant removal in water at room temperature: case study of the total degradation of the insecticide fenitrothion (phosphorothioic acid O, O-dimethyl- O-(3-methyl-4-nitro-phenyl) ester). Catal Today 1996;27:215–20.
[342] Okonomoto K, Yamamoto Y, Tanaka H, Tanaka M, Itaya A. Heterogeneous photocatalytic decomposition of phenol over TiO_2 powder. Bull Chem Soc Jpn 1985;58: 2015–22.

[343] Martin ST, Lee AT, Hoffmann MR. Chemical mechanism of inorganic oxidants in the TiO_2/UV process: increased rates of degradation of chlorinated hydrocarbons. Environ Sci Technol 1995;29:2567–73.
[344] Zhu H, Zhang M, Xia Z, Low GKC. Titanium dioxide mediated photocatalytic degradation of monocrotophos. Water Res 1995;29:2681–8.
[345] Mu Y, Yu HQ, Zheng JC, Zhang SJ. TiO_2-mediated photocatalytic degradation of Orange II with the presence of Mn2. in solution. J Photochem Photobiol A Chem 2004;163:311–6.
[346] Burns R, Crittenden JC, Hand DW, Sutter LL, Salman SR. Effect of inorganic ions in heterogeneous photocatalysis. J Environ Eng 1999;125:77–85.
[347] Rizzo L, Koch J, Belgiorno V, Anderson MA. Removal of methylene blue in a photocatalytic reactor using polymethylmethacrylate supported TiO_2 nanofilm. Desalination 2007;211:1–9.
[348] Diebold U. The surface science of titanium dioxide. Surf Sci Rep 2003;48:53–229.
[349] Matthews RW, McEnvoy SR. Photocatalytic degradation of phenol in the presence of near-UV illuminated titanium dioxide. J Photochem Photobiol A Chem 1992;64: 231.
[350] Lindner M, Bahnemann DW, Hirthe B, Griebler WD. Novel TiO_2 powders as highly active photocatalysts. In: Stine WR, Tanaka T, Claridge DE, editors. Solar water detoxification; solar eng. New York: ASME; 1995. p. 339.
[351] Kabra K, Chaudhary R, Sawhney RL. Treatment of hazardous organic and inorganic compounds through aqueous-phase photocatalysis: a review. Ind Eng Chem Res 2004;43:7683–96.
[352] Prairie MR, Evans LR, Martinez SL. Destruction of organics and removal of heavy metals in water via TiO_2 photocatalysis. In: Eckenfelder WW, Roth JA, Bowers AR, editors. Chemical oxidation: technologies for the nineties, vol. 2. Pennsylvania (US): Technomic Publishing Company Inc.; 1994. p. 428–41.
[353] Herrmann JM, Disdier J, Pichat P. Photocatalytic deposition of silver on powder titania: consequences for the recovery of silver. J Catal 1988;113:72–81.
[354] Huang M, Tso E, Datye AK. Removal of silver in photographic processing waste by TiO_2-based photocatalysis. Environ Sci Technol 1996;30:3084–8.
[355] Angelidis TN, Koutlemani M, Poulios I. Kinetic study of the photocatalytic recovery of Pt from aqueous solution by TiO_2, in a closed-loop reactor. Appl Catal B Environ 1998;16:347–57.
[356] Muñoz I, Peral J, Ayllón JA, Malato S, Passarinho P, Domènech X. Life cycle assessment of a coupled solar photocatalytic-biological process for wastewater treatment. Water Res 2006;40:3533–40.
[357] Gernjak W, Fuerhacker M, Fernandez-Ibáñ ez P, Blanco J, Malato S. Solar photo-Fenton treatment process parameters and process control. Appl Catal B Environ 2006;64:121–30.
[358] Lapertot M, Pulgaín C, Fernandez-Ibáñez P, Maldonado MI, Pérez-Estrada L, Oller I, Gernjak W, Malato S. Enhancing biodegradability of priority substances (pesticides) by solar photo-Fenton. Water Res 2006;40:1086–94.
[359] Muñoz I, Rieradevall J, Torrades F, Peral J, Domenech X. Environmental assessment of different solar driven advanced oxidation processess. Sol Energy 2005;79:369–75.
[360] Consoli F, Allen D, Boustead I, de Oude N, Fava J, Franklin W, Quay B, Parrish R, Perriman R, Postlethwaite D, Seguin J, Vigon B. Guidelines for life-cycle assessment: a code of practice. Brussels (Belgium): Society of Environmental Toxicology and Chemistry; 1993.
[361] Azapagic A, Clift R. Life cycle assessment and multiobjective optimisation. J Cleaner Prod 1999;7:135–43.
[362] Reinhard A. Strongly correlated photons on a chip. Nat Photonics 2011.

FURTHER READING

[1] Fu J, Zhao Y, Wu Q. Optimising photoelectrocatalytic oxidation of fulvic acid using response surface methodology. J Hazard Mater 2007;144:499−505.
[2] Hossain MF. In situ geothermal energy technology: an approach for building cleaner and greener environment. J Ecol Eng 2016;77:49−55 [Polish Society of Ecological Engineering].
[3] Hossain MF. Production of clean energy from cyanobacterial biochemical products. Strateg Plan Energy Environ 2016;3:6−23 [Taylor and Francis].
[4] Hossain MF. Solar energy integration into advanced building design for meeting energy demand. Int J Energy Res 2016;40:1293−300 [Wiley].
[5] Hossain MF. Theoretical modeling for hybrid renewable energy: an initiative to meet the global power. J Sustain Energy Eng 2016;4:5−36 [Wiley].
[6] Hossain MF. Theory of global cooling. Energy Sustain Soc 2016;6:24 [Springer].
[7] Hossain MF. Application of advanced technology to build a vibrant environment on planet Mars. Int J Environ Sci Technol 2017;14(12):2709−20 [Springer].
[8] Hossain MF. Design and construction of ultra-relativistic collision PV panel and its application into building sector to mitigate total energy demand. J Build Eng 2017; 9:147−54 [Elsevier].
[9] Hossain MF. Green science: independent building technology to mitigate energy, environment, and climate change. Renew Sustain Energy Rev 2017;73:695−705 [Elsevier].
[10] Hossain MF. Invisible transportation infrastructure technology to mitigate energy and environment. Energy Sustain Soc 2017;7:27 [Springer], https://doi.org/10.1186/s13705-017-0128-x.
[11] Hossain MF. Green science: advanced building design technology to mitigate energy and environment. Renew Sustain Energy Rev 2018;81(2):3051−60 [Elsevier].
[12] Hossain MF. Green science: decoding dark photon structure to produce clean energy. Energy Rep 2018 [Elsevier], https://doi.org/10.1016/j.egyr.2018.01.001.
[13] Hossain MF. Transforming dark photon into sustainable energy. Int J Energy Environ Eng 2018. https://doi.org/10.1007/s40095-017-0257-1 [Springer].
[14] Hossain MF, Nowshin F. Integration of wind into running vehicles to meet its total energy demand. Energy Ecol Environ 2017;2(1):35−48 [Springer].
[15] Hossain MF, Hiroshi M. Importance of nutrients (N, P, K and NO_3 +NO_2) in the development of *Phylospdix iwatensis*, Makino. Benthos Res 2000;55:23−35 [The Japanese Association of Benthology].
[16] Hristovski KD, Nguyen H, Westerhoff PK. Removal of arsenate and 17-ethinyl estradiol (EE2) by iron (hydr)oxide modified activated carbon fibers. J Environ Sci Health Part A Toxic Hazard Subst Environ Eng 2009;44(4):354−61.
[17] Evaristo J, Jasechko S, McDonnell JJ. Global separation of plant transpiration from groundwater and streamflow. Nature 2015;525:91−4.
[18] Masle J, Gilmore SR, Farquhar GD. The ERECTA gene regulates plant transpiration efficiency in Arabidopsis. Nature 2005;436:866−70.
[19] Langer L, Poltavtsev SV, Yugova IA, Salewski M, Yakovlev DR, Karczewski G, Wojtowicz T, Akimov IA, Bayer M. Access to long-term optical memories using photon echoes retrieved from semiconductor spins. Nat Photonics 2014;8:851−7.
[20] Yang L, Wang S, Zeng Q, Zhang Z, Pei T, Y.L., Peng L-M. Efficient photovoltage multiplication in carbon nanotubes. Nature Photonics 2011:672−6.
[21] Qiong Li, Xu DZ, Cai CY, Sun CP. Recoil effects of a motional scatterer on single-photon scattering in one dimension. Sci Rep 2013.
[22] Pignatello JJ. Dark and photoassisted iron(3/=)-catalyzed degradation of chlorophenoxy herbicides by hydrogen peroxide. Environ Sci Technol 1992;26:944−51.

[23] Pignatello JJ, Liu D, Huston P. Evidence of an additional oxidant in the photoassisted Fenton reaction. Environ Sci Technol 1999;33:1832–9.
[24] Maxwell1 RM, Condon LE. Connections between groundwater flow and transpiration partitioning. Science July 22, 2016;353(6297):377–80.
[25] Jasechko S, Sharp ZD, Fawcett PJ. Terrestrial water fluxes dominated by transpiration. Nature 2013;496:347–50.
[26] Pregnolato T, Lee EH, Song JD, Stobbe S, Lodahl P. Single-photon non-linear optics with a quantum dot in a waveguide. Nat Commun 2015.
[27] Tame MS, McEnery KR, Özdemir ŞK, Lee J, Maier SA, Kim MS. Quantum plasmonics. Nat Phys 2013.
[28] Wheeler TD, Stroock AD. The transpiration of water at negative pressures in a synthetic tree. Nature 2008;455:208–12.
[29] Yan W-B, Fan H. Single-photon quantum router with multiple output ports. Sci Rep 2014.
[30] Wang Y, Zhang Y, Zhang Q, Zou B, Schwingenschlogl U. Dynamics of single photon transport in a one-dimensional waveguide two- point coupled with a Jaynes-Cummings system. Sci Rep 2016.
[31] Zhu Y, Hu X, Yang H, Gong Q. On-chip plasmon-induced transparency based on plasmonic coupled nanocavities. Sci Rep 2014.

CHAPTER SEVEN

Best Management Practices

A best management practice (BMP) is a method or technique that has been generally accepted to any alternatives because it produces results that are superior to those achieved by other means or because it has become a sustainable way of doing things [1,2]. Thus, the BMPs are used to maintain quality as an alternative to mandatory technical and legislative standards that can be based on self-assessment accredited sustainable research and technology. Necessary to building a better Earth by good operating practice, six major actions must be executed with the fullest extension of efforts, enforcement, and laws which are discussed in the following sections.

7.1 ENVIRONMENTAL MANAGEMENT

Based on the present information, the worldwide carbon dioxide (CO_2) discharges increased more rapidly ever since 2000 than in the course of past years [3–5]. In 2007, the Intergovernmental Panel on Climate Change (IPCC) reported that this increase in CO_2 emissions is beyond the worst scenario case [6–8]. As compared with developed countries, developing countries are releasing more CO_2 [9–11]. Over the last 50 years, the efficiency of natural sinks such as the oceans, which absorb CO_2, has declined, indicating that human efforts to reduce these emissions from human undertakings will have to be increasingly effective in case they are to stabilize the CO_2 level within the atmosphere [12–14]. Acquiring a carbon-balanced environment requires implementation of sustainable transport, accommodation, and consumer habits to create a basic communal alteration in human lifestyles.

Because air pollution is severely affecting human health and completely damaging the ecosystem, the priority task is to improve the quality of air for all living things and the environment [15,16]. Applying clean technology, involving proper conservation and recycling of water, within every sector, where most significantly recycling the industrial, municipal, and domestic waste and converting it into valuable products will be the best decision for environmental management, can manage the practice for reducing air pollution in the atmosphere [17,18]. Considering the environmental and social benefits, the adoption of measures such as low consumption of water,

Sustainable Design and Build
ISBN: 978-0-12-816722-9
https://doi.org/10.1016/B978-0-12-816722-9.00007-0

reduction of sources of water, an obligation to application of more effectual energy systems, and cleaner vehicles promotion shall firmly be upheld.

An additional most common source of human stress in both rural and urban areas is environmental noise [19–21]. Based on research findings, noise pollution results in several health issues including psychological problems, raised hormone levels, cardiovascular disease, and even untimely death [22,23]. The levels of noise continue to increase with the increase in demand for mobility, which causes increase in use of planes, trains, and cars [24,25]. Accordingly, the mobility consequence with both rural and urban regions should be controlled including, but not limited to, unloading and loading activities, sirens, outdoor activities, and public road works. Moreover, to drastically reduce noise pollution, rural and urban authorities must draft a plan to shift the heavy industries to a well-planned new industrial zone that is distant from residential zones. The authorities must put in place "Noise and Vibration Regulations" rule to set a limit for maximum noise levels, as well as for the acoustic measures of all construction, motilities, industries, and their corresponding offices [26,27]. Furthermore, amid the upcoming residential areas and industrial area, a transitional area must be established with a scope to include less polluting and less dense industrial activity. This will enable both the rural and urban authorities to carry on with their plans of economic development which certainly will alleviate noise pollution.

Nearly four billion people staying within the urban areas, yearly, dispose waste of more than two billion tons globally in an unmannered way of waste disposal [28,29]. It is necessary for waste management principles such as "polluter pays principle" to be practiced and for the Waste Framework Directive to set out to alleviate the pollution of the environment. In order of priorities, the waste management approach must be set with waste prevention as the preferred option, followed by reuse, recycling, other recovery forms such as energy from waste, and finally, improved waste disposal that is followed by enforcing strict monitoring. Clearly, it is essential to appropriately manage waste so as it does not adversely disturb places of spatial interest; be a nuisance via noise or odors; pose risk to animals, plants, soil, or water; cause harm to the environment; or endanger the health of human beings [30–32]. Consequently, establishing environmental compatibility examination and effect estimates of utilized products for the whole life cycle is important. Moreover, development, marketing, usage, and disposal stages should carefully be measured to attain the best environmental management and ultimately secure a sustainable Earth.

7.2 ENERGY MANAGEMENT

Management of energy involves production of both renewable alternative or conventional resources and appropriate salvage and reuse of energy that otherwise would be wasted. To reduce energy demand and acquire environmental and societal benefits, it is necessary to determine and enforce energy-efficiency and conservation measures. Energy is used by societies for heating and air-conditioning, illumination, manufacturing, transportation, and communication and also for other domestic, commercial, and industrial purposes [33–35]. Use of energy is classified as primary resource with used resources being renewable or nonrenewable sources of energy. Although the renewable resources are created by continuing processes that are likely to withstand infinite time, nonrenewable resources are considerably exhausted by human usage. The conventional industry that encompasses the nuclear industry, electrical power industry, natural gas industry, and petroleum industry causes serious ecological disaster. Energy management according to the USDOE (United States Department of Energy)-guideline 4602 refers to the proactive prearranged and methodical coordination of energy procurement, distribution, and usage to meet the necessities, considering the economic and environmental objectives [36–38]. For energy management to be implemented, it is necessary to incorporate the energy management within the structure of an organization. Decision-makers' interaction and responsibilities must be regularized. Functions and skills spread to the executive worker via delegations from the top management. Also, task fulfillment can be guaranteed by a comprehensive coordination. As a result, within huge or energy-intensive firms, it is recommended to create a distinct organizational unit, that is, the energy management unit, which keeps track and supports management. This relies on the fundamental organizational structure form to which this particular unit is linked. For instance, in the case of a functional institution, the energy management unit is situated directly amid the first and the second hierarchical level. Central and numerous sector-particular energy units must exist within a divisional company so that the individual sectors' needs, as well as the coordination between the head office and branches, can be meet [5,39,40]. Energy management within a matrix company can be incorporated as a matrix function and hence approaches a number of functions directly.

The amount of energy needed to provide goods and services can be reduced through efficient use of energy. For example, house insulation permits less use of cooling and heating energy to sustain a comfy temperature. Accordingly, building constructions promote less use of energy for cooling, heating, and lighting through more refined usage of materials, innovative airflow systems, and facade design, all of which are essential elements of the effort to attain clean and efficient cities. Correspondingly, buildings that proactively integrated with lower carbon energy formation are also capable of making a substantial contribution to the general energy performance of metropolises. Afterward, establishment of incorporated and compatible energy infrastructure on an entire metropolitan basis shall create the platform above which smart metropolises can enhance their energy performance to achieve environmental sustainability [41—43]. Having a strategy for the "energy optimization of buildings" would be the most interesting step for encouraging the rational usage of clean energy by which constructed buildings can make itself sustainable energy, such as exterior curtain wall skin Photovoltaic (PV) panel system as a source of application of renewable energy for a novel inventive industry.

Application of renewable energy thus encompasses the sustainable energy industry consisting of sustainable and alternative manufacture, sale, and distribution of substitute fuels that would certainly be the finest choice for the Earth's sustainability [44,45]. Essentially, management of energy including energy production planning and operation, as well as energy consumption, should be set to use prudently putting in mind logistics, production management, environmental management, and other recognized business functions for creating an improved planet Earth.

7.3 BUILDING MANAGEMENT

Building management system (BMS) also known as building automation system (BAS) refers to a computer-based control system that needs to be installed within buildings to monitor and regulate the building's electrical and mechanical equipment such as power system, lighting, and ventilation to confirm sustainability [46—48]. Given that the systems connected to a BMS usually represent a building's energy use of 40%, this percentage will approach 70% in case lighting is included [49,50]. It is essential for a BMS to be set up because they are critical elements to intelligently manage the energy demand, for example, electrical system, plumbing, fire alarm system, heating ventilation and air-conditioning (HVAC), electric power control, and illumination control. As a result, through BMS, building management requires to have electronic centralized regulation of a building's air-conditioning, ventilation and heating, lighting, and other building systems [51,52]. Bearing in mind enhanced utilities' life cycle, decrease in energy operating costs and consumption, building systems' efficient operation, and comfort of the occupant, the BMS objectives require to be enhanced. Consequently, distributed control system's great work on the computer networking of automation instruments should be structured to control and monitor the HVAC and humidity ventilation, control systems, lighting, flood and fire safety, security, and mechanical systems within a building [53–55]. Therefore, the core functionality of a BAS should be retained within the building climate in a specified range, offer malfunction alarms to the maintenance staff of a building, monitor device failure and performance in all building systems, and lighting to rooms based on an occupancy schedule. Compared to uncontrolled buildings, a BAS reduces building energy and costs of maintenance, hence most industrial, institutional, and commercial buildings constructed after 2000 include a BMS that should be restructured for a timeworn building [56–58]. Normally, several timeworn buildings have been retrofitted with a novel BAS, financed via insurance and energy savings, as well as other savings linked with fault detection and preemptive maintenance. So, it is regarded that a building regulated by a BMS is frequently termed as an intelligent building, a "smart home" in case of residence or just a "smart building." When proprietary procedures were used in homes, industrial and commercial buildings must have in history depended on healthy recognized procedures such as BACnet. Current IEEE standards and groups' efforts have offered a standard-based grounds for different networking of several instruments in several physical networks for varied purposes and service quality and failover, which promises proper support to human safety and health [59,60].

Designing building to accommodate a BAS for energy, water, and air preservation characteristics, as well as appropriate electrical instrument demand response in a typical BAS function, is necessary. This is due to the fact that it is the more refined ventilation and moisture monitoring needed of "tight" insulation for achieving environmental sustainability.

7.4 WATER MANAGEMENT

Having a sustainable and healthy rural and urban community makes the residents to be certain of accessing safe water for municipal, industries, and household use. The significance of acquiring clean drinking water via the urban authorities is the fact that they make sure that the water supplied to residents through water supply systems is nontoxic to drink. It is the duty of the rural and urban authorities to often monitor the water quality supplied at the tap and constantly inform the consumers of the quality. Concerning the quality of water, the rural and urban authorities should always inspire citizens to drink the tap water rather than bottled water based on quality and environmental effect [61,62].

Among the wastewater treatment directive goals that should have to be adopted appropriately, a clean rural and urban environment is one. The wastewater treatment directive calls for all the rural establishments, towns, and cities to acquire clean sewage systems in site and/or treat the sewage water to particular minimum standards before discharging it into the environment [57,63]. Through this approach, coastal areas, lakes, and rivers can be cleaned up to meet the health-based minimum standards for fauna, flora, and human health.

For mega cities, if the connection of sewer system should have to avert leakages, the better option could be the incorporation of density polyethylene pipes. It is necessary to carry out investigations within the catchment area of water that accounts for 10%—15% of the yearly water distribution [38,64]. To avert pressure surges, all pipe-work workers are trained to handle valves as well as other appliances in the right way.

Essentially, the wastewater treatment plants are to take good care of the removal of prolonged nitrogen, in case it is over 10 years old. Occasionally, new and strict necessities to deal with the elimination of nitrogen, phosphorus, and organic materials should be integrated. Besides the customary measures to oversee wastewater, rural and urban authorities must have as well-analyzed different climate-change settings to handle the impacts that change in climate will have and is presently having because of increased storm levels, water situation, and water supply. As a result, effective management of storm water turns out to be more and more important and hence requires active work to separate storm water from joint system, thereby causing a reduction in the overflow risk, and support smoother flow. This eventually will avert overflow of sewage system facilities or drainage system. Afterward, the rural and urban authorities must manage rainfall in a more natural way, which will finally take a central role in climate change.

7.5 INFRASTRUCTURE AND TRANSPORTATION MANAGEMENT

Besides serving the nations, infrastructure, which is a basic facility and system, serves world's movement to facilitate the essential routes, to help the economy function properly, and to comfort the day-to-day lives [48,62]. Usually, infrastructure describes technical structures such as telecommunications, electrical grids, tunnels, bridges, and roads, and as defined, the interconnected systems physical components offering goods and services vital to facilitate, sustain, or improve social condition of living. On the other hand, transportation refers to the movement of goods, animals, and humans from one place to another. Space, pipeline, cable, water, land, and air are the modes of transportation. Because transport facilitates trade amid people, which is necessary for the civilization development, it is important. Transport infrastructure is composed of the fixed installations of canals, waterways, airways, railways, roads, and terminals, as well as pipelines such as seaports, refueling depots, trucking terminals, warehouses, bus stations, railway station, and airports. Terminuses can be used both for maintenance and exchange of cargo and passengers. Aircrafts, automobiles, spacecraft, watercraft, helicopters, people, trucks, trains, buses, and bicycles are the vehicles moving along these networks.

Ownership and operations within the transportation industry is either private or public, depending on the country as well as the mode playing an essential part in globalization and economic growth, though most kinds use huge amounts of land and cause air pollution [11,65]. Essentially, operations handle the way in which vehicles are functioned, and to acquire sustainability to preserve biodiversity, processes set for this particular purpose, including policies, legalities, and financing, should be followed. Besides making cities pleasant, habitable, and green places, biodiversity also provides valued services to urban residents, as a result, it is essential for infrastructure to be structured and constructions to be nontoxic to the biodiversity when building infrastructure for transportation [14,23]. When building infrastructure to facilitate transport for sport and health, relaxation, social activities, and nature watching, recreational parks, quiet streets, and green spaces should simply be given the priority. This will definitely serve as the building blocks for consoling the ecological crisis and encouraging quality of life.

Consequently, through the use of enhanced technology, the approach must strongly be applied to build clean transport and infrastructure plans. The aim of these plans will include working toward acquiring zero-emission vehicles and improving traffic around and within capitals, by considering economic interest and encouraging the public transport use, walking, and cycling within towns and cities, and, above all, balancing the environmental effects.

7.6 GENERAL MANAGEMENT

The general management is based primarily on the overall management system that includes organizational structure, planning activities, responsibilities, practices, procedures, processes, and resources for developing, implementing, achieving, reviewing, and maintaining the environmental policy". It is a simple definition of "a management tool designed to help the urban and rural authority to manage the environmental aspects by conducting five principles: (1) commitment and policy, (2) -planning, (3) measurement and evaluation, (4) review and improvement, and (5) implementation.

Rural and urban councils therefore should conform to the sustainability commitment to realize the total management because it has a significant citizen and social participation dimension for the rural and urban environment improvement [19,30]. The use of resources should be characterized by rural and urban council, considering continuous thinking of the environment, so as to practice sustainability. Afterward, the city, industry, and residents will find it easier to make sustainable selections concerning waste management, travel, services, and commodities. The following objectives will need to be enforced, precise to its own responsibilities and undertakings, if the aforementioned goal is to be attained: (1) through the increase in the quantities of waste management waste, they should be recycled in an operative manner to minimize waste creation. Waste sorting and recycling system will be more easily reachable and artless; (2) initiating to reduce the usage of dangerous substances by acquiring information, collaboration, as well as supervision, which will contribute to stage out the hazardous substances through circulation. At the acquiring stage, both the rural and urban rural authorities will require to eliminate commodities with harmful matters, and sustainable substances should be selected within all sectors; and (3) clean procurement where the rural and city authorities will have to review its purchase and procurement procedure, as well the use of resources by the city. Finally, the following three steps must be taken care of to confirm a sustainable planet.

7.6.1 Green Public Procurement

Given that in the entire world, the public authorities are the main consumers, they must use their purchasing power to select works, services, and goods that are environmentally friendly and are be able to make a significant contribution to clean production and consumption. This is what is referred to as Green Public Procurement (GPP). GPP should be enforced as a major rule to show the efforts to turn into a more efficient resource within the economy. This can assist in stimulating a serious mass of demand for increasingly sustainable services and goods for a robust inducement in ecoinnovation states and cities, as well as the world. Rural and city authority should therefore be having a volunteer process enforced to start purchasing environmental and social necessities in markets, converse with suppliers, as well as strong engagements with the environmental consultant for sustainable/clean public contracting. Consequently, the rural and city authorities' need for ecolabeling should be added to the best environmental procedure in tendering services and commodities such as buying printing services, furniture, and waste-collection lorries by taking necessary action.

7.6.2 Green Procurement

Based on the rural and urban procurement policy, workers purchasing for rural and city authorities should consider environmental factors. Having the vision to make green procurement a principle within the whole procurement process is one of the major objectives. This vision should have the frame in such a manner that the buying party needs to assess the effects of the purchase on the environment; in case rural and city authorities' policy on green procurement is not observed, reasoned argument should support this particular decision. Certainly, this process will enable green procurement monitoring to sustain the rural and city authorities' green procurement compliance.

7.6.3 Domestic and Global Collaboration

It is necessary for the rural and city authorities to collaborate with the global sustainable and/or appropriate authorities, Regulated Ministry for the

Environment and/or appropriate authorities, and State Trading Centre to make people know about the GPP that has been carried out in their own countries, rural areas, and cities. Essentially, the rural and urban authorities must have to play a part, for example, in creating a new buying policy for its government, within which is the green procurement, which is the major pillar of the National Action Plan for Green Public Procurement to secure clear and greener globe. Subsequently, the city and rural agencies should have a representative on the steering committee on green procurement, set up a website on GPP, prepare criteria documents for green procurement, and organize conferences on GPP to encourage citizens about the green purchase and provide the knowledge about the benefit of the green product for sustainability.

7.7 CONCLUSION

BMPs that are broadly applicable to local public administrations in different areas of their activities have been identified. These range from the management of the public administration to individual activities to achieve sustainability. Because the engagement of the public sector is the driving force for securing sustainability, public administrations can take a leading role to show the way to conform the sustainability. Indeed, municipality and rural authority can influence and enforce the citizens for management of appropriate energy technology, air quality, land use, noise pollution, and green urban areas to secure a better environment. Subsequently, at the local level, municipalities and local authorities must take the responsibilities for activities ranging from energy and building sectors, land use planning to wastewater treatment facilities, and transport and mobility and infrastructure sectors. Within its work on identification of BMPs and enforcement of the same, BMPs must be developed as a sectoral reference document for the best environmental management practices by conducting advanced research, planning, application, and using management strategy to adopt sustainable design and build to confirm a cleaner and green Earth.

ACKNOWLEDGMENTS

This research was supported by Green Globe Technology under the grant RD-02018-03. Any findings, conclusions, and recommendations expressed in this chapter are solely those of the author and do not necessarily reflect those of Green Globe Technology.

REFERENCES

[1] Ballantyne AP, et al. Audit of the global carbon budget: Estimate errors and their impact on uptake uncertainty. Biogeosciences 2015;12:2565−84.

[2] Box P. Spatial Units as Agents: making the landscape an equal player in Agent-based Simulations. In: Gimblett RH, editor. Integrating Geographic information systems and Agent-based Modeling techniques for Simulating social and ecological processes. New York: Methuen; 2002.

[3] Achard F, et al. Determination of tropical deforestation rates and related carbon losses from 1990 to 2010. Glob. Change Biol 2014;20:2540−54.

[4] Bao H, Niggemann J, Luo L, Dittmar T, Kao S-J. Aerosols as a source of dissolved black carbon to the ocean. Nat Commun 2017.

[5] Hossain MF. Green science: Independent building technology to mitigate energy, environment, and climate change. Renew Sustain Energy Rev 2017;73:695−705. Elsevier.

[6] Andres RJ, Boden TA, Higdon D. A new evaluation of the uncertainty associated with CDIAC estimates of fossil fuel carbon dioxide emission. Tellus B Chem. Phys. Meteorol 2014;66:23616.

[7] Betts RA, Jones CD, Knight JR, Keeling RF, Kennedy JJ. El Nino and a record CO2 rise. Nat Clim Change 2016;6:806−10.

[8] Ciais P, Sabine C. Chapter 6: carbon and other biogeochemical cycles in climate change. In: Stocker T, Qin D, Platner GK, editors. The physical science basis. Cambridge University Press, 2013; 2013.

[9] Andreozzi R, Caprio V, Ciniglia C, De Champdor_e M, Lo Giudice R, Marotta R, Zuccato E. Antibiotics in the environment: occurrence in Italian STPs, fate, and preliminary assessment on algal toxicity of amoxicillin. Environ Sci Technol 2004;38: 6832−8.

[10] Barbier EB. The economic determinants of land degradation in developing countries. Phil. Trans. R Soc. London 1997;352:891−9.

[11] Bebbington AJ, Batterbury SPJ. Transnational livelihoods and landscapes: political ecologies of globalization. Ecumene 2001;8(4):369−80.

[12] Ballantyne AP, Alden CB, Miller JB, Tans PP, White JWC. Increase in observed net carbon dioxide uptake by land and oceans during the past 50 years. Nature 2012;488: 70−2.

[13] Bridges EM, Oldeman LR. Global assessment of human-induced soil degradation. Arid Soil Res Rehabil 1999;13(4):319−25.

[14] Bauer JE, et al. The changing carbon cycle of the coastal ocean. Nature 2013;504: 61−70.

[15] Chevallier F. On the statistical optimality of CO2 atmospheric inversions assimilating CO2 column retrievals. Atmos Chem Phys 2015;15:11133−45.

[16] Le Quéré C, Andrew RM, Friedlingstein P, Sitch S, et al. Global carbon budget 2017. Earth Syst Sci Data Discuss 2017.

[17] Cleaver K, Schreiber G. Reversing the Spiral: the Population, Agriculture and environment Nexus in Sub-Saharan Africa. Washington, D.C.: World Bank; 1994.

[18] Denman KL, et al. Couplings between changes in the climate system and Biogeochemistry. Cambridge University Press; 2007.

[19] Hossain MF, Mukai H. Importance of Nutrients (N, P, and NO3+NO2) in growth of the Surfgrass, Phyllospadix iwatensis Makino. 2000.

[20] Houghton R. Balancing the global carbon budget. Annu Rev Earth Planet Sci 2007; 35:313−47.

[21] Li W, et al. Reducing uncertainties in decadal variability of the global carbon budget with multiple datasets. Proc Natl Acad Sci USA 2016;113:13104−8.

[22] Conroy MJ. Mapping of species Richness for conservation of Biological diversity: Conceputal and Methodological issues. Ecol Appl 1996;6(3):763–73.
[23] Davis SJ, Caldeira K. Consumption-based accounting of CO2 emissions. Proc Natl Acad Sci USA 2010;107:5687–92.
[24] Dietzenbacher E, Pei J, Yang C. Trade, production fragmentation, and China's carbon dioxide emissions. J. Environ. Econ. Manage 2012;64:88–101.
[25] Duce RA, et al. Impacts of atmospheric anthropogenic nitrogen on the open ocean. Science 2008;320:893–7.
[26] Houghton RA. Balancing the global carbon budget. Annu Rev Earth Planet Sci 2007.
[27] Karl DM, Church MJ. Microbial oceanography and the Hawaii ocean time-series programme. Nat Rev Microbiol 2014;12:699–713.
[28] de Sherbinin A. Land-use and land-Cover change. In a CIESIN Thematic Guide. Palisades, NY: Center for International Earth Science Information Network of Columbia University; 2002.
[29] Hyland KC, Dickenson ERV, Drewes JE, Higgins CP. Sorption of ionized and neutral emerging trace organic compounds onto activated sludge from different wastewater treatment configurations. Water Res 2012;46:1958–68.
[30] Diouf A, Lambin EF. Monitoring land-cover changes in semi-arid regions: Remote sensing data and field observations in the Ferlo, Senegal. J Arid Environ 2001;48: 129–48.
[31] Ewel KC. Natural resource management: the need for interdisciplinary collaboration. Ecosystems 2001;4:716–22.
[32] Fairhead J, Leach M. Misreading the African Landscape: Society and Ecology in a Forest-Savanna Mosaic. Cambridge: Cambridge University Press; 1996.
[33] Donald CE, Elie MR, Smith BW, Hoffman PD, Anderson KA. Transport stability of pesticides and PAHs sequestered in polyethylene passive sampling devices. Environ Sci Pollut Control Ser 2016.
[34] Gibson CC, Ostrom E, Ahn TK. The concept of scale and the human dimensions of global change: a survey. Ecol Econ 2000;32:217–39.
[35] Gonzalez-Gaya B, et al. High atmosphere-ocean exchange of semivolatile aromatic hydrocarbons. Nat Geosci 2016;9:438–42.
[36] Garcia-Robledo E, Padilla CC, Aldunate M, Stewart FJ, Ulloa O, Paulmier A, Gregori G, Revsbech NP. Cryptic oxygen cycling in anoxic marine zones. Proc Natl Acad Sci Unit States Am 2017;114(31):8319–24.
[37] Erb K-H, et al. Bias in the attribution of forest carbon sinks. Nat Clim Change 2013;3: 854–6.
[38] Schwietzke S, et al. Upward revision of global fossil fuel methane emissions based on isotope database. Nature 2016;538:88–91.
[39] Helbling DE, Hollender J, Kohler H-PE, Singer H, Fenner K. High-throughput identification of microbial transformation products of organic micropollutants. Environ Sci Technol 2010;44:6621–7.
[40] Tengberg A, Ellis-Jones J, Kiome R, Stocking M. Applying the concept of Agrodiversity to Indigenous soil and water conservation practices in Eastern Kenya. Agric Ecosyst Environ 1998;70:259–72.
[41] Hossain MF. Green science: advanced building design technology to mitigate energy and environment. Renew Sustain Energy Rev 2018;81(2):3051–60 [Elsevier].
[42] Hossain MF. Photonic Thermal energy control to naturally Cool and Heat the building. Advanced Thermal Engineering 2018;131:576–86 [Elsevier].
[43] Hossain MF. Theoretical Modeling for Hybrid renewable energy: an initiative to meet the global power. Journal Sustainable Energy Engineering 2016;4:5–36 [Wiley].

[44] Hossain MF. Design and construction of Ultra-Relativistic Collision PV panel and its application into building sector to mitigate total energy demand. Journal of Building Engineering 2017;9:147–54 [Elsevier].
[45] Hossain MF. Theory of global cooling. Energy, Sustainability, and Society 2016;6:24 [Springer].
[46] Hossain MF. Application of advanced technology to build a Vibrant environment on planet Mars. Int J Environ Sci Technol 2017;14(12):2709–20 [Springer].
[47] Hossain MF. Invisible transportation infrastructure technology to mitigate energy and environment. Energy, Sustainability, and Society 2017;7:27 [Springer].
[48] Redford KH, Dinerstein E, Groves C, Mace G, Maginnis S, Mittermeier RA, Noss R, Olson D, Robinson JG, Vedder A, Wright M, Coppolillo P, Sanderson EW, Da Fonseca GAB. Mapping the conservation landscape. Conserv Biol 2003;17(1): 116–31.
[49] Parazoo, et al. Contrasting carbon cycle responses of the tropical continents to the 2015–2016 El Niño. Science 2017.
[50] Borrelli P, Robinson DA, Fleischer LR, Lugato E, et al. An assessment of the global impact of 21st century land use change on soil erosion. Nat Commun 2017.
[51] Hossain MF, Fara N. Integration of Wind into Running vehicles to meet its total energy demand. Energy, Ecology, and Environment 2017;2(1):35–48 [Springer].
[52] Hossain MF. Production of clean energy from Cyanobacterial Biochemical products. Strat Plann Energy Environ 2016;3:6–23 [Taylor and Francis].
[53] Hossain MF. Solar energy integration into advanced building design for meeting energy demand. Int J Energy Res 2016;40:1293–300 [Wiley].
[54] Irwin EG, Geoghegan J. Theory, data, methods: developing spatially explicit economic models of land use change. Agric Ecosyst Environ 2001;85:7–23.
[55] Liu Z, et al. Reduced carbon emission estimates from fossil fuel combustion and cement production in China. Nature 2015;524:335–8.
[56] Kinzig AP. Bridging disciplinary divides to address environmental and intellectual challenges. Ecosystems 2001;4:709–15.
[57] Stephens BB, et al. Weak northern and strong tropical land carbon uptake from vertical profiles of atmospheric CO2 science. Science 2007;316:1732.
[58] Yin Y, et al. Variability of fire carbon emissions in equatorial Asia and its nonlinear sensitivity to El Niño. Geophys Res Lett 2016;43:10,472–410,479.
[59] Kok K, Farrow A, Veldkamp A, Verburg PH. A method and application of multi- scale validation in spatial land use models. Agric Ecosyst Environ 2001;85:223–38.
[60] Ogunlaja OO, Parker WJ. Impact of activated sludge process configuration on removal of micropollutants and estrogenicity. Water Sci Technol 2015.
[61] Mason Earles J, Yeh S, Skog KE. Timing of carbon emissions from global forest clearance. Nat Clim Change 2012;2:682–5.
[62] Nagendra H, Munroe DK, Southworth J. From pattern to process: landscape fragmentation and the analysis of land use/land cover change. Agric Ecosyst Environ 2004;101: 111–5.
[63] Metzger KL, Boone RB, Coughenour MB, Reich RM. Effects of seasonal grazing on plant species diversity and vegetation structure in a semi-arid ecosystem. J Arid Environ 2005;61(1):147–60.
[64] Marti E, Variatza E, Balcazar JL. The role of aquatic ecosystems as reservoirs of antibiotic resistance. Trends Microbiol 2014;22:36–41.
[65] Bagnall JP, Malia L, Lubben AT, Kasprzyk-Hordern B. Stereoselective biodegradation of amphetamine and methamphetamine in river microcosms. Water Res 2012b;47: 5708–18.

INDEX

'*Note:* Page numbers followed by "f" indicate figures, "t" indicate tables.'

A

D

E

F

H

I

K

L

Q

R

T

U

Printed in the United States
By Bookmasters